Oxford Medical Publications

Psychiatric genetics and genomics

Psychiatric Genetics and Genomics

Edited by

Peter McGuffin
Social, Genetic and Developmental Psychiatry Centre,
Institute of Psychiatry,
Denmark Hill,
London, UK

Michael J. Owen
Department of Psychological Medicine,
University of Wales College of Medicine,
Cardiff, UK

Irving I. Gottesman
University of Minnesota, Minneapolis, USA

OXFORD
UNIVERSITY PRESS

OXFORD

UNIVERSITY PRESS

Great Clarendon Street, Oxford OX2 6DP

Oxford University Press is a department of the University of Oxford.
It furthers the University's objective of excellence in research, scholarship,
and education by publishing worldwide in

Oxford New York
Auckland Bangkok Buenos Aires Cape Town Chennai
Dar es Salaam Delhi Hong Kong Istanbul Karachi Kolkata
Kuala Lumpur Madrid Melbourne Mexico City Mumbai Nairobi
São Paulo Shanghai Taipei Tokyo Toronto

Oxford is a registered trade mark of Oxford University Press
in the UK and in certain other countries

Published in the United States
by Oxford University Press, Inc., New York

A catalogue record for this title is available from the British Library

Library of Congress Cataloging in Publication Data
(Data available)

ISBN 0-19-263148-9 (Hbk.)

10 9 8 7 6 5 4 3 2 1

Typeset by Newgen Imaging Systems (P) Ltd., Chennai, India
Printed in Great Britain
on acid-free paper by
Biddles Ltd., Guildford & King's Lynn

Preface

It is over 30 years since Oxford University Press published the first textbook of psychiatric genetics in the English language, Eliot Slater and Valerie Cowie's *The Genetics of Mental Disorders*. In the intervening years the field has changed dramatically, with the advent of molecular genetics and rapid progress in the human genome project culminating in the publication of the draft human genome sequence. There have also been profound, if less publicized advances in statistical and computational methods that have aided our understanding of the genetic and environmental causes of a number of psychiatric disorders. Yet in other respects the bedrock of psychiatric genetics remains strikingly constant and unchanging. Much of what we know comes from carrying out family, twin and, where possible, adoption studies. A consistent theme has been one of healthy scepticism about genetics and abnormal behavior which has meant that few within or outside the field have been willing to accept that mental disorders or symptoms have genetic underpinnings only on the basis that they cluster in families. Consequently the quantity and quality of twin and adoption data available from studies of psychiatric disorders exceeds that for nearly all other forms of common disease. For this reason, and in spite of the challenges posed by genetic and phenotypic complexity, this is a time of optimism that the application of molecular genetics to psychiatric disorders will ultimately bear fruit. Indeed, genetics has become one of the most vital and productive areas of psychiatric research and publication of a major new text evaluating the field is timely.

The book starts with chapters on the basic principles of molecular and quantitative genetics aimed at readers without much expertise in these areas, such as those with their primary backgrounds in psychiatry and psychology. As such, this section of the book is not extensively referenced but the reader is given suggestions for further reading. The next section deals with the genetics of normal cognitive ability and personality followed by developmental and childhood disorders. This is followed by a section that deals with the major psychiatric syndromes of adult life in detail, with coverage of both quantitative and molecular findings. In each case the latest research is critically evaluated and the prospects for future advances described. Finally, we consider the application and implications of likely advances in psychiatric genetics from clinical, academic and ethical perspectives.

Most of the authors of this book work at, or have worked at, the Institute of Psychiatry London or at the University of Wales College of Medicine, Cardiff or both and these two centres have had substantial collaborations in genetic research and training over the past 15 years. We hope that this helps give a coherence of outlook without narrowing too much the book's breadth of vision.

Peter McGuffin
Michael J Owen
Irving I Gottesman
June 2002

Contents

List of Contributors

Dr Maria J. Arranz, PhD
Clinical Neuropharmacology
Division of Psychological Medicine
Institute of Psychiatry, King's College
London UK

Dr David Ball, MA, BM, BCh, MRCPsych
Social, Genetic and Developmental
Psychiatry Research Centre
Institute of Psychiatry, King's College
London UK

Dr Alastair Cardno, MB PhD MRC Psych
Department of Psychological Medicine
University of Wales College of Medicine,
Cardiff UK

Dr Avshalom Caspi, MA PhD
Social, Genetic and Developmental
Psychiatry Research Centre
Institute of Psychiatry, King's College
London UK

Dr Jill Clayton-Smith MB ChB,
MRCP, MD
The University Department of Medical
Genetics
St. Mary's Hospital, Manchester, UK

Dr David Collier, BSc PhD
Division of Psychological Medicine and
Social, Genetic and Developmental
Psychiatry Research Centre
Institute of Psychiatry, King's College
London UK

Professor Nick Craddock, MA, MB,
ChB, MMedSci, PhD MRCPsych
Department of Psychiatry

University of Birmingham,
Birmingham UK

Dr Thalia C. Eley
Social, Genetic and Developmental
Psychiatry Research Centre
Institute of Psychiatry, King's College
London UK

Professor Anne Farmer, MD FRCPsych
Social, Genetic and Developmental
Psychiatry Research Centre
Institute of Psychiatry, King's College
London UK

Professor Irving I. Gottesman, BSc, PhD,
Hon. FRCPsych
Departments of Psychiatry and
Psychology
University of Minnesota,
Minneapolis USA

Dr Francesca Happé, BA, PhD
Social, Genetic and Developmental
Psychiatry Research Centre
Institute of Psychiatry, King's College
London UK

Dr Ian Jones, BSc, MB BS, MSc,
MRCPsych
Department of Psychiatry
University of Birmingham,
Birmingham UK

Dr Lindsey Kent, MB ChB, PhD,
MRCPsych
ILTM Department of Psychiatry
University of Birmingham,
Birmingham UK

Professor Rob W. Kerwin, MB, PhD,
FRCPsych
Division of Psychological Medicine
Institute of Psychiatry, King's College
London UK

Dr Malcolm B. Liddell, MB PhD
MRCPsych
Department of Psychological Medicine
University of Wales College of Medicine
Cardiff UK

Professor Peter McGuffin, MB PhD
FRCP FRCPsych FMedSci
Social, Genetic and Developmental
Psychiatry Research Centre
Institute of Psychiatry, King's College
London UK

Dr Terrie Moffitt, MA PhD FMedSci
Social, Genetic and Developmental
Psychiatry Research Centre
Institute of Psychiatry, King's College
London UK

Professor Michael C. O'Donovan, MB,
PhD. FRCPsych
Department of Psychological Medicine
University of Wales College of Medicine,
Cardiff UK

Professor Michael J. Owen, PhD
FRCPsych FMedSci
Professor of Psychological Medicine

Department of Psychological Medicine
University of Wales College of Medicine,
Cardiff UK

Dr Robert Plomin, BA PhD FMedSci
Social, Genetic and Developmental
Psychiatry Research Centre
Institute of Psychiatry, King's College
London UK

Dr Jane Scourfield, MA, BMBch
MRCPsych PhD
Department of Psychological Medicine
University of Wales College of Medicine,
Cardiff UK

Dr Pak Sham, BM, BCh, MA, MSc, PhD,
MRCPsych
Social, Genetic and Developmental
Psychiatry Research Centre
Institute of Psychiatry, King's College
London UK

Professor Anita Thapar, MB BCh,
MRCPsych, PhD
Department of Psychological Medicine
University of Wales College of Medicine,
Cardiff UK

Professor Julie Williams, BSc PhD
Department of Psychological Medicine
University of Wales College of Medicine,
Cardiff UK

Part 1

Basic principles

Chapter 1

Basic molecular genetics

Michael C. O'Donovan and Michael J. Owen

Introduction

The purpose of this chapter is to enable readers who have little knowledge of biology to understand the rest of the book. Thus this chapter covers the organization and structure of DNA, and how information is encrypted within, decoded from and transmitted by this molecule. We describe how alteration and variation in DNA may lead to disease, and how these alterations may be detected in a laboratory.

Given its purpose, our emphasis has been to try and make this chapter accessible rather than comprehensive, although we hope it might also be a worthwhile read for a genetics-literate readership with a non-molecular background. Given the goal, this text is not extensively referenced, nor should it be viewed by the more highly motivated or demanding readership as a replacement for the excellent textbooks on molecular genetics that are recommended at the end of this chapter.

DNA, Chromosomes, and Genetic Information

Organization of DNA

Genetic information is encrypted in a macromolecule called *deoxyribonucleic acid* (DNA). In eukaryotes (that is organisms in which the cells have nuclei containing the main genetic material), linear DNA molecules are complexed with proteins to form a material called *chromatin*. This is organized in a highly structured and condensed way to form the *chromosomes* that can be seen in the nucleus of dividing cells. *Gametes* (sperm and eggs) from different eukaryotic species have a characteristic number of chromosomes. Normal human male and female gametes (sperm and eggs respectively), contain 23 chromosomes and are termed *haploid*. Since we are the result of fertilization of an egg by a sperm, most of our cells are *diploid*, that is, they contain twice this number of chromosomes.

When dividing cells are treated with particular enzymes and dyes, chromosomes can be visualized under the microscope as long, thin banded rectangular structures that are pinched in at one point along their length (Fig. 1.1) and each chromosome has a unique appearance. Human chromosomes are classified under the International System for Human Cytogenetic Nomenclature (ISCN). Two of the diploid complement of 46 chromosomes are called *sex chromosomes*, and these come in two forms, a large X chromosome and a much smaller Y chromosome (Fig. 1.1). These are called sex chromosomes because females carry two copies of the X form, while males carry one X and one Y. The Y form determines gender, with possession of a Y chromosome resulting in male development and its absence, female development. As only males carry Y chromosomes, it is the sperm (which can contribute either an X or Y to the embryo) that determine gender. The other 44 chromosomes are called *autosomes*.

The autosomes are numbered from 1 through to 22. The 'pinched in' region of each chromosome is known as the *centromere* and the ends are called *telomeres*. Centromeres are not positioned exactly in the middle of the chromosome, which therefore appears to have a long and a short arm. These are called q and p respectively. The arms are further divided according to banding patterns, with those nearest the centromere (*cen*) having smaller numbers than those near the telomere (*tel*). Thus, 22q11 designates a region on the long arm of chromosome 22 that is nearer the centromere than 22q12.

The *karyotype* is a shorthand description of the chromosomal content of a cell, consisting of the total number of chromosomes, followed by a description of the sex chromosome content. Thus, the normal male karyotype is 46,XY, the female 46,XX. However, occasionally there are abnormalities of chromosome number (see below). Thus, Turner's syndrome (females with only a single X chromosome) karyotype is 45,X while Klinefelter's (males with an extra X) is 47,XXY. The usual karyotype of males with Downs syndrome is 47,XY,+21, where +21 indicates an extra copy of chromosome 21.

The 22 autosomes are each present in duplicate, with one complete set inherited from each parent. Pairs of the same type are called *homologous* chromosomes, and each

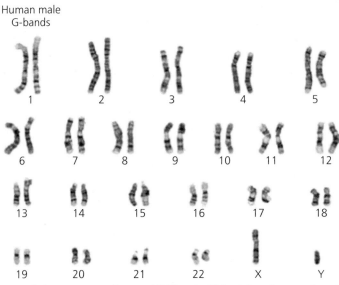

Fig. 1.1 Image of chromosomes from a 46XY male. This picture is reproduced from *http://www.pathology.washington.edu/Cytogallery/* by permission and is copyright University of Washington Pathology.

member of a pair is virtually identical at the DNA level. Since it is DNA that carries genetic information, we therefore contain duplicate copies of the same information, one coming from our mother, one from our father. However, the situation is different for the sex chromosomes. The X chromosome is considerably larger than the Y chromosome, and therefore female cells contain more genetic information than the cells of a male. However, this excess is compensated for by the fact that in female cells, one of the X chromosomes is randomly compressed into a structure that can be seen in cells, for example taken from the lining of the mouth, as the *Barr body*. This chromosome is almost completely inactivated and thus in some female cells, the paternal X chromosome is active, in others the maternal. The importance of this mechanism, known as *lyonisation* after Mary Lyons who proposed it, will be seen later in the sections dealing with patterns of inheritance (Chapter 2) and X-linked genetic diseases that result in mental retardation (Chapter 5).

The Structure of DNA

At the macro level, DNA is organized into chromosomes. However, it is the microstructure of DNA that is largely responsible for carrying genetic information. To understand genetics, it is therefore important to consider the nature and structure of DNA. DNA is a polymeric macromolecule constructed from monomeric *deoxynucleotides* (Fig. 1.2). Deoxynucleotides consist of a modified 5-carbon sugar molecule (deoxyribose). The first of the carbon molecules in the sugar ring (called 1′) carries a *base* molecule, the third (3′) caries a hydroxyl group, and the fifth (5′) carbon molecule carries a triphosphate group. The polymer is constructed from the individual monomers by the formation of phosphodiester bonds between the 5′ phosphates and the 3′ hydroxyl group (Fig. 1.2). DNA polymerization occurs in a 5′ to 3′ direction, that is, deoxynucleotide molecules are added to the 3′ OH group of the terminal nucleotide. Thus, one end of a DNA molecule is called the *5′ end*, the other is called the *3′ end*. This terminology will be important in discussing gene structure and molecular genetic techniques.

There are four possible bases attached to the 1′ carbon. These are:

The two purines adenine (A) and guanine (G) and the two pyrimidines cytosine (C), and thymine (T). The primary structure, or *sequence,* of DNA is usually described by the order of bases in a 5′ —3′ direction. Thus a stretch of DNA may be represented as:

5′—AGCTTTGGCA—3′

The DNA content of an organism is called its *genome*, and in humans is defined by approximately 3.3×10^9 bases. However, DNA in its natural state generally exists as a double stranded structure (dsDNA) consisting of two DNA molecules held together by hydrogen bonding between the bases in the opposing strands. This is represented in Fig. 1.3. Pairs of bases in opposite strands do not align randomly. Thus, the base A is always paired with T while the base C will always bind with G. This is called base *complementarity* and two strands displaying this property are said to be *complementary*.

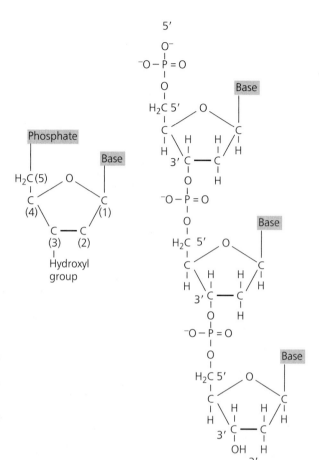

Fig. 1.2 Representation of monomeric deoxynucleotide and polymer DNA. During DNA synthesis, deoxynucleotides are linked by phosphodiester bonds between the 3′ hydroxyl group of the terminal deoxynucleotide in the DNA molecule and the 5′ phosphate of the next monomer to be added.

Consequently, if the sequence of bases in one strand of DNA is known, the complementary sequence of bases in the other strand is also known according to the above simple rule. This permits the dsDNA sequence to be described by one of the two strands. Thus, the sequence above would be:

5′—AGCTTTGGCA—3′
3′—TCGAAACCGT—5′

The complementary strands of DNA are said to be *antiparallel*, that is the 5′ ends of each strand are at opposite ends (Fig. 1.3). Finally, DNA is described as an *antiparallel double helix* because the molecule *in vivo* adopts a helical structure consisting of the pairs of antiparallel strands (Fig. 1.3).

DNA Replication

At conception, each of us consists of a single cell containing 46 chromosomes; yet by adulthood, each of us consists of around 10^{14} cells, most of which contain at least one

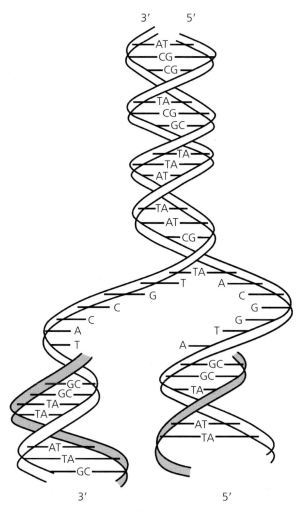

Fig. 1.3 Double stranded helical DNA held together by hydrogen bonds between complementary bases. The strands are antiparallel because each strand runs in the opposite direction (5′ to 3′ and 3′ to 5′). During DNA replication and cell division, new double helices are formed, each of which contains one original strand of DNA and one newly synthesized complementary copy (shaded). Synthesis is 5′ to 3′.

complete diploid genome. This increase in cell number occurs by a process of cell division in somatic cells called *mitosis* which has many stages, the details and names of which are not relevant to understanding this book. During mitosis, each chromosome is first replicated to form pairs of identical *sister chromatids* joined at the centromere. These are separated to form pairs of identical chromosomes (the cell is now briefly tetraploid, i.e. has two copies of each of the original diploid complement of chromosomes), and then the cell divides in such a way as to separate the sister chromatids, with each daughter cell retaining a single copy of each sister chromatid, and thus a full complement of diploid chromosomes.

The production of gametes is a rather different procedure as DNA must be both replicated, *and* the diploid number of chromosomes must be reduced to haploid. This is called *meiosis*. Pairs of homologous chromosomes migrate within the cell to lie together side by side to form a structure called a *bivalent*. Each chromosome is replicated to form pairs of sister chromatids. Because homologues are aligned, and each homologue consists of a pair of sister chromatids, the bivalent consists of four dsDNA molecules. Chromosomal material is then almost randomly exchanged between homologues, a process called *recombination* (Fig. 1.4). After recombination, the chromosomes are no longer identical to those inherited from the parents. Furthermore, pairs of sister chromatids are not identical. Instead, the bivalent represents four recombinant strands of DNA, each containing some material originating from the father and some from the mother. Recombination is central to understanding of much of the rest of this book, particularly of studies that make use of the phenomena of linkage and association, and further details on its implications for genetics are given in Chapter 3.

   After recombination, cellular division occurs, with a single representative of each of the 23 chromosomes (in the form of paired of chromatids) being randomly assigned to each daughter cell. This is then followed *without further DNA replication* by a second cell division in which each daughter cell receives a single member of each of the 23 sister chromatid pairs. Thus, in spermatogenesis, instead of a diploid 46,XY karyotype, each of the four spermatozoa have a haploid 23,X or 23,Y chromosome complement. Furthermore, because of recombination, each sperm is genetically distinct. In fact the random nature of recombination ensures that every gamete is genetically different. The

Fig. 1.4 Recombination and cell division during spermatogenesis. Pairs of homologous chromosomes migrate to lie together side by side in a structure called a bivalent. Each chromosome is replicated to form pairs of sister chromatids thus the bivalent contains four dsDNA molecules. Chromosomal material is almost randomly exchanged between homologues during recombination. After recombination, the chromosomes are no longer identical to those inherited from the parents. Furthermore, pairs of sister chromatids are not identical. After the second cell division, each sperm therefore contains different chromosomes.

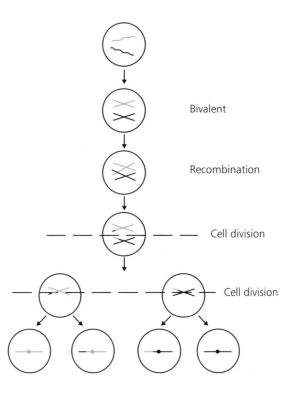

Bivalent

Recombination

Cell division

Cell division

process is similar in females except that at each division, most of the cytoplasm is assigned to one of the two daughter cells. The smaller of these are known as polar bodies, and these do not divide. Thus, the product of oogenesis is one large haploid 23,X egg and two small polar bodies.

Regardless of whether we consider meiosis or mitosis, if genetic information is to be accurately transmitted between generations (of cells or organisms) DNA must be replicated faithfully since it is this molecule that carries the information. How is this achieved?

As we have seen, it follows from the law of complementarity that if the sequence of one strand in the DNA helix is known, so is the sequence of the other. It is this property of DNA that permits faithful duplication. During DNA replication, the original (*parent*) strands of each chromosome are unwound and partially separated (Fig. 1.3). Each parent strand then serves as a template for the synthesis of a complementary *daughter* DNA sequence. At the end of this process, two double helix molecules are generated. Each helix contains one parental strand and one daughter strand. Since the daughter strand in one of the new pair of helices has effectively replaced the parental strand which is now in the other new helix, by the rules of complementarity, the two must be identical. This type of replication is called *semi-conservative* because each helix, while identical to the original helix, contains one strand from the original chromosome and a newly synthesized strand (Fig. 1.3).

Storage of information in DNA

DNA molecules are organized into stretches of sequence called *genes* (Fig. 1.5) which are the sequences encoding the information required to synthesize RNA molecules and thus proteins. During 2001 the Human Genome Project, which involves many centres worldwide, delivered an almost complete sequence of the human genome (The International Human Genome Mapping Consortium, 2001). Of the approximately 3.2×10^9 bases, only ~3 per cent is thought to encode proteins and this is estimated to be distributed across 30,000–40,000 genes (The International Human Genome Mapping Consortium, 2001). There are, however, other important functional elements to DNA including sequences that may alter the three-dimensional structure of the DNA and thus regulate transcription (this will be discussed in more detail later), centromeric elements necessary for DNA replication, telomeres, and stretches rich in the sequence CG (called CpG islands) which may activate or inactivate genes. These are often separated by long sequences that appear to serve no function or, perhaps more correctly, no known function.

In order to express the information encoded in genes, DNA must first be *transcribed*. In the *transcription* process a strand of DNA acts as a *template* for the synthesis of another macro-molecule called ribonucleic acid (RNA). RNA is similar to DNA except that the sugar backbone has a hydroxyl group on the 2' carbon and is called ribose rather than deoxyribose. With the exception of the replacement of thymine (T) with the base uracil (U), the bases in RNA are the same as in DNA. Furthermore, RNA and

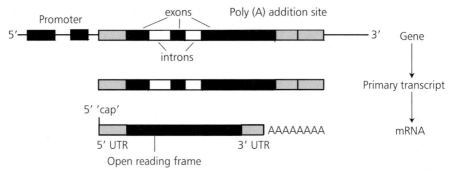

Fig. 1.5 The structure of a gene encoding messenger RNA (mRNA). Genes contain a promoter at the 5' end which binds RNA polymerase and initiates transcription. Although the whole gene sequence is transcribed into a primary transcript, only a proportion of the gene is represented in mRNA. Sequences present in mRNA (grey and black) are called exons. The intervening sequences are called introns (white). In producing a mature mRNA, the primary transcript undergoes RNA splicing to remove intronic sequence, is cleaved at its 3' end and a poly A sequence added (polyadenylation) while the 5' end is capped. The mature RNA sequence contains a stretch of sequence (black) encoding protein that is not interrupted by a stop codon. This is called the open reading frame. The residual sequences and both 5' and 3' ends are called untranslated regions because they do not encode protein.

DNA molecules can form stable structures by base pairing. This allows the information stored in a stretch of DNA to be copied to RNA (as for DNA replication) which can then carry the information to where it is needed in the cell. When the sequence of DNA that encodes an RNA molecule is given, it is by tradition the sequence of the DNA strand that is similar to the RNA molecule. This is called the *sense strand*. However, since the RNA is synthesized by the rules of base pairing, it is actually the complementary DNA strand, called the *anti-sense strand,* that acts as the template for RNA synthesis, and therefore strictly speaking, just as in DNA replication, the DNA molecule is not copied.

Genes encode several different species of RNA. Most encode RNA molecules called *messenger RNA* (mRNA) that contain information for making polypeptides. Others encode *ribosomal RNA* (rRNA) which forms ribosomes, *transfer RNA* (tRNA) which translate the information held in mRNA, and yet others encode a group of small cytoplasmic and nuclear RNA molecules. The functions of the latter two RNA molecules are not fully understood but involve a range of processes including regulating gene activation and mRNA processing.

Accessing the genetic code

mRNA

In order to understand the potential effects of DNA sequence diversity on function and disease, we must briefly consider how the information contained in DNA is processed. Genes that encode polypeptides are first transcribed into mRNA molecules. In order to initiate transcription, a number of *transcription factors* bind to the anti-sense strand of DNA at a rather ill-defined region called the *promoter* located at the 5′ end of the gene (Fig. 1.5). After the required transcription factors have bound to the promoter region, an enzyme called RNA polymerase II binds to the transcription factor complex and initiates transcription.

Whereas the promoter region is located in the immediate region flanking the 5′ end of the gene, there are other regions which can either enhance or repress the rate of transcription. These are called *enhancers* and *silencers* respectively and unlike promoter elements, their position relative to the gene is unpredictable. Thus, they may be located at the 5′ end, the 3′ end, within genes, or even several tens of thousands of bases away from the beginning of transcription. After initiation of transcription, the complete gene sequence is transcribed until the polymerase recognizes a transcription termination site. However, the immediate product of transcription does not constitute a functional mRNA. Instead, this *primary transcript* must be subjected to *post-transcriptional processing* (Fig. 1.5). There are several steps in post-transcriptional processing, but for the purposes of understanding the remainder of this book, the most important of these is *RNA splicing*. In most genes in man and other vertebrates, gene sequence comprises sequence that is represented in the mature mRNA (called *exons*), interrupted by stretches of DNA sequence that is not (called *introns*). Intron sequence is removed from the primary transcript by a process called RNA splicing to produce a contiguous mRNA species comprised only of exons (Fig. 1.5).

RNA splicing is of importance for several reasons. First, by altering the pattern of splicing (called *alternative splicing*), a single gene may result in several different mRNAs which in turn encode different polypeptides (many genes encoding neuro-receptors display this). Second, mutations in the gene sequence may cause either failure of splicing (*read through*), or excessive splicing (*exon skipping*). Clearly, this can alter the sequence of the mature mRNA, and as we shall see later, may result in disease.

Other post-transcriptional processes include (i) the addition of a cap to the 5′ end of the mRNA, (ii) cleavage of the RNA at a specified point or points in the 3′ end of the primary transcript, and (iii) addition of a hundred or so A monomers to the cleaved 3′ end to create a 3′ poly A tail (Fig. 1.5). Such processing is of importance to the mature mRNA, affecting mRNA stability, transport out of the nucleus, efficiency of translation, and possibly cellular differentiation.

Sometimes, both *in vivo* and in the laboratory, mRNA sequences are used as a template to synthesize DNA molecules by a process known as *reverse transcription*. Reverse transcription is particularly important to RNA viruses whose infective form consists of a genome of RNA rather than DNA. After infection, the RNA genome is reverse transcribed to DNA before replication. DNA molecules arising by reverse transcription are called *complementary DNA*, or more commonly, *cDNA*. From what has gone before, it should be clear that cDNA sequence will differ from that of the corresponding gene (the *genomic sequence*) by the absence of intronic sequence.

Codons, amino acids and polypeptides

Like nucleic acids, polypeptides are polymers, but in this case, the monomeric units are amino acids. The properties of a mature protein are dependent upon the sequence of amino acids, and it is mRNA molecules that carry the information specifying this sequence. Information is encrypted linearly in the sequence of mRNA, where groups of three adjacent bases serially code for each amino acid. These groups of three bases are called *codons* and, since there are 4 bases there are 4^3, or 64 possible codons. However, there are only 20 different amino acids and therefore most amino acids are coded for by more than one codon (Fig. 1.6).

Because there is not a unique codon-amino acid relationship, the DNA code is *degenerate*. Some bases (for example the third base in codons beginning AC) are called fourfold degenerate sites because any of the four possibilities encode the same amino acid. In contrast, some bases if changed will inevitably change the encoded amino acid. These are called non-degenerate.

The process of assembling a polypeptide based upon the information encoded in mRNA is called *translation*. How this decoding is achieved is described in the next section.

Reading the genetic code

Mature mRNA molecules move to the cytoplasm where protein synthesis takes place. mRNA molecules are first bound to ribosomes, which are large complexes composed of rRNA and ribosomal proteins. Yet another species of RNA called tRNA is then responsible for actually reading the genetic code.

tRNA molecules are small clover-leaf shaped molecules (Fig. 1.7). Specific species of tRNA covalently bind specific amino acids, a process that is catalysed by different

Fig. 1.6 The genetic code in eukaryotes chromosomal DNA. Note that the code is degenerate, that is several codons can specify the same amino acid. Codons are given as in DNA rather than RNA sequence. In RNA, T is replaced with U (Fig. 1.7).

AAA AAG } Lys	CAA CAG } Gln	GAA GAG } Glu	TAA TAG } STOP
AAC AAT } Asn	CAC CAT } His	GAC GAT } Asp	TAC TAT } Tyr
ACA ACG ACC ACT } Thr	CCA CCG CCC CCT } Pro	GCA GCG GCC GCT } Ala	TCA TCG TCC TCT } Ser
AGA AGG } Arg	CGA CGG } Pro	GGA GGG } Gly	TGA STOP
AGC AGT } Ser	CGC CGT	GGC GGT	TGG Trp TGC TGT } Cys
ATA ATG } Met	CTA CTG } Leu	GTA GTG } Val	TTA TTG } Leu
ATC ATT } Ile	CTC CTT	GTC GTT	TTC TTT } Phe

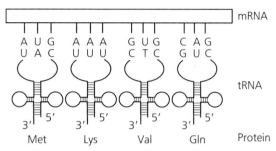

Fig. 1.7 Translation. Specific anti-codons in transfer RNA (tRNA) recognize mRNA codons by base pairing. Each tRNA brings a specific amino acid to the translational machinery consisting of an mRNA/ribosomal complex. In this way, the codon sequence is translated to an amino acid sequence.

specific *aminoacyl tRNA synthetases*. As well as binding specific amino acids, each species of tRNA has a specific three base *anti-codon sequence* (Fig. 1.7) which reads the genetic code by base pairing with the appropriate codon in the mRNA. Thus, specific tRNA molecules base pair with specific sequences in the messenger RNA, where they give up their specific amino acid for inclusion in the lengthening polypeptide. This is the mechanism by which linear genetic information encoded in DNA results in linear information encoded in a polypeptide. The flow of information from DNA to RNA by transcription and from RNA to polypeptide by translation is sometimes called the *central dogma* of molecular genetics. While such flow of information is the general rule, it is transgressed by DNA sequences that encode reverse transcriptase enzymes which permit information flow in the opposite direction, from RNA templates to DNA synthesis.

The codon AUG is the *initiation* codon for polypeptide synthesis. As it also specifies the amino acid methionine, all newly synthesized polypeptides start with methionine. After initiation, translation proceeds until a stop codon is reached (UAA, UAG or UGA). The sequence corresponding to that part of the mRNA that specifies polypeptide sequence is often called the *coding sequence*. However, in most mRNA species, the initiation codon is preceded by a number of bases that are not translated. This is called the 5′ UTR (*untranslated region*). Most mRNAs also contain several hundred bases at the 3′ end which are also not translated (3′ UTR), in addition to the poly A tail. Whilst 5′ and 3′ UTRs are not translated, they may have important roles in determining the efficiency of mRNA interaction with ribosomes, translation, and the half-life of the mRNA.

The codon sequence that is translated is called the *reading frame*. Any single-stranded molecule of RNA has three potential reading frames depending upon whether the first complete codon begins at base 1, 2 or 3 (see example below). Clearly, the order of amino acids specified is critically dependent upon which base forms the start of a codon unit. An *open reading frame* is a translation sequence uninterrupted by a stop codon.

> ACC AGG AUG GGG CAG = threonine arginine methionine glycine glutamate
> A CCA GGA UGG GGC AG = proline glycine tryptophan glycine

This is an important concept to grasp because, as we shall see later, mutations that alter the reading frame can have profound functional consequences.

Post-translational modification of polypeptides

Just as primary transcripts are modified to produce mature mRNA, so polypeptide products of translation are often modified before a functional protein is produced. Common modifications include the removal of an N-terminal signal sequence, glycosylation, phosphorylation, and cleavage. An example of latter is the large precursor polypeptide pro-opiomelanocortin (POMC), which is cleaved to produce ACTH (corticotropin), alpha-, beta-, and gamma-MSH (melanocyte stimulating hormones), beta- and gamma-LPH (lipotropins), CLIP (corticotropin-like intermediate lobe peptide), and beta-endorphin.

Variation

As we have seen earlier, most *somatic* (i.e. non germ-line) human cells are diploid and therefore contain two homologous copies of each autosome. Because the sequence of homologous pairs is almost identical, each homologue carries the same genes. However, while they are similar, because of *mutation* events that change the DNA sequence they are not identical. The science of genetics is essentially the study of this variation. Without DNA variation, every human would begin life as a clone, each with identical genetic information. All *observable* variation (which is called *phenotypic* variation) would therefore be due to the effects of the environment.

Before we discuss molecular genetics further, it is necessary to introduce some more terminology. The term *locus* is used to describe a sequence of DNA that is situated at a defined position on a chromosome. If the DNA sequence at a given locus (which may or may not contain a gene) shows variation in the population, that locus is said to show *allelic* variation with each DNA variant being called an *allele*. If the DNA sequence at a given locus on both homologous chromosomes is identical in an individual, then that person is said to be *homozygous* at that locus. If, however, the locus shows allelic variation and an individual carries two different allelic forms at a given locus, then that individual is described as *heterozygous* at that locus. Females are also homozygous or heterozygous at X chromosomal loci because they have two copies of the X chromosome. Males are *hemizygous* because they only carry a single copy of each sex chromosome.

Classes of variant

Mutation is the process of change in the sequence of DNA. Where allelic variants exist in DNA sequence with at least two alleles at the locus having frequencies of more than 0.01, the variants are called *polymorphisms*. Where the allele frequency is lower than this, the variants are often referred to as *mutations*. There are several classes of DNA variation.

1 *Base substitution*: a replacement of DNA sequence with a different sequence. This is the most common type of variation found in the human genome and is usually the result of a change in the sequence of a single base (e.g. C is replaced by a T).

2 *Deletion*: the loss of one or more bases relative to a reference sequence.

3 *Insertion*: a gain of one or more bases relative to a reference sequence.

4 *Chromosomal abnormalities*: includes abnormalities of chromosome number, structure, or parental origin, and will be considered below.

The effects of DNA variation

The effect of DNA variation depends on its nature, location and sequence context. The most common types are single base substitutions or single base insertions and deletions. These are often termed *single nucleotide polymorphisms* or SNPs. Such changes are extremely common in the human genome. On average, the sequence in a pair of homologous chromosomes carried by any individual is expected to show such variation around once every 1,500 bases or so, and since the vast majority occur in 'junk DNA', most have no phenotypic consequences. Mutations in coding sequence are much more likely to have phenotypic effects. These can be classified as follows:

1 *Nonsense mutations*: single base substitutions may result in a codon specifying termination of polypeptide synthesis instead of an amino acid. This is called *nonsense mutation*. Such changes can often be expected to have a dramatic effect, at least on the protein.

2 *Missense mutations*: single base substitutions may result in a change in amino acid. Such variants are called *missense mutations*, or *non-synonymous substitutions*. Non-synonymous substitutions are sometimes described as *conservative* where the amino acids specified by both variants are similar in chemical and physical properties (e.g. hydrophobic versus hydrophilic), and *non-conservative* where they are not. Because they are likely to result in a greater change in the properties of the protein, non-conservative substitutions are generally expected to have greater pathogenic potential than conservative ones.

3 *Silent substitutions*: many single base substitutions even within coding sequence are not expected to have a phenotypic consequence. This is likely if substitution occurs at a degenerate base, (particularly the third base in a codon) and therefore does not result in a change in amino acid. Changes in coding sequences that do not result in altered amino acid sequence are often called *synonymous* or *silent* substitutions.

4 *Frameshift mutations*: small insertions and deletions in coding sequences may have dramatic phenotypic consequences depending upon the number of bases inserted or deleted. This is because the genetic code is based upon the triplet codon system. Consequently, a single base deletion will cause the entire downstream sequence to be read out of frame resulting in a dramatically different polypeptide. Note that a deletion/insertion of a complete codon (or any number of complete codons) may actually have a lesser impact on the mature protein than deletion of one or two bases.

Most variation outside coding sequence is expected to have no phenotypic effect, but this is not always so. As we have seen, sequences at the 5′ end of genes (promoters) but also elsewhere (enhancers and repressors) are important in regulating gene expression. Sequences in introns may serve similar functions and semi-conserved sequences near intron/exon boundaries are also important in determining gene splicing. Furthermore, sequences in the UTRs are important in stabilizing mRNA and altering its translational efficiency. Thus, it should not be assumed that just because a variant does not result in a change in amino acid, it has no phenotypic consequences. Even so-called silent substitutions in coding sequence may alter gene splicing or transcriptional/translational efficiency. Furthermore, some species of mRNA are prone to a phenomenon known as *RNA editing*. RNA editing is the process of changing nucleotides in RNA after transcription. This may result in a change in the encoded protein. Examples of interest to psychiatric geneticists include the mRNA species encoding several glutamate receptors and the amyloid precursor protein. Silent polymorphisms that alter mRNA sequence may alter editing and have a downstream effect on encoded proteins.

At present, the majority of known pathogenic mutations *do* involve amino acid changes, but there are already examples of relatively simple phenotypes that are caused by DNA variation affecting gene expression. Classic examples include the thalassaemias, haemophilias, and, at a slightly more complex level, Fragile X and Friedreich's Ataxia. Furthermore, such variants have also been implicated in susceptibility to complex diseases such as diabetes and epilepsy and, more controversially, Alzheimer's disease, schizophrenia and bipolar disorder.

Chromosomal abnormalities

Chromosomal number

Rarely, individuals are found containing three copies of autosomes as well as an extra sex chromosome (karyotypes 69,XXX, 69,XXY, or 69,XYY). This is called *polyploidy* and may be caused by simultaneous fertilization of an egg by two sperm or alternatively, fertilization that involves a diploid gamete. More commonly, numerical increases in chromosomal number result from possession of an extra copy of a single chromosome. This is called *trisomy* and the best known example of this is trisomy 21 or Down syndrome (male karyotype 47,XY,+21) in which individuals carry an extra copy of chromosome 21. The converse situation where an entire chromosome is missing is called *monosomy* and, with the exception of Turner's syndrome females who carry only a single X chromosome (45,X), monosomy is lethal and therefore is not found.

If the chromosomal abnormality arises by failure of separation of chromosomes during mitosis rather than at conception, some cells in the body will have a normal chromosomal number, while others will not. This is called *mosaicism*, and the phenotypic consequences can vary dramatically depending upon the proportion of cells with the abnormal number of chromosomes. (Note it is normal for some cells types to lack any chromosomes, for example mature red blood cells, while some rapidly dividing cells are polyploid, for example hepatocytes and megakaryocytes.)

Structural rearrangements

Other chromosomal abnormalities occur as a result of breakage of chromosomes. These include:

1 *Deletion* where a large stretch of DNA is lost.

2 *Inversion* where a piece of chromosome breaks and becomes reattached in the opposite orientation. Consequently there is no loss of DNA although individual genes may be interrupted.

3 *Duplication* where a segment of chromosome is duplicated.

4 *Translocation* of segments from one chromosome to another (non-homologous) chromosome. Translocations may also produce highly disorganized chromosomes, for example, a *Robertsonian* translocation when two different chromosomes fuse at or near the centromere, or *ring* chromosomes where a single chromosome breaks at two positions and the ends of the middle segment join to form a circular fragment.

Translocations can be categorized into *balanced* or *unbalanced*. If an individual has part of chromosome 1 attached to chromosome 2 and vice versa, the offspring may be phenotypically normal if both abnormal chromosomes are simultaneously transmitted. This is said to be a balanced translocation. However, if the abnormal chromosome 1 is transmitted with the normal chromosome 2, the offspring will receive an extra dosage of chromosome 2 (having a normal chromosome 2 plus the chromosome 2 fragment on the abnormal chromosome 1) and a deficit of chromosome 1 (since part of this is lost to the untransmitted abnormal chromosome 2). The offspring with this unbalanced translocation will have both a *partial monosomy* and a *partial trisomy*.

Parental origin

Occasionally, both homologous chromosomes originate from the same parent. If both copies of the chromosome are different (i.e. they represent the simultaneous transmission of both homologous chromosomes carried by one parent) the situation is termed *uniparental heterodisomy*. However, if they are identical copies of a single parental chromosome, the situation is called *uniparental isodisomy*. It might be thought that if a child possesses a pair of chromosomes from a phenotypically normal parent, this should not pose a problem, however, this is not the case. As we will discuss later (see section on imprinting) chromosomes are sometimes 'marked' or *imprinted* as they are transmitted from a parent. Where this occurs, chromosomes from both parents are not equivalent, and abnormalities can occur when an individual does not receive an imprinted sequence from each of his parents. Classic examples of relevance to psychiatry are Prader–Willi and Angelman syndromes which result when both copies of chromosome 15 originate from the mother or the father respectively. This will be discussed further in Chapter 5. Uniparental isodisomy also carries an increased risk of recessive disorders (see Chapter 2), since the child will be homozygous for all variants on that chromosome.

The mitochondrial genome

So far in this chapter, we have focused on chromosomal DNA. However, mitochondria contain their own small genome, and also have their own machinery for protein synthesis. The mitochondrial genome consists of a single circular double helix of 16,500 bases which encodes a small number of proteins and RNA molecules required for oxidative phosphorylation. It should be noted that the majority of proteins and RNA molecules required for mitochondrial function are actually encoded in chromosomal DNA and have to be imported from the cell nucleus. The biology of the mitochondrial genome is essentially the same as that of the nuclear genome. However, although both are based upon the codon system, there is not an exact match between the two codes. Also, mitochondrial genes have no introns, and genetic information is extremely densely packed with more than 90 per cent of the sequence representing coding DNA.

Sperm only very rarely contribute mitochondria during fertilization and consequently, diseases of the mitochondrial genome are almost always maternally transmitted. Furthermore, because there are several thousand mitochondrial DNA molecules per cell, the concepts of homo-, hemi- or and heterozygosity do not apply. Instead, when considering mitochondrial variation, the issue is the proportion of DNA molecules carrying a variant mutation. This may be 100 per cent, 0 per cent (*homoplasmy*) or any intermediate ratio (*heteroplasmy*). Furthermore, because mitochondria are randomly dispersed into daughter cells during cell division, the ratios of heteroplasmy can vary between mother and child and between individual cells. This all has an impact on the severity of the phenotype of diseases caused by mitochondrial mutations.

Unstable DNA

We tend to think of DNA as a relatively stable structure that, with the exception of uncommon mutation, is transmitted faithfully. However, it is clear that for DNA sequences containing repetitive sequences, this is not always the case. Such repeats are often unstable, that is they are prone to extremely high rates of mutation, and in some cases these may be pathogenic. For example an unstable 12 bp repeat in the 5′ flanking region of the cystatin B gene is the most common pathogenic mutation in type 1 progressive myoclonic epilepsy (Lafreniere *et al.* 1997). However, in recent years, a particular class of repeat—the trinucleotide repeat—has emerged as playing an important role in human diseases, particularly neurological and psychiatric disorders, and we will focus in this section on this class of unstable DNA sequence.

Trinucleotide repeats are tandem repeat elements where the unit that is repeated consists of three bases, for example CAGCAGCAG. At a given locus, trinucleotide repeat sequences are often polymorphic in repeat numbers making them useful genetic markers for molecular genetic studies (see below). However, when they are located near or within genes, possession of a relatively large copy number of repeats (called an *expanded trinucleotide repeat*) can sometimes lead to disease. Expanded trinucleotide repeats were first identified as pathogenic in classic fragile X syndrome (FRAXA, Yu *et al.* 1991). Normal individuals carry up to fifty copies of the trinucleotide sequence CCG in the FMR-1 gene, whereas affected individuals carry more than two hundred copies of the same motif. Since this discovery many other disorders have been attributed to expanded trinucleotide repeats, including myotonic dystrophy (Brook *et al.* 1992), Huntington's disease (Huntington's Disease Collaborative Research Group 1993), spino–bulbar–muscular atrophy (SMA) (La Spada *et al.* 1991), and numerous forms of ataxia.

The mechanisms by which trinucleotide repeats lead to disease are still not understood. Some with very large repeats in the promoter (FRAXE), untranslated region (FRAXA) or intronic regions (Friedreich's ataxia) of genes appear to exert their effects by silencing transcription. However, the largest group of *identified* trinucleotide repeat diseases are neurodegenerative disorders (e.g. HD) caused by moderate sized CAG repeats encoding polyglutamine sequences. The pathogenic mechanisms of this group are essentially unknown, and although neurones in people with these diseases have characteristic protein deposits called nuclear inclusions (NI) it now appears that these are unlikely to provide a primary explanation (Cummings and Zoghbi 2000).

Anticipation

It has been known for many years that myotonic dystrophy tends to get worse from one generation to the next. Progression in pedigrees occurs from mild, late-onset illness in a grandparent through offspring with an earlier-onset severe phenotype, to, in the youngest generation, an extremely severe congenital form of myotonic dystrophy. This progression in age at onset and/or severity is called *anticipation*, and is seen in some of the other trinucleotide repeat disorders, albeit generally to a lesser degree.

Anticipation is at least partly explained by (a) the propensity of trinucleotide repeats to increase in size during transmission from parent to child, an instability that has led to their description as 'dynamic mutations' and (b) correlation between age at onset and/or severity with repeat size. Anticipation was first described as a phenomenon related to severe mental disorder by Mott (O'Donovan and Owen 1996) not long after

Mendel's laws had been 'rediscovered' (see Chapter 2) but because there was no known mechanism by which anticipation could be explained and because the phenomenon appeared to violate Mendelian principles, it was dismissed by many geneticists as an artefact (O'Donovan and Owen 1996). However, the discovery of the association between anticipation and trinucleotide repeats has brought about renewed interest in the phenomenon of anticipation in psychiatric disorders and has led to the hypothesis that schizophrenia, bipolar affective disorder, and even anxiety disorders, may at least in part be attributed one or more pathogenic trinucleotide repeat. This will be discussed further in Chapters 9 and 10.

Methylation, imprinting and epigenetic inheritance

So far, our discussion has concerned the genetic information carried by DNA sequence. However, there are mechanisms by which variation in the rate at which a gene is expressed can be inherited in the absence of a change in the primary DNA sequence. This is often referred to as *epigenetic* inheritance.

Although our understanding of epigenetic phenomena is far from complete, we do have a limited understanding of several important mechanisms. One mechanism that is known to silence gene expression involves chemical modification (methylation) of cytosine molecules in stretches of DNA that are rich in the dinucleotide sequence CpG (called CpG islands). CpG islands are found at the 5' end of around half of all genes, and methylation here leads to changes in chromatin structure, which makes the DNA unavailable for transcription and thus silences the genes. This is an important mechanism in cell differentiation. Tissue development requires cells with specific patterns of gene expression, and, moreover, that daughter cells stably inherit these patterns after mitosis.

Methylation and gene silencing are also important in X inactivation (see above), Fragile X syndrome, and also *genomic imprinting*. The latter arises because some sequences of DNA are differentially methylated in male and female germ cell lines. This results in variation in the transcriptional activity of genes depending upon the transmitting parent, and is one of the mechanisms that is responsible for the problems that can arise as a consequence of uniparental disomy (e.g. Prader–Willi syndrome). Another example of the importance of methylation in human disease is that of Rett syndrome, one of the most common causes of mental retardation in females. At least in some patients, Rett syndrome is caused by mutations that disrupt the MECP2 gene (Amir *et al.* 1999), which encodes a protein that binds CpG dinucleotides and mediates gene silencing. Rett syndrome is almost exclusively found in females, presumably because mutations in this X-linked gene are almost invariably lethal in males.

Genetic information can also be altered by other influences on chromatin structure. For example gene activity can be dependent upon the precise chromosomal localization of the gene (*position effects*) and therefore the activity of a gene may be altered by translocation, even if the gene itself is not disrupted by the chromosomal breakage.

It appears that in some species, repetitive stretches of DNA on different chromosomes can interact in a way that also results in methylation and gene silencing. Furthermore there is accumulating evidence that RNA molecules transcribed at one locus may interact with homologous DNA sequences elsewhere in the genome, again resulting in stable gene silencing. These mechanisms, and others that are still unknown may contribute to the poorly understood phenomenon called *paramutation* (Wolffe and Matzke 1999). This describes an interaction between different alleles or even different loci, which results in a stable alteration in their functional state. Interaction may occur between different alleles which are subsequently separated during gametogenesis. Consequently, the properties of an inherited gene may in part be dependent on a gene sequence that is not actually co-inherited. Clearly, this flouts what we generally think of as genetic inheritance. Furthermore, if parental experiences affect the expression of RNA molecules involved in RNA induced DNA silencing, it is conceivable that heritable changes in gene activity might result from environmental stimuli.

The result is that inheritance patterns of disorders caused in part by epigenetic events will not necessarily display classical patterns of inheritance (see Chapter 2). Although as yet there are no examples of human diseases with epigenetic inheritance, a phenotype in mice has recently been described (agouti) in which the phenotype of offspring carrying the agouti genotype is dramatically affected by the maternal phenotype (Morgan *et al.* 1999). Although the mechanism is still unknown, the variable phenotype in animals with identical genotypes has been shown to be the result of epigenetic factors rather than environment. It remains to be seen if similar mechanisms are common in humans, but the agouti phenomenon provides a clear example of how epigenetic inheritance can result in a phenotype with variable expressivity and incomplete penetrance, both of which are features of neuropsychiatric phenotypes (see Chapter 2).

Basic techniques in molecular genetics

The key technologies in molecular genetic analysis are those that aim to detect and measure DNA sequence variation. Possessed with such tools, the task of geneticists is then to correlate the possession of DNA variants with different aspects of the phenotype either in closely related individuals (usually by *linkage*) or in individuals who are so distantly related that they are generally considered unrelated (*association*). How measurements at the molecular level are correlated with phenotypes in these studies will be discussed in detail in Chapter 3. This section will simply give an outline of how the measurements are made.

The polymerase chain reaction

Most modern molecular genetic studies employ the *polymerase chain reaction* (PCR) to identify smaller fragments of DNA that can be conveniently manipulated in the laboratory. It is not necessary to understand the technical details, but for the reader to gain a better appreciation of subsequent chapters, a brief outline of PCR now follows.

PCR is basically a process that allows exponential replication of a small sequence of DNA, typically 200–1000 bases in size. The first step is to identify the DNA sequence that defines each end of the fragment of interest. A pair of synthetic, single stranded DNA molecules, usually 20 bases or so in length, are then commercially purchased. These are called *primers* because they are used to prime synthesis of DNA. One primer is identical to the 5′ end of one of the strands of genomic DNA to be amplified, the other is identical to 5′ end of the complementary strand. Genomic DNA is mixed with the primers, a heat stable DNA polymerase (e.g. *Taq* polymerase) and a mixture of the individual nucleotides (A, T, C and G, referred to often as dNTPs) from which DNA is manufactured. The mixture is heated, which causes the double stranded DNA to *denature*, that is, to become single stranded. Upon cooling, the primers bind to the single stranded genomic DNA by base pairing, a process that is called *hybridization*. The primers are then extended from their 3′ ends, resulting in a larger DNA molecule that is complementary to the DNA strand to which the primer is bound, a process not dissimilar to chromosomal replication but over a very small region. At the beginning of this process, the reaction mix contains equal molar representations of all fragments of genomic DNA. At the end, because it has been replicated, the molarity of the target sequence is around double this. The whole process is then repeated a second time, but on this occasion, both genomic DNA and the newly synthesized DNA molecules are duplicated, and therefore there are four copies of the target sequence. This process is repeated many times, with each cycle exponentially increasing the number of copies of the target sequence. At the end of the process (which is automated on a machine called a *thermocycler*) the starting concentration of the target fragment has been increased by $\sim 10^5 - 10^6$ to a concentration that can easily be detected by several techniques (see below).

Genotyping polymorphisms that alter the size of fragments

Genotyping is the process of determining which alleles are carried by an individual at a given locus (called the genotype). The polymorphisms most commonly used in genetic studies are called short tandem repeat polymorphisms (STRPs). These are polymorphisms where a certain sequence is repeated in tandem several times but the exact number of times a sequence is repeated varies in the population. The most commonly used class of STRPs at present are dinucleotide repeats, where the repeat unit consists of multiple copies of the sequence CA (or any other dinucleotide). However, other STRPs consist of differences in the number of repeats of a single nucleotide (mononucleotide repeat), three nucleotides (trinucleotide repeat), four nucleotides (tetranucleotide repeat) and so on. In all STRPs, the fundamental difference between alleles is the number of repeats and therefore the size of the target molecule. Genotyping is simply a case of measuring the size of the repeat.

This is generally achieved by first amplifying genomic DNA by PCR in the presence of primers that flank the STRP. One of the primers is usually tagged with a fluorescent dye. The PCR products are then subjected to gel electrophoresis, a process where the

DNA molecules are placed at one end of a gel-like substance and subjected to a voltage gradient. The PCR products, being negatively charged, migrate along the gradient towards the positive terminal at a rate that is inversely related to their size. The DNA fragments in the gel can then be visualized by the use of one of several DNA staining techniques, or if fluorescent dyes have been used, by a fluorescence scanning device. This allows the size of the amplified molecules to be determined and the alleles identified.

Genotyping single nucleotide polymorphisms

SNPs are the most common types of polymorphism in the genome and are increasingly thought of as the polymorphisms of the future for genetic marker studies. At the time of writing, the most common method for genotyping SNPs is based upon enzymes called *restriction endonucleases*. Such enzymes recognize and cleave DNA at specific stretches of DNA sequence of usually between four and eight bases. If a polymorphism changes the recognition sequence for one of these enzymes, one allele will be cut, the other will not. After PCR, the reaction products are digested with an appropriate enzyme which will result in two smaller fragments in the presence of the allele that is cleaved by the enzyme, or one large fragment in the presence of an allele that is not. Polymorphisms like these are called *restriction fragment length polymorphisms* (RFLPs). As the assay is based upon fragment size, the genotype can be determined by electrophoresis.

This PCR based process has, in most instances, replaced a previous method for genotyping RFLPs called Southern blotting. Southern blotting is now restricted to fairly specific applications and will not be considered further in this chapter. The interested reader is directed to either of the two recommended textbooks for further details.

Emerging methods

There are numerous new techniques for genotyping SNPs on the horizon, some of which have already been installed in a limited number of laboratories. It is not yet clear which technique will emerge as the method of choice. Here, we consider a few of the most promising.

Mini-sequencing/primer extension

After PCR, a third primer (the extension primer) is added. This is designed so that it anneals to one of the strands of the PCR product with its 3′ end immediately adjacent to the polymorphic base. A second reaction is then performed in the presence of specially designed nucleotides, which permit the primer to be extended by only a single base. The specific base that is added to the primer will be the base that is complementary to the base at the polymorphic site. If the polymorphism at a given site is either an A or a C, then by complementarity, either a T or a G will be incorporated respectively. The extended primers differing at one end by a single base can then be detected by a variety of methods. For example the different nucleotides can each be labelled with one of four different fluorescent dyes, each of which emits light of a different colour upon laser stimulation. Thus, one colour indicates extension by a T, another indicates extension by a G, and simultaneous emission of both colours upon laser stimulation indicates extension of some molecules by a T and some by a G (and therefore the person genotyped is heterozygous for the polymorphism).

There are several other promising detection methods. One depends upon measuring the mass of the extended molecule via matrix-assisted laser desorption/ionization time-of-flight mass spectrometry (MALDI-TOF MS). This is possible because each nucleotide has a different molecular weight. Regardless of the detection method, because this reaction is based upon directly determining the sequence of the PCR product, the assay is sometimes called *mini-sequencing*, although it also has a variety of other names such as *primer extension* or *single base extension*.

Micro-arrays

Micro-arrays are also sometimes called *DNA chips*. SNP genotyping on DNA chips may be based upon primer extension, or *allele specific hybridization*. In general terms, a DNA sequence that is perfectly complementary to another DNA sequence will bind (*hybridize*) more efficiently to that sequence than to a very similar sequence where there is a small mismatch in complementarity. In the case of the SNP above, amplified alleles containing an A at the polymorphic site will hybridize to a complementary DNA molecule containing a T in the correct position in preference to a molecule that is identical in all respects other than that it contains a G instead of a T. The opposite is true for alleles containing a C.

In one application of micro-arrays, PCR products of a single locus from thousands of subjects are individually spotted out (or arrayed) in an area approximately the size of a postage stamp. This micro-array can then be hybridized with two short synthetic DNA molecules which are essentially the same as primers, but which in this case are called *probes*. Each probe is perfectly complementary to one of the two alleles, and each is tagged with a different fluorescent dye. After hybridization, the micro-array is analysed by a confocal laser system to determine the colour of the probe(s) bound to each spot that represents a PCR reaction from an individual. The colour of the bound probes then defines the genotype of that individual at that locus.

Alternatively, instead of arraying PCR products from thousands of individuals, each spot on the micro-array can contain a different probe, with pairs of spots representing a pair of possible alleles at a given locus. Under this scenario, a chip can be constructed that will allow thousands of loci to be arrayed on a single micro-array. Different PCR products representing thousands of different loci from an individual can then be used to probe the micro-array, and the pattern of binding at each pair of spots then yields the genotype at each locus.

Both types of method, where numerous pieces of genetic information are simultaneously assayed in a single reaction, are sometimes referred as 'massively parallel'.

Regardless of the technology, be it chip based, MALDI-TOF, or otherwise, all available SNP genotyping assays still require loci to be amplified by PCR. In order to advance genotyping forward for some of the applications discussed in the chapter dealing with association studies (Chapter 3) this is a step that will have to be overcome, though it is not at all clear how.

Detecting unknown DNA variation

The term 'mutation detection' describes the process of identifying unknown variation in a defined stretch of DNA. There are currently numerous mutation detection techniques available. The definitive method is *DNA sequencing*, which is the process of identifying the complete sequence in a fragment of DNA. Thus, one method for looking for population variation is to sequence the same stretch of DNA in many different people. However, sequencing is expensive, and outside the major public and private genome centres, labour intensive. For this reason, prior to sequencing, most laboratories undertake mutation-scanning first to identify fragments that are likely to contain variants.

One of the simplest and most widely used of these is *single strand conformation analysis* (SSCA). Even a single base change in a single stranded PCR product can alter its three-dimensional folding pattern (or its secondary structure), particularly if the DNA molecule is small (around 200 bases). This can be detected as altered migration of single stranded PCR products during electrophoresis. *Heteroduplex analysis* (HA) is another simple widely-used method. After amplification, the PCR products are denatured and allowed to slowly re-anneal. In heterozygotes (who carry allele 'A' and its variant 'a'), some of the re-annealed double stranded DNA molecules will carry sequence corresponding to allele 'A' on both strands, some will carry allele 'a' on both, and some will have allele 'A' sequence on one strand and 'a' sequence on the other. Perfectly complementary sequences (AA, aa) are called *homoduplexes* while imperfectly matched ones (Aa, aA) are called *heteroduplexes*. Heteroduplexes, indicating the presence of a sequence variant, can often be distinguished from homoduplexes because the latter migrate faster during electrophoresis. There are many more complex mutation scanning techniques (see Cotton 1997). In general the relatively simple methods (SSCA, HA) tend to lack sensitivity whereas the more sensitive methods (for example denaturing gradient gel electrophoresis (DGGE), dideoxy fingerprinting (ddF)) are often relatively labour intensive, expensive, and time consuming.

However, again, these technologies are rapidly evolving, and the degree of automation and sensitivity improving. For example HA analysis can be now performed using a high performance liquid chromatography (HPLC) column rather than a gel. This method, called denaturing HPLC (DHPLC) along with other semi-automated technologies, mean that whereas a few years ago, screening a single gene for polymorphisms was a fairly labour-intensive process, now it is possible to screen hundreds, possibly even thousands of genes, in a single laboratory.

Information technology and positional cloning

Positional cloning refers to a set of techniques by which disease related genes are identified on the basis of their chromosomal location. Classically, positional cloning is achieved by first identifying the chromosomal location of a disease gene by identifying polymorphic markers whose alleles tend to be inherited along with the disease in families with more than one affected member. Cosegregation of marker alleles with disease alleles occurs when the two loci are so close together on the same chromosome as to cause departures from independent assortment; a phenomenon known as genetic linkage (see Chapter 3). This process is greatly simplified by the availability of fairly detailed 'maps' of polymorphisms called *DNA markers* which are sequence variants known to reside at particular locations in the genome. By studying the segregation of a series of markers of known location it is possible systematically to search the genome for genetic linkage.

After a manageable region of interest containing a disease gene has been identified, the next step is to identify the specific sequence variations in that region that confer susceptibility to disease. A number of complementary strategies may be used. However, most first require the creation of a complete or almost complete representation of the candidate region as *cloned* DNA. In brief, DNA from the whole genome, a single chromosome or even a fragment of a chromosome is broken into fragments. These are then randomly inserted into *cloning vectors* that allow replication of the human fragment in a host organism, usually a bacterium or yeast. For example, genomic DNA may be cut or physically broken into pieces of approximately 10^6 bases in length, which are then randomly inserted into *yeast artificial chromosomes* (YACs). The YACs are then inserted into a host strain of yeast which replicates the human DNA along with its own. After many cycles of reproduction, each single yeast organism that received a YAC will have grown into a yeast colony containing many copies of that single YAC. The YAC DNA can then be extracted, yielding large amounts of that specific human DNA fragment for further laboratory manipulation. A number of cloning systems are available, each with their own advantages and disadvantages; for example, YACs and *bacterial artificial chromosomes* (BACs) carry relatively large fragments of DNA. Intermediate sizes of DNA are carried by P_1 artificial chromosomes (PACs), *cosmids* and some *phage*, and relatively small fragments of DNA are carried by phage and *plasmids*.

The next step is to isolate a series of individual clones containing DNA fragments originating from the candidate region until the whole region of interest is represented in the test tube by a series of contiguous fragments of DNA, abbreviated to a *contig*.

The final stages are to identify the relatively small proportion of the DNA within the region corresponding to genes, screen these for sequence mutations/polymorphisms, and then test whether any of the detected variants are more common in patients with the disease compared with people who are unaffected. This whole process involves a great deal of cost and labour such that analyses even of relatively small genomic regions of one to two million bases would be massive undertakings. Indeed, although the approximate chromosomal location of Huntington's disease gene was identified in 1983, it took a further 10 years to actually identify the gene itself.

Many of these steps have now been largely replaced by cloning 'in silico'. Increasingly, raw genomic sequence, gene sequence, and even the identification of at least common polymorphisms can all be achieved by searching appropriate computer databases. Thus, the genetic researchers of the future will increasingly depend at least as much upon expertise with bioinformatics as they do 'wet' laboratory experiments. The interested reader is directed to *http://www.ncbi.nlm.nih.gov/*, *http://www.ebi.ac.uk/*, *http://www.ensembl.org/*, which are useful entry points to the electronic data, and to *http://www.ncbi.nlm.nih.gov/entrez/query.fcgi?db=OMIM* which contains summary information about the genetic aspects of most diseases.

Candidate gene and positional candidate gene analysis

A major benefit of the positional cloning approach is that genes can be identified without any knowledge of disease pathophysiology. However, where there are good clues to the pathophysiology of disease, or where the genetic effects are so small that they are difficult to detect by (at least today's) positional cloning technologies, hypothesis driven approaches are appropriate. This will be discussed in detail in the chapters on linkage and association. Essentially the task of the candidate gene approach is to select genes with plausible a priori roles in the disease pathogenesis (called functional *candidate genes*), screen the candidate gene sequences for DNA sequence variation and test populations of affected and unaffected subjects for evidence that variants in the candidate gene are more commonly associated with affected status. As mentioned above, the technologies that make this approach possible (availability of gene sequence, mutation detection and genotyping) are all undergoing rapid improvements, and whereas testing a single candidate gene was rather time consuming in the past, it is now possible to test many candidate gene hypotheses in relatively short periods of time. Furthermore, although it is some considerable distance in the future, a catalogue of every human polymorphism in every gene will eventually become available, making the mutation screening step at least partially obsolete. Ultimately, one can envisage the possibility of buying micro-arrays with either every known common polymorphism represented, or alternatively, specific chips which will allow whole systems of candidate genes (e.g. the genes known to be involved in neurotransmission or neurodevelopment) to be tested for association with disease. However, with multiple testing on such massive scales on the horizon, appropriate statistical techniques and sample sizes will be required. This will also be discussed further in Chapter 3.

Gene expression

The function of a cell depends upon its pattern of gene expression and it follows that altered function as part of a disease process is likely to be accompanied by alteration in gene expression profile. This offers the possibility of identifying the pathophysiological pathways or genes involved in a disease process by looking for differences in gene expression in tissues from individuals with and without that disease. Most readers of this book will already be familiar with this broad approach. For example, there are numerous published studies looking at differences in metabolic enzyme activities or neuro-receptor densities in psychiatric patients versus controls.

Gene expression can be quantitated at the level of mRNA or the level of translated protein. It is not necessary in order to understand this book to know the detail of the techniques used, but a small amount of terminology might be helpful. Individual mRNA species can be assayed by techniques such as *Northern blotting*, PCR amplification after reverse transcription of the mRNA (called *RT-PCR*) or a variety of 'solution hybridization' methods. However, it is now possible to measure the expression levels of thousands of genes simultaneously using so called 'gene expression chips'. These are simply micro-arrays containing thousands of probe DNA sequences, each of which is individually complementary to a different species of mRNA.

In order to measure numerous species of mRNA simultaneously, the total RNA content of a cell or a tissue is first reverse transcribed in the presence of a fluorescent dye (or some other labelled molecule) to create labelled cDNA. This is then hybridized to the micro-array. The amount of labelled cDNA that binds to each arrayed probe sequence is then measured, giving an indication of the abundance of each specific mRNA complementary to each of the arrayed sequences. By comparing the patterns of hybridization from either different tissues or different individuals, it is possible to create a profile of the genes that are differentially expressed either in different tissues or between subjects. Analogous technologies have been developed for the study of proteins. These will allow differential gene expression to be quantitated at the level of protein abundance, and will even allow the function of thousands of proteins to be analysed simultaneously.

Given that many of our clues to the aetiology of mental disorders have been provided by a range of neuroscientists measuring gene expression at the level of enzyme activity, receptor abundance and so on, the development of technology allowing virtually every protein or mRNA in a tissue to be assayed is likely to provide important opportunities for developing new hypotheses concerning the aetiology of mental disorders.

Further reading

Strachan T. and Reid A. P. (1999) *Human Molecular Genetics*, 2nd edn. Oxford, BIOS Scientific Publishers Ltd.

Lewin B. (2000) *Genes VII*, Oxford, Oxford University Press.

References

Amir R. E., Van den Veyver I. B., Wan M. *et al.* (1999) Rett syndrome is caused by mutations in X-linked MECP2, encoding methyl-CpG-binding protein 2. *Nature Genetics* **23**:127–8.

Brook J. D., McCurrach M. E., Harley H. G. *et al.* (1992) Molecular basis of myotonic dystrophy: expansion of a trinucleotide (CTG) rpeat at the 3′ end of a transcript encoding a protein kinase family member. *Cell* **68**:799–808.

Cotton R. G. H. (1997) Slowly but surely towards better scanning for mutations. *TIGS* **13**:43–7.

Cummings J. and Zoghbi H. Y. (2000) Fourteen and counting: unraveling trinucleotide repeat diseases. *Human Molecular Genetics* **9**:909–16.

Lafreniere R. G., Rochefort D. L., Chretien N. *et al.* (1997) Unstable insertion in the 5′ flanking region of the cystatin B gene is the most common mutation in progressive myoclonus epilepsy type 1, EPM1. *Nature Genetics* **15**:298–302.

La Spada A. R., Wilson E. M., Lubahn D. B. *et al.* (1991) Androgen receptor gene mutations in X-linked spinal and bulbar muscular atrophy. *Nature* **352**:77–9.

Morgan H. D., Sutherland H. G., Martin D. I. *et al.* (1999) Epigenetic inheritance at the agouti locus in the mouse. *Nature Genetics* **23**:314–8.

O'Donovan M. C. and Owen M. J. (1996) Dynamic mutations and psychiatric genetics. *Psychological Medicine* **26**:1–6.

The Huntington's Disease Collaborative Research Group (1993) A novel gene containing a trinucleotide repeat that is expanded and unstable on Huntington's disease chromosomes. *Cell* **72**:971–83.

The International Human Genome Mapping Consortium (2001) Initial sequencing and analysis of the human genome. *Nature* **409**:860–921.

Wolffe A. P. and Matzke M. A. (1999) Epigenetics: regulation through repression. *Science* **286**:481–6.

Yu S., Pritchard M., Kremer E. *et al.* (1991) Fragile X genotype characterized by an unstable region of DNA. *Science* **52**:1179–91.

Chapter 2

Quantitative genetics

Alastair Cardno and Peter McGuffin

Introduction

Quantitative genetics has traditionally been based on investigations of continuous or 'metric' phenotypes, where there are measurable differences between individuals, rather than on phenotypes that are simply present or absent (Falconer and Mackay 1996). Most characteristically, this involves phenotypes that tend to show a normal distribution in the population, such as height, weight or IQ-test scores. However, a similar approach can be taken to studying phenotypes that occur as an ordered series of classes (e.g. graded severity of a disorder), or as dichotomous (present/absent) phenotypes that do not show classical Mendelian inheritance, and this includes the presence or absence of all common psychiatric disorders.

Quantitative genetic investigations of psychiatric disorders are primarily concerned with the following questions:

1　Do genetic factors contribute to the aetiology of the disorder?; if so,

2　What is the relative contribution of genetic and environmental factors?; and

3　What is the mode of inheritance?

These questions are addressed by investigating the pattern of resemblance for the disorder in various classes of relatives who differ in their degree of genetic relatedness or degree of environmental sharing (Neale and Cardon 1992; McGuffin *et al.* 1994a; Falconer and Mackay 1996; Plomin *et al.* 2001; Sham 1998).

Searching for genetic effects

Family studies

Family studies investigate the degree of familial clustering of a disorder. The procedure is as follows:

First, a systematic collection of families is made via a sample of index cases or probands. For studies of relatively infrequent disorders, such as psychoses, probands are usually ascertained from a series of consecutive hospital admissions, or a register of hospital admissions linked to a population register. However, even when systematically collected, probands ascertained via psychiatric services can be subject to sampling biases.

Compared with cases from the general population, such samples will be selected for (a) more severe forms of the disorder, (b) comorbidity with other disorders, and (c) clinical profiles that are more likely to elicit acute treatment, such as suicidal thoughts or acts (Neale and Kendler 1995). Therefore, particularly for relatively common milder disorders, ascertainment at a point as close to the general population base as possible is often desirable, for example, via general practice attenders or household surveys.

More general ascertainment biases include (d) cohort effects, and (e) volunteer bias. Cohort effects refer to changes in the characteristics of a disorder over time, which may be relevant if probands have a wide range of birth years. Volunteer bias tends to result in studies including probands with less severe disorders, and fewer of the sorts of symptoms that may be associated with declining to participate in research, such as paranoid symptoms. Volunteers also tend to be better educated than non-volunteers and are less likely to come from socially deprived backgrounds.

In addition, there may be interviewer bias, and also measurement error that shows up as unreliability of clinical ratings.

Second, the lifetime expectation, or morbid risk (MR) of the disorder is calculated in relatives of probands. The MR is calculated from the number of affected relatives divided by the total number of relatives, with an adjustment made for variable age of onset, and the fact that not all relatives will have passed through the period of risk. Various methods can be used to make this adjustment, including the Weinberg, Strömgren–Slater, and Kaplan–Meier methods (Sham 1998). The uncorrected lifetime prevalence of the disorder is sometimes calculated instead. The lifetime prevalence is the number of relatives who have ever been affected divided by the total number of relatives. This is obviously simple to calculate, but it is inevitably lower than the true lifetime risk of disorder unless all of the relatives have completely passed through the period of risk, which is unlikely to be the case for adult onset disorders such as schizophrenia or affective illness. Also, lifetime prevalence is not useful for comparing risks across generations, for example, in the parents versus offspring.

A further complication is that even if one does go to the trouble of calculating MRs for different classes of first-degree relative, the results may be perplexing. For example, the risk of schizophrenia in the parents, siblings and children of probands is around 6 per cent, 10 per cent and 13 per cent, respectively (Gottesman 1991). The relatively low MR in parents who, like siblings and offspring share half of their genes with the probands, may be a consequence of the low rate of reproduction associated with active schizophrenia. This means that most schizophrenics who successfully reproduce, do so before the onset of their illness. On the other hand, individuals who carry a genetic liability to schizophrenia but who never manifest the disorder are less likely to be reproductively impaired and may thus be over-represented among the parents of schizophrenics compared with other classes of first degree relative. Parents of schizophrenics who themselves have the illness are likely to have become ill after having had their children.

Third, the MR of the disorder in relatives of probands is compared with the MR in the general population or, preferably, in relatives of a demographically-matched control sample. In practice, control probands may be screened for psychiatric illness or unscreened. Using unscreened controls is a more conservative approach, because some control probands may have a psychiatric disorder; unscreened controls sampled from the general population also allow estimates of population MRs to be made.

Familial aggregation is indicated by a significantly higher MR for the disorder in relatives of probands than controls. However, family studies do not distinguish between genetic and shared-environmental effects. This distinction can be addressed in twin and adoption studies.

Twin studies

The basic premise of twin studies is that if pairs of monozygotic (MZ) twins, who inherit all of their genes in common, are more similar for a phenotype than pairs of dizygotic (DZ) twins, who on average inherit 50 per cent of their genes in common, a genetic contribution to the phenotype can be inferred.

This premise is based on a number of assumptions:

The equal environments assumption

MZ and DZ twin pairs are assumed to share environmental risk factors for the disorder to the same degree. The assumption would be invalid if, for example, MZ twins shared environmental risk factors to a greater degree than DZ twins; in this case, a greater resemblance for the phenotype in MZ than DZ twin pairs could not be attributed solely to genetic factors.

Studies that have attempted to measure environmental sharing provide some evidence that pairs of MZ twins have more sharing than DZ twins for example MZ pairs tend to socialize together more, and their parents emphasise their similarities more (McGuffin et al. 1996; Kendler and Gardner 1998). However, the degree of environmental sharing has not been shown to predict concordance rates for most common psychiatric disorders (McGuffin et al. 1996; Kendler and Gardner 1998; Cannon et al. 1998; Cardno et al. 1999). The effects of more similar MZ environments can theoretically be avoided by studying MZ twins who have been reared apart. However, such twins are very rare, and probably not representative of twins in general (Gottesman 1991).

The effects of environmental sharing in utero are unclear because they are difficult to measure. Such effects could potentially make MZ twins either more or less similar (Martin et al. 1997). About 65 per cent of MZ twins share a common chorion, while DZ twins never do. A consequence of this shared blood supply could be to increase the chances that a viral infection affects both twins; on the other hand, competition for nutrition and sharing a crowded chorion could result in differences in growth rates. Comparison of monochorionic and dichorionic MZ twins could help to unravel these effects (Rose et al. 1981; Martin et al. 1997).

The risk of the disorder is the same in MZ and DZ twins

This assumption would be invalid if MZ twins had a higher risk of the disorder than DZ twins. In this case, a higher concordance in MZ than DZ twin pairs could not be interpreted as simply being due to MZ twins sharing a greater proportion of susceptibility genes; they would also resemble each other because of factors related to being a monozygotic twin. However, most studies have found no significant differences in the rates of psychotic or affective disorders in MZ versus DZ twins (Kendler *et al.* 1996).

The risk of the disorder is the same in twins and singletons

If this assumption is invalid the results of twin studies could not be extrapolated to the general population in a straightforward way. Again, most studies have not found an excess of psychotic or affective disorders in twins versus singletons (Kendler *et al.* 1996). The generalizability of twin studies can also be tested by extending studies to include other classes of relatives, and comparing with the results of the twin data alone (Martin *et al.* 1997). This has been done by McGue *et al.* (1983) for schizophrenia.

MZ twins are genetically identical

A range of non-inherited or epigenetic influences could potentially make MZ twins genetically discordant, for example, differential trinucleotide repeat expansion, or skewed X-inactivation (McGuffin *et al.* 1994b; Martin *et al.* 1997; Petronis *et al.* 1999). A number of molecular genetic investigations are currently underway to investigate these phenomena in more detail (Petronis *et al.* 1999).

The procedure for the most common twin strategy is as follows:

Twin probands are ascertained systematically. Ideally, this involves direct assessment of twins ascertained from a population-based twin register. However, for less common disorders such as psychoses ascertainment may need to be via a twin register based on hospital attendance, or a register of hospital treatment linked to a population-based twin register. When ascertainment is based on clinical referral, the same potential ascertainment biases apply as for clinically-ascertained family studies. In addition, there may be selection for (a) twin status, (b) concordant pairs, and (c) MZ pairs, because such factors make twins stand out as being unusual, and therefore potentially more likely to be referred for specialist hospital treatment.

Twin pairs are most usefully counted probandwise, that is, if both members of an affected twin pair are ascertained independently they are counted as two pairs. This approach allows comparison between the general population MR for the disorder and the rate of the disorder in co-twins of probands, which is a prerequisite for model fitting and calculating heritability. The alternative approach is to count pairwise, where each pair is counted once. However, this rate cannot be directly compared with the population MR.

The resemblance for the disorder in twin pairs is initially expressed as the probandwise concordance rate. This is the number of concordant probandwise pairs divided by the total number of probandwise pairs. A higher concordance rate in MZ than DZ pairs suggests a genetic contribution to the disorder, but care is required to avoid errors of interpretation as would happen in a twin study of infectious disease such as tuberculosis where data from spouse pairs is also critical (McGue *et al.* 1985). There is no standard adjustment for variable age of onset in twin studies of psychiatric disorders; age correction is made problematic by the high correlation for age of onset in MZ pairs. However, in twin studies of other phenotypes an adjustment has sometimes been made using survival analysis (Meyer *et al.* 1991).

Beyond investigating whether a genetic effect is present, twin studies are particularly informative for investigating the relative contribution of genetic and environmental factors to the aetiology of a disorder (see below), and for investigating epigenetic and environmental effects (Cardno and Gottesman 2000).

Adoption studies

Adoption studies commonly take one of three formats:

1 Ill biological parents as probands who have adopted away offspring (the adoptee strategy). The risk of the disorder in their adopted children, with whom they have no shared environment or experiences, is compared with the risk in adopted offspring of unaffected biological parents. The work by Heston (1966) and Rosenthal *et al.* (1968) in schizophrenia used this approach.

2 Ill adoptee as proband (the adoptee's family strategy). Adoptees with the disorder are ascertained. The risk of the disorder is compared in their biological relatives and their adoptive relatives, and in the biological and adoptive relatives of unaffected adoptees (who form the control or comparison group). This approach has been used by Kety and colleagues (Ingraham and Kety 2000) in schizophrenia, and Mendlewicz and Rainer (1977) in bipolar disorder.

3 Cross-fostering design. The risk of the disorder is compared in adoptees who have affected biological parents but unaffected adopting parents, and in adoptees with unaffected biological parents but affected adopting parents. Such a sample is particularly difficult to ascertain, but this approach has been used by Wender *et al.* (1974) for schizophrenia.

In each case, if the risk of the disorder is higher in biological relatives of probands (where genes are shared but not environments and experiences) than in the other control groups, a genetic contribution to the disorder is suggested.

A high proportion of adoption studies have been carried out in Scandinavia, where systematic ascertainment can be achieved by linking population registers with adoption registers, and then with registers of hospital treatment. There may be ascertainment biases relating to the identification of probands via hospital treatment, as with family studies. In addition, adoptees and their biological and adoptive parents may not be representative of the general population; for example the circumstances leading to adoption may include parental illness or social disadvantage and adoptees frequently have a family history of psychiatric disorders. Other potentially biasing factors include the following: adoptive rearing practices by screened parents may be biased towards health; correlations have been found between the social classes of biological and adoptive parents; and the partners of parent probands (often unmarried adolescents, like the parent proband themselves) whose child is put up for adoption may not be representative of partners of people who have psychiatric disorders in general (Gottesman and Shields 1982).

Thus, family, twin and adoption studies each have particular strengths and limitations. If biases are minimized by methodological rigour, a convergent pattern of results from all three types of study provides strong evidence concerning whether or not there is a genetic contribution to the aetiology of a psychiatric disorder.

Patterns of genetic transmission

The types of study we have just considered allow us to determine whether genetic factors contribute to psychiatric disorders, and as such can be understood without much formal knowledge of genetics. In order to more fully understand the ways in which characteristics are passed on from one generation to the next it is necessary to discuss the patterns and mechanisms of genetic transmission.

Mendelian inheritance

The principle laws of inheritance were first reported by the Augustinian monk, Gregor Mendel, in 1866. These were based on his studies of the effects of cross-breeding on present or absent characteristics of pea plants. The significance of Mendel's work was not initially recognized by the scientific community, but his studies were rediscovered in 1900 and brought to the attention of the world as fundamental to genetics. Mendel's laws for the inheritance of qualitative characteristics by hereditary elements (later to be termed alleles) are summarized in Table 2.1.

The traits studied by Mendel were qualitatively distinct because of the phenomenon of dominance/recessivity. Thus if A is dominant and *a* recessive, then homozygotes *AA* and heterozygotes *Aa* have identical phenotypes, and the trait coded for by *a* is only expressed in the homozygotes *aa*. Hence, the offspring of two unaffected 'carriers' of a recessive disease gene will on average show a ratio of unaffected/affected individuals of 3:1 (25 per cent of offspring affected). Phenylketonuria and Wilson's disease are examples of diseases showing recessive inheritance. For dominant disorders, the most common situation is that an affected heterozygote parent marries an unaffected homozygote. When offspring are then produced, the ratio of affected/unaffected is 1:1 (50 per cent of offspring affected). Unlike recessive conditions, dominant traits do not therefore 'skip' generations, and in the case of a typical dominant condition arising with unaffected parents either a new mutation or a *phenocopy* (a phenotypically indistinguishable

Table 2.1 Mendel's laws

Laws	Comments
Uniformity	
Parent AA × aa	Two alternative alleles A, a at one locus. Each parent has double dose of different allele (homozygotes).
Offspring Aa	All offspring are of the same type (heterozygotes).
Segregation	
Parents Aa × Aa	Parents both heterozygotes.
Offspring AA Aa Aa aa	Three possible types of offspring with probability of occurrence = 1/2/1.
Independent assortment	
Parents AaBb x aabb	Two loci with alleles A, a and B, b, double heterozygote × double homozygote parents.
Offspring AaBb aaBb Aabb aabb	Four possible types of offspring, each with equal probability of occurrence.

disorder in the absence of the genotype) must be inferred. Huntington's disease and acute intermittent porphyria are examples of dominant conditions which may be seen in psychiatric practice.

The patterns of *segregation* just described do not apply to those conditions carried on the X chromosome. Whereas normal females have a pair of X chromosomes, normal males have one X and one Y and hence are effectively *hemizygous* for any characteristic transmitted on the X chromosome. Most X-linked characters are recessive (e.g. colour-blindness, non-specific X-linked mental retardation) in that heterozygous females are unaffected. However, half of their sons (i.e. all of those who receive the X chromosome bearing the specific genes from their mother) are affected. On the other hand, none of the sons of an affected male will be affected (because only a Y chromosome is transmitted) but all of his daughters will be carriers.

Non-Mendelian patterns of inheritance

Many inherited human characteristics (e.g. blood pressure, height, IQ) are *quantitative*, or continuously distributed. Others, such as the presence or absence of a common disease, are *qualitative*, or *discrete*, but do not conform to Mendelian segregation ratios within families. The mechanisms of inheritance in these cases are not radically different from those in the case of Mendelian characters but do require some conceptual modifications. It is important to realize that dominance is really a property of the phenotype rather than of the gene. For example sickle cell anaemia is clear-cut Mendelian recessive disorder if we define it according to the full-blown clinical phenotype. This results from affected homozygotes producing virtually all of the haemoglobin in their red blood cells in the form of an abnormal variant called haemoglobin S. However 'unaffected' heterozygotes have around half of their haemoglobin in the form of the S variant. Thus if we were to redefine affected status as having any haemoglobin S in red blood cells we would have a dominant trait.

We also need to consider the phenomenon called *variable expressivity*. For example, the dominant gene for neurofibromatosis may be expressed in a parent only as *café au lait* spots on the skin but may manifest as the full-blown disorder with CNS involvement in the offspring. One molecular mechanism underlying variable expression is the existence of unstable or dynamic mutations resulting from trinucleotide repeat expansions. These lead to anticipation, a systematic form of variable expression where a disorder becomes more severe and tends to have an earlier onset over successive generations (see Chapter 1).

Another cause of departure from classic Mendelian segregation ratios is *incomplete penetrance*. Penetrance can be defined as the probability of a certain phenotype given a particular genotype. Regular Mendelian traits always have penetrances of zero or one. Thus if we consider a single locus with two alleles G_1 and G_2, where G_2 is the 'disease allele', there are three possible genotypes—$G_1 G_1$, $G_1 G_2$ and $G_2 G_2$. The penetrance values for a recessive disorder would be 0, 0 and 1 respectively and for a dominant disorder the penetrance values would be 0, 1, 1. If we consider a more general model, the general single major locus (SML), the three penetrance values may lie anywhere between 0 and 1. Models of this type have been invoked to explain the transmission of schizophrenia and bipolar affective disorder but have been found to be unsatisfactory as the sole explanation of either (McGue *et al.* 1985; Craddock *et al.* 1995). Nevertheless this does not rule out the possibility that genes of large enough effect to be detected by linkage may be found in some families with these disorders, and the general SML model allows the implementation of linkage analysis in traits that do not show simple dominant or recessive patterns (see Chapter 3).

Continuous traits

Let us consider a single locus with two alleles, A, a. As we can see in Fig. 2.1, we have three genotypes aa, Aa, and AA whose phenotypic values on some particular scale are x_1, x_2, x_3 respectively. Depending on the value of x_2 we can have three possible situations. When x_2 is equal to x_3 we have classical dominance of A with respect to a. If, however, x_2 is equal to x_1 we have recessivity.

But if the value of x_2 is exactly midway between x_1 and x_3, we have a purely additive gene effect. Suppose the two alleles have frequencies in the population of p and q

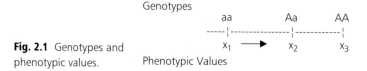

Fig. 2.1 Genotypes and phenotypic values.

(equal to $1 - p$) and assuming there is no migration, mutation or selection against a genotype, the genotype would be distributed in the population as follows:

Genotypes: aa Aa AA

Frequency: p^2 $2pq$ q^2

This is known as the *Hardy–Weinberg equilibrium*. If, for the sake of simplicity we take, $p = q = 0.5$, and suppose that we have variations about the mean value in each genotype due to environmental factors, we might see three phenotypic distributions as in Fig. 2.2(a) occurring in the proportions $1:2:1$. For a trait controlled by two loci with each having two alleles of equal frequency and additive effect, we would see, as in Fig. 2.2(b), five phenotypic distributions occurring in the proportions $1:4:6:4:1$. In general then, the relative proportions of phenotypes controlled by N loci will be given by $(p + q)^{2N}$. As the size of N increases the overall phenotypic distribution more closely approximates a normal distribution which is the limiting case when N becomes very large. It is thought that most continuously distributed hereditary traits are due to *polygenic* mechanisms of this kind, with each individual gene still being transmitted according to Mendel's laws. When the phenotype is due to the combination of multiple genes as well as environmental contributors, the term *multifactorial* is used.

Non-additive effects

In practice the interplay between genes and environment is often more complex than the simple additive situation just described. Non-additive polygenic effects may be present due to *dominance* or due to *epistasis* (i.e. interactions between different genes). Similarly there may be gene $\times$ environment *interactions* as opposed to simple additive

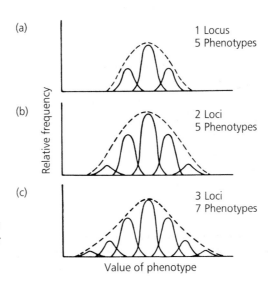

Fig. 2.2 A trait measured on a continuous scale coded for by (a) one (b) two and (c) three loci.

gene + environment coactions. The phenomenon of gene–environment interaction is well described in experimental animals. For example pure-bred strains of mice differ markedly on certain tasks, such as the time taken to find food in a maze. Henderson (1970) has shown that all of several strains performed better when reared in an enriched environment than when reared in a standard environment. However, the degree of improvement differed greatly. Some genotypes showed a pronounced difference in performance in the two environments; in other words they exhibited a large phenotypic *reaction range* (Gottesman 1963). Other strains showed only a modest improvement in the enriched environment. If the gene–environment combination was purely additive, a parallel improvement across all genotypes should have been seen. Gene–environment interactions in humans also occur. For example an adoption study (Cadoret *et al.* 1995) found that adverse environmental factors increased the risk of antisocial behaviour in adoptees at high genetic risk but not in adoptees whose biological parents did not show antisocial tendencies (see Chapter 8 for further details). *Gene–environment correlations* may also occur. Three types have been described: *passive, evocative* and *active* (Plomin *et al.* 1977). Passive gene–environment correlation results from the fact that children not only receive their genes form their parents but parents also provide them with the environment in which they are reared. Plomin *et al.* (2001) give the example of musical ability. Supposing that being a gifted musician is heritable, a child who is talented in this way is also likely to have musically talented parents who will tend to provide an environment that encourages the development of musical skills. An example of evocative gene–environment correlation would be that when the gifted child goes to school, the musical talents are recognized and the child is provided with opportunities to develop them. Finally active gene–environment correlation would occur if the talented young musician deliberately seeks out the circumstances that allow their abilities to improve.

Investigating the relative contribution of genetic and environmental factors

We can summarize all that we have discussed so far regarding complex traits as:

$$\text{Phenotype} = \text{genotype} + \text{environment} + \text{gene–environment correlation}$$
$$+ \text{gene–environment interaction}$$

Furthermore we can consider the environment as consisting or shared (or familial) environment that tends to make relatives similar to each other and non-shared (or unique) environment that differs between family members. We have already covered the principles of twin and adoption studies which help us to tease out whether family resemblance can be attributed, at least in part, to genes rather than just shared environment, but how do we go further and quantify the extent to which a phenotype, or more precisely the variance in a phenotype within the population, is explained by each of these?

In essence this is accomplished by investigating patterns of covariance or correlation (Neale and Cardon 1992; Falconer and Mackay 1996) between twins and sometimes other classes of family member. In psychiatric research it may be useful to investigate continuous measures such as questionnaire scores on the assumption that test scores throughout the entire range can tell us something about the extremes, and subsequent chapters of this book provide many examples of this approach. However, most research in psychiatry takes a categorical approach in which diseases are considered to be present or absent. Liability–threshold models enable a reconciliation between dimensional and categorical approaches.

The multifactorial liability–threshold model

Liability–threshold models (Falconer 1965; Gottesman and Shields 1967) require the assumption of a continuous liability distribution in the general population (Fig. 2.3(a)), contributed to by multiple genes and environmental factors. The liability distribution is set to have mean = 0, and variance = 1. Individuals below a particular threshold on this liability distribution are unaffected, while those above the threshold are affected.

The position of the threshold is determined by the lifetime morbid risk of the disorder in the population. In Fig. 2.3(a) the population risk is 3 per cent (for severe affective

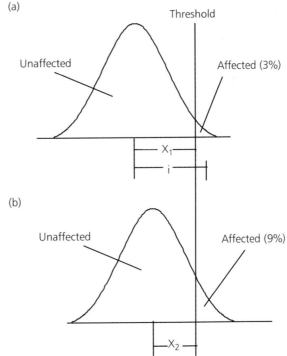

Fig. 2.3 Liability–threshold model. (a) population liability distribution for an illness with a 3% morbid risk. (b) liability distribution for relatives of affected individuals (9% recurrence risk). X_1, distance between threshold and mean population liability; X_2, distance between threshold and mean liability of relatives; i, distance between mean liability for affected individuals and mean population liability. Distances are expressed in standard deviations. Risks here are based on severe affective disorders (McGuffin and Katz, 1986).

disorder requiring inpatient treatment), so 3 per cent of the distribution is to the right of the threshold. Distances (e.g. X_1 between the threshold and the mean liability of the population) are measured in standard deviations (z-scores), with the threshold designated as the origin (zero).

Figure 2.3 (b) represents the liability distribution for relatives of affected individuals, whose recurrence risk is 9 per cent, in this example (McGuffin and Katz 1986). Thus 9 per cent of the distribution is to the right of the threshold.

Correlations in liability

From data on the population risk for the disorder, and the risk in a particular class of relatives of probands, a correlation in respect of liability (or correlation in liability) for the disorder can be calculated. Correlations in liability can be most accurately calculated using computer algorithms that carry out numerical integration as implemented in programs such as PRELIS (Jöreskog and Sörbom 1986a) and Mx (Neale et al. 1999). However an approximation derived by Falconer (1965) and modified by Reich et al. (1972) provides an easier to calculate alternative that is accurate except at more extreme values. If ascertainment is via a population-based twin register, a two-by-two contingency table can be constructed, as shown in Table 2.2.

Separate tables are constructed for MZ and DZ pairs. The frequencies A, B, C and D are assumed to represent proportions of a bivariate normal liability distribution. In this example, where the population risk is 50 per cent (and so the threshold divides the distribution into two equal halves), if there is no correlation in the liabilities of probands and co-twins (Fig. 2.4 (a)) it is expected that there will be equal numbers of concordant and discordant pairs, so A = B = C = D. As the correlation in liability increases, the proportion of concordant pairs increases, so (A = D) > (B = C) (Fig. 2.4 (b)). Numerical integration is used to calculate the expected values of A, B, C and D for a given correlation in liability.

If probands are ascertained via hospital attendance, all probands are affected, so there is no sampling from the parts of the distribution represented by A and B. In this case, a correction for non-random ascertainment is made which is given by Neale et al. (1999). Correlations in liability can be used to estimate the relative contribution of

Table 2.2 A contingency table for a twin study of a dichotomous illness phenotype

| | | Co-twin | |
		Unaffected	Affected
Twin proband	Unaffected	A	B
	Affected	C	D

The cell frequencies A, B, C and D represent the numbers of twins in the four alternative categories of affected and unaffected probands and co-twins.

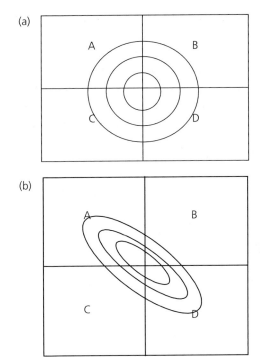

Fig. 2.4 Representation of a bivariate normal liability distribution viewed from above. (a) Correlation in liability of 0.0. A=B=C=D. (b) Correlation in liability of 0.9. (A=D)>(B=C).

genetic and environmental factors to the aetiology, or more specifically, to the variance in liability to a psychiatric disorder.

Multiple thresholds

Threshold models can also be used to explore sex differences in the prevalence of a disorder or the relationship between associated disorders which differ in prevalence and severity. It has been pointed out (Carter 1969) that where a familial disorder is more common in one sex than the other, we might expect to find relatives of probands of the less commonly affected sex more frequently affected than the relatives of probands of the more commonly affected sex. Thus we might describe the more commonly affected sex as the 'broad form' and the less commonly affected sex as the 'narrow form' of the disorder. We can consider this in terms of the tetrachoric correlations between relatives as shown in Table 2.3.

The model we have just outlined has been called the *isocorrelational* model since it requires that

$$r_{11} = r_{12} = r_{21} = r_{22}.$$

The transmission of antisocial personality in the relatives of male (broad form) and female (narrow) probands has been shown to conform to this pattern (see Chapter 8).

Table 2.3 Correlations between probands and relatives for two forms of disorder or disorders with differing sex specific prevalence or morbid risks

		Relative	
		1	2
Proband	1	r_{11}	r_{12}
	2	r_{21}	r_{22}

1=broad form of illness or illness in the more commonly affected sex.
2=narrow form of illness or illness in the less commonly affected sex.

This type of model has also been applied to disorders that can be conceptualized in broad/less severe and narrow/more severe forms. For example attention deficit hyperactivity disorder(ADHD) accompanied by conduct disorder (CD) has sometimes been considered to be more severe than 'pure' ADHD without accompanying CD. A recent analysis has shown that twin data on ADHD with or without CD are well explained by an isocorrelational model (Thapar *et al.* 2001).

Under an alternative model termed the 'environmental' or *isoproportional* model (Reich *et al.* 1979) the differences in prevalence of the disorder in men and women is due to systematic non-familial, environmental or biological differences between the sexes. The tetrachoric correlations are no longer equal and the ratio of affected males/affected females approximates to the ratio in the general population. Data from family studies on alcoholism show this pattern (Reich *et al.* 1975). It should be noted that this model allows that the transmission of the trait may be contributed to by genetic factors and the term 'environmental' signifies only that the across-sex difference in prevalences are due to non- familial causes.

A third alternative is that two disorders which have been observed to occur in the same families may actually be *independent*. Under the independent model, the correlations r_{12} and r_{21} will tend to 0, or at least be substantially lower than the correlations r_{11} and r_{22}, signifying that the transmissible components of the two types of illness are distinct.

Partitioning components of phenotypic variance

The quantitative genetic investigation of continuous phenotypes is generally based on patterns of phenotypic variance, and covariances or correlations. The total phenotypic variance (Vp) is decomposed into genetic (Vg) and environmental (Ve) components, as follows:

$$Vp = Vg + Ve$$

Vg can be further subdivided into additive genetic variance (Va) and dominance variance (Vd). Where multiple genes are involved, there may also be gene–gene interaction (epistasis). However, in most human studies epistasis and dominance variance are difficult to distinguish.

The environmental component can be subdivided into variance due to shared or common family environment (Vc), and variance due to non-shared or individual–specific environment (Ve). No assumptions are made about the nature of the environmental factors, except that shared factors make individuals living together resemble one another, and non-shared factors make individuals differ from each other.

Genetic and environmental components of variance cannot be measured directly in a group of individuals. Instead, they are estimated indirectly from patterns of resemblance between relatives, as we will describe.

Heritability

Heritability is a central concept in quantitative genetics. The heritability of a quantitative phenotype is the proportion of phenotypic variance that is accounted for by genetic effects; for a phenotype that is present or absent it is the proportion of variance in liability (based on the liability–threshold model) that is accounted for by genetic effects.

The broad sense heritability (h^2_b) is the proportion of total phenotypic variance (Vp) (or variance in liability) accounted for by total genetic variance (i.e. additive and non-additive genetic effects) (Vg):

$$h^2b = Vg/Vp.$$

The narrow sense heritability (h^2 or a^2) refers only to additive genetic variance (Va):

$$h^2(\text{or } a^2) = Va/Vp.$$

Heritability refers to variance in a population; it has no straightforward meaning for an individual. Also, heritability estimates may vary between populations, and over time for a single population, because they are influenced by anything that can affect the denominator of phenotypic variance, including secular or cultural changes in longevity or the 'stress of industrialization'.

In a similar manner to heritability, the proportion of phenotypic variance due to shared environmental factors (c^2) can be written as:

$$c^2 = Vc/Vp$$

and for non-shared environmental factors (e^2):

$$e^2 = Ve/Vp.$$

Path analysis and structural equation modelling

The relationships between correlations or covariances between pairs of relatives and components of variance (a^2 etc) can be derived from path analysis (Li 1975; Neale and Cardon 1992). This involves constructing a diagram of the unobserved (latent) genetic

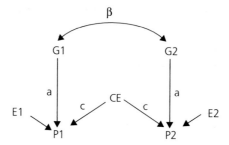

Fig. 2.5 Path analysis. Example of using path analysis to obtain expected covariances or correlations between pairs of relatives.

and environmental effects contributing to observed phenotypes. A relatively simple example is given in Fig. 2.5. These formulae can then be used in model fitting.

To obtain the expected correlation (r) (or covariance) between a pair of relatives we

1 Trace paths between phenotypes (P1 and P2);

2 Multiply path coefficients for each connecting path; and

3 Add these products together.

Thus:

$$r(P1)\,(P2) = (a \times \beta \times a) + (c \times c) = \beta a^2 + c^2$$

For MZ twins $\beta = 1$, and for DZ twins $\beta = 1/2$
So:

$$rMZ = a^2 + c^2, \text{ and } rDZ = 1/2\, a^2 + c^2$$

Model fitting

Biometrical model fitting allows formal testing of (a) whether and (b) to what extent various genetic and environmental effects (a^2, d^2, c^2 and e^2) contribute to variation in a quantitative phenotype (or to variance in liability for a categorical phenotype). The process of biometrical model fitting for the analysis of data on MZ and DZ twins generally proceeds as follows:

1 A set of models is chosen that includes various combinations of genetic and environmental effects. Commonly, models specify phenotypic variance contributed by

  (a) non-shared environmental factors only (E model)

  (b) shared and non-shared environmental factors (CE model)

  (c) additive genetic and non-shared environmental factors (AE model)

  (d) additive genetic, shared, and non-shared environmental factors (ACE model)

  (e) additive genetic, genetic dominance, and non-shared environmental factors (ADE model).

2 Equations are derived, e.g. from path analysis, showing the relationships between (a) correlations in liability for twins and (b) the parameters included in the model

(e.g. a^2, c^2, and e^2 in the ACE model). This sets up a series of simultaneous equations to be solved. For example, under the ACE model rMZ = $a^2 + c^2$, and rDZ = $1/2a^2 + c^2$.

3 The correlations in liability are inspected. Patterns of correlations that suggest particular effects are as follows:

(a) additive genetic, rMZ = 2rDZ

(b) dominance, rMZ > 2rDZ

(c) shared environment, rMZ = rDZ

(d) non-shared environment—degree to which rMZ < 1.0

If there are mixed effects, for example, additive genetic and shared environment, the expected pattern is between (a) and (c), i.e. rMZ between 1 and 2 times rDZ. This gives a general idea of what to expect from the model fitting.

4 The fit of each model is then formally tested using a computer program, such as Mx (Neale *et al.* 1999) or LISREL (Jöreskog and Sörbom 1986b). This involves an iterative process, in which the probability of obtaining the observed data is calculated for a range of estimates of the parameters in the model (e.g. a^2, c^2 and e^2 for the ACE model), until convergence is reached, that is, when the parameter estimates have been found that give the best agreement between the observed and expected correlations. This is usually done by maximizing a likelihood function or minimizing a goodness-of-fit χ^2. A good model has a fit with a high (non-significant) P-value.

5 The fits of nested models are compared. Thus, models can be compared where one model is a submodel of the other, e.g. ACE can be compared with AE, CE and E. However, e.g. CE and AE cannot be directly compared (Neale and Cardon 1992).

6 The best model is chosen on the grounds of goodness-of-fit and parsimony. Thus, if two models do not differ significantly in goodness-of-fit, the model with fewer parameters is chosen. Sometimes the Akaike information criterion (AIC) is used to assess the best model; this is an index of the balance between goodness-of-fit and parsimony and is calculated as χ^2-twice the degrees of freedom. The preferred model has the lowest AIC.

This type of approach has been used, for example, to investigate genetic and environmental influences on quantitative symptom scores in children and adolescents (e.g. Thapar and McGuffin 1994; Eaves *et al.* 1997), liability to major depressive disorder (Kendler *et al.* 1992; McGuffin *et al.* 1996) and liability to psychotic disorders (Cannon *et al.* 1998; Cardno *et al.* 1999) and many more examples are given elsewhere in this book.

More complex model fitting

Recent studies using structural equation modelling and allied approaches can go beyond merely estimating genetic and an environmentally explained variance to cover more complex issues. For example, the question of developmental change and continuity of behaviour has recently received considerable attention. Much of this has been focused on cognitive ability, where there is no considerable evidence that heritability of IQ increases from childhood to adulthood and actually appears to go on increasing into old age (McClearn *et al.* 1997 and see also chapter 4). There is also evidence of similar phenomenon with regards to psychopathology. In particular, depressive symptoms in childhood are fairly infrequent and appear to have low or negligible heritability with an increasing genetic effect evident during adolescence, similar to that found in overt clinical depression in adults (Thapar and McGuffin 1994). Similarly, delinquent behaviour in adolescence shows low heritability, whereas antisocial behaviour in adults shows substantial genetic affects (Lyons *et al.* 1995).

Multivariate structural models allow exploration of the genetic and environmental underpinnings of disorders that frequently co-occur. For example, there is evidence both in twin studies of adults (Kendler *et al.* 1987) and children (Thapar and McGuffin 1997) of a large overlap between the genes that affect anxious and depressive symptomatology. The details of model fitting to explore the genetic and environmental overlap of comorbid disorders is discussed in greater detail elsewhere, for example by Neale and Cardon (1992), but the broad principles are illustrated in Fig. 2.6.

The figure shows two phenotypes that overlap clinically: here they are unipolar depression and anxiety disorder. The question is, to what extent are the two aetiologically related? Each condition has its additive genetic (A), common environmental (C) and non-shared environmental (E) sources of variation as in the simple univariate model discussed earlier and here these are labelled A1, C1 and E1 for depression, and A2, C2 and E2 for anxiety. The paths from these unobserved (or latent) variables are denoted by h, c and e with the subscripts d or a. The questions of aetiological overlap can be tested by comparing the fit of models where the genetic (rg) common environmental (rc) and non-shared environmental (re) correlations are freely estimated compared with models where they are either set at zero (no aetiological overlap) or one (complete aetiological overlap).

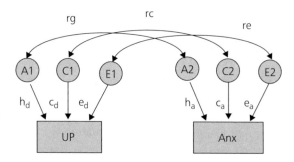

Fig. 2.6 A bivariate path model to explore the causal overlap between a pair of comorbid phenotypes, eg unipolar depression (UP) and anxiety (Anx). See text for further explanation.

Extremes analysis

An alternative method of analysing twin and family data, extremes analysis, was devised by De Fries and Fulker (1985) (often called DF extremes analysis after these authors). Here a regression method is used to analyse data where the probands have a present/absent trait or disorder but where the co-twins (or other relatives) are measured on some continuous scale. One of its first applications was to data on twins ascertained via probands who had marked reading disability. The co-twins were tested and assigned reading scores (De Fries *et al.* 1987). The broad principle is that if there is a genetic influence on the trait the MZ co-twins will, on average, show less regression toward the population mean than the DZ co-twins. This was found to be the case for reading disability. The difference between the average MZ and DZ co-twins' scores regression to the mean provides a measure of the size of the genetic effect. One attraction of the DF extremes method is that a modified form of the original method can be used to test whether the heritability of extremely low (or extremely high) scores differs from that in the rest of the range. For example Dale *et al.* (1998) showed that the genetic influence on language delay in two-year-olds was significantly higher than the genetic influence on the normal development of language.

Complex segregation analysis and related methods

If evidence for a genetic contribution to the aetiology of a psychiatric disorder is found, approaches such as complex segregation analysis (Lalouel and Morton 1981; Lalouel *et al.* 1983) can be used to attempt to identify the most likely mode of inheritance, based on the pattern of inheritance in families. Commonly a mixed model (Morton and MacLean 1974), in which there are both major gene and polygenic effects, is compared with the submodels of a single major locus, and polygenic inheritance. Large sample sizes are required to distinguish between models, especially between the polygenic and mixed models, which has severely limited the usefulness of complex segregation analysis for psychiatric genetics to date. Furthermore complex segregation analysis has been shown to be fallible (McGuffin and Huckle 1990) in suggesting the presence of a major gene effect in a familial present/absent trait, 'attending medical school', that seems highly likely to have a more complicated causation! Unfortunately it therefore seems that more recent sophisticated statistical methods are susceptible to error in the analysis of present/absent traits (although they may have better discriminatory power with continuous traits). This is a problem that Edwards (1960) long ago referred to as the 'simulation of Mendelism'.

Ultimately the test of whether a major gene contributes to a disorder is whether such a gene can be detected and localized by linkage analysis (see Chapter 3). Nevertheless, before embarking on linkage studies it might seem reasonable to use statistical approaches to explore whether major gene inheritance is a possibility. One such is a graphical approach (Suarez *et al.* 1976), in which the range of familial morbid risks compatible with an SML model is calculated. If the morbid risks for an illness lie outside the range that is defined by constraining the model parameters, the gene frequency and the three penetrances, between 0 and 1, it implies more complex inheritance.

Studies applying this method or an extension of it strongly suggest that there are multiple susceptibility loci for schizophrenia (O'Rourke *et al.* 1982) and bipolar disorder (Craddock *et al.* 1995). On the other hand, polygenic and mixed models cannot be differentiated by this type of analysis. In addition such studies do not tell us about the effect size of individual genes, or the total number of susceptibility loci that are involved. However attempts have been made (e.g. Risch 1990) to use the pattern of recurrence risk in first degree relatives, compared with more distant classes of relative, to infer the number of loci and the size of their effect in disorders such as schizophrenia. Here Risch (1990) has pointed out that just two loci each of only moderate effect could explain the available data if epistasis is present.

Conclusions

Quantitative genetics provides the bedrock of psychiatric genetic research. Quantitative methods have provided answers to the questions of whether a disorder or trait is influenced by genes and what is the size of the influence. They also provide important information on environmental effects and the interplay between genes and environment. As such, quantitatively based studies can be seen both as ends in themselves, many examples of which are given elsewhere in this volume, and as necessary preludes to molecular genetic studies.

Further reading

Falconer D. S. and Mackay T. F. C. (1996) *Introduction to Quantitative Genetics*, 4th edn. Harlow: Longman.

Neale M. C. and Cardon L. R. (1992) *Methodology for Genetic Studies of Twins and Families*. Dordrecht: Kluwer Academic Publishers.

Neale M. C., Boker S. M., Xie G. *et al.* (1999) Mx: Statistical Modelling Box 126 MCV, Richmond Virginia 23298: Department of Psychiatry, 5th edn (can be down loaded from *http://views.vcu.edu/mx/documentation.html*)

Sham P. (1998) *Statistics in Human Genetics*. London: Arnold.

References

Cadoret R. J., Yates W. R., Troughton E. *et al.* (1995) Genetic–environmental interaction in the genesis of aggressivity and conduct disorders. *Archives of General Psychiatry* 52:916–24.

Cannon T. D., Kappio J., Lonnqvist J. *et al.* (1998) The genetic epidemiology of schizophrenia in a Finnish twin cohort. *Archives of General Psychiatry* 55:67.

Cardno A. G., Coid B., MacDonald A. M. *et al.* (1999) Heritability estimates for psychotic disorders: The Maudsley Twin Psychosis series. *Archives of General Psychiatry* 56:162–8.

Cardno A. G. and Gottesman I. I. (2000) Twin studies of schizophrenia: from bow-and-arrow concordances to Star Wars Mx and functional genomics. *Am J Med Genet (Semin. Med. Genet.)* 97(12):12–17.

Carter C. O. (1969) Genetics of common disorders. *British Medical Bulletin* **25**:52–7.

Craddock N., Khodel V., Van Eerdewegh P. *et al.* (1995) Mathematical limits of multilocus models: the genetic transmission of bipolar disorder. *Am J Hum Genet* **57**:690–702.

Dale P. S., Simonoff E., Bishop D. V. M. *et al.* (1998) Genetic influence on language delay in two-year-old children. *Nature Neuroscience* 1(4):324–8.

De Fries J. C. and Fulker D. W. (1985) Multiple regression analysis of twin data. *Behavior Genetics* **15**:467–73.

De Fries J. C., Fulker D. W., LaBuda M. C. (1987) Evidence for a genetic aetiology in reading disability of twins. *Nature* **329**:537–9.

Eaves L. J., Silberg J. L., Meyer J. M. *et al.* (1997) Genetics and developmental psychopathology: 2. The main effects of genes and environment on behavioral problems in the Virginia Twin Study of Adolescent Behavioral Development. *J Child Psychol Psychiatry* **38**:965–80.

Edwards J. H. (1960) The simulation of Mendelism. *Acta Genetica* **10**:63–70.

Falconer D. S. (1965) The inheritance of liability to certain diseases, estimated from the incidence among relatives. *Annals of Human Genetics* **29**:51–76.

Falconer D. S. and Mackay T. F. C. (1996) *Introduction to Quantitative Genetics*, 4th edn. Harlow, Longman.

Gottesman I. I. (1991) *Schizophrenia Genesis*. New York, W. H. Freeman.

Gottesman I. I. and Shields J. (1967) A polygenic theory of schizophrenia. *Proceedings of the National Academy of Sciences of the United States of America* 1:199–205.

Gottesman I. I. and Shields J. (1982) *The Epigenetic Puzzle*. Cambridge: Cambridge University Press.

Gottesman I. I. (1963) Genetic aspects of intelligent behavior, in N. Ellis (ed.) *The Handbook of Mental Deficiency: Psychological Theory and Research*. New York: McGraw-Hill.

Henderson N. (1970) Genetic influences on the behavior of mice can be observed by laboratory rearing. *Journal of Comparative Physiological Psychology*, **72**:505–11.

Heston L. L. (1966) Psychiatric disorders in foster home reared children of schizophrenic mothers. *British Journal of Psychiatry* **112**:819–25.

Ingraham L. J. and Kety S. S. (2000) Adoption studies of schizophrenia. *Am J Med Genet* **97**:18–22.

Jöreskog K. G. and Sörbom D. (1986a) *Prelis: A preprocessor for LISREL*. Mooresville, Indiana: Scientific Software.

Jöreskog K. G. and Sörbom D. (1986b) *LISREL 7. A Guide to the Program and its Applications*, 2nd edn. Chicago, IL: SPSS Inc.

Kendler K. S. and Gardner C. O. Jr. (1998) Twin studies of adult psychiatric and substance dependence disorders: are they biased by differences in the environmental experiences of monozygotic and dizygotic twins in childhood and adolescence? *Psychol Med* **28**:625–33.

Kendler K. S., Heath A. C., Martin N. G. *et al.* (1987) Symptoms of anxiety and symptoms of depression. *Archives of Surgery* **122**:451–7.

Kendler K. S., Neale C., Kessler C. *et al.* (1992) A population-based twin study of major depression in women. *Archives of General Psychiatry* **49**:257–66.

Kendler K. S., Pedersen N. L., Farahmand B. Y., Persson P. G. (1996) The treated incidence of psychotic and affective illness in twins compared with population expectation: a study in the Swedish Twin and Psychiatric Registries. *Psychol Med* **26**:1135–44.

Lalouel J. M. and Morton N. E. (1981) Complex segregation analysis with pointers. *Hum Hered* **31**:312–21.

Lalouel J. M., Rao D. C., Morton N. E. *et al.* (1983) A unified model for complex segregation analysis. *Am J Hum Genet* **35**:816–26.

Li (1975) *Path Analysis. A Primer.* Pacific Grove, California: Boxwood Press.

Lyons M. J., True W. R., Eisen A. *et al.* (1995) Differential heritability of adult and juvenile antisocial traits. *Archives of General Psychiatry* 52:906–15.

Martin N., Boomsma D., Machin G. (1997) A twin-pronged attack on complex traits. *Nat Genet* 17:387–92.

McClearn G. E., Johansson B., Berg S. *et al.* (1997) Substantial genetic influence on cognitive abilities in twins 80 or more years old. *Science* 276:1560–3.

McGue M., Gottesman I. I., Rao D. C. (1983) The transmission of schizophrenia under a multifactorial threshold model. *Am J Human Genet* 35:1161–78.

McGue M., Gottesman I. I., Rao D. C. (1985) Resolving genetic models for the transmission of schizophrenia. *Genetic Epidemiology* 2:99–110.

McGuffin P., Owen M. J., O'Donovan M. C. *et al.* (1994a) *Seminars in Psychiatric Genetics.* London, Gaskell.

McGuffin P., Asherson P., Owen M., Farmer A. E. (1994b) The strength of the genetic effect—is there room for an evironmental influence in the aetiology of schizophrenia? *British Journal of Psychiatry* 164:593–9.

McGuffin P. and Huckle P. (1990) Simulation of Mendelism revisited: The recessive gene for attending Medical School. *Am J Hum Genet* 46:994–9.

McGuffin P. and Katz R. (1986) Nature, nurture and affective disorders, in J. W. K. Deakin *The Biology of Depression.* London: Gaskell Press.

McGuffin P., Katz R., Rutherford J. *et al.* (1996) A hospital based twin register of the Heritability of DSM-IV Unipolar Depression. *Archives of General Psychiatry* 53:129–36.

Mendlewicz J. and Rainer J. D. (1977) Adoption study supporting genetic transmission in manic-depressive illness. *Nature* 268:327–

Meyer J. M., Eaves L. J., Heath A. C., Martin N. G. (1991) Estimating genetic influences on the age–at–menarche: a survival analysis approach. *Am J Med Genet* 39:148–54.

Morton N. E. and Maclean C. J. (1974) Analysis of family resemblance. III Complex segregation of quantitative traits. *Am J Hum Genet* 26:489–503.

Neale M. C. and Cardon L. R. (1992) *Methodology for Genetic Studies of Twins and Families.* Dordrecht, Kluwer Academic Publishers.

Neale M. C. and Kendler K. S. (1995) Models of comorbidity for multifactorial disorders. *Am J Hum Genet* 57:935–53.

Neale M. C., Boker S. M., Xie G. *et al.* (1999) Mx: Statistical Modelling Box 126 MCV, Richmond Virginia 23 298: Department of Psychiatry. 5th edn.

O'Rourke D. H., Gottesman I. I. *et al.* (1982) Refutation of the general single-locus model for the etiology of schizophrenia. *Am J Hum Genet* 34:630–49.

Petronis A., Paterson A. D., Kennedy J. L. (1999) Schizophrenia: an epigenetic puzzle? *Schizophr Bull* 25:639–55.

Plomin R., DeFries, J. C., Loehlin, J. C. (1977) Genotype-environment interaction and correlation in the analysis of human behavior. *Psychological Bulletin* 84:309–322.

Plomin R., De Fries J. C., McClearn G. E. *et al.* (2001) *Behavioral Genetics,* 4th edn. New York: Worth Publishers.

Reich T., James J. W., Morris C. A. (1972) The use of multiple thresholds in determining the mode of transmission of semi-continuous traits. *Annals of Human Genetics* 36:163–84.

Reich T., Winokur G., Mullaney J. (1975) The transmission of alcoholism, in D. Rosenthal, R. R. Fieve, H. Brill (eds) *Genetic Research in Psychiatry.* Baltimore and London: The John Hopkins University Press.

Reich T., Rice J., Cloninger C. R. *et al.* (1979) The use of multiple thresholds and segregation analysis in analyzing the phenotypic heterogeneity of multifactorial traits. *Ann Hum Genet* **42**:371–90.

Risch N. (1990) Linkage strategies for genetically complex traits. III: the effect of marker polymorphism analysis on affected relative pairs. *Am J Hum Genet* **46**:242–53.

Rose R. J, Uchida I. A, Christian J. C. (1981) *Placentation effects on cognitive resemblance of adult monozygotes. Twin Research 3: Intelligence, Personality, and Development.* New York: Alan R. Liss, Inc.

Rosenthal D., Wender P. H., Kety, S. S. *et al.* (1968) Schizophrenic's offspring reared in adoption homes. In D. Rosenthal and S. S. Kety (eds) *The Transmission of Schizophrenia* (pp. 377–91). Oxford: Pergaman.

Sham P. (1998) *Statistics in Human Genetics.* London, Arnold.

Suarez B. K., Reich T., Trost J. (1976) Limits of the general two-allele single locus model with incomplete penetrance. *Annals of Human Genetics* **40**:231–44.

Thapar A. and McGuffin P. (1994) A twin study of depressive symptoms in childhood. *British Journal of Psychiatry* **165**:259–65.

Thapar A. and McGuffin P. (1997) Anxiety and depressive symptoms in childhood—a genetic study of comorbidity. *Journal of Child Psychology and Psychiatry* **38**(6):651–6.

Thapar A., Harrington R., McGuffin P. (2001) Examining the Comorbidity of ADHD related behaviours and conduct problems using a twin study design. *British Journal of Psychiatry* **179**:224–9.

Wender P. H., Rosenthal D., Kety S. S. *et al.* (1974) Crossfostering. A research strategy for clarifying the role of genetic and experimental factors in the etiology of schizophrenia. *Archives of General sychiatry* **30**:121.

Chapter 3

Linkage and association

Pak Sham and Peter McGuffin

Introduction

Linkage and association analyses are methods for mapping disease loci. Both methods depend on the existence of polymorphic genetic markers and can be used to test specific candidate genes, or in a systematic scan of a chromosome or the entire genome. When used systematically, linkage and association analyses do not require knowledge of the pathogenesis of the disease. Linkage and association have complementary properties, in that linkage is only able to detect genes of major effect, but at large distances, whereas association can detect genes of minor effect, but only at small distances. Systematically screening the whole genome requires only several hundred markers for linkage analysis, but many thousands of markers for association analysis.

Linkage and recombination

In order to understand the principles of linkage analysis, it is necessary to appreciate how genetic material is transmitted from parent to offspring. As described in Chapter 1, each individual's genetic material is made up of a set of 23 chromosomes inherited from the father and a set of 23 chromosomes inherited from the mother. These chromosomes form 23 homologous pairs, so that each gene is present in duplicate (except those on the X and Y chromosomes). When the individual reproduces, it passes a set of 23 chromosomes to an offspring (through a gamete), where each chromosome is a 'hybrid' of two homologous chromosomes. To be more precise, each chromosome contains a number of segments alternating between a chromosome inherited from the individual's father, and the homologous chromosome inherited from the individual's mother. Each of these hybrid chromosomes is generated by crossing over, or recombination, between homologous chromosomes during meiosis. Each crossover point, or chiasma, causes a switch from paternal to maternal chromosome, or vice versa (see Fig. 1.4).

The crossover points of a chromosome appear to occur in a nearly random fashion. The genetic distance between two points on a chromosome, in units of Morgan (named after T.H. Morgan who first described the phenomenon of linkage following his experiments with fruit flies [Morgan *et al.* 1915]), is defined as the average number of crossover points between the two points. Morgans are further subdivided into centimorgans (cM), where 1 cM corresponds to the occurrence of crossing over one time in a hundred. The relationship between genetic distance defined in terms of Morgans, and physical distance as measured by the number of DNA base pairs, is not constant. The rate of crossover points as a function of physical distance varies between species, between male and female gametes (usually it is more in females), between different chromosomes, and between different positions on a chromosome. In humans, on average, one (cM) of genetic distance corresponds approximately to a million base pairs of DNA, i.e. one megabase, (Mb). There are on average a total of about 35 crossover points on the chromosomes of a gamete. In other words, the human genome is approximately 35 Morgans in genetic length.

It has recently been estimated that there are approximately 32,000 functional genes in the human genome. Since the genetic material is transmitted from parent to offspring in about 35 chromosomal segments, it is obvious that each segment must contain multiple genes, and that the genes on these segments will be transmitted together as a group. Two genes that are on the same chromosome, transmitted from either the individual's father or the individual's mother, will be transmitted together unless they are separated by a crossover point. The nearer the two genes, the less likely that a crossover point will occur between them, and the more likely that they will be transmitted together. This represents departure from Mendel's law of independent assortment, and is the phenomenon known as linkage.

When a parent transmits two alleles to an offspring, the two alleles may have the same parental origin (i.e. both from the parent's father or both from the parent's mother) or have different parental origins (i.e. one from the parent's father and the other from the parent's mother). In the latter case, the transmitted gamete consists of two alleles that were transmitted through two separate gametes in the previous generation, and is therefore known as a *recombinant* (Fig. 3.1). The proportion of gametes that are recombinant with respect to two loci is defined as the *recombination fraction* (θ) of the two loci. Loci on different chromosomes segregate independently, so that the recombination fraction

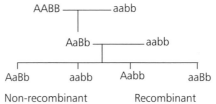

Fig. 3.1 Recombinant and non-recombinant gametes. The parent with AaBb genotype is doubly heterozygous. It can be deduced that the alleles transmitted from the two grandparents to this parent are AB and ab. The 'coupling' of A with B, and that of a with b, are known as the phase of the genotype. The first and second children inherit AB and ab; the same combination as those transmitted in the previous generation. In contrast, the third and fourth children inherit Ab and aB, and those combinations are unlike those transmitted in the previous generations. These latter gametes are known as recombinants.

is 0.5. The recombination fraction between two loci on the same chromosome (i.e. *syntenic loci*) depends on the genetic distance between them, being 0 at a genetic distance of 0 Morgan, and approaches 0.5 as genetic distance increases. The exact relationship between recombination fraction and genetic distance depends on the underlying assumptions about the crossover process. It is usual in linkage analysis to assume that crossover events occur at random without interfering with each other. In this case the relationship between recombination fraction (θ) and genetic distance (m) can be described mathematically by the Haldane mapping function $\theta = (1 - e^{-2\,m})/2$.

In reality, the occurrence of a crossover reduces the probability of further crossovers in the immediate vicinity. This phenomenon is called *interference*, and can be modelled in linkage analysis by the use of appropriate map functions.

Parametric linkage analysis

The phenomenon of linkage means that, if we can demonstrate non-independent segregation between a disease and a genetic marker, then there must be a gene for the disease in the region of the chromosome containing the marker. Furthermore, if additional markers in the region are examined, then the gene for the disease can be narrowed down to a small region where the markers show no recombination with the disease locus. Genotyping of additional individuals may identify recombination between some of these markers and disease, further narrowing the region containing the disease locus. The resolution of linkage mapping is inherently limited by the frequency of recombination events. For example, it is necessary to study a sample involving more than 100 transmission events to have above a 50 per cent chance of observing a recombination between two loci separated by a recombination fraction of 0.01. The narrowing down of a disease locus to a region of less than 1 cM therefore requires a large sample of pedigrees.

In order for a recombination event to be identified with certainty, it is necessary that the genotypes of both loci are directly measurable, or can be inferred from the phenotype. In addition, three-generation pedigree data are desirable in order to establish *linkage phase*. Having at least three generations of data allows us to compare the gametes transmitted from the second generation to the third generation with the gametes transmitted from the first to the second generation. For example, let us consider a heterozygous parent who has alleles *G* and *g* at one locus and alleles *T* and *t* at another locus, so that the genotype can be summarized as *GgTt*. Let us suppose that their spouse has the genotype *ggtt*. There are then four different possible types of offspring: *GgTt*, *ggTt*, *Ggtt*, and *ggtt*. If the heterozygous parent carries alleles *G* and *T* on the same chromosome and *g* and *t* on the other (classically referred to as the coupling linkage phase and written as *GT/gt*) then the offspring of the types *Ggtt* and *ggtt* must be recombinants. If however the heterozygous parent is in the opposite linkage phase, *Gt/gT* (known as repulsion) then offspring of the types *GTgt* and *ggtt* must be recombinants. Unfortunately, no matter how many children the couple have it is impossible to tell the linkage phase of the heterozygous parent if there is no grandparental information, hence in nuclear or two generation families one has to assume that there is a equal chance that a heterozygous parent is in coupling as in repulsion. Consequently nuclear families contain less information than families with the same number of offspring but with complete genotypic data on grandparents.

Even in Mendelian diseases, highly informative families may be difficult to obtain. For example, in the case of adult-onset disorders grandparental information is usually incomplete or missing altogether, while in other families a parent transmitting the disorder may be homozygous at the marker, rendering the transmitted and non-transmitted marker alleles indistinguishable. Since it is often not possible to identify recombination events with certainty, a more sophisticated method of statistical analysis is necessary.

The standard method of linkage analysis for general pedigree data is based on the theory of likelihood, which requires the formulation of a model that contains certain unknown parameters (such as the recombination fraction between two loci). The relative values of the probabilities of the observed data under different parameter values can be used as measures of relative support for these parameter values provided by the data. Such probabilities, which are calculated for a fixed data set under different parameter values, are formally called likelihoods. For mathematical convenience it is usual to take the logarithm of the likelihoods (so that the product of the likelihoods of multiple independent families translates to the sum of the log-likelihoods) The difference between the log-likelihood at a particular recombination fraction and the log-likelihood at a recombination fraction of 1/2; is defined as the lod (log of the odds) score at that recombination fraction (Morton 1955).

$$Lod(\theta = \theta_1) = \log_{10} \frac{likelihood \; (\theta = \theta_1)}{likelihood \; (\theta = 0.5)}.$$

In principle, lod score can be calculated from data on a single marker locus or multiple marker loci (Elston and Stewart 1971; Ott 1974; O'Connell and Weeks 1995). Efficient algorithms for likelihood calculations have been developed that allow *multipoint* (i.e. multiple marker loci) lod scores to be calculated in small to moderate-sized pedigrees (Kruglyak *et al.* 1996). In principle, multipoint lod score analysis is more powerful than single-marker lod score analysis. However, multipoint analysis can be misleading if the positions of the markers are mis-specified. For this reason, it is customary to report both single-marker and multipoint results from linkage studies.

For historical reasons (the method was introduced in the days of books of mathematical tables and long before electronic calculators or desktop personal computers became available) it is customary to use common logarithm (i.e. base 10) in genetic linkage analysis. Thus, a lod score of 3 at recombination fraction 0.1 means that the data are 1,000 times more likely to have arisen given recombination fraction 0.1 than recombination 0.5. A plot of lod score against recombination fraction (from 0 to 0.5) provides a summary of the degree of support for different values of recombination fraction, against the baseline of no linkage. Figure 3.2 gives a typical lod score curve.

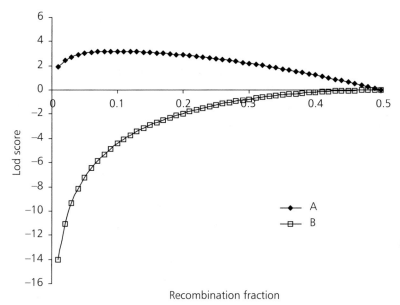

Fig. 3.2 Illustrative lod score curves. Lod score curves for data containing 20 fully informative gametes. Curve A: 2 recombinants and 18 non-recombinants. Curve B: 10 recombinants and 10 non-recombinants. Note that for curve A the maximum likelihood estimate of the recombination fraction is 0.1.

Summarizing the evidence for linkage using lod score tables has the important advantage that the lod scores at any particular recombination fraction are additive across independent samples. It is therefore easy to combine results across a number of studies to provide an overall summary of the evidence for or against linkage. Morton (1955) suggested that a cumulative lod score exceeding 3 should be regarded as strong evidence for linkage, while a cumulative lod score below -2 should be regarded as strong evidence against linkage. These criteria were based on the consideration that the prior probability of linkage to any particular chromosomal region is low.

Another way of interpreting lod scores is through significance levels. Since a lod score can be regarded as a log-likelihood ratio, and twice such a ratio expressed in natural logs is approximately a chi square, a lod score of 3 corresponds to a chi square of $3 \times 2\ln 10 = 13.8$. In large samples this has a one-tailed significance of approximately 0.0001. However, taking the multiple testing involved in a whole genome scan into account, a lod score of 3 corresponds to a p value of only 0.05, the conventional criterion for statistical 'significance'. In fact, in a survey of linkage studies on Mendelian diseases, 98 per cent of studies showing a lod scores of over 3 were subsequently replicated (Rao et al. 1979). Therefore in Mendelian diseases parametric linkage analysis using the lod score method is an extremely powerful method for mapping genes and is unlikely to throw up type I errors.

Complex modes of transmission

The lod score method was developed for the analysis of Mendelian disorders. It was soon recognized that many disorders are not strictly Mendelian. A simple form of 'non-Mendelian' or 'complex' transmission is locus heterogeneity, where the disease is caused by a mutation at one locus in some families but by a mutation at another locus in other families. Failure to recognize the possibility of locus heterogeneity may lead to the false inference of some gametes as recombinants, increasing the estimate of the recombination fraction, and possibly excluding linkage a true disease locus. Smith (1959) introduced the 'admixture model' in which only a proportion (α) of families shows linkage between disease and marker. The hypothesis of no linkage, H_1, corresponds to either $\theta = 0.5$, or $\alpha = 0$, that of homogeneous linkage, H_2, corresponds to $\theta < 0.5$ and $\alpha = 1$, and that of locus heterogeneity, H_3, corresponds to $\theta < 0.5$ and $\alpha > 0$. A test for locus heterogeneity is obtained by the ratio of the likelihood of H_3 (linkage in some families) to that of H_2 (linkage in all families). A test for linkage that allows for the possibility of locus heterogeneity is obtained by the ratio of the likelihood of H_3 (linkage in some families) to that of H_1 (no linkage).

Other potential complexities are reduced penetrance or absence of disease in some individuals with a high-risk genotype (see also Chapter 2) and phenocopies (presence of disease in some individuals with a low risk genotype). Allowance for these complexities has been made possible by extending the Mendelian model to have arbitrary penetrance parameters, so that each genotype can be specified to have any risk of disease ranging from 0 to 1. This is known as the single major locus (SML) or generalized single locus (GSL) model. In order to carry out a parametric linkage analysis, it is usually necessary to pre-specify the allele frequencies and the penetrances of the disease locus.

For most diseases, the risk of disease will depend not only on genotype but also on other variables such as age and sex. In these cases the penetrances are likely to be dependent on these other variables. Variable penetrance is allowed for in parametric linkage analysis by defining two or more liability classes differing from each other in age, sex, or other personal characteristics. Each liability class is then assigned a specific set of penetrance values that are likely to reflect the risk of that class.

The SML model remains a simplification for complex traits, which are likely to be determined by multiple genes and environmental factors. In principle, the SML model can be extended to an oligogenic model, by including additional loci that have an influence on risk. In practice, however, this does not appear to increase power appreciably, except when the two susceptibility loci demonstrate strong gene–gene interaction or epistasis. Simulation studies have shown that parametric linkage analysis under an SML model will extract most of the linkage information even when the true model is oligogenic, if the analysis is repeated over several sets of penetrance parameters designed to cover both dominant and recessive gene action (Greenberg *et al.* 1998). A lod score maximized over several models is sometimes called a mod score, and requires an adjustment for multiple testing before it can be interpreted in the same way as a simple lod score calculated under a known model. A straightforward adjustment for multiple tests is to subtract $\log_{10}$ M from the mod score where M is the number of tests.

Non-parametric linkage analysis

The hallmark of parametric linkage analysis is that it assumes an explicit mode of inheritance for the disease, usually the SML model. In contrast, non-parametric linkage analysis, first introduced by Penrose (1935), does not require an explicit model of the disease. Instead, evidence of linkage between disease and marker is based simply on an excess of marker-allele sharing between relative pairs concordant for disease, and a deficit of marker-allele sharing between relative pairs discordant for disease. Allele sharing is defined technically either as 'identity-by-state' (IBS) or 'identity-by-descent' (IBD). Two alleles are defined as IBS if they have the same DNA sequence at the polymorphic site. If, in addition to being IBS, two alleles are both descended from a single allele in a recent common ancestor, then they are said to be IBD. The difference between IBS and IBD is demonstrated in Fig. 3.3. Although both IBD and IBS have been used to define allele sharing in non-parametric linkage analysis, it is now accepted that IBD is the superior measure because it is more informative and less dependent on correct specification of marker allele frequencies.

The most popular design for non-parametric linkage analysis is the affected sib pairs (ASP) method. Under the null hypothesis of no linkage, sib pairs are expected to share on average half their genes IBD. If IBD sharing can be unambiguously assigned from marker genotypes, then the analysis simply involves testing whether the proportion of alleles IBD among affected sib pairs is greater than half. In the more realistic scenario that IBD sharing cannot always be unambiguously assigned, the data can be analysed by a likelihood method that treats the proportion of alleles IBD as the unknown parameter to be inferred from the data. This leads to a 'maximum lod score' (MLS)

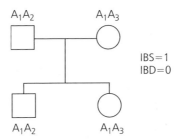

Fig. 3.3 Identity-by-descent (IBD) and identity-by state (IBS). Males are drawn in squares and females in circles. The two sibs are IBS for one allele (A_1) but IBD for 0 allele. The latter is because the A_1 allele in the son is transmitted from the mother, whereas the A_1 in the daughter is transmitted from the father.

(Risch 1990; Holmans 1993) that summarizes the evidence for linkage. The estimated proportion of alleles IBD depends both on the magnitude of effect of the disease locus, and how tightly linked the marker is to the disease locus.

The ASP method focuses on affected sib pairs because for complex disorders unaffected siblings may well be non-penetrant carriers of the susceptibility allele. For the same reason, extensions of the ASP method to pedigree data have also concentrated on affected family members. In general, these methods examine the marker genotype data to see if there is evidence for increased allele sharing among the affected members of a pedigree. The simplest measure of the extent of IBD sharing among a set of affected relatives is the average IBD sharing between all possible pairs of affected relatives. The observed values of this measure, over a sample of pedigrees, can be assessed for evidence of increased sharing over the level predicted under the hypothesis of no linkage. This method has been implemented in the Genehunter program (Kruglyak *et al.* 1996).

Linkage analysis of quantitative traits

Linkage analysis, originally developed for Mendelian disorders, has traditionally used a categorical approach to phenotype definition, with each individual being classified as being either affected or normal. A quantitative approach to phenotype definition may be more appropriate for some complex disorders. Some common disorders are likely to represent the extreme of a continuum of symptom severity in the population, the cut-off that defines 'caseness' being somewhat arbitrary. For example, scales that measure anxiety and depression symptoms in community do not show any evidence of discontinuity in terms of the frequency distribution of symptom levels, nor in terms of their correlations with other variables such as sociodemographic risk factors and service usage. Other disorders do not so obviously represent the extreme of a continuum, but quantitative 'intermediate phenotypes' may be identifiable. For example, efforts have been made to characterize the neurological or cognitive abnormality of schizophrenia in terms of evoked potentials and various neuropsychological tests.

The loci that contribute significantly to the variation of a quantitative trait in a population have become known as quantitative trait loci (QTLs). Variance–components models, which were developed for the partitioning of phenotypic variation into genetic and environmental components from correlational data from pairs of relatives (Fisher 1918), have been extended for QTL analysis (Schork 1993; Amos 1994; Eaves *et al.* 1996; Fulker and Cherny 1996; Almasy and Blangero 1998). A QTL component of variance is characterized by being correlated between relatives to the extent that they share alleles IBD at the QTL. In other words, the component contributes no covariance for relative pairs who share 0 allele IBD, half the QTL component for relative pairs who share 1 allele IBD, and the entire QTL component for relative pairs who share two alleles IBD. Linkage to a chromosomal position is indicated when such a relationship exists between the covariance in the phenotype and the IBD sharing at that position. The magnitude of the QTL variance is estimated by maximum likelihood. The hypothesis that a QTL component exists at a particular location is tested by likelihood ratio statistics.

One problem in the use of the variance–components model is the assumption of multivariate normality. While obvious outliers can be readily detected and then either removed or adjusted, more subtle deviations from multivariate normality can inflate the test statistics and lead to false positive linkage results (Allison *et al.* 1999). Similarly, when applied to the analysis of samples selected for phenotypic extremes, standard variance–components models have been shown to produce inflated test statistics and an increase in false positive linkage findings (Dolan *et al.* 1999). One way of making variance–components models more robust to non-normality and phenotypic selection is to use the conditional likelihood of the genotypes given the trait values, rather than the usual likelihood which is that of the trait values given the genotypes (Sham *et al.* 2000).

Another method of QTL linkage analysis that is applicable to sib pairs is a regression of the squared difference of sib pair trait values on the proportion of alleles IBD (Haseman and Elston 1972). The expected value of the regression coefficient of such an analysis is minus twice the QTL variance. This method has been improved by including the squared sum of sib pair trait values in the regression (Sham and Purcell 2001).

Significance and power of linkage tests

Linkage analysis of Mendelian disorders is both robust and powerful. The vast majority of linkage findings where the lod score exceeded 3 have been subsequently replicated and confirmed, and to date over a thousand genes have been mapped using this method. For complex disorders, however, replication has been much less consistent, even for findings that meet the lod score of 3 criterion. When the mode of inheritance is uncertain, a lod score of 3 no longer implies a very high probability of true linkage. Lander and Kruglyak (1995) suggested guidelines for the interpretation of linkage studies when the mode of inheritance is uncertain. According to these criteria, a finding can be said to be significant if its probability of occurring by chance is 0.05 per genome scan, but only 'suggestive' if its probability of occurring by chance is 1 per genome scan. How these guidelines correspond to lod scores depends on family structure; for large pedigrees significant and suggestive linkage correspond to lod scores of 3.3 and 1.9 respectively, whereas for sib pairs the corresponding lod scores are 3.6 and 2.2.

The above guidelines are meaningful only if the lod score is calculated correctly. Very large lod scores can be produced in the absence of linkage, when the frequencies of marker alleles are mis-specified for samples that consist mostly of sibships without genotyped parents. Mis-specification of the disease model, on the other hand, does not increase the rate of false positive linkage (although it does reduce power when linkage is present) (Clerget-Darpoux *et al.* 1986). Lod scores can be inflated by selecting the largest value from multiple methods of analysis, multiple disease models, or multiple phenotypic definitions. In these cases an adjustment for multiple testing is necessary as mentioned earlier in connection with using the mod score approach, the simplest of which is to subtract $\log_{10}M$ from the maximum lod score, where M is the number of different analyses performed.

Simulation studies comparing non-parametric to the optimal parametric analyses (conducted under the true disease model) have shown non-parametric methods to be almost fully efficient for the detection of low-penetrance genes. However, whatever statistical method is used, the power of linkage analysis decreases rapidly with diminishing effect size. The minimum effect size of an allele that can be detected by linkage with a feasible sample size is an odds ratio of around 1.5 (Risch and Merikangas 1996). Because of this, evidence for a substantial genetic contribution, in terms of the sibling recurrence ratio (λ_S) is now considered a prerequisite for embarking on gene mapping studies. Furthermore, because investigators of complex disorders are equally concerned with false negative as with false positive results, it is standard practice to perform both a non-parametric analysis, and at least two parametric analyses (under dominant and recessive models), to ensure that no linkage signal is missed.

For a diallelic QTL, effect size can be measured by the difference in average trait value between the two homozygous genotypes. The contribution of the QTL to the population phenotypic variance is called QTL heritability, and is proportional to the square of the effect size for any given allele frequencies. The information for detecting a QTL is proportional to the square of QTL heritability for linkage analysis. In other words, halving the effect size of a QTL will increase the required sample size by 16 fold for linkage analysis.

The most informative families for linkage analysis are those that are large and have multiple individuals on the extremes of the phenotypic dimensions (or affected individuals in the case of a rare disorder). Extreme discordant sib pairs are particularly informative. The efficiency of linkage studies can be maximized by choosing families with high information content, which can be calculated from the genetic relationships and the trait values of the family members.

Allelic association

Allelic association refers to the co-occurrence of an allele at a particular locus and a disease, above the level to be expected by chance (Edwards 1965). In general association between any two variables suggests, but does not necessarily indicate, causation. Allelic association is no exception. Non-causal explanations for allelic association include linkage disequilibrium (also known as indirect association) and population substructure (resulting in a phenomenon called 'spurious associations').

Linkage disequilibrium (LD) refers to the association between two alleles at different loci. The reshuffling of genes in meiosis will tend to reduce the level of LD between all pairs of loci from one generation to the next. However, alleles that are tightly linked on the same chromosome are unlikely to be separated in meiosis, so that the degree of LD between such alleles will decay at a slow rate over time. Other factors that influence the extent of LD between two loci include the size and histories of the population, and the mutational and recombinational histories of the loci. The theoretical prediction that the strength of LD will tend to be strong for tightly linked loci, and diminish with increasing genetic distance, has been borne out by empirical studies. These studies have also demonstrated great variability, with level of LD ranging from being almost complete to almost absent, for loci separated by the same distance (Abecasis et al. 2001). Generally speaking, the chance of appreciable LD is quite high at distances of less than 50 kb, but very low at distances of above 500 kb.

This frequent presence of LD between tightly-linked loci is an important indirect cause of allelic association. The associated allele has no direct effect on the risk of disease but is in LD with an allele at a nearby locus that does have a causal influence. This phenomenon means that allelic association may be observed not only at the causal locus itself, but also at neighbouring polymorphisms. It is therefore possible to screen a genomic region for association by genotyping a small proportion of the polymorphisms in the region. This leads to the possibility of using allelic association for systematic genetic mapping.

Another important cause of allelic association is population stratification. Basically this means that the population consists of several subgroups, and that there is variation in the frequency of both the disease and a particular allele among these subgroups, such that the frequency of both is higher in some subgroups than others. This will generate a 'spurious' association between the disease and the allele, when there is no 'true' association between the two within any of the subgroups. This is a potentially serious cause of false positive association findings.

It is also possible for population stratification to mask a real association. This occurs when the direction of the spurious association is opposite to the direction of the true association. In other words, there is a tendency for subpopulations with a higher frequency of a certain allele also to have a higher prevalence of the disease, but within each subpopulation there is a negative association between the allele and the disease. Indeed, such opposing effects can 'reverse' the direction of association, so that the apparent association is in the opposite direction of the true association. In the statistical literature this phenomenon is known as the 'Simpson paradox'.

Case-control association studies

The simplest type of association involves comparing individuals with a disease (cases) with unaffected subjects from the same population (controls) for differences in genotype or allele frequencies. This is analogous to case-control studies in epidemiology, with exposure being defined as the presence of a genotype, allele or haplotype (a combination of alleles inherited from the same parent, possibly on the same chromosome). Because family data are not required, case-control samples are relatively easy to collect.

Single-locus genotype data from a case-control study can be summarized in a contingency table. This tabulates the observed counts of the alternative alleles in cases and controls (Table 3.1). From the observed counts it is possible to calculate various measures of the magnitude of association (e.g. odds ratio, risk ratio and population attributable risk) with confidence intervals, and a chi-squared statistic to test the significance of the association.

If there are potential confounding variables (e.g. age, sex, ethnicity) then standard epidemiological methods of adjustment such as stratified analysis or logistic regression can be used. Logistic regression is particularly convenient in also allowing the analysis of potential interaction between genotype and demographic or environmental factors. There are several ways of coding the genotypes of a locus, and the choice of coding depends on the hypothetical model for gene action. For example, the three genotypes (aa, Aa and AA) at a diallelic locus might be assigned the values (0, 1, 2) under a 'gene-dosage' model, (0, 1, 1) under a dominant model, and (0, 0, 1) under a recessive model. Genotype data from multiple unlinked loci can also be conveniently analysed using logistic regression for gene–gene interaction.

The association analysis of multiple tightly linked loci requires special treatment because it is possible that the disease is associated with a particular haplotype (that is a set of alleles at two or more adjacent loci on the same region of a chromosome) more strongly than with any single allele. This can arise, for example, when a substantial number of high-risk alleles in the current population are descended from a single mutation that occurred on an ancestral chromosome, and that each of these high risk alleles are still flanked by a variable-length segment of that ancestral chromosome. In this case the disease will be associated with the haplotype around the disease mutation on the ancestral chromosome. An analysis of haplotype frequency differences between cases and controls is therefore expected to be more informative than an analysis of allele frequency differences (assuming that the actual causal polymorphism is not one

Table 3.1 Case-control contingency table and measures of association

Marker	Cases	Controls
Present	a	b
Absent	c	d

Tests of significance: chi square (1 degree of freedom) or Fisher's exact test if any expected frequency <5.
Measures of strength of association:
Relative risk (RR) = P1/P2, where P1 = a/(a + b) and P2 = c/(c + d).
Odds ratio (OR) = O1/O2 where O1 = a/b, O2 = c/d.
Attributable risk (AR) = F/(1 + F), where F = P(RR − 1), and P = exposure prevalence.
Above definition of RR is valid only when the proportions of cases and controls in the sample are the same as those in the population.
2 Where the proportion of affecteds in the population is small OR approximates to RR.
3 AR is dependent on the frequency of the marker/risk factor in the population. High frequency markers (or risk factors with high exposure prevalence) may therefore be associated with high attributable risks even when the RR is low (McGuffin and Gottesman 1999).

of the markers). Analysis of haplotype frequencies is complicated by the fact that the haplotypes making up a person's genotype cannot always be inferred with certainty, when parental data are not available or sometimes even when grandparental data are not available. Methods for statistical estimation of haplotype frequencies, and for testing haplotype frequency differences, are available (Zhao *et al.* 2000).

Hidden population substructure

When population substructure is apparent and each individual can be classified accurately into a subpopulation, standard epidemiological methods of stratified analysis or logistic regression modelling can be used to adjust for any potential confounding effects. However, population substructure may not be obvious, and it may be impossible to categorize many individuals accurately into a subpopulation. It is not unusual to come across individuals of mixed or uncertain ethnic origin in multi-ethnic communities. Since the confounding variable is not directly measurable, it is not possible to use standard methods of statistical adjustment.

There are two main methods for dealing with hidden population substructure. The first is to use related instead of unrelated control subjects. Choosing controls related to the cases ensures matched comparisons between individuals from the same subpopulation. The second is to use random genetic markers to reveal the substructure and to assign each individual to each subpopulation. If the hidden substructure is revealed and measured, then standard methods of statistical adjustment can be applied.

A popular design for family-based association is case-parent triads (affected cases and both parents). The transmitted and non-transmitted alleles at a locus for each parent are identified and tabulated (Fig. 3.4). The counts of transmitted and non-transmitted alleles can be analysed by one of the two methods: the haplotype-based haplotype relative risk (HHRR) (Terwilliger and Ott 1992) and the transmission/disequilibrium test (TDT) (Spielman *et al.* 1993). The two methods will usually give similar results, although TDT is preferred since it constitutes a matched-pair analysis appropriate for the design. The TDT is applicable to families with multiple cases.

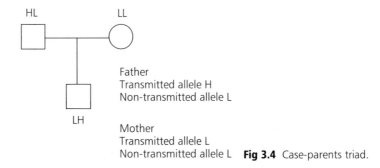

HL   LL

LH

Father
Transmitted allele H
Non-transmitted allele L

Mother
Transmitted allele L
Non-transmitted allele L   **Fig 3.4** Case-parents triad.

For late-onset diseases parents may be often unavailable and an alternative design is to compare cases and unaffected siblings (Curtis 1997). This discordant sibling design is however less powerful than the case-parent triad design because, for a low penetrance gene, many unaffected siblings are expected to be non-penetrant gene carriers. The power of discordant sibling design approaches the power of the case-parent design as the number of unaffected siblings increases, effectively allowing the genotypes of the two parents to be reconstructed (Whitaker and Lewis 1999).

The use of random genetic markers to overcome to problem of hidden population substructure can be loosely labelled as 'genomic control' (Devlin and Roeder 1999). One form of genomic control is to use random markers to estimate the average inflation of the chi-squared test statistic, and to use this factor to adjust the results of subsequent analyses (Pritchard and Rosenberg 1999). The disadvantage of this method is that a uniform adjustment may be too much for some analyses and too little for others. Furthermore, the adjustment does not deal with the problem that population substructure can mask real associations.

The inference of population substructure from the genotypes of random markers is based on the allele frequency differences between the subpopulations at a number of loci, which lead to associations between these alleles (Pritchard *et al.* 2000). A subpopulation might therefore be revealed by a pattern of positive correlations between alleles that are more common in the subpopulation than in other subpopulations. The statistical problem is therefore one of revealing a latent structure from the correlations

Table 3.2 Haplotype-based haplotype relative risk (HHRR) and transmission/disequilibrium test (TDT). Contingency table of allele transmission to affected individuals

Transmitted allele		Non-transmitted allele	
	H	L	Total
H	a	b	a + b
L	c	d	c + d
Total	a + c	b + d	a + b + c + d

HHRR Method

Estimated frequency of H among cases = (a + b)/(a + b + c + d)
Estimated frequency of H in general population = (a + c)/(a + b + c + d)
Estimated odds ratio = [(b + a)(b + d)]/[(c + a)(c + d)]
Chi-squared test statistic (one degree of freedom)

$$HHRR = \frac{2(a+b+c+d)(b-c)^2}{(2a+b+c)(2d+b+c)}$$

TDT Method

Estimated odds ratio = b/c
Chi-squared test statistic (one degree of freedom)

$$TDT = \frac{(b-c)^2}{(b+c)}$$

between observed variables. In the case of continuous variables factor analysis would be applicable, but because genotypes are categorical, a more appropriate statistical method is latent class analysis. The aim of the latent class analysis is to reveal the subpopulations and to assign each individual to a subpopulation. The success of such an approach will depend on the information (the number of genetic markers and whether they discriminate between the subpopulations), and on the complexity of the population and of the sample. If successful, then subpopulation membership can be used as a confounding variable in subsequent association analyses between disease and genetic markers.

Association analysis using pooled DNA samples

The cost of association analysis may be drastically reduced by performing genotyping not on individual DNA samples, but on pools made up of DNA from multiple individuals (Daniels *et al.* 1998). Thus, the allele frequencies in a sample of 200 cases and 200 controls can be measured from two pooled samples, rather than 400 individual samples.

There are some problems in the use of pooled DNA for association analysis. Significant covariates will need to be balanced in the design of the pools, because of the inability for individual adjustment in the analysis. Similarly, there will be limited ability to delineate hidden population stratification, and to adjust for this. Analyses will be restricted to the main effects of alleles; it will be impossible to examine for multilocus haplotype effects or interactions. Finally, allele frequency estimation from pooled DNA is subject to both bias and random measurement errors. Because of these problems, association analysis of pooled DNA samples should be viewed as an initial screen, with any positive findings followed up by further analysis using individual genotype data.

Association analysis of quantitative traits

For quantitative traits association analysis can be carried out on a simple random sample of unrelated individuals. Simple linear regression can be performed, with the trait as dependent and the genotype as independent variable. Different coding schemes for the genotype correspond to different hypothetical modes of inheritance. For example, the additive effects of the 3 genotypes of a diallelic locus can be coded -1, 0 and 1, with the heterozygous being coded 0 and the two homozygous coded -1 and $+1$. Dominance can be included as a second independent variable coded 1 for the heterozygous and 0 for the homozygous. The regression model can accommodate potential confounding variables, as well interactions between the genotype and other variables (which may include the genotype at another locus).

For samples with related family members, the model must also incorporate any residual covariances between family members. This can be achieved in a general variance–components framework, in which effects can be modelled either in the mean vector or the residual variance–covariance matrix. Effects of variables that are measured directly, such as genotype, age and sex, are modelled in the mean vector. Effects of variables that are

Table 3.3 A Comparison of linkage and association

Linkage	Association
Uses families	Uses cases and controls or families with 'internal controls'
Detectable over large distances >10 cM	Detectable only over very small distances, 1 cM
Can usually only detect large effects, e.g. relative risk >2 or >10% of variance	Capable of detecting small effects, e.g. odds ratio <2 or <1% of variance

not measured directly, but have predictable consequences on residual phenotypic correlation, are modelled in the variance–covariance matrix. A QTL on which IBD information is available can be included here, as a component whose correlation between relatives is equal to the proportion of alleles IBD. Similarly, residual polygenes can be included as a component whose correlation between relatives is equal to the expected proportion of alleles IBD given the genetic relationship. Such a model can simultaneously model allelic association in the mean vector and linkage in the variance–covariance matrix (Fulker *et al.* 1999).

The information for detecting a QTL by association analysis is proportional to the QTL heritability for association analysis (Sham *et al.* 2000). In other words, halving the effect size of a QTL will increase the required sample size by fourfold (as compared to 16 fold for linkage analysis). Association analysis is therefore potentially much more powerful than linkage for the detection of QTLs of small effect.

Conclusion

Our ability to map genes by positional cloning derives from the linear arrangement of coding DNA sequences on chromosomes, and the occurrence of crossovers in meiosis. The two main methods of positional cloning–linkage and association—have complementary properties. Linkage can be detected at long genetic distances but has low power to detect genes of small effect. Conversely, association is limited to short genetic distances but is able to detect minor-effect QTLs (see Table 3.3). Both linkage and association rely on the availability of polymorphic genetic markers throughout the genome. Systematic linkage analysis of the entire genome was already feasible in the 1990s. With the completion of the Human Genome Project and the development of high-throughput genotyping, systematic association analysis on a large scale is likely to become feasible in the near future.

Further reading

Ott (1999) is the standard reference work of human genetic linkage analysis. Sham (1997) gives an introduction to the principles of the statistical principles involved in linkage and association analysis. Bishop and Sham (2000) provides a review of the current status of the application of linkage and association analysis to complex disorders.

References

Abecasis G. R., Noguchi E., Heinzmann A. *et al.* (2001) Extent and distribution of linkage disequilibrium in three genomic regions. *Am J Hum Genet* **68**:191–8.

Allison D. B., Heo M., Kaplan N. *et al.* (1999) Sibling-based tests of linkage and association for quantitative traits. *Am J Hum Genet* **64**:1754–63.

Allison D. B., Neale M. C., Zannolli R. *et al.* (1999) Testing the robustness of the likelihood ratio test in a variance component quantitative trait loci mapping procedure. *Am J Hum Genet* **65**:531–44.

Almasy L. and Blangero J. (1998) Multipoint quantitative-trait linkage analysis in general pedigrees. *Am J Hum Genet* **62**:1198–211.

Amos C. I. (1994) Robust variance–components approach for assessing genetic linkage in pedigrees. *Am J Hum Genet* **54**:535–43.

Bader J. S., Bansal A., Sham P. C. (2001) Efficient SNP-based tests of association for quantitative phenotypes using pooled DNA. *Genescreen* **1**:143–50.

Bishop T. and Sham P. C. (2000) *Analysis of Multifactorial Disease.* London: Bios.

Clerget-Darpoux F., Bonaiti Pellie C., Hochez J. (1986) Effects of misspecifying genetic parameters in lod score analysis. *Biometrics* **42**:393–9.

Curtis D. (1997) Use of siblings as controls in case-control association studies. *Annals of Human Genetics* **61**:319–33.

Daniels J., Holmans P., Williams N. *et al.* (1998) A simple method for analysing microsatellite allele image patterns generated from DNA pools and its applications to allelic association studies. *Am J Hum Genet* **62**:1189–97.

Devlin B, and Roeder K. (1999) Genomic control for association studies. *Biometrics* **55**:997–1004.

Dolan C. V., Boomsma D. I., Neale M. C. (1999) A simulation study of the effects of assignment of prior identity-by-descent probabilities to unselected sib pairs, in covariance-structure mapping of a quantitative trait locus. *Am J Hum Genet* **64**:268–80.

Eaves L. J., Neale M. C., Maes H. (1996) Multivariate multipoint linkage analysis of quantitative trait loci. *Behavior Genetics* **26**:519–25.

Edwards J. H. (1965) The meaning of associations between blood groups and disease. *Annals of Human Genetics* **29**:77–83.

Elston R. C. and Stewart J. (1971) A general model for the analysis of pedigree data. *Human Heredity* **21**:523–42.

Fisher R. A. (1918) The correlation between relatives on the supposition of Mendelian Inheritance. *Transactions of the Royal Society of Edinburgh* **52**:399–433.

Fulker D. W. and Cherny S. S. (1996) An improved multipoint sib-pair analysis of quantitative traits. *Behavior Genetics* **26**:527–32.

Fulker D. W., Cherny S. S., Sham P. C. *et al.* (1999) Combined linkage and association analysis for quantitative traits. *Am J Hum Genet* **64**:259–67.

Greenberg D. A., Abreu P. C., Hodge S. E. (1998) The power to detect linkage in complex disease using simple genetic models. *Am J Hum Genet* **63**:870–9.

Haseman J. K. and Elston R. C. (1972) The investigation of linkage between a quantitative trial and a marker locus. *Behavior Genetics* **2**:3–19.

Holmans P. (1993) Asymptotic properties of affected-sib-pair linkage analysis. *Am J Hum Genet* **52**:362–74.

Knapp M. (1999) The transmission/disequilibrium test and parental-genotype reconstruction: the reconstruction-combined transmission/disequilibrium test. *Am J Hum Genet* **64**:861–70.

Kruglyak L., Daly M. J., Reeve-Daly M. P. *et al.* (1996) Parametric and nonparametric linkage analysis: a unified multipoint approach. *Am J Hum Genet* **58**:1347–63.

Lander E. S. and Kruglyak L. (1995) Genetic dissection of complex traits: guidelines for interpreting and reporting linkage results. *Nature Genetics* **11**:241–7.

McGuffin P. and Gottesman I. I. (1999) Risk factors for schizophrenia. *New England Journal of Medicine* **341**:370–1.

Morgan T. H., Sturvetant A. H., Muller H. J. *et al.* (1915) *The Mechanism of Mendelian Heredity*. New York: Holt.

Morton N. E. (1955) Sequential tests for the detection of linkage. *Am J Hum Genet* **7**:277–318.

O'Connell J. R. and Weeks D. E. (1995) The VITESSE algorithm for rapid exact multilocus linkage analysis via genotype set recoding and fuzzy inheritance. *Nature Genetics* **11**:402–8.

Ott J. (1974) Estimation of the recombination fraction in human pedigrees: efficient computation of the likelihood for human linkage studies. *Am J Hum Genet* **26**:588–97.

Ott J. (1999) *Analysis of Human Genetic Linkage*. Baltimore, MD: Johns Hopkins University Press.

Penrose L. S. (1935) The detection of autosomal linkage in data which consist of pairs of brothers and sisters of unspecified parentage. *Annals of Eugenics* **6**:133–8.

Plomin R., Hill L., Craig I. *et al.* (2002) A genome-wide scan of 1847 DNA markers for allelic associations with general cognitive ability: a five-stage design using DNA pooling. *Behavior Genetics* **31**:497–509.

Pritchard J. K. and Rosenberg N. A. (1999) Use of unlinked genetic markers to detect population stratification in association studies. *Am J Hum Genet* **65**:220–8.

Pritchard J. K., Stephens M., Rosenberg N. A. *et al.* (2000) Association mapping in structured populations. *Am J Hum Genet* **67**:170–81.

Purcell S., Cherny S. S., Hewitt J. K. *et al.* (2001) Optimal sibship selection for genotyping in quantitative trait locus linkage analysis. *Human Heredity* **52**:1–13.

Rao D. C., Keats B. J. D., Morton N. E. *et al.* (1979) Variability of human linkage data. *Am J Hum Genet* **30**:516–29.

Risch N. (1990) Linkage strategies for genetically complex traits. III. The effect of marker polymorphism on analysis of affected relative pairs. *Am J Hum Genet* **46**:242–53.

Risch N. and Merikangas K. (1996) The future of genetic studies of complex human diseases. *Science* **273**:1516–7.

Risch N. and Zhang H. (1995) Extreme discordant sib pairs for mapping quantitative trait loci in humans. *Science* **268**:1584–9.

Schork N. J. (1993) Extended multipoint identity-by-descent analysis of human quantitative traits: efficiency, power, and modeling considerations. *Am J Hum Genet* **55**:1306–19.

Sham P. (1997) *Statistics in Human Genetics*. London: Arnold.

Sham P. C. and Curtis D. (1995) An extended transmission/disequilibrium test (TDT) for multiallele marker loci. *Annals of Human Genetics* **59**:323–36.

Sham P. C. and Purcell S. (2001) Equivalence between Haseman–Elston and variance–components linkage analyses for sib pairs. *Am J Hum Genet* **68**:1527–32.

Sham P. C., Cherny S. S., Purcell S. *et al.* (2000) Power of linkage versus association analysis of quantitative traits, by use of variance–components models, for sibship data. *Am J Hum Genet* **66**:1616–30.

Sham P. C., Lin M. W., Zhao J. H. *et al.* (2000) Power comparison of parametric and non-parametric linkage tests in small pedigrees. *Am J Hum Genet* **66**:1661–8.

Sham P. C., Sterne A., Purcell S. *et al.* (2001) Creating a composite index of the vulnerability to anxiety and depression in a community sample of siblings. *Twin Research* **3**:316–22.

Sham P. C., Zhao J. H., Cherny S. S. *et al.* (2000) Variance components QTL linkage analysis of selected and non-normal samples: conditioning on trait values. *Genet Epi* 19:S22–8.

Smith C. A. B. (1959) Some comments on the statistical methods used in linkage investigations. *Am J Hum Genet* **11**:289–304.

Spielman R. S., McGinnis R. E., Ewens W. J. (1993) Transmission test for linkage disequilibrium: the insulin gene region and insulin-dependent diabetes mellitus (IDDM). *Am J Hum Genet* **52**:506–16.

Suarez B. K., Rice J. P., Reich T. (1978) The generalised sib-pair IBD distribution: its use in the detection of linkage. *Annals of Human Genetics* **42**:87–94.

Terwilliger J. and Ott J. (1992) A haplotype-based 'haplotype relative risk' approach to detecting allelic associations. *Human Heredity* **42**:337–46.

Whittaker J. C. and Lewis C. M. (1999) Power comparisons of the transmission/disequilibrium test and sib-transmission/disequilibrium-test statistics. *Am J Hum Genet* **65**:578–80.

Zhao H. (2001) Family-based association studies. *Statistical Methods in Medical Research* **9**:563–87.

Zhao H., Zhang S., Merikangas K. R. *et al.* (2000) Transmission/disequilibrium tests for multiple tightly linked markers. *Am J Hum Genet* **67**:936–46.

Zhao J. H., Curtis D., Sham P. C. (2000) Model-free analysis and permutation tests for allelic associations. *Human Heredity* **50**:133–9.

Part 2

The genetics of normal and abnormal development

Chapter 4

Personality and cognitive abilities

Robert Plomin, Francesca Happé, and
Avshalom Caspi

Introduction

This chapter reviews genetic research on personality and cognitive abilities, the two major domains of behavioural genetic research on normal variation in human behaviour. Rather than considering the two domains separately, they are reviewed together because the contrasts between them are interesting in relation to quantitative and molecular genetic research.

Why include a chapter on normal personality and cognitive abilities in a book on psychiatric genetics? One reason is that it seems likely that some psychopathology such as personality disorders or depression might represent the quantitative extreme of the normal range of genetic and environmental variation in personality (Cloninger 2002; Nigg and Goldsmith 1998). Similarly, some cognitive disabilities such as learning disabilities might be the quantitative extreme of the normal range of genetic and environmental variation in cognitive abilities. This hypothesis follows from the perspective of quantitative trait loci (QTLs), genes in multiple-gene systems that contribute quantitatively to the variance of a trait (Plomin *et al.* 1994). If there are multiple genes that affect complex heritable traits such as personality and cognitive abilities, it is likely that the traits are distributed quantitatively as dimensions rather than qualitatively as disorders. This was the essence of a 1918 paper that forms the cornerstone of quantitative genetic theory (Fisher 1918).

To the extent that disorders are the quantitative extreme of the same genetic and environmental factors that contribute to variability throughout the distribution, it could be argued that there are no disorders, just dimensions. For example, reading disability could be the genetic extreme of reading ability. Depression might be the quantitative extreme of mood or the personality trait of neuroticism. Quantitative genetic research mentioned later tends to support the hypothesis that common disorders such as these represent the genetic extremes of quantitative traits. The definitive test of this QTL hypothesis will come when genes are identified that are associated with behaviour. The main prediction arising from this hypothesis is that genes associated with disorders will also be associated with dimensions and vice versa. In other words, when a gene is identified that underlies the well-replicated QTL linkage for reading disability on chromosome 6 (see Chapter 6), the prediction is that this is not a gene conferring susceptibility to reading disability per se. Rather, the gene will be associated with reading ability throughout the distribution. That is, having a particular allele of the gene will decrease reading ability for children of high ability and average ability as well as those of low reading ability.

In addition to the implications for design and analysis of molecular genetic studies, there is an important conceptual aspect to the QTL perspective. A common mistake is to think that we are all basically the same genetically except for a few rogue mutations that lead to disorders. In contrast, the QTL perspective suggests that genetic variation is widespread and normal. Many genes affect most complex traits and, together with environmental variation, these QTLs are responsible for normal variation as well as for the abnormal extremes of these quantitative traits. This QTL perspective has some implications for thinking about mental illness because it blurs the aetiological boundaries between the normal and the abnormal. That is, we all have many alleles that contribute to mental illness, but some of us are unluckier in the hand that we draw at conception from our parents' genetic decks of cards.

A more subtle conceptual advantage of a QTL perspective is that it frees us to think about both ends of the normal distribution—the positive as well as the problem end, abilities as well as disabilities, and resilience as well as vulnerability. It has been proposed that we move away from an exclusive focus on pathology towards considering positive traits that improve the quality of life and perhaps prevent pathology (Seligman and Csikszentmihalyi 2000).

In this chapter, brief discussions of background issues are followed by an overview of research on the heritability of personality and cognitive abilities. Although much remains to be learned about whether and how much genetic factors influence these domains, quantitative genetic research is going beyond heritability to ask more interesting questions such as genetic links between the normal and abnormal, genetic and environmental contributions to the covariance between traits, developmental change and continuity, the importance of non-shared environment, and the interplay between nature and nurture. The future of genetic research lies with molecular genetic attempts to identify specific genes responsible for the ubiquitous heritability of personality traits and cognitive abilities, which we discuss in the last section.

Concepts of personality and cognition

Personality traits are relatively enduring individual differences in behaviour that are stable across time and across situations (Pervin and John 1999). In the 1970s, there was an academic debate about whether personality exists, a debate reminiscent of the nature–nurture debate. Some psychologists argued that behaviour is more a matter of the situation than of the person, but it is now generally accepted that both are important and can interact (Kenrick and Funder 1988; Rowe 1987). Another definitional issue concerns temperament, personality traits that emerge early in life and, according to some researchers (e.g. Buss and Plomin 1984), may be more heritable. However, there are many different definitions of temperament (Goldsmith *et al.* 1987), and exploring the supposed distinction between temperament and personality is beyond the scope of this chapter.

Cognitive abilities have traditionally been considered separately from other personality traits. One reason is that cognitive abilities are assessed as maximal performance whereas personality involves typical performance. Most theories and research about cognitive processing are at the level of species universals rather than individual differences. For example, brain imaging research typically asks what part of the brain is engaged in a particular task on average across a sample rather than focusing on differences between individuals in the sample. The focus of this review is on individual differences in cognitive performance, the level of analysis that is amenable to genetic research.

Measurement

The vast majority of genetic research on personality relies on self-report questionnaires. There are dozens of such questionnaires assessing hundreds of personality dimensions. Although one might question the validity of these self-perceptions, they are remarkably reliable and stable over decades (Roberts and Del Vecchio 2000). They also show reasonable validity in relation to ratings by peers and, for some personality dimensions, by observers rating behaviour in the laboratory. Although structured interviews could be used to assess personality as they typically are used to diagnose psychiatric disorders, there is no evidence that interviews do a sufficiently better job than questionnaires to justify their much greater cost. Because self-report questionnaires cannot be used in infancy and early childhood, ratings by parents, testers and teachers have been used.

In genetic research, cognitive abilities have typically been assessed using so-called psychometric tests. These are often paper-and-pencil tests but also include individually administered IQ tests such as the Wechsler series of tests developed for children, adolescents, and adults. Both types of tests are highly reliable and stable (Gottfredson 1997). For example, a recent 60-year follow-up of IQ scores from age 11 to age 77 yielded a correlation of 0.63 (Deary *et al.* 2000). Cognitive psychologists prefer to focus on cognitive processes such as short-term memory, often using reaction time measures rather than behavioural performance in substantive domains such as vocabulary. The level of analysis preferred by neuroscientists is brain function, using neuroimaging techniques in the human species and even more specific measures in animal models, such as recording the activity of a particular synapse. A future direction for genetic research is to address all of these levels of analysis from an individual differences perspective.

Factor structure

Both personality and cognitive abilities yield a hierarchical factor structure. For personality, there is no '*g*' at the top of the hierarchy. However, progress in developing a taxonomy of personality traits has been facilitated by the recognition that personality is organized hierarchically. At the highest level are broad traits (e.g. extraversion) representing the most general dimensions of individual differences in personality. At successively lower levels are more specific traits (e.g. sociability, activity, dominance) that are, in turn, composed of more specific responses (e.g. talkative, energetic, enthusiastic). In this hierarchical scheme, higher-order constructs can be shown to account for the observed covariation among lower-order constructs. The search for the structure of personality yields somewhat different results depending at which level personality factors are derived, how factors are extracted, and how they are rotated, but during the past two decades consensus has been growing towards a Five-Factor Model (FFM) (John and Srivastava 1999). These Big Five personality factors are identified in Table 4.1. The FFM is the focus of current genetics research (Beer and Rose 1995; Bergeman *et al.* 1993; Jang *et al.* 1996; McCrae *et al.* 2000) as well as research on links between normal personality variants and psychopathology (Jang and Livesley 1999; Widiger *et al.* 1999).

For cognitive abilities, one general factor ('g' or general cognitive ability) is on top of the hierarchy. 'g' represents the fact that all tests of cognitive ability intercorrelate, tests

Table 4.1 Examples of trait adjectives and Q-sort items defining the Five-Factor Model (FFM)

Factor	Example of factor definers from different instruments	
	Adjective checklist items	**Q-sort items**
Extraversion vs. introversion	Active	Skilled in play, humour
	Assertive	Behaves assertively
	Enthusiastic	Facially, gesturally expressive
	Outgoing	Gregarious
Agreeableness vs. antagonism	Generous	Warm, compassionate
	Kind	Arouses liking
	Sympathetic	Sympathetic, considerate
	Trusting	Basically trustful
Conscientiousness vs. lack of direction	Organized	Not self-indulgent
	Planful	Able to delay gratification
	Reliable	Dependable, responsible
	Responsible	Behaves ethically
Neuroticism vs. emotional stability	Anxious	Basically anxious
	Self-pitying	Thin-skinned
	Tense	Concerned with adequacy
	Worrying	Fluctuating moods
Openness/intellect vs. closedness to experience	Artistic	Aesthetically reactive
	Curious	Introspective
	Imaginative	Values intellectual matters
	Wide interests	Wide range of interests

as diverse as vocabulary, spatial ability, and memory. The term *general cognitive ability* is to be preferred to the word *intelligence* because intelligence has come to mean too many different things to different people. Although the existence of 'g' is well documented, it is less clear what 'g' is—whether 'g' is due to a single general process such as speed of information processing, or whether it represents overlap among components of more specific cognitive processes such as executive functions (Mackintosh 1998). The next step down in the hierarchy divides ability into verbal and non-verbal (performance) skills, and the next lower step involves so-called specific cognitive abilities such as spatial, memory and processing speed (Carroll 1993). Each of these specific cognitive abilities is indexed by several individual tests.

Heritability

Personality

Genetic research on personality is extensive, with several books devoted to the topic (Cattell 1982; Eaves *et al.* 1989; Loehlin 1992; Loehlin and Nichols 1976) and hundreds of research papers. Loehlin (1992) fitted different gene–environment models to a meta-analysis of twin and adoption studies of personality traits. He accomplished this in two steps. First, he organized according to the FFM the many different personality measures that have been administered in twin and family studies. Second, he fitted different gene–environment models to the data about each of the five broad traits. Figure 4.1 summarizes the results. It shows that genetic effects are important for all traits and account for 22–46 per cent of the total variation in each of the five factors. Model-fitting also points to a second factor that may indicate either a special monozygous (MZ) environmental effect or non-additive genetic effects. Special MZ environmental effects refer to the possibility that MZ twins' resemblance is increased because their environments are more similar than those of other first-degree relatives. Non-additive genetic effects refer to the possibility that MZ twins' resemblance is more than double that for other first-degree relatives because they share the entire configuration of their genes. This includes within-locus interactions (dominance) and between-locus interactions (epistasis) (see Chapter 2). To the extent that non-additive genetic effects are important, it may be that altogether between 40 per cent to 50 per cent of the variation in most personality traits is broadly heritable. The implications of non-additivity for molecular genetics research are discussed later. The results in Fig. 4.1 also highlight important environmental influences on personality traits, and show that the contribution of the shared family environment is small. The largest factor—accounting for 50 per cent or more of the variation in most personality traits—is nonshared, or person-specific, environmental variation. This category of environmental effects is discussed later.

Heritabilities in the 30–50 per cent range are typical of personality results. It is surprising that studies have not found any personality traits assessed by self-report questionnaire that consistently show no heritability. Can this be true? One way to explore

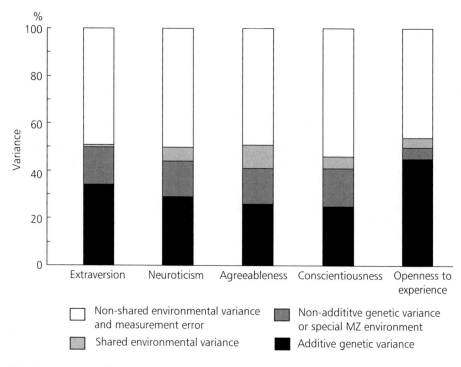

Fig. 4.1 Summary of behavioural genetics research and the Five-Factor Model (Loehlin 1992).

this issue is to use measures of personality other than self-report questionnaires to investigate whether this result is somehow a consequence of the use of self-report measures.

A recent study of adult twins in Germany and Poland compared twin results for self-report questionnaires and ratings by peers for the FFM for nearly a thousand pairs of twins (Riemann *et al.* 1997). Each twin's personality was rated by two different peers. The average correlation between the two peer ratings was 0.61, a result indicating substantial agreement concerning each twin's personality. The averaged peer ratings correlated 0.55 with the twins' self-report ratings, a result indicating moderate validity of self-report ratings. Figure 4.2 shows the results of twin analyses for self-report data and peer ratings averaged across two peers. The results for self-report ratings are similar to other studies. The interesting result is that peer ratings also show significant genetic influence, although somewhat less than self-report ratings. Moreover, multivariate genetic analysis indicates that the same genetic factors are largely involved in self-report and peer ratings, which provides strong evidence for genetic validity of self-report ratings.

Do these five broad factors represent the best level of analysis for research in genetics and psychiatry? Higher-order dimensions, such as those in the FFM, are summaries of specific-lower order traits. For example, the higher-order trait of neuroticism or negative emotionality subsumes propensities toward anger, guilt, self-criticism and other specific

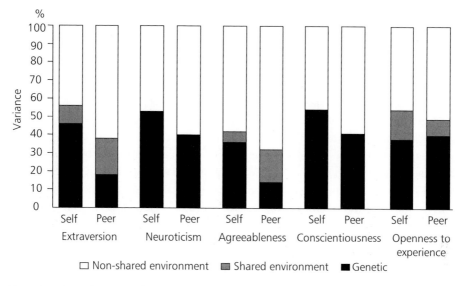

Fig. 4.2 Twin study using self-report and peer ratings of personality traits (Riemann *et al.* 1997).

negativistic biases. Analysis at these lower levels of the personality hierarchy may offer additional useful information for describing the factors of personality associated with psychopathology (Krueger *et al.* 2000). As discussed later, multivariate genetic research suggests that a narrower focus will add to our understanding because subtraits within each broad factor show some unique genetic variance not shared with other traits in the factor (Loehlin 1992).

Cognitive abilities

Although the genetic contribution to 'g' has produced controversy in the media, especially following the publication of *The Bell Curve* (Herrnstein and Murray 1994), there is considerable consensus among researchers, even those who are not geneticists, that 'g' is substantially heritable (Brody 1992; Mackintosh 1998). Correlations for first-degree relatives living together average 0.43 for more than 8000 parent-offspring pairs and 0.47 for more than 25,000 pairs of siblings. However, 'g' might run in families for reasons of nurture or of nature. In studies involving more than 10,000 pairs of twins, the average 'g' correlations are 0.85 for identical twins and 0.60 for same-sex fraternal twins. All the data converge on the conclusion that the heritability of 'g' is about 50 per cent, that is, genes account for about half of the variance in 'g' scores (Bouchard and McGue 1981; Plomin *et al.* 1997a). Even an attempt to explain as much of the variance of 'g' as possible in terms of prenatal effects yielded a heritability estimate of 48 per cent (Devlin *et al.* 1997; McGue 1997). Heritability is specific to an environment and so could differ in different cultures (see Chapter 2). However, moderate heritability of g has been found, not only in twin studies in North American and western European countries, but also in urban Russia, former East Germany, Japan, and both urban and rural India.

Although far fewer studies have focused on specific cognitive abilities, genetic influence is also substantial for specific cognitive abilities but somewhat less than for '*g*' (Plomin and DeFries 1998). The results of dozens of twin studies of specific cognitive abilities are summarized in Table 4.2 (Nichols 1978). Doubling the difference between the correlations for identical and fraternal twins as a rough estimate of heritability, these results suggest that specific cognitive abilities show slightly less genetic influence than general cognitive ability. Memory and verbal fluency show heritabilities of about 30 per cent and the other abilities yield heritabilities of 40 to 50 per cent. Although some larger twin studies do not consistently find that some cognitive abilities are more heritable than others (Bruun *et al.* 1966; Schoenfeldt 1968), it has been suggested that verbal and spatial abilities in general show greater heritability than do perceptual speed and memory abilities (Plomin 1988). Earlier twin studies of specific cognitive abilities have been reviewed in detail elsewhere (DeFries *et al.* 1976). A recent study of 160 pairs of twins aged 15 to 19 found similar results for tests of verbal and spatial ability. This study is notable because the sample population was Croatian (Bratko 1997), thus broadening the population base of observations on this topic.

Two studies of identical and fraternal twins reared apart provide additional support for genetic influence on specific cognitive abilities. One is a US study of 72 reared-apart twin pairs of a wide age range in adulthood (McGue and Bouchard 1989), and the other is a Swedish study of older twins (average age of 65 years), including 133 reared-apart twins and 142 control twin pairs reared together (Pedersen *et al.* 1992). Both studies show significant heritability estimates for all four specific cognitive abilities. As shown in Table 4.3, the heritability estimates are generally higher than implied by the twin results summarized in Table 4.2. This discrepancy may be due to the trend, discussed later, for heritability for cognitive abilities to increase during the lifespan. In both studies, the lowest heritability is found for memory.

Specific cognitive abilities are central to an ongoing longitudinal adoption study called the Colorado Adoption Project (DeFries *et al.* 1994). Figure 4.3 summarizes parent-offspring results for verbal, spatial, perceptual speed, and recognition memory abilities

Table 4.2 Average twin correlations for tests of specific cognitive abilities (Nichols 1978)

Ability	Number of studies	Twin correlations	
		Identical twins	**Fraternal twins**
Verbal comprehension	27	0.78	0.59
Verbal fluency	12	0.67	0.52
Reasoning	16	0.74	0.50
Spatial visualization	31	0.64	0.41
Perceptual speed	15	0.70	0.47
Memory	16	0.52	0.36

Table 4.3 Heritability estimates for specific cognitive abilities in two studies of twins reared apart

Ability	Heritability estimate (%)	
	McGue and Bouchard (1989)	**Pedersen et al. (1992)**
Verbal	57	58
Spatial	71	46
Speed	53	58
Memory	43	38

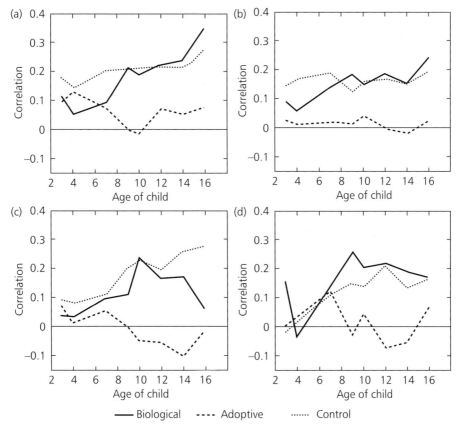

──── Biological   ---- Adoptive   ·········· Control

Fig. 4.3 Colorado Adoption Project: parent–offspring correlations for factor scores for specific cognitive abilities for adoptive, biological, and control parents and their children at 3, 4, 7, 9, 10, 12, 14, and 16 years of age. (a) verbal, (b) spatial, (c) speed of processing, (d) recognition memory. Parent–offspring correlations are weighted averages for mothers and fathers. (Reprinted with permission from Plomin et al. 1997).

from early childhood through adolescence (Plomin *et al.* 1997b). Mother–child and father–child correlations were averaged for both adoptive and control (nonadoptive) families. For each ability, biological mother–adopted child and control parent–control child correlations tend to increase as a function of age. In contrast, adoptive parent–adopted child correlations do not differ substantially from zero at any age. These results indicate increasing heritability and no shared environment. Although it has been argued that adoption studies overestimate heritability and underestimate the influence of shared environment due to a reduced range of environments in adoptive families (Stoolmiller 2000; cf. Loehlin and Horn 2000), the three types of parents (adoptive, biological, and control) in the Colorado Adoption Project show similar variances, which preserves the internal integrity of comparisons between offspring and their environmental (adoptive) parents, genetic (biological) parents, and genetic plus environmental (control) parents.

The results for family, twin, and adoption studies of verbal and spatial ability are summarized in Fig. 4.4 (Plomin and Craig 1998). These results converge on the conclusion that both verbal and spatial ability show substantial genetic influence but only modest influence of shared environment.

An interesting difference between personality and cognitive abilities is that assortative mating is much greater for cognitive abilities than for personality. Spouses correlate about 0.10 for personality traits but about 0.45 for 'g' and even higher for verbal ability (Jensen 1978). Unless assortative mating is taken into account in model-fitting analyses, it will be read as shared environmental influence in twin studies because it raises fraternal twin correlations but not identical twin correlations. Identical twins are genetically identical and thus their genetic similarity is not increased by assortative mating. This means that twin studies can underestimate heritability if assortative mating is not taken into account (see Chapter 2 for further discussion).

At a population level, assortative mating increases additive genetic variance cumulatively generation after generation (Plomin *et al.* 2001). Additive genetic variance involves genetic effects that add up rather than interact across loci. A hallmark of non-additive genetic variance is that identical twins (who are similar for all genetic effects no matter how complexly interactive the genes happen to be) are more than twice as similar as fraternal twins. Fraternal twins are said to be 50 per cent similar genetically but this is only in relation to additive genetic effects—they are hardly similar at all for epistatic effects (higher-order interactions among multiple genes). The typical twin study correlations of 0.85 for identical twins and 0.60 for fraternal twins for 'g' suggest that genetic variance for 'g' is largely additive. In contrast, some personality traits show non-additive genetic influence (Loehlin 1992). The issue of additive versus non-additive genetic variance is important in relation to attempts to identify specific genes responsible for genetic influence. If genetic effects involve interactions among genes, it will be much more difficult to identify individual genes in such complex systems. With 30,000 genes expressed in the brain, two- and three-way gene interactions alone would yield millions of possible interactions.

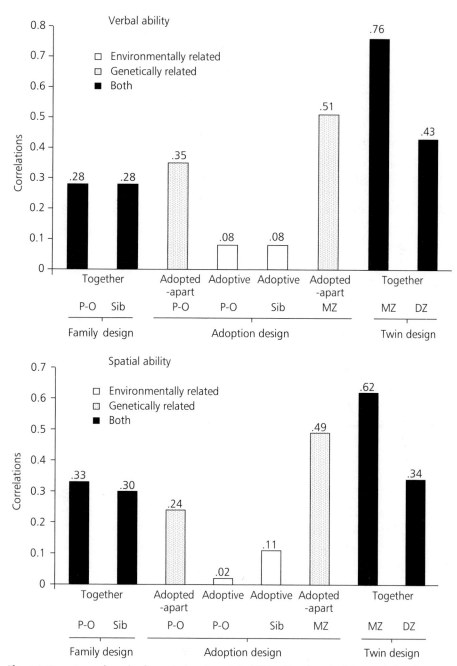

Fig. 4.4 Summary of results for verbal and spatial abilities (Reprinted with permission from Plomin and Craig 1998).

Beyond heritability

Quantitative genetics is hardly needed any longer merely to ask whether and how much genetic factors influence behavioural traits because the answers are 'yes', and 'a lot', respectively, for nearly all traits that have been studied, including personality and cognitive abilities. New quantitative genetic techniques make it possible to go beyond estimating heritabilities to investigate more interesting questions such as the genetic links between normal variation and disorders, the genetic contribution to covariance between traits, developmental changes and continuities in genetic effects, the importance of non-shared environment, and correlations between genetic propensities and experience.

Genetic links between the normal and abnormal

A quantitative genetic method called DeFries-Fulker (DF) extremes analysis makes it possible to investigate the genetic and environmental links between the normal and abnormal (DeFries and Fulker 1985, 1988). For example, to what extent does reading disability represent the quantitative extreme of the same genetic and environmental factors responsible for normal variation in reading ability, as predicted by a QTL perspective? DF extremes analysis addresses this issue by assessing relatives of diagnosed probands on a quantitative measure. To the extent that genetic factors involved in reading disability are also responsible for normal variation as assessed by a quantitative measure of reading ability, the reading ability scores of MZ co-twins of reading disabled probands will be more similar to those of the probands than will the scores of DZ co-twins. That is, the mean reading score of the co-twins will be closer to the probands for MZ than for DZ co-twins. This is the case for quantitative reading scores of co-twins of reading disabled probands, which suggests that genes associated with reading disability will also be associated with normal variation in reading ability. DF extremes analysis can yield evidence for genetic influence only to the extent that the quantitative trait is genetically linked to the disorder, that is, to the extent that the same genes affect the disorder and the dimension. DF extremes analysis also led to sib pair QTL analysis which is in essence the same analysis except that, instead of comparing MZ and DZ twins, siblings are compared who are identical for a DNA marker (that is, they have the same two alleles), who are 50 per cent similar (sharing just one allele), or who are zero per cent similar (sharing neither of their two alleles) (Cardon *et al.* 1994).

In the case of reading disabled probands and a quantitative measure of reading ability, DF extremes analysis estimates that half of the mean difference between the probands and the population is heritable (DeFries and Gillis 1993). This is called 'group heritability' to distinguish it from the usual heritability estimate, which refers to differences between individuals rather than to mean differences between groups. DF extremes analysis is similar to the liability–threshold model which is often used to analyse dichotomous data such as twin concordances (Falconer 1965; Smith 1974). The major difference is that the liability–threshold model assumes a continuous dimension even though it assesses a dichotomous disorder. The liability–threshold analysis converts dichotomous diagnostic data to a hypothetical construct of a threshold with an underlying continuous liability. In contrast, DF extremes analysis assesses rather than assumes a continuum. If all of the assumptions of the liability–threshold model are correct for a particular disorder, it will yield results similar to DF extremes analysis to the extent that the quantitative dimension assessed underlies the qualitative disorder. In the case of reading disability, a liability–threshold analysis of twin data yields an estimate of heritability similar to that of the DF extremes analysis (Plomin 1991).

Similar analyses have been conducted at the low extreme of 'g' (Petrill *et al*. 1997; Saudino *et al*. 1994) and language development (Dale *et al*. 1998). In general, research of this sort in the cognitive realm yields results similar to those for reading disability in suggesting strong genetic links between the normal and abnormal. A possible exception involves low 'g' late in life which has been reported to show no group heritability (Petrill *et al*. 2001). A clear exception is severe retardation. Using a method similar conceptually to DF extremes analysis, a large study of siblings addressed the issue of the familial links between normal and abnormal cognitive development (Nichols 1984), although a sibling study cannot disentangle genetic and environmental sources of familial resemblance. In a sample of over 17,000 children, 0.5 per cent had IQs below 50. Surprisingly, the siblings of these retarded children were not retarded. The siblings' average IQ was 103 and ranged from 85 to 125. In other words, moderate to severe mental retardation showed no familial resemblance. This result suggests that there is no genetic link between severe mental retardation and normal variation in IQ. In contrast, siblings of mildly retarded children yielded familial resemblance in that they had lower than average IQ scores. The average IQ for these siblings of mildly retarded children (1.2 per cent of the sample) was 85. This finding suggests that mild mental retardation is familial, unlike severe retardation. Similar results emerged from the largest family study of mild mental retardation, which considered 80,000 relatives of 289 mentally retarded individuals (Johnson *et al*. 1976; Reed and Reed 1965). Together, these results are consistent with the view that mild mental retardation, but not moderate or severe retardation, represents the lower end of normal variation in 'g'.

To date there has been little research using DF extremes analysis to address the extent to which normal variation in personality is linked genetically to the extremes of personality or psychopathology. However, one example was a twin study of attention deficit hyperactivity disorder (ADHD) symptoms suggesting that extreme and 'normal' or mid-range ADHD scores reported by questionnaire are influenced by the same genes (Levy *et al*. 1997). Another study along these lines found that neuroticism is strongly predictive of an onset of major depression in a one-year period and that approximately 70 per cent of the correlation between major depression and neuroticism was estimated to be genetic (Kendler *et al*. 1993).

Therefore it seems likely that the extremes of some normal personality traits are associated genetically with certain disorders. Some extremes of normal cognitive traits also clearly correspond to disorders, for example reading ability and reading disability. However, not all normal dimensions of personality and cognitive abilities completely match psychiatric disorders, and vice versa. This lack of correspondence suggests directions for future research. For example, some dimensions of personality and cognitive abilities that are not represented in current diagnostic classifications of disorders might usefully be considered in this light, often for both the low and high ends of the normal distribution. Activity level in the sense of vigour or energy level is not clearly represented in terms of pathology. For children, high activity can be a problem that is only tangentially considered in diagnosis of hyperactivity, which primarily focuses on attention and impulsivity. More tellingly, for both children and adults, low activity level can be a problem that only surfaces as one of several symptoms of depression or chronic fatigue syndrome. Sociability in the sense of gregariousness is another example. Both low sociability (little satisfaction in interacting with others) and high sociability (lack of independence) can be problems. A third example involves emotionality. Although the fearfulness component of emotionality contributes heavily to anxiety, the anger component of emotionality does not have a clear diagnostic analogue, despite the problems caused by quick-tempered people.

For cognitive abilities and disabilities, the links are also slightly askew. For example, specific cognitive ability factors include verbal, spatial, memory, and processing speed, but these abilities are not closely aligned with what are studied as cognitive disabilities. Although verbal ability is somewhat involved in communication disorders, there is much to learn about the relationship between low functioning and the development of disorder. Poor memory is a core feature of dementia but poor memory exists throughout life whereas dementia is acquired late in life. Low spatial ability is not even recognized as a problem, although it is probably included in non-verbal learning disability. Diagnostic classifications of cognitive disorders have focused on functional cognitive problems such as communication and language, reading, and mathematics. Although these behaviours are not represented in traditional genetic research on specific cognitive abilities, they could easily be considered in relation to dimensions of normal variation. Research along these lines has made good progress for reading, which shows strong genetic links between the normal and abnormal (DeFries *et al.* 1999). One study of mathematics disorder also shows genetic links between the normal and abnormal (Alarcón *et al.* 1997) and, more surprisingly, substantial genetic overlap with reading disability (Light and DeFries 1995). Genetic research has been slow in coming to the field of normal and abnormal language development but it is making up for lost time (Gilger 1997; Plomin and Dale 2000; Stromswold, 2001).

Multivariate genetic analysis

To what extent do the same sets of genes affect different personality traits or cognitive abilities? In broader terms, to what extent does the genetic factor structure of personality and cognitive abilities correspond to their phenotypic structures? A set of techniques called multivariate genetic analysis examines covariance among measures and yields a statistic called the *genetic correlation* (see Chapter 2 for further discussion). The genetic correlation is an index of the extent to which genetic effects on one trait correlate with genetic effects on another trait independent of the heritability of the two traits and independent of the magnitude of their phenotypic correlation. That is, although personality traits and cognitive abilities are moderately heritable, the genetic correlations between them could be anywhere from 0.0, indicating complete independence of their genetic effects, to 1.0, indicating that the same genes are involved.

For personality, the few reported multivariate genetic analyses suggest that genetic factor structures generally correspond to phenotypic factor structures. Most importantly, multivariate genetic analyses of FMM in two twin samples found that the genetic structure of the five factors largely corresponds to the phenotypic structure of the five factors (Jang *et al.* 1998). This finding suggests that genes found to be associated with neuroticism, for example, are not likely to be associated with extraversion, two of the major FFM traits. It also suggests that introversion items and extraversion items are opposite ends of the same continuum, genetically as well as phenotypically.

Earlier analyses focused on neuroticism, extraversion, and psychoticism as assessed by the Eysenck Personality Questionnaire (EPQ). A multivariate genetic analysis of the EPQ items yielded independent genetic factors of neuroticism and extraversion (Heath *et al.* 1989). However, correspondence between the phenotypic and genetic factor structures for psychoticism was not as clear. The genetic structure of psychoticism showed two distinct genetic factors: paranoid attitudes and hostility. (The psychoticism scale of the EPQ is not closely related to conventional clinical concepts of psychotic symptoms.) Another multivariate genetic analysis comparing the EPQ and Cloninger's Tridimensional Personality Questionnaire (TPQ) indicated that the two systems of personality did not show much genetic overlap (Heath *et al.* 1994). In other words, from a genetic perspective, the EPQ and TPQ assess somewhat different traits.

As mentioned in the introduction to this chapter, considerable genetic overlap has been found between the FFM and a measure of personality dysfunction (Jang and Livesley 1999; Widiger *et al.* 1999). The best documented example is the relationship between neuroticism and depressive symptoms (Eaves *et al.* 1989). Another strategy is the use of profile analysis (DiLalla *et al.* 1996).

Multivariate genetic analyses of cognitive abilities generally support the widely accepted hierarchical phenotypic factor structure, but they emphasize the genetic importance of 'g'. That is, although phenotypic correlations among specific cognitive abilities are only moderate, genetic correlations among specific cognitive abilities are very high, even though some unique genetic variance for specific cognitive ability remains that is not explained by 'g' (Petrill 1997). The high genetic 'g' correlations suggest that genes found to be associated with one cognitive ability, such as spatial ability, are also likely to be associated with other cognitive abilities such as verbal ability or memory. A related interesting finding is that the 'g'-loadings of cognitive tests on a first unrotated principal component are highly correlated with their heritabilities. That is, the higher a test's 'g' loading, the higher its heritability (Jensen 1998). This finding suggests that a large part of the heritability of individual cognitive capacities has to do with the heritability of 'g'.

In addition to indicating that 'g' is a good target for molecular genetic research, the multivariate genetic finding of high genetic correlations among diverse cognitive abilities also has important implications for understanding the brain mechanisms that mediate genetic effects on cognitive ability. In contrast to the prevalent modular view of cognitive neuroscience that assumes that many cognitive processes are independent, these results suggest that genetic effects are general (Plomin 1999). However, it is important to note that the perspective of cognitive neuroscience largely focuses on species-typical phenomena, and these processes are assumed to show little variation. In contrast, genetic research focuses on individual differences within a species. An emerging area of interest is the relationship between individual differences in reaction time measures of mental speed and 'g', which generally show substantial genetic overlap (Spinath and Borkenau 2000).

Development

Does heritability change during development? For personality it is clear that twins become less similar as time goes by, but this decreasing similarity occurs for identical twins as much as for fraternal twins for most personality traits, suggesting that heritability does not change (McCartney *et al.* 1990). In contrast, genetic research on 'g' indicates that heritability increases steadily from infancy (when it is about 20 per cent) to childhood (40 per cent) to adulthood (60 per cent) (McGue *et al.* 1993b; Plomin 1986). Why does heritability of 'g' increase during the lifespan? It is possible that completely new genes come to affect 'g' as more sophisticated cognitive processes come on line. A more likely hypothesis is that relatively small genetic effects early in life snowball during development, creating larger phenotypic effects as individuals select or create environments that foster their genetic propensities. Another developmental finding about 'g' concerns environmental influences, which is discussed in the following section.

In addition to investigating developmental changes in components of variance, longitudinal genetic analyses can assess genetic contributions to continuity and change from age to age. For cognitive ability, genetic factors largely contribute to stability from age to age rather than to change, although some evidence can be found for genetic change during the transition from infancy to early childhood and from early childhood to middle childhood (Fulker *et al.* 1993; Fulker *et al.* 1988) and perhaps in adulthood (Loehlin *et al.* 1989). Unlike genetic effects, which contribute to change as well as to continuity, longitudinal analysis suggests that shared environmental effects contribute only to continuity. That is, the same shared environmental factors affect 'g' in infancy and in early and middle childhood (Fulker *et al.* 1993). Socio-economic factors, which remain relatively constant, might account for this shared environmental continuity.

Although less well studied longitudinally than cognitive ability, developmental genetic findings for personality appear to be similar. In childhood there is evidence for genetic change (Goldsmith and Gottesman 1981; Loehlin 1992; Wilson and Matheny 1986), although some traits such as shyness show little evidence of genetic change even during childhood (Cherny *et al.* 1994; Saudino *et al.* 1996). During adulthood, there is little evidence for genetic influence on change from age to age (e.g. Loehlin *et al.* 1990; McGue *et al.* 1993a; Pedersen *et al.* in press; Pogue-Geile and Rose 1985).

The importance of non-shared environment

Genetic research is changing the way we think about the environment. For example, two of the most important discoveries from genetic research in psychology are about nurture rather than nature. The first discovery is that environmental influences do not tend to make children growing up in the same family similar. Because environmental influences that affect psychological development are not the same for different children in the same family, they are called *non-shared environment*. The second topic, genotype–environment correlation, is discussed in the next section.

From Freud onward, most theories about how the environment works in development implicitly assumed that offspring resemble their parents because parents provide the family environment for their offspring and that siblings resemble each other because they share that family environment. Twin and adoption research during the past two decades has dramatically altered this view. The reason why genetic designs such as twin and adoption methods were devised was to address the possibility that some of this widespread familial resemblance may be due to shared heredity rather than to shared family environment. The surprise is that genetic research consistently shows that family resemblance is almost entirely due to shared heredity rather than to shared family environment. The message is not that family experiences are unimportant, but that the effects of environmental influences are specific to each child, not general to an entire family.

How do genetic designs estimate the net effect of non-shared environment? In terms of the components-of-variance approach of quantitative genetics, environmental variance is variance not explained by genetics. The effect of shared family environment is estimated as family resemblance not explained by genetics. Non-shared environment is the rest of the variance, variance not explained by genetics or by shared family environment. The conclusion that environmental variance is largely non-shared refers to this residual component of variance, usually estimated by model-fitting analyses (see Chapter 2 for further discussion). However, simple tests of shared and nonshared environments make it easier to understand how they can be estimated (Plomin *et al.* 1994a). A direct test of shared family environment is resemblance among adoptive relatives. For personality, adoptive 'siblings' (genetically uncorrelated children adopted into the same adoptive family) correlate near zero, a value implying that shared environment is unimportant and that environmental influences, which are substantial for personality, are of the nonshared variety (Plomin *et al.* 1994a).

Just as genetically unrelated adoptive siblings provide a direct test of shared family environment, identical twins reared together provide a direct test of nonshared environment. Because they are identical genetically, differences within pairs of identical twins can only be due to non-shared environment plus error of measurement. For example, for self-report personality questionnaires, identical twins typically correlate about 0.40. This value means that about 60 per cent of the variance is due to non-shared environment plus error of measurement. For self-report personality questionnaires, genetics typically accounts for about 40 per cent of the variance, shared family environment for 0 per cent, and non-shared environment plus error for 60 per cent of the variance. Such questionnaires are usually at least 80 per cent reliable, which means that about 20 per cent of the variance is due to error of measurement. In other words, systematic non-shared environmental variance excluding error of measurement accounts for about 40 per cent of the variance (i.e. 60 per cent minus 20 per cent).

Estimates of shared environmental influence for cognitive abilities change over development. Up until adolescence, about a quarter of the variance of general cognitive ability is due to shared family environment. For example, in childhood adoptive siblings correlate about 0.25 for general cognitive ability, suggesting that about a quarter of the variance of general cognitive ability in childhood is due to shared family environment. However, by adolescence, the correlation for adoptive siblings is negligible. This developmental finding is the basis for the conclusion that shared family environment has negligible impact in the long run, even for 'g'. A complexity for 'g' is that twins share more family environment than do non-twin siblings. Identical twins reared together correlate about 0.85, a result that does not seem to leave much room for non-shared environment (i.e. $1 - 0.85 = 0.15$). However, fraternal twins correlate about 0.60, whereas non-twin siblings correlate about 0.40, implying that twins have a special shared environment that accounts for as much as 20 per cent of the variance as compared to non-twin siblings. For this reason, the identical twin correlation of 0.85 may be inflated by 0.20 by this special shared twin environment. In other words, about a third of the variance of general cognitive ability may be due to nonshared environment [i.e. $1 - (0.85 - 0.20) = 0.35$]. Another example of underestimating the importance of non-shared environment in twin studies is assortative mating. 'g' shows substantial assortative mating which artificially inflates estimates of shared family environment unless assortative mating is taken into account (Plomin *et al.* 2001).

In summary, environmental influence is important, accounting for as much as half of the variance for personality and cognitive abilities, but the salient environmental influences are not shared by family members. This remarkable finding means that environmental influences that affect development operate to make children growing up in the same family no more similar than children growing up in different families. Shared and non-shared environment are not limited to family environments. Experiences outside the family can also be shared or not shared by siblings, such as peer groups, life events, and educational, and occupational experiences.

Current research is trying to identify specific sources of non-shared environment and to investigate associations between non-shared environment and psychological traits. That is, the components-of-variance approach of quantitative genetics described above assesses the net effect of shared and non-shared environment on behaviour rather than using specific measures of the environment, just as quantitative genetic designs estimate the net effect of genetic influence without assessing specific DNA markers. In order to identify non-shared environmental factors, it is necessary to begin by assessing aspects of the environment specific to each child, rather than aspects shared by siblings. Many measures of the environment used in studies of psychological development are general to a family rather than specific to a child. For example, asking whether or not their parents have been divorced gives the same answer for two children in the family. Assessed in this family-general way, divorce cannot be a source of differences in siblings' outcomes, because it does not differ for two children in the same family. However, research on divorce has shown that divorce affects children in a family differently (Hetherington and Clingempeel 1992). If divorce is assessed in a child-specific way (for example, asking about the children's different perceptions about the stress caused by the divorce), divorce could well be a source of differential sibling outcome.

Even when experiences are assessed in a child-specific manner, they can be shared by two children in a family. For example, to what extent are maternal vocalizing and maternal affection towards their children shared by siblings in the same family? Observational research on maternal interactions with siblings assessed separately when each child was one and two years old indicates that mothers' spontaneous vocalizing correlates substantially across the siblings (Chipuer and Plomin 1992). This research implies that maternal vocalizing is an experience shared by siblings. In contrast, mothers' affection yields negligible correlations across siblings, indicating that maternal affection is not shared and is thus a better candidate for non-shared influence.

Some family structure variables, such as birth order and sibling age spacing, are by definition non-shared environmental factors. However, these have generally been found to account for only a small portion of variance in psychological outcomes, although birth order has been resurrected as a possible non-shared environmental influence for personality (Sulloway 1996). Research on more dynamic aspects of non-shared environment such as parenting has found that children growing up in the same family lead surprisingly separate lives (Dunn and Plomin 1990).

A decade-long study called the Nonshared Environment and Adolescent Development (NEAD) project systematically explored family environment as a source of non-shared environment (Reiss et al. 2000). During two, two-hour visits to 720 families with same-sex sibling offspring ranging in age from 10 to 18 years, a large battery of questionnaire and interview measures of the family environment was administered to both parents and offspring, and parent–child interactions were videotaped during a session when problems in family relationships were discussed. Sibling correlations for children's reports of their family interactions (for example, children's reports of their parents' negativity) were modest, as they were for observational ratings of child-to-parent interactions and parent-to-child interactions. This finding suggests that these experiences are largely non-shared. In contrast, parent reports yielded high sibling correlations, for example, when parents reported on their own negativity toward each of the children. Although this may be due to a rater effect in that the parent rates both children, the high sibling correlations indicate that parent reports of children's environments are not good sources of candidate variables assessing non-shared environmental factors.

Once child-specific environmental factors are identified, the next question is whether these non-shared experiences relate to psychological outcomes. For example, to what extent do differences in parental treatment account for the non-shared environmental variance known to be important for personality? Although research in this area has only just begun, some success has been achieved in predicting differences in adjustment from sibling differences in their experiences (Hetherington *et al.* 1994). The NEAD project provides an example, in that negative parental behaviour directed specifically to one adolescent sibling (controlling for parental treatment of the other sibling) is highly correlated with that child's antisocial behaviour (Reiss *et al.* 2000). Most of the associations found in NEAD and other studies involve negative aspects of parenting such as conflict, and negative outcomes such as antisocial behaviour. Associations are generally weaker for positive parenting such as affection and positive outcomes such as prosocial behaviour.

When associations are found between non-shared environment and outcome, the question of direction of effects is raised. That is, is differential parental negativity the cause or the effect of sibling differences in antisocial behaviour? Genetic research is beginning to suggest that most differential parental treatment of siblings is in fact the effect rather than the cause of sibling differences. One of the reasons why siblings differ is genetics. Siblings are 50 per cent similar genetically, but that means that siblings are also 50 per cent different. Research on non-shared environment needs to be embedded in genetically sensitive designs in order to distinguish true non-shared environmental effects from sibling differences due to genetics. For this reason, the NEAD project included identical and fraternal twins, full siblings, half siblings, and genetically unrelated siblings. Multivariate genetic analysis of associations between parental negativity and adolescent adjustment yielded an unexpected finding: most of these associations were mediated by genetic factors (Reiss *et al.* 2000). In other words, the apparent non-shared environmental associations between differential parental treatment of siblings and differential outcomes largely disappear when genetic mediation is taken into account in a genetically sensitive design. This finding implies that differential parental treatment of siblings to a substantial extent reflects genetically influenced differences between the siblings, such as differences in personality. This is an example of genotype–environment correlation which is the topic of the next section.

No matter how difficult it may be to find specific non-shared environmental factors within the family, it should be emphasized that non-shared environment is generally the way the environment works for personality and cognitive abilities. Although NEAD focused on the family environment, it seems reasonable that experiences outside the family, such as experiences with peers and other life events, might be richer sources of non-shared environment as children make their own way in the world (Harris 1998). It is also possible that chance contributes to non-shared environment in the sense of random noise, idiosyncratic experiences or the subtle interplay of a concatenation of events (Dunn and Plomin 1990; Turkheimer and Waldron 2000). Francis Galton, the founder of behavioural genetics, suggested that non-shared environment is largely due to chance: 'The whimsical effects of chance in producing stable results are common enough. Tangled strings variously twitched, soon get themselves into tight knots' (Galton 1889: 195). Compounded over time, small differences in experience might lead to large differences in outcome.

The critical question for understanding how the environment influences psychological development is why children in the same family are so different. To address this question, it is obviously necessary to study more than one child per family in order to identify sibling differences in experience, and to investigate the relationship between these different experiences and differences in their psychological outcomes. Answers to the question why children in the same family are so different pertain not only to sibling differences. This is a key to unlocking the environmental origins of psychological development for all children.

Nature–nurture interplay

Another new direction for genetic research on personality and cognitive abilities involves their role in explaining a fascinating finding: environmental measures widely used in psychological research show genetic influence. An example mentioned in the previous section was that the relationship between parental negativity and adolescents' antisocial behaviour was largely mediated genetically. This implies that parental negativity as well as adolescents' antisocial behaviour are both influenced by genetic factors. Genetic research consistently finds that measures of the environment show genetic influence when they are analysed as dependent measures in twin and adoption studies. Indeed, measures of family environment, peer groups, social support, and life events often show as much genetic influence as behavioural measures (Plomin 1994). Most of this research involves parent and child ratings on self-report questionnaires assessing family environment. However, a few observational studies suggest that genetic effects on family interactions are not solely in the eye of the beholder. For example, a twin and step-family study of videotaped parent–child interactions found significant heritability for all measures (O'Connor et al. 1995; Reiss et al. 2000). A widely used measure of the home environment relevant to cognitive development that combines observations and interviews is called the Home Observation for Measurement of the Environment (HOME) (Caldwell and Bradley 1978). In an adoption study of the HOME, correlations for non-adoptive and adoptive siblings were compared when each child was one year old and again when each child was two years old (Braungart *et al.* 1992b). HOME scores were significantly more similar for non-adoptive siblings than for adoptive siblings at both one and two years, suggesting genetic influence on the HOME. Other observational studies of mother–infant interaction in infancy using the adoption design (Dunn *et al.* 1986) and the twin design (Lytton 1977, 1980) also indicate genetic influence.

This research suggests that people create their own experiences in part for genetic reasons. Although this topic has been labelled the *nature of nurture* (Plomin and Bergeman 1991), it is better described by the more prosaic term, *genotype–environment correlation*, because it refers to experiences that are correlated with genetic propensities (Plomin *et al.* 1977). These findings are not as paradoxical as it might seem at first: How can the environment be influenced by DNA? The answer of course is that the environment independent of the individual is not heritable—what is heritable is the extent to which individuals are involved in their experiences. DNA influences personality and cognitive processes of individuals, who in turn influence their environment. Three types of genotype–environment correlation have been identified—passive, evocative, and active—and three analytic methods are available to assess their effects (Plomin 1994).

Measures of psychological environments in part assess genetically influenced characteristics of the individual, such as personality and cognitive abilities. For example, genetic influence on perceptions of life events can be accounted for by the Five-Factor Model (FFM) of personality (Saudino *et al.* 1997). That is, using multivariate genetic model fitting, there is no residual genetic variance on perceptions of life events once the genetic variance of the FFM is taken into account. Genetic influence on personality has also been reported to contribute to genetic influence on parenting in two studies (Chipuer *et al.* 1993; Losoya *et al.* 1997) although not in another (Vernon *et al.* 1997). These findings are not limited to self-report questionnaires. For example, genetic influence found on the HOME mentioned earlier can be explained entirely by genetic influence on a tester-rated measure of attention called task orientation (Saudino and Plomin 1997) and by children's '*g*' (Braungart *et al.* 1992a).

This research does not mean that experience is entirely driven by genes. Nearly all environmental measures studied to date show some significant genetic influence, but most of the variance in these measures is not genetic. On the other hand, this research suggests that environmental measures cannot be assumed to be entirely environmental just because they are called environmental measures. Research to date suggests that it is safer to assume that measures of the environment include some genetic effects. Moreover, especially in families of genetically related individuals, associations between measures of the family environment and children's developmental outcomes cannot be assumed to be purely environmental in origin. Taking this argument to the extreme, some authors have concluded that socialization research is fundamentally flawed because it has not considered the role of genetics (Harris 1998), but these attacks have met with stiff resistance (Collins *et al.* 2000; Vandell 2000).

These findings support a current shift from thinking about passive models of how the environment affects individuals toward models that recognize the active role we play in selecting, modifying, and creating (and recreating in memory) our own environments. Personality and cognitive factors shape both our choice of environment and our construal and representation (hence experience) of that environment. Progress in this field depends on developing measures of the environment that reflect the active role we play in constructing our experience.

Molecular genetics

The two worlds of genetic research, quantitative genetics and molecular genetics, both have their origins early in the twentieth century (Plomin *et al.* 2001). Their ideas and research grew apart as quantitative geneticists focused on naturally occurring genetic variation and quantitative traits and complex multifactorial disorders, using experiments of nature such as twinning, and experiments of nurture such as adoption, in order to disentangle the effects of nature and nurture. In contrast, the progenitors of modern molecular genetics analysed single-gene effects, often focusing on mutations created artificially by chemicals or X-radiation.

During the past decade, quantitative genetics and molecular genetics have begun to come together to identify genes that contribute to complex, quantitative traits. One of the most exciting directions for research on behavioural dimensions and disorders is to identify specific genes responsible for their substantial heritability. For complex traits and common disorders, the goal is not to find *the* gene for a particular trait but rather to find the many genes that contribute to the variance of the trait. In order to emphasize this distinction, genes in multiple-gene systems have been called *quantitative trait loci* (QTLs). As mentioned at the beginning of this chapter, the name indicates that complex traits influenced by multiple genes are thought to be distributed as continuous, quantitative dimensions rather than as discontinuous, qualitative disorders. Unlike single-gene effects that are necessary and sufficient for the development of a disorder, QTLs act as probabilistic risk factors. Although QTLs are inherited in the same Mendelian manner as single-gene effects, if many genes affect a trait then each gene is likely to have a relatively small effect. Moreover, genetic effects may be nonadditive. This makes it much more difficult to detect QTLs than single-gene effects.

Other chapters in this volume review the issues involved in finding QTLs that contribute to complex traits. There are too many unknowns in this area at this time to be dogmatic about what will and what will not work. The issues will be resolved empirically. For example, a few years ago, linkage was deemed to be the only acceptable approach for identifying genes, but now allelic association is widely accepted. Candidate gene approaches came and went as the flavour of the year. Today some argue that only within-family association designs are acceptable in order to control for ethnic stratification, while others point out that until replicable results emerge, there is little need to worry about ethnic stratification and that it makes more sense to find replicable associations using more powerful case-control designs, matching cases and controls as epidemiologists have done for decades. After replicable associations are found, it is a relatively simple matter to assess the extent to which such associations can be explained by ethnic stratification.

If additive genetic factors are important for complex traits like personality and cognitive abilities, it should be possible to identify specific genes responsible if their effect sizes are large enough to be detected with the power provided by our designs. The critical question is the distribution of effects sizes of these QTLs: What is the average effect size of QTLs and how are QTL effect sizes distributed? Extrapolating from work so far in plants and animals, it seems reasonable to assume that the distributions of QTL effect sizes for complex traits are positively skewed, with few QTLs that account for more than 5 per cent of the variance and with very long tails that fade out with QTL effect sizes so small that they will be very difficult to detect. If the average effect size is 1 per cent, we will eventually need to detect many QTLs in order to account for typical heritabilities of around 50 per cent. However, if the average QTL effect size is 0.1 per cent, we will detect very few QTLs and they will be difficult to replicate. Time and much more empirical research will tell.

Personality

In contrast to molecular genetic research on psychopathology, molecular genetic research on personality has just begun but has led to the first reports of associations between variation in candidate genes and normal variation in behaviour (Benjamin *et al.* 2002; Hamer and Copeland 1998). In 1996 two papers reported an association between the dopamine D4 receptor gene (*DRD4*) and novelty seeking in unselected samples (Benjamin *et al.* 1996; Ebstein *et al.* 1996). Novelty seeking is one of the four traits included in a theory of temperament developed by Cloninger (Cloninger *et al.* 1993), although it is very similar to the impulsive sensation-seeking dimension studied by Zuckerman (1979; 1994). Individuals high in novelty seeking are characterized as impulsive, exploratory, fickle, excitable, quick-tempered and extravagant. Cloninger's theory predicts that novelty seeking involves genetic differences in dopamine transmission.

The polymorphism in *DRD4* that has been studied consists of seven alleles involving 2, 3, 4, 5, 6, 7, or 8 repeats of a 48-base pair sequence in a coding region. The number of repeats changes the structure of the receptor by altering the length of the third intracytoplasmic loop and appears to affect the receptor's efficiency *in vitro*. The shorter alleles (2, 3, 4, or 5 repeats) code for a receptor variant that is more efficient in binding dopamine than the larger alleles (6, 7, or 8 repeats). For this reason, the *DRD4* alleles are usually grouped as *short* (about 85 per cent of alleles) or *long* (15 per cent of alleles). It has been suggested that there is a reward mechanism whereby novelty seeking promotes dopamine release, and that individuals with the less efficient long-repeat *DRD4* allele have to 'work harder' to seek novelty and increase dopamine release.

In both studies (Benjamin *et al.* 1996; Ebstein *et al.* 1996), individuals with the longer *DRD4* alleles (6–8 repeats) had significantly higher novelty-seeking scores than individuals with the shorter alleles (2–5 repeats). This association was also found within families, a result indicating that the association is not due to ethnic differences. That is, within the same families, individuals with the longer *DRD4* alleles had significantly higher novelty-seeking scores than their siblings with the shorter *DRD4* alleles. The reported effect size accounted for about 4 per cent of the variance in this trait. As would be expected for an association of small effect, some studies have failed to replicate the association, but after a dozen studies there appears to be an association of small effect size (Prolo and Licinio 2002; Plomin and Caspi 1999). Interestingly, there also appears to be an association between the long *DRD4* allele and attention deficit hyperactivity disorder in children (see Chapter 7).

Other candidate genes studies of personality have focused on neuroticism—especially a functional serotonin transporter promoter polymorphism (Lesch *et al.* in press; Osher *et al.* 2000)–shyness (Fox and Schmidt in press), and aggression (Siever and New in press). QTL research in personality so far has been limited to screening candidate genes. No systematic genome scans for linkage or association have been reported.

Cognitive abilities

Dozens of single-gene causes of mental retardation are known, such as phenylketonuria (PKU) and fragile-X syndrome. Moreover, heterozygotes for PKU show slightly lowered IQ scores (Bessman *et al.* 1978; Propping 1987). Differences in the number of fragile X repeats in the normal range, on the other hand, do not relate to differences in '*g*' (Daniels *et al.* 1994). The two strongest examples of behavioural QTLs to date are cognitive examples: dementia (*APOE*) and reading disability (6p21 linkage). In addition, it has been reported that the allele of *APOE* that is associated with dementia is also associated with greater cognitive decline in an unselected population of elderly men (Feskens *et al.* 1994) with supportive results found in other studies (Bartres-Faz *et al.* 1999; Plomin *et al.* 1995).

Only one project has focused on finding QTLs for cognitive ability rather than disability. The IQ QTL Project compares high '*g*' and control groups (Plomin in press). The first phase of the project employed an allelic association strategy using DNA markers in or near candidate genes likely to be relevant to neurological functioning, such as genes for neuroreceptors. Allelic association results were reported for 100 DNA markers for such candidate genes (Plomin *et al.* 1995). Although several significant associations were found in an original sample, only one association was replicated in an independent sample. This finding might well have been a chance result because 100 markers were investigated and a follow-up study failed to replicate the result (Petrill *et al.* 1998). Negative findings for isolated candidate genes have led to interest in systems of candidate genes and in systematic genome scans.

Systematic genome scans for QTL associations can be conducted using dense maps of DNA markers. As part of the IQ QTL Project, a first attempt to take a systematic approach to identify QTLs associated with IQ focused on the long arm of chromosome 6 and found replicated associations for a DNA marker that happened to be in the gene for insulin-like growth factor-2 receptor (IGF2R) (Chorney *et al.* 1998), which has been shown to be especially active in brain regions most involved in learning and memory (Wickelgren 1998). However, although the original finding was replicated, a new sample has not replicated the association (Hill *et al.* in press).

The problem with using a dense map of markers for a genome scan for allelic association is the amount of genotyping required. In order to scan the entire genome at 1 million DNA base pair intervals (1Mb), about 3500 DNA markers would need to be genotyped. This would require 700,000 genotypings in a study of 100 high '*g*' individuals and 100 controls. With markers at 1Mb intervals, no QTL would be farther than 500,000 base pairs from a marker. However, empirical data indicate that at least 10 times as many markers would be needed in order to detect most QTLs (Cambien *et al.* 1999). Despite the daunting amount of genotyping for such a systematic genome scan, this approach has been fuelled by the promise of single nucleotide polymorphisms (SNPs)—'SNPs on chips'—which can quickly genotype thousands of DNA markers (see Chapter 1).

DNA pooling, developed for use in the IQ QTL Project, provides a low-cost and flexible alternative to SNPs on chips for screening the genome for QTL associations (Daniels *et al.* 1998). DNA pooling greatly reduces the need for genotyping by pooling DNA from all individuals in each group (e.g. a high '*g*' group and a control group) and comparing the pooled groups so that only 14,000 genotypings are required to scan the genome in the previous example involving 3500 DNA markers. DNA pooling is being used to scan the genome for QTLs in the IQ QTL Project. A proof of principle paper examining 147 markers on chromosome 4 using DNA pooling reported three replicated QTLs (Fisher *et al.* 1999). Figure 4.5 illustrates DNA pooling results for one of the markers (*D4S2943*) that showed a significant difference in allele frequency for a high '*g*' group and a control group. Because the DNA is pooled, all six alleles are seen rather than just one or two alleles which would be seen when individuals rather than pools of individuals are genotyped. The relative area under the curve for each allele indicates its frequency. The overlaid allele image patterns (AIPs) for the original high '*g*' group and the original control group indicate that differences between the AIPs for the two groups are due primarily to the fourth allele. This result from DNA pooling was verified using traditional individual genotyping. The IQ QTL Project is currently applying this DNA pooling approach to examine 2000 DNA markers throughout the genome (Plomin and Crabbe 2000). It should be emphasized that this approach attempts to find *some*—certainly not *all*—of the QTL associations responsible for the heritability of this complex trait. The hypothesis that such QTLs will also be associated with normal variation as well as mild mental retardation remains to be tested.

Identifying chromosomal neighbourhoods of QTLs is just the first step in gene mapping. Localizing the specific address of the functional polymorphism, sometimes called

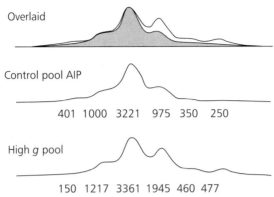

Overlaid

Control pool AIP

401 1000 3221 975 350 250

High *g* pool

150 1217 3361 1945 460 477

Fig. 4.5 DNA pooling results for *D4S2943* for a high *g* group (bottom), a control group of average *g* (middle), and their overlaid images (top). The bumps represent alleles (consisting of different numbers of two base-pair repeats) and the area of each bump represents the allele's frequency. AIP refers to this allele image pattern. The difference in the height of the bumps for the high *g* and average *g* groups for allele 4 suggests a QTL association with *g*. (Reprinted with permission from Plomin and Craig, British Journal of Psychiatry Supplement, 2001.)

quantitative trait nucleotides (QTNs), is a critical next step. Localizing QTNs will be greatly aided by knowing the entire DNA sequence of the human genome, from an effort to identify hundreds of thousands of SNPs in these DNA sequences especially SNPs in coding regions (cSNPs), and from efforts to map gene expression which will indicate which genes are expressed in any given brain region.

The ultimate scientific goal is not just finding genes but understanding how they function, and this is an even more daunting task. Functional genomics is usually discussed in terms of bottom-up molecular biology analyses of cellular function. For example, studies of coordinated patterns of gene expression using DNA chips for thousands of genes—like functional imaging at a cellular level—will make major contributions to the study of gene function. However, other levels of analysis are also important in understanding how genes work. A top-down behavioural level of analysis will also contribute to functional genomics, for example, by means of psychological theories of cognitive processing and by investigating interactions and correlations between individuals and their environment. For instance, psychological theories suggest how different components of information processing are related and the role of genes in these cognitive systems can be examined. Such top-down strategies can yield equally important information to the bottom-up approach in which the products of these genes are studied at a cellular level of analysis. As an antidote to the tendency to define functional genomics at the cellular level of analysis, the phrase behavioural genomics has been proposed (Plomin and Crabbe 2000). Behavioural genomics is likely to pay off more quickly than other levels of analysis in terms of prediction, diagnosis, and intervention.

The brain is clearly where bottom-up molecular levels of analysis will eventually meet top-down behavioural levels of analysis. Studies of brain functioning, for example as assessed by neuroimaging, will foster this integration. For humans, the expense of neuroimaging presently restricts sample sizes and, for this reason, differences between individuals are rarely considered. However, should imaging become more available, this technology could play an important part in establishing links between genes, brain, and cognition. In the meantime, mouse models will be valuable, especially so given current large-scale behavioural screens of mutagenized mice. Finding genes associated with personality and cognitive abilities in the human species will provide discrete windows through which brain pathways between genes and behaviour can be observed using animal models. Although animal models are not appropriate for some distinctly human aspects of personality (e.g. sympathy, aesthetics) and cognition (e.g. language), some of the most exciting areas of genetic research involve mouse models of personality (e.g. fearfulness; Flint 2002) and cognition (e.g. learning and memory). Although research on synaptic mechanisms of learning and memory using animal models is leading the way in genetic research in cognitive neuroscience, genetic research in this area has been limited to gene knockouts in which a gene is altered so that it is no longer expressed rather than studying naturally occurring genetic variation that might underlie individual differences in learning, memory and cognitive abilities. Moreover, commonalities across cognitive processes that are indicative of 'g' have not yet been explored or even considered systematically using animal models.

The scientific impact of finding genes associated with personality and cognitive abilities will affect all aspects of behavioural research. Perhaps one day behavioural and social scientists will routinely collect DNA using cheek swabs (no blood needed) in order to investigate, or at least control for, genes associated with personality or cognition, as is happening now for research on dementia and the apolipoprotein E gene. Even if hundreds of genes contribute to the heritability of these traits, finding genes associated with them will make it possible to investigate long-standing scientific issues with much greater precision, such as the developmental, multivariate, and gene–environment interplay issues discussed earlier.

One of the most exciting questions is whether genes associated with normal variation in personality and cognitive performance are related to traditional diagnoses of psychiatric disorders. Indeed, from a QTL perspective, common disorders may be the quantitative extreme of normal genetic variation rather than qualitatively different, as mentioned earlier. Finding genes associated with disorders as well as dimensions may lead to gene-based diagnoses and treatment programmes. Gene-based classification of disorders may bear little resemblance to our current symptom-based diagnostic systems. The most exciting potential is for secondary prevention because DNA can be used to predict genetic risk for an individual, offering the hope for behavioural intervention before disorders create cascades of complications. The exemplar is phenylketonuria (PKU), a single-gene disorder that leads to mental retardation unless it is detected early in life and a dietary intervention is used to ameliorate its effects on the developing brain.

Finding QTLs involved in personality and cognition has important implications for society as well as science (Plomin 1999). In terms of implications for society, it should be emphasized that no policies necessarily follow from finding such genes because policy involves values. For example, finding genes for '*g*' does not mean that we ought to put all of our resources into educating the brightest children. Depending on our values, we might invest more effort in helping children of lower ability. Appreciating the genetic contribution to individual differences will allow us to assess, for example, the differential effects of possible environmental and educational interventions.

Many ethical issues related to DNA are being broached at the level of single-gene disorders that are hard-wired in the sense that a single gene is necessary and sufficient for the development of the disorder. Ethical deliberations in the genetics of complex traits such as personality and cognition may be eased by the sheer complexity of the developmental process. (See Chapter 16.) Genetic effects of these sorts of traits are probabilistic rather than deterministic, and attempts to select for intelligence or personality are likely to fail for two reasons. First, heritability is closer to 50 per cent than to 100 per cent, which means that non-genetic factors make a major contribution. Second, because many genes contribute to the heritability of these traits, the system is inherently probabilistic. Potential problems related to finding genes associated with '*g*' have been discussed, such as prenatal and postnatal screening, discrimination in education and employment, and group differences (Newson and Williamson 1999). In the shadows of such discussions lurks the fear that that finding genes will limit our freedom and our free will. In large part such fears involve misunderstandings about how genes affect complex traits (Rutter and Plomin 1997). An important job for the future is to communicate to the public the complex and fascinating interplay between genes and environment, so that the wider community is able to engage in informed debate.

We need to be cautious and to consider societal implications and ethical issues carefully, but there is much to celebrate as well. The greatest implication for science is that the functional genomics of complex traits such as personality and cognitive abilities will serve as an integrating force across diverse disciplines with DNA as the common denominator, opening up new scientific horizons for understanding behaviour.

References

Alarcón M., DeFries J. C., Light J. G. *et al.* (1997) A twin study of mathematics disability. *Journal of Learning Disabilities* **30**:617–23.

Bartres-Faz D., Clemente I. and Junque C. (1999) Cognitive changes in normal aging: nosology and current status. *Revista de Neurologia* **29**:64–70.

Beer, J. M. and Rose, R. J. (1995) The five factors of personality. *Behav Genet* **25**:254 (Abstract).

Benjamin, J., Ebstein, R. and Belmaker, R. H. (2002) *Molecular Genetics of Human Personality*, American Psychiatric Press.

Benjamin J., Li L., Patterson C. *et al.* (1996) Population and familial association between the D4 dopamine receptor gene and measures of novelty seeking. *Nat Genet* **12**:81–4.

Bergeman C. S., Chipuer H. M., Plomin R. *et al.* (1993) Genetic and environmental effects on openness to experience, agreeableness, and conscientiousness: an adoption/twin study. *J Pers* **61**:159–179.

Bessman S. P., Williamson M. L., Koch, R. (1978) Diet, genetics, and mental retardation interaction between phenylketonuric heterozygous mother and fetus to produce non-specific dimunution of IQ: Evidence in support of the justification hypothesis. *Proceedings of the National Academy of Sciences* **78**:1562–6.

Bouchard T. J., Jr and McGue M. (1981) Familial studies of intelligence: a review. *Science* **212**:1055–9.

Bratko D. (1997) Twin studies of verbal and spatial abilities. *Person Individ Diff* **23**:365–9.

Braungart J. M., Fulker D. W., Plomin R. (1992a) Genetic mediation of the home environment during infancy: A sibling adoption study of the HOME. *Developmental Psychology* **28**:1048–55.

Braungart J. M., Plomin R., DeFries J. C. *et al.* (1992b) Genetic influence on tester-rated infant temperament as assessed by Bayley's Infant Behavior Record: Nonadoptive and adoptive siblings and twins. *Developmental Psychology* **28**:40–7.

Brody N. (1992) *Intelligence*, 2nd edn. New York: Academic Press.

Bruun K., Markkananen T., Partanen J. (1966) *Inheritance of Drinking Behaviour: a Study of Adult Twins*. Helsinki, Finland: Finnish Foundation for Alcohol Research.

Buss A. H. and Plomin R. (1984) *Temperament: Early developing personality traits*, Hillsdale, NJ: Lawrence Erlbaum.

Caldwell B. M. and Bradley R. H. (1978) *Home Observation for Measurement of the Environment*, Little Rock: University of Arkansas.

Cambien F., Poirier O., Nicaud V. *et al.* (1999) Sequence diversity in 36 candidate genes for cardiovascular disorders. *Am J Hum Genet* **65**:183–91.

Cardon L. R., Smith S. D., Fulker D. W. *et al.* (1994) Quantitative trait locus for reading disability on chromosome 6. *Science* **266**:276–9.

Carroll J. B. (1993) *Human Cognitive Abilities*, New York: Cambridge University Press.

Cattell R. B. (1982) *The Inheritance of Personality and Ability*, New York: Academic Press.

Cherny S. S., Fulker D. W., Emde R. N. *et al.* (1994) Continuity and change in infant shyness from 14 to 20 months. *Behav Genet* **24**:365–79.

Chipuer H. M. and Plomin R. (1992) Using siblings to identify shared and non-shared HOME items. *Br J Dev Psych* **10**:165–78.

Chipuer H. M., Plomin R., Pedersen N. L. *et al.* (1993) Genetic influence on family environment: The role of personality. *Developmental Psychology* **29**:110–18.

Chorney M. J., Chorney K., Seese N. *et al.* (1998) A quantitative trait locus (QTL) associated with cognitive ability in children. *Psych Sci* **9**:1–8.

Cloninger C. R. (2002) The relevance of normal personality for psychiatrists, in J. Benjamin and R. H. Belmaker (eds) *Molecular Genetics and Human Personality* (pp. 33–42). New York: American Psychiatric Press.

Cloninger C. R., Svrakic D. M., Przybeck T. R. (1993) A psychobiological model of temperament and character. *Archives of General Psychiatry* **50**:975–90.

Collins W. A., Maccoby E. E., Steinberg L. *et al.* (2000) Contemporary research in parenting: The case for nature *and* nurture. *Am Psychol* **55**:218–32.

Dale P. S., Simonoff E., Bishop D. V. M. *et al.* (1998) Genetic influence on language delay in 2-year-olds. *Nature Neuroscience* **1**:324–8.

Daniels J., Holmans P., Plomin R. *et al.* (1998) A simple method for analyzing microsatellite allele image patterns generated from DNA pools and its application to allelic association studies. *Am J Hum Genet* **62**:1189–97.

Daniels J. K., Owen M. J., McGuffin P. *et al.* (1994) IQ and variation in the number of fragile X CGG repeats: No association in a normal sample. *Intelligence* **19**:45–50.

Deary I. J., Whalley L. J., Lemmon H. *et al.* (2000) The stability of individual differences in mental ability from childhood to old age: Follow-up of the 1932 Scottish Mental Survey. *Intelligence* **28**:49–55.

DeFries J. C., Knopik V. S., Wadsworth S. J. (1999) Colorado Twin Study of reading disability, in D. D. Duane (ed.) *Reading and Attention Disorders: Neurobiological Correlates.* Baltimore, MD: York Press.

DeFries J. C. and Fulker D. W. (1985) Multiple regression analysis of twin data. *Behav Genet* **15**:467–73.

DeFries J. C. and Fulker D. W. (1988) Multiple regression analysis of twin data: etiology of deviant scores versus individual differences. *Acta Genet Med Gemellol* **37**:205–16.

DeFries J. C. and Gillis J. J. (1993) Genetics and reading disability, in R. Plomin and G. E. McClearn (eds) *Nature, Nurture and Psychology.* Washington, DC: American Psychological Association.

DeFries J. C., Plomin R., Fulker D. W. (1994) *Nature and Nurture During Middle Childhood*, Oxford, UK: Blackwell.

DeFries J. C., Vandenberg S. G. and McClearn G. E. (1976) Genetics of specific cognitive abilities. *Annu Rev Genet* **10**:179–207.

Devlin B., Daniels M., and Roeder K. (1997) The heritability of IQ. *Nature* **388**:468–71.

DiLalla D. L., Carey G., Gottesman I. I. *et al.* (1996) Heritability of MMPI personality indicators of psychopathology in twins reared apart. *J Abnorm Psychol* **105**:491–9.

Dunn J. F. and Plomin R. (1990) *Separate Lives: Why Siblings are so Different*, New York: Basic Books.

Dunn J. F., Plomin R., Daniels, D. (1986) Consistency and change in mothers' behavior toward young siblings. *Child Development* **57**:348–56.

Eaves L. J., Eysenck H., Martin N. G. (1989) *Genes, Culture, and Personality: An Empirical Approach.* London: Academic Press.

Ebstein R. P., Novick O., Umansky R. *et al.* (1996) Dopamine D_4 receptor (D_4DR) exon III polymorphism associated with the human personality trait novelty-seeking. *Nat Genet* **12**:78–80.

Falconer D. S. (1965) The inheritance of liability to certain diseases estimated from the incidence among relatives. *Ann Hum Genet* **29**:51–76.

Feskens E. J., Havekes L. M., Kalmijn S. *et al.* (1994) Apolipoprotein e4 allele and cognitive decline in elderly men. *BMJ* **309**:1202–6.

Fisher P. J., Turic D., McGuffin P. *et al.* (1999) DNA pooling identifies QTLs for general cognitive ability in children on chromosome 4. *Hum Mol Genet* **8**:915–22.

Fisher R. A. (1918) The correlation between relatives on the supposition of Mendelian inheritance. *Transactions of the Royal Society of Edinburgh* **52**:399–433.

Flint J. (2001) Molecular genetic studies in animal models: relevance for studies of human personality, in J. Benjamin, R. Ebstein, R. H. Belmaker (eds) *Molecular Genetics and Human Personality* (pp. 63–90). New York: American Psychiatric Press.

Fulker D. W., Cherny S. S., Cardon L. R. (1993) Continuity and change in cognitive development, in R. Plomin R. and G. E. McClearn (eds) *Nature, Nurture, and Psychology.* Washington, DC, Psychological Association.

Fulker D. W., DeFries J. C., Plomin R. (1988) Genetic influence on general mental ability increases between infancy and middle childhood. *Nature* **336**:767–9.

Galton F. (1889) *Natural Inheritance*. London: Macmillan.

Gilger J. W. (1997) How can behavioral genetic research help us understand language development and disorders? In M. L. Rice (ed.) *Towards a Genetics of Language*. Hillsdale, NJ: Lawrence Erlbaum.

Goldsmith H. H., Buss A. H., Plomin R. *et al.* (1987) Roundtable: what is temperament? Four approaches. *Child Development* **58**:505–29.

Goldsmith H. H. and Gottesman I. I. (1981) Origins of variation in behavioral style: A longitudinal study of temperament in young twins. *Child Development* **52**:91–103.

Gottfredson L. S. (1997) Why g matters: the complexity of everyday life. *Intelligence* **24**:79–132.

Hamer D. and Copeland P. (1998) *Living with our Genes*, New York: Doubleday.

Harris J. R. (1998) *The Nurture Assumption: Why Children turn out the way they do*. New York: The Free Press.

Heath A. C., Cloninger C. R., Martin, N. G. (1994) Testing a model for the genetic structure of personality: a comparison of the personality systems of Cloninger and Eysenck. *J Pers Soc Psychol* **66**:762–75.

Heath A. C., Eaves L. J., Martin N. G. (1989) The genetic structure of personality: III. Multivariate genetic item analysis of the EPQ scales. *Person Individ Diff* **10**:877–88.

Herrnstein R. J. and Murray C. (1994) *The Bell Curve: Intelligence and Class Structure in American Life*. New York: Free Press.

Hetherington E. M. and Clingempeel W. G. (1992) Coping with marital transitions: a family systems perspective. *Monogr Soc Res Child Development* 2–3, Serial No. 227.

Hetherington E. M., Reiss D., Plomin R. (1994) *Separate Social Worlds of Siblings: Impact of Nonshared Environment on Development*. Hillsdale, NJ: Lawrence Erlbaum Assoc.

Hill L., Chorney M. J., and Plomin R. (in press) A quantitative trait locus (not) associated with cognitive ability in children? *Psychological Science*.

Jang K. L. and Livesley W. J. (1999) Why do measures of normal and disordered personality correlate? A study of genetic comorbidity. *J Pers Disorders* **13**:10–17.

Jang K. L., Livesley W. J., Vernon P. A. (1996) Heritability of the Big Five dimensions and their facets: a twin study. *J Pers* **64**:577–591.(Abstract)

Jang K. L., McCrae R. R., Angleitner A. *et al.* (1998) Heritability of facet-level traits in a cross-cultural twin sample: support for a hierarchical model of personality. *J Pers Soc Psychol* **74**:1556–65.

Jensen A. R. (1978) Genetic and behavioural effects of nonrandom mating, in R. T. Osbourne, C. E. Noble, N. Weyl (eds) *Human Variation: The Biopsychology of Age, Race, and Sex*. New York: Academic Press.

Jensen A. R. (1998) *The g Factor: The Science of Mental Ability*. Wesport: Praeger.

John O. P. and Srivastava S. (1999) The big five trait taxonomy: history, measurement, and theoretical perpectives, in L. A. Pervin and O. P. John (eds) *Handbook of Personality*, 2nd edn. New York: Guildford.

Johnson C. A., Ahern F. M., and Johnson R. C. (1976) Level of functioning of siblings and parents of probands of varying degrees of retardation. *Behav Genet* **6**:473–7.

Kendler K. S., Neale M. C., Kessler R. C. *et al.* (1993) A longitudinal twin study of personality and major depression in women. *Archives of General Psychiatry* **50**:853–62.

Kenrick D. T. and Funder D. C. (1988) Profiting from controversy: lessons from the person-situation debate. *Am Psychol* **43**:34.

Krueger R. F., Caspi A., and Moffitt T. E. (2000) Epidemiological personology: the unifying role of personality in population-based research on problem behaviors. *J Pers* **68**:967–98.

Lesch K.-P., Greenberg B. D., Higley J. D. *et al.* (2002) Serotonin transporter, personality, and behavior: toward dissection of gene-gene and gene-environment interaction, in J. Benjamin, R. Ebstein, R. H. Belmaker (eds) *Molecular Genetics and Human Personality* (pp. 109–35). New York: American Psychiatric Press.

Levy F., Hay D. A., McStephen M. *et al.* (1997) Attention-deficit hyperactivity disorder: A category or a continuum? Genetic analysis of a large-scale twin study. *Journal of the American Academy of Child and Adolescent Psychiatry* **36**:737–44.

Light J. G. and DeFries J. C. (1995) Comorbidity of reading and mathematics disabilities: genetic and environmental etiologies. *J Learning Disabilities* **28**:96–106.

Loehlin J. C. (1992) *Genes and Environment in Personality Development.* Newbury Park, California: Sage Publications Inc.

Loehlin J. C. and Horn J. M. (2000) Stoolmiller on restriction of range in adoption studies: a comment. *Behav Genet* **30**:245–7.

Loehlin J. C., Horn J. M., Willerman L. (1989) Modeling IQ change: evidence from the Texas Adoption Project. *Child Development* **60**:993–1004.

Loehlin J. C., Horn J. M., Willerman L. (1990) Heredity, environment, and personality change: evidence from the Texas Adoption Study. *J Pers* **58**:221–43.

Loehlin J. C. and Nichols J. (1976) *Heredity, Environment and Personality.* Austin: University of Texas.

Losoya S. H., Callor S., Rowe D. C. *et al.* (1997) Origins of familial similarity in parenting: a study of twins and adoptive siblings. *Developmental Psychology* **33**:1012–23.

Lytton H. (1977) Do parents create or respond to differences in twins? *Developmental Psychology* **13**:456–9.

Lytton H. (1980) *Parent-child Interaction: The Socialization Process Observed in Twin and Singleton Families.* New York: Plenum.

Mackintosh N. J. (1998) *IQ and Human Intelligence.* Oxford: Oxford University Press.

McCartney K., Harris M. J., Bernieri F. (1990) Growing up and growing apart: a developmental meta-analysis of twin studies. *Psychological Bulletin* **107**:226–37.

McCrae R. R., Costa-P. T. J., Ostendorf F. *et al.* (2000) Nature over nurture: temperament, personality, and life span development. *J Pers Soc Psychol* **78**:173–86.

McGue M. (1997) The democracy of the genes. *Nature* **388**:417–18.

McGue M., Bacon S., Lykken D. T. (1993a) Personality stability and change in early adulthood: a behavioral genetic analysis. *Developmental Psychology* **29**:96–109.

McGue M. and Bouchard T. J., Jr (1989) Genetic and environmental determinants of information processing and special mental abilities; a twin analysis, in R. J. Sternberg (ed.) *Advances in the Psychology of Human Intelligence*, Vol. 5. Hillsdale, NJ: Erlbaum.

McGue M., Bouchard T. J., Jr, Iacono W. G. *et al.* (1993b) Behavioral genetics of cognitive ability: A life-span perspective, in R. Plomin and G. E. McClearn (eds) *Nature, Nurture, and Psychology.* Washington, DC: American Psychological Association.

New A., Goodman M., Mitropoulon V. *et al.* (2002) Genetic polymorphisms and aggression, in J. Benjamin, R. Ebstein, R. H. Belmaker (eds) *Molecular Genetics and Human Personality* (pp. 231–44). New York: American Psychiatric Press.

Newson A. and Williamson R. (1999) Should we undertake genetic research on intelligence? *Bioethics* **13**:327–42.

Nichols P. L. (1984) Familial mental retardation. *Behav Genet* **14**:161–70.

Nichols R. C. (1978) Twin studies of ability, personality, and interests. *Homo* **29**:158–73.

Nigg J. T. and Goldsmith H. H. (1998) Developmental psychopathology, personality, and temperament: reflections on recent behavioral genetics research. *Hum Biol* **70**:387–412.

O'Connor T. G., Hetherington E. M., Reiss D. *et al.* (1995) A twin-sibling study of observed parent-adolescent interactions. *Child Development* **66**:812–29.

Osher Y., Hamer D., Benjamin J. (2000) Association and linkage of anxiety-related traits with a functional polymorphism of the serotonin transporter gene regulatory region in Israeli sibling pairs. *Mol Psychiatry* **5**:216–19.

Pedersen N. L., Harris J. R., Plomin R. *et al.* (In Press) Genetic and environmental stability for personality. *Psychology and Aging*

Pedersen N. L., McClearn G. E., Plomin R. *et al.* (1992) Effects of early rearing environment on twin similarity in the last half of the life span. *Br J Dev Psych* **10**:255–67.

Pervin L. A. and John O. P. (1999) *Handbook of Personality: Theory and Research*, 2nd edn. New York: Guilford Press.

Petrill S. A. (1997) Molarity versus modularity of cognitive functioning? A behavioral genetic perspective. *Current Directions in Psychological Science* **6**:96–9.

Petrill S. A., Ball D. M., Eley T. C. *et al.* (1998) Failure to replicate a QTL association between a DNA marker identified by EST00083 and IQ. *Intelligence* **25**:179–84.

Petrill S. A., Saudino K. J., Cherny S. S. *et al.* (1997) Exploring the genetic etiology of low general cognitive ability from 14 to 36 months. *Developmental Psychology* **33**:544–8.

Petrill S. A., Saudino K. S., Wilkerson B. *et al.* (2001) Genetic and environmental molarity and modularity of cognitive functioning in two-year-old twins. *Intelligence* **29**:31–43.

Plomin R. (1986) *Development, Genetics, and Psychology*, Hillsdale, NJ: Erlbaum.

Plomin R. (1988) The nature and nurture of cognitive abilities, in R. J. Sternberg (ed.) *Advances in the Psychology of Human Intelligence*, Vol. 4. Hillsdale, NJ: Lawrence Erlbaum Associates.

Plomin R. (1991) Behavioral genetics, in P. R. McHugh V. A. McKusick (eds) *Genes, Brain and Behavior*. New York: Haven Press.

Plomin R. (1994) *Genetics and Experience: The Interplay Between Nature and Nurture.* Thousand Oaks, California: Sage Publications.

Plomin R. (1999) Genetics and general cognitive ability. *Nature* 402:C25–9.

Plomin R. (2002) Quantitative trait loci and general cognitive ability, in J. Benjamin, R. Ebstein, R. H. Belmaker (eds) *Molecular Genetics of Human Personality* (pp. 211–30). New York: American Psychiatric Press.

Plomin R. and Bergeman C. S. (1991) The nature of nurture: genetic influences on 'environmental' measures. *Behavioral and Brain Sciences* **14**:373–427.

Plomin R. and Caspi A. (1999) Behavioral genetics and personality, in L. A. Pervin and O. P. John (eds) *Handbook of Personality: Theory and Research*, 2nd edn. New York: Guilford Press.

Plomin R., Chipuer H. M., Neiderhiser J. M. (1994a) Behavioral genetic evidence for the importance of nonshared environment, in E. M. Hetherington, D. Reiss, R. Plomin (eds) *Separate Social Worlds of Siblings: The Impact of Non-shared Environment on Development*. New Jersey: Lawrence Erlbaum Assoc.

Plomin R. and Crabbe J. C. (2000) DNA. *Psychological Bulletin* **126**:806–28.

Plomin R. and Craig I. (1998) Human behavioural genetics of cognitive abilities and disabilities. *BioEssays* **19**:1117–24.

Plomin R. and Craig I. (2001) Genetics, environment and cognitive abilities. *British Journal of Psychiatry Supplement* **178**:541–48.

Plomin R. and Dale P. S. (2000) Genetics and early language development: a UK study of twins, in D. V. M. Bishop and B. E. Leonard (eds) *Speech and Language Impairments in Children: Causes, Characteristics, Intervention and Outcome*. Hove, UK: Psychology Press.

Plomin R., DeFries J. C., McClearn G. E. *et al.* (2001) *Behavioral Genetics*, 4th edn. New York: Worth Publishers.

Plomin R. and DeFries J. C. (1998) Genetics of cognitive abilities and disabilities. *Sci Am* May:62–9.

Plomin R., DeFries J. C., Loehlin J. C. (1977) Genotype–environment interaction and correlation in the analysis of human behaviour. *Psychological Bulletin* **85**:309–22.

Plomin R., DeFries J. C., McClearn G. E. *et al.* (1997a) *Behavioral Genetics*, 3rd edn. New York: W. H. Freeman.

Plomin R., Fulker D. W., Corley R. *et al.* (1997b) Nature, nurture and cognitive development from 1 to 16 years: a parent-offspring adoption study. *Psych Sci* **8**:442–7.

Plomin R., McClearn G. E., Smith D. L. *et al.* (1995) Allelic associations between 100 DNA markers and high versus low IQ. *Intelligence* **21**:31–48.

Plomin R., Owen M. J., McGuffin P. (1994b) The genetic basis of complex human behaviors. *Science* **264**:1733–9.

Pogue-Geile M. F. and Rose R. J. (1985) Developmental genetic studies of adult personality. *Developmental Psychology* **21**:547–57.

Prolo P. and Licinio J. (2002) D4DR and novelty seeking, in J. Benjamin, R. Ebstein, R. H. Belmaker (eds) *Molecular Genetics and Human Personality* (pp. 91–107). New York: American Psychiatric Press.

Propping P. (1987). Single gene effects in psychiatric disorders, in F. Vogel and K. Sperling (eds) *Human Genetics: Proceedings of the 7th International Congress, Berlin*. New York: Springer.

Reed E. W. and Reed S. C. (1965) *Mental Retardation: A Family Study*. Philadelphia: Saunders.

Reiss D., Neiderhiser J. M., Hetherington E. M. *et al.* (2000) *The Relationship Code: Deciphering Genetic and Social Patterns in Adolescent Development*. Cambridge, MA: Harvard University Press.

Riemann R., Angleitner A., Strelau J. (1997) Genetic and environmental influences on personality: a study of twins reared together using the self- and peer report NEO-FFI scales. *J Pers* **65**:449–76.

Roberts B. W. and Del Vecchio W. (2000) The rank-order consistency of personality traits from childhood to old age: a quantitative review of longitudinal studies. *Psychological Bulletin* **126**:3–25.

Rowe D. C. (1987) Resolving the person-situation debate: invitation to an interdisciplinary dialogue. *Am Psychol* **42**:218–27.

Rutter M. and Plomin R. (1997) Opportunities for psychiatry from genetic findings. *Br J Psychiat* **171**:209–19.

Saudino K. J., Pedersen N. L., Lichtenstein P. *et al.* (1997) Can personality explain genetic influences on life events? *J Pers Soc Psychol* **72**:196–206.

Saudino K. J. and Plomin R. (1997) Cognitive and temperamental mediators of genetic contributions to the home environment during infancy. *Merrill-Palmer Quarterly* **43**:1–23.

Saudino K. J., Plomin R., DeFries J. C. (1996) Tester-rated temperament at 14, 20, and 24 months: environmental change and genetic continuity. *Br J Dev Psych* **14**:129–44.

Saudino K. J., Plomin R., Pedersen N. L. *et al.* (1994) The etiology of high and low cognitive ability during the second half of the life span. *Intelligence* **19**:353–71.

Schmidt L. A. and Fox N. A. (2002) Molecular genetics of temperamental differences in children, in J. Benjamin, R. Ebstein, R. H. Belmaker (eds) *Molecular Genetics and Human Personality* (pp. 245–55). New York: American Psychiatric Press.

Schoenfeldt L. F. (1968) The hereditary components of the project TALENT two-day test battery. *Measurement and Evaluation in Guidance* **1**:130–40.

Seligman M. E. P. and Csikszentmihalyi M. (2000) Positive psychology: an introduction. *Am Psychol* **55**:5–14.

Smith, C. (1974) Concordance in twins: Methods and interpretation. *Am J Hum Genet* **26**:454–66.

Spinath F. M. and Borkenau P. (2000) Genetic and environmental influences on reaction times: evidence from behavior-genetic research. *Psychologische Beiträge* **42**:201–12.

Stoolmiller M. (2000) Implications of the restricted range of family environments for estimates of heritability and nonshared environment in behavior-genetic adoption studies. *Psychological Bulletin* **125**:392–409.

Stromswold K. (2001) The heritability of language: a review of twin and adoption studies. *Language* **77**:647–723.

Sulloway F. J. (1996) *Born to Rebel: Family Conflict and Radical Genius.* New York: Pantheon.

Turkheimer E. and Waldron M. (2000) Nonshared environment: a theoretical, methodological, and quantitative review. *Psychological Bulletin* **126**:78–108.

Vandell D. (2000) Parents, peer groups, and other socializing influences. *Developmental Psychology* **36**:699–710.

Vernon P. A., Jang K. L., Harris J. A. *et al.* (1997) Environmental predictors of personality differences: A twin and sibling study. *J Pers Soc Psychol* **72**:177–83.

Wickelgren I. (1998) Tracking insulin to the mind. *Science* **280**:517–19.

Widiger T., Verheul R., van den Brink W. (1999) Personality and psychopathology, in L. A. Pervin, and O. P. John (eds) *Handbook of Personality* 2nd edn. New York: Guilford.

Wilson R. S. and Matheny A. P., Jr (1986) Behavior genetics research in infant temperament: The Louisiville Twin Study, in R. Plomin and J. F. Dunn (eds) *The Study of Temperament: Changes, Continuities, and Challenges.* Hillsdale, NJ: Erlbaum.

Zuckerman M. (1979) *Sensation Seeking: Beyond the Optimal Level of Arousal.* Hillsdale, NJ: Erlbaum.

Zuckerman M. (1994) *Behavioral Expression and Biosocial Bases of Sensation Seeking.* New York: Cambridge University Press.

Chapter 5

Genetics of mental retardation

Anita Thapar and Jill Clayton-Smith

Definitions and categorization

Mental retardation, known as learning disability in the UK, is defined as early onset intellectual deficit (IQ test scores of below 70), accompanied by impairment of functioning. Although this is a widely accepted definition, the term mental retardation describes a group of individuals who show an extremely wide range of abilities and adaptive skills and who differ enormously in terms of clinical presentation. Given that mental retardation represents such a broad category, it is not surprising that there have been many attempts to subdivide the group at a phenotypic and aetiological level.

Mental retardation has most commonly been subdivided into two groups according to IQ level. Those who show IQ scores of between 50 and 70 are categorized as having mild mental retardation (MMR) and those with IQ scores of below 50 are considered as having moderate-severe mental retardation (SMR). Most recent population studies suggest population prevalence estimates of approximately 20 per 1000 for MMR and 3.8 per 1000 for SMR (Roeleveld *et al.* 1997). This two group approach has been influential for many years (Penrose 1938) and originally it was presumed that there was an aetiological as well as phenotypic distinction between these two groups. Whereas MMR was considered as representing the lower end of the normal IQ distribution (see chapter 4) with psychosocial disadvantage having a major influence, SMR was thought to represent a qualitatively different group, mainly accountable for by identifiable causes such as environmental insults and specific genetic anomalies. There was some empirical support for this separation, in that the recurrence risk of mental retardation is considerably higher amongst relatives of those with MMR whereas families of those with SMR do not show this increased risk (Bundey *et al.* 1989; Costeff and Weller 1987). A definite cause is also found much more frequently for SMR (65 per cent) (Raynham *et al.* 1996), than for MMR (30 per cent) (Knight *et al.* 1999). Nevertheless the distinction between MMR and SMR at an aetiological level is not clear-cut, given that the cause remains unknown in approximately 40 per cent of those with SMR (Knight *et al.* 1999) and specific causes of mental retardation have increasingly been recognized in those with MMR, for example between 4–19 per cent show gross chromosomal anomalies (Hagberg *et al.* 1981; Gostason *et al.* 1991).

The genetic basis of mental retardation

Undoubtedly there has been remarkable progress in the mapping and localization of an increasing number of single gene defects, and in the recognition of chromosomal anomalies. Chromosomal and genetic disorders account for between 30–40 per cent of SMR and about 15 per cent of MMR. However, it is likely that many defects remain undetected. Indeed new techniques involving molecular analysis and flourescent *in situ* hybridization (FISH) have recently enabled the identification of previously unrecognized subtle chromosomal abnormalities in a proportion of those with mental retardation (Flint *et al.* 1995; Knight *et al.* 1999; Slavotinek *et al.* 1999). The most recent of these studies (Knight *et al.* 1999) reported that in a sample of 466 children with unexplained mental retardation, subtle chromosomal rearrangements in telomeric regions were found in 0.5 per cent of children with mild retardation and 7.4 per cent of those with moderate-severe retardation, with half of these rearrangements being familial. This suggests that such anomalies are the second most common cause, after Down syndrome, of moderate-severe mental retardation.

Although single gene disorders and chromosomal anomalies are relatively rare, the advances in understanding their genetic basis nevertheless have an impact on the field of psychiatry in two important ways. First, mental retardation is a major clinical manifestation of many of these recognized genetic anomalies, and it is clearly clinically important to recognize specific genetic causes of mental retardation. Second, where a recognized and mapped genetic defect is more commonly associated with a particular behavioural phenotype, this could offer a clue to the location of potential susceptibility genes for a given type of behaviour or psychiatric disorder. Thus, for example as described in chapters 7 and 10, it is recognized that the rate of autism is increased in tuberous sclerosis and psychotic illness more commonly occurs in those with velocardiofacial syndrome (see later). This has led to interest in examining the relevant candidate regions for autism and schizophrenia. It remains to be seen to what extent the study of behavioural phenotypes for specific genetic defects has an impact in identifying susceptibility genes for much more common psychiatric disorders (Flint 1999).

Although there have been such major advances in the identification of specific genetic anomalies and the mapping of single gene disorders, it has to be recognized that most mental retardation is not accounted for by these syndromes. Although there are clearly many different causes of mental retardation including environmental insults such as birth trauma and infections, in most instances, no specific cause can be identified. Thus it is worthwhile emphasizing that individuals with so-called idiopathic mental retardation constitute much the largest group.

In this chapter we will begin by considering what is known about the genetics of idiopathic mental retardation, and then move onto discuss single gene defects and chromosomal anomalies.

Idiopathic mental retardation

In contrast to the wealth of research on the genetics of intelligence within the normal range (see chapter 4), until recently there has been remarkably little interest in the genetic epidemiology of idiopathic mental retardation, with no properly designed family and twin studies published so far. Forthcoming results from a new twin study of young children should help remedy this shortcoming (see chapter 4). The results of such studies are dependent both on sample ascertainment, given that findings are likely to differ depending on whether probands are identified from the community, special schools or institutions, and on the rigour of screening for identifiable causes.

Until recently it would have been reasonable to use the term idiopathic mental retardation to refer to those individuals who showed no evidence of gross chromosomal defects or single gene anomalies. However older studies will not have carried out molecular testing for fragile X or checked for subtle chromosomal anomalies, and clearly other specific genetic defects that we cannot yet detect will in time be recognized. Nevertheless it seems likely that a substantial proportion of individuals for whom mental retardation is truly idiopathic will remain. Here aetiology may still best be explained in terms of the traditional polygenic multifactorial model. The difficulty is that there are too few studies to even properly consider the likely role of genes and environment in determining idiopathic mental retardation. Given the dearth of published literature we are reliant on attempting to draw conclusions from family and twin studies, most of which are very old now and which report widely varying recurrence risks (Crow and Tolmie 1998). Nevertheless, methodological considerations aside, a recent review summarizes reported recurrence risks in siblings as varying between 2.4 per cent and 19.8 per cent for MMR and 2 per cent to 14 per cent for SMR (Crow and Tolmie 1998).

The familial clustering of MMR is perhaps not surprising given that genes and psychosocial adversity, which are thought to be influential, are shared by family members. Unfortunately so far there have only been three published small twin studies of mental retardation: although they showed higher concordance rates in the MZ than DZ, twin pairs were not large enough to draw definite conclusions about the contribution of genetic influences (Rosanoff *et al.* 1937; Gottesman, 1971). However there have been twin studies that have focused on cognitive ability (see chapter 4), but not necessarily mild mental retardation. Those that have used the DeFries and Fulker regression method (see chapter 2; DeFries and Fulker 1985, 1988) to examine the aetiology of low cognitive ability suggest that it is influenced by the same genetic and environmental factors that influence cognitive ability within the normal range (Petrill *et al.* 1997; Saudino *et al.* 1994).

Single gene defects

Single gene defects account for only a small proportion of mental retardation, and are more likely to be seen in individuals with moderate to severe retardation. Bundey *et al.* (1989)

estimated that single gene defects accounted for 7–14 per cent of moderate to severe retardation and 5 per cent of mild retardation. This figure did not include individuals with multiple congenital anomaly syndromes, some of which are now known to result from single gene defects. The group of single gene disorders can be divided into autosomal disorders (dominant and recessive) and X-linked mental retardation (XLMR). A detailed family history may provide clues as to the mode of inheritance of a single gene disorder. When the diagnosis of a single gene disorder is made it is appropriate to offer the families concerned genetic counselling, as recurrence risks may be significant.

In many disorders which are due to single genes there are associated phenotypic features such as physical, radiological or behavioural abnormalities, in addition to the mental retardation (Table 5.1). Recognition of these patterns facilitates diagnosis and has allowed discrete phenotypes to be delineated. Syndrome recognition may be aided by reference to a dysmorphology database (Winter 2000; http://dhmhd.mdx.ac.uk/LDDB.html) where mental retardation may be entered as one of the features in a syndrome search. One large group of single gene disorders causing mental retardation is the inborn errors of metabolism which have been well documented by Scriver *et al.* (1995). The underlying genetic defects often involve genes encoding enzymes or co-factors within metabolic pathways. Biochemical defects may occur at different points along the pathway, giving rise to variation in clinical features (phenotypic variability) e.g. phenylketonuria may be due to mutations within the phenylalanine hydroxylase gene or within the gene encoding dihydropteridine reductase, and there are subtle differences between the clinical features in both. In some circumstances different mutations occurring within the same gene may cause different clinical phenotypes, depending on how the mutation affects the production and function of the protein involved. This phenomenon is known as allelic heterogeneity. The majority of inborn errors of metabolism follow an autosomal recessive pattern of inheritance, with some notable exceptions e.g. Hunter syndrome. Diagnosis is most often by biochemical methods, although direct mutation testing is now gradually becoming the method of choice for some of these conditions.

Several of the single gene disorders are associated with specific behavioural patterns and abnormalities on neuropsychological testing, and interest in the genetic contribution to behaviour has escalated in recent years. These features are proving to be very relevant to the diagnosis and management of these disorders. Noonan syndrome is a relatively common autosomal dominant disorder (1 in 2000 livebirths). The classical clinical features include intrauterine oedema, pulmonary stenosis and short stature. The facies are dysmorphic with wide-spaced eyes, ptosis and down-slanting palpebral fissures. The helix of the ear is often overturned and the earlobes are prominent and forward facing. There may be nuchal oedema or frank neck-webbing. Although many individuals with Noonan syndrome have normal intellectual development, mild to moderate retardation may be seen along with a variety of neuropsychological abnormalities including a discrepancy in verbal and performance IQ, poor communication skills and autistic features

Table 5.1 Autosomal single gene disorders causing mental retardation

Disorder features	Inheritance	Gene	Location	Population frequency	Degree of MR	Clinical features
Tuberous sclerosis	AD	TSC1 TSC2	9q34 16p13.3	1 in 30,000	Ranges from normal IQ to severe retardation	Adenoma sebaceum, Hypopigmented patches, hamartomata of brain, heart and kidneys
Apert Syndrome	AD	FGFR2	10q	1 in 150,000	Mild to severe	Coronal craniosynostosis, shallow orbits, trapezoid mouth, 'mitten' hands and feet
Treacher Collins Syndrome	AD	Treacle	5q31–33	1 in 40,000	Mild to moderate	Maxillary and mandibular hypoplasia, absent or malformed external ears, downslanting palpebral fissures
Noonan Syndrome	AD	unknown	12	1 in 1,500	Mild	Short stature nuchal oedema/webbing, pulmonary stenosis, cryptorchidism
Bardet-Biedl	AR	BBS1,2,3 etc	Various	Unknown	Mild	Post-axial polydactyly, obesity, hypogonadism, pigmentary retinopathy
Hurler Syndrome	AR	IDUA	4p16.3	1 in 100,000	Gradual deterioration after age of two years	Coarse facies, corneal clouding, stiff joints, skeletal abnormalities
Seckel Syndrome	AR	Unknown		1 in 40,000	Moderate to severe	Severe microcephaly ('bird' head), beaked nose, extreme short stature

AD = autosomal dominant
AR = autosomal recessive

(Ghaziuddin 1994; Wood *et al.* 1995; Cornish 1996). Cohen syndrome is an autosomal recessive disorder in which moderate to severe mental retardation is associated with truncal obesity, joint hypermobility and a pigmentary retinopathy (Cohen *et al.* 1973). Affected individuals have a cheerful, sociable disposition (Kivitie-Kallio *et al.* 1999). Tuberous sclerosis is an autosomal dominant disorder which may be caused by mutations in two separate genes on chromosomes 9 and 16 (an example of genetic heterogeneity). These genes both act as tumour suppressor genes, yet specific behavioural features and cognitive deficits are known to be associated with this condition (Gillberg *et al.* 1994; Harrison *et al.* 1999). It has been suggested that these may correlate with the number and positions of the hamartomas found within the brain in this condition.

X-linked mental retardation

In 1938 Penrose recognized that mental retardation was commoner among males than females, and initially this was attributed to a greater susceptibility of males to environmental effects. In later years several large pedigrees of families with mental retardation were published (Martin and Bell 1943; Borjeson *et al.* 1962). It was clear that these were following an X-linked pattern of inheritance because of the increased frequency of affected males, lack of male to male transmission and an increased likelihood that the transmitting parent was the mother. Turner *et al.* (1971) and Lehrke (1972) demonstrated the significant contribution of X-linked genes to mental retardation, and the prevalence of this condition, has been estimated at 1.83 per thousand (Glass 1991) with a carrier frequency in females of 2.44 per thousand. In 2000 an update on XLMR reported that 202 disorders had been identified (Chiurazzi *et al.* 2001). Almost a third of these fell into the category of 'non-specific XLMR', with no other features present apart from mental retardation. The commonest cause of XLMR is the Fragile X syndrome, which is discussed in more detail later. Manifesting females have been described in several of the XLMR disorders. In some conditions, such as Coffin–Lowry syndrome (Young 1988) females always manifest, whereas in other disorders carrier females are usually phenotypically normal but may occasionally manifest symptoms as a result of altered Lyonisation (Lyon 1993) or the presence of an X-autosome translocation. In the latter case, the breakpoint on the X-chromosome affects the function of the XLMR gene concerned. Some forms of XLMR which follow an X-linked dominant pattern of inheritance affect females exclusively and usually lead to lethality in males. Perhaps the best example of such a disorder is Rett syndrome, a neurodevelopmental disorder which occurs almost exclusively in females. Development is normal for the first 6–18 months of life and then there is a period of regression with loss of acquired skills, especially of purposeful hand movement. There is deceleration of head growth and development of stereotypic hand-wringing or patting movements and hyperventilation. Rett syndrome affects 1 in 10,000 females and most cases are sporadic, occurring as a result of a new mutation in the MECP2 gene (Amir *et al.* 1999). MECP2 encodes a protein MeCP2 which is thought to be involved in transcriptional silencing of other genes by means of methylation. Faulty expression of MeCP2 may lead to a failure to silence other genes throughout the genome at critical times. Truncating mutations which cause a failure in production of normal amounts of MeCP2 cause a more severe phenotype than missense mutations where an altered protein is produced in relatively normal amounts.

Several of the XLMR disorders map to the same locus, and the final number of XLMR genes will therefore be less than the number of conditions described. In some cases the genetic abnormality will be specific to just a single family and these genes will be harder to identify. Of particular interest has been the finding that mutations within some of the genes causing syndromic mental retardation such as Coffin–Lowry syndrome (Trivier *et al.* 1996) may also be responsible for non-specific XLMR (Touraine *et al.* 2000). The human genome mapping project has recently resulted in the sequencing of almost all of the human genome, and in its wake many genes involved in the aetiology of mental retardation will be identified. The majority of these will be expressed within the brain, and many different types of gene will be involved including cell adhesion molecules, receptors and genes involved in the process of ubiquitination which affects protein turnover (Table 5.2).

Fragile X syndrome

Fragile X syndrome is the single commonest known inherited cause of mental retardation and originally derived its name from the characteristic non-staining band or fragile site on the X chromosome. It affects between 1 in 4000 and 1 in 6000 males (DeVries *et al.* 1998) and accounts for approximately 7 per cent of moderate and 4 per cent of mild retardation in males, and approximately 2.5 per cent of moderate and 3 per cent of mild retardation in females (Webb and Thake 1991).

The key clinical characteristics of fragile X syndrome are mental retardation, large ears and a long face and macro-orchidism. Many affected individuals show higher rates of speech and language problems, attentional difficulties and hyperactivity and autistic-like

Table 5.2 X-linked recessive disorders associated with mental retardation

Disorder	Associated clinical features	Gene and location
X-linked hydrocephalus	Adducted thumbs, spasticity aphasia	L1CAM Xq28
Fragile X syndrome	Macrocephaly, large ears, macro-orchidism	FMR1 Xq27–28
ATR-X syndrome	Microcephaly, dysmorphic facies, genital abnormalities	XNP Xq13
Coffin–Lowry syndrome	Coarse face, hypertelorism, tapering fingers	RSK2 Xp22
Simpson–Golabi-Behmel syndrome	Tall stature, polydactyly supranumerary nipples	GPC3 Xq25–27
Rett syndrome	Normal development followed by regression acquired microcephaly, hand-wringing hyperventilation, seizures	MECP2 Xq28

features such as gaze avoidance and hand flapping (Hagerman and Sobesky 1989; Brown *et al.* 1991). The observation of this type of behavioural phenotype in conjunction with early reports of increased rates of fragile X in autism led to interest in the relationship between fragile X and autism. Although it is clearly important to screen children who have autism, the rate of fragile X is probably little higher than amongst children with mental retardation (Fisch 1992; Bailey *et al.* 1993).

The diagnosis of fragile X was originally made cytogenetically when the fragile site on Xq 27.3 could be detected in 4 per cent or more lymphocytes in certain culture media. This method of diagnosis has been replaced by much more accurate molecular testing for the mutation which has now been identified (Verkerk *et al.* 1991). Although fragile X syndrome is an X-linked disorder the pattern of inheritance is unusual in two major ways. First it has long been noted that the clinical severity of the disease increases with each successive generation; this phenomenon is known as anticipation. Second, it had been observed that disease expression seemed to depend on whether the defect was transmitted by mother or father. These observations remained an unexplained puzzle for many years but now have a molecular explanation. The mutation for fragile X is a heritable unstable sequence of trinucloeotide CGG repeats within the first exon of the FMR1 gene (Oberle *et al.* 1991; Yu *et al.* 1991). This sequence contains between approximately 6 and 53 repeats and is stable in the normal population. A premutation of 54–200 repeats is unstable and expands when transmitted by females. The full mutation, when the repeat sequence reaches a critical length of about 200 copies, is associated with hypermethylation of the repeat and adjacent region. This results in the failure of FMR1 transcription and an absence of the FMR1 gene protein product (FMRP) which is responsible for the characteristic clinical features of fragile X syndrome. Heritable unstable repeat sequences within other genes have also been found for other disorders such as Huntington's chorea and myotonic dystrophy (see chapter 13).

Two other folate sensitive fragile sites have also been detected at nearby sites. The fragile site at the first exon of FMR1 gene is termed FRAXA. The second fragile site FRAXE, which is situated at Xq28, also appears to be associated with mental retardation and loss of expression of the FMR2 gene, whereas so far FRAXF does not seem to be associated with any clinical features. It is now possible to detect these mutations using PCR based assays (Strelnikov *et al.* 1999).

Chromosomal abnormalities

Chromosomal abnormalities account for 35–40 per cent of severe mental retardation and 10 per cent of mild mental retardation (Bundey *et al.* 1989; Raynham *et al.* 1996). These figures are likely to be an underestimate, as recent developments in techniques of chromosome analysis have enabled the detection of many small chromosome abnormalities which were not visible with previous techniques. The commonest autosomal abnormalities are trisomies, particularly involving chromosomes 13, 18 and 21, and these are all associated with increased maternal age. Almost all chromosomal aneuploidies which involve an alteration in the amount of chromosome material are associated with mental retardation. It is not certain whether this is due to dosage effects of specific genes within the duplicated or deleted segments, or to a more general effect of aneuploidy per se. It is certainly true that individuals with chromosomal aneuploidy share some non-specific features in common such as poor growth, microcephaly, epicanthic folds and unusual palmar creases, in addition to features more specific to the chromosome involved. Some chromosome abnormalities occur in the mosaic form, and for disorders which are usually seen in the full form, mosaicism will confer a milder phenotype. However, there are some conditions which are lethal in the full form and are therefore found only in the mosaic form in surviving individuals. The mental retardation in these conditions, which include trisomy 8 mosaicism (Gagliardi *et al.* 1988) and 12p tetrasomy (Schinzel 1991) is usually severe. Mosaic karyotypes are more likely to be identified within cultured fibroblasts than in blood cells.

Down syndrome

Trisomy 21 (Down syndrome) is the commonest chromosomal cause of mental retardation, with an overall incidence of 1 in 700 pregnancies. The incidence is greatly influenced by maternal age. In the majority of cases the extra chromosome arises as a result of failure of separation (non-dysjunction) of the two maternal chromosome 21s in the oocyte during meiosis, giving rise to three separate copies of chromosome 21 in the affected individual. In 5 per cent of cases the extra copy of chromosome 21 is attached at the centromere to one of the other chromosomes, usually chromosome 14. This is termed a Robertsonian translocation. If one of the parents carries a balanced form of the same translocation, there will be a significant risk of recurrence of Down syndrome during a future pregnancy. The clinical features of Down syndrome include neonatal hypotonia, a flat occiput, flat facial profile, small simply formed ears, protrusion of the tongue, single palmar crease, incurved fifth fingers and a wide gap between the first and second toes (sandal gap). Congenital heart defects occur in 40 per cent of infants and are a common cause of death. Other structural abnormalities include duodenal or oesophageal atresia and imperforate anus. There is a predisposition to Hirschprung's disease, hypothyroidism and leukaemia and to the early onset of Alzheimer's disease. The latter is likely to be related to the fact that the gene for amyloid precursor protein, known to be involved in familial Alzheimer's disease (Goate *et al.* 1991) lies on chromosome 21. The majority of individuals have moderate to severe mental retardation, and some of those with a milder phenotype have been shown to have a mosaic karyotype. The critical region for the Down syndrome phenotype is in the region of bands 21q21.3 to 21q22 (Epstein *et al.* 1991). The specific region associated with mental retardation was defined further by Delabar *et al.* in 1993, and now that the entire chromosome sequence has been identified we can expect further elucidation of the mechanisms giving rise to the features of this condition.

Microdeletion syndromes

Improvements in cytogenetic analysis using techniques such as high resolution chromosome banding and fluorescent *in situ* hybridization (FISH) together with the molecular genetic techniques of microsatellite analysis (Trask 1991; Strachan and Read 1996) have enabled the detection of increasingly small chromosome abnormalities (see chapter 1). The microdeletions are usually small – 4kb or less – and encompass multiple genes which may all contribute to the phenotype. Those microdeletions which are observed most commonly tend to have similar breakpoints, occurring in regions of the chromosome where there is a repetitive DNA sequence. The detection of submicroscopic chromosome rearrangements in individuals with idiopathic mental retardation has been mentioned previously. Several of these microdeletions occur with a frequency significant enough to merit special mention.

Di George syndrome (Velocardiofacial syndrome)

This is the commonest microdeletion syndrome and involves a deletion of chromosome 22q11. It gives rise to features of the Di George syndrome (Ryan *et al.* 1997), also known as the velocardiofacial syndrome (VCFS). In addition to mild to moderate mental retardation there may be a cleft palate, hypocalcaemia, congenital heart defects and speech and swallowing difficulties. The facies are dysmorphic with a broad nasal bridge, narrow alae nasae which are often notched, small mouth and chin and overfolded helices. A large number of other structural abnormalities have been described in this condition, which appears to have protean manifestations. The features vary considerably from person to person, with the speech and swallowing difficulties being the most consistent features. As mentioned earlier, there is an excess of psychotic disorders in patients who have VCFS. The most recent study found that more than one quarter fulfilled diagnostic criteria for schizophrenia (Murphy *et al.* 1999) (see also chapter 10).

Williams syndrome

Williams syndrome, which is characterized by infantile hypercalcaemia, supravalvular aortic stenosis and a happy, sociable disposition is associated with moderate mental retardation and is caused by a microdeletion of chromosome 7q11. The behavioural features of this condition are quite distinctive and consist of moderate-severe learning difficulties, superficially fluent language skills, a propensity to be overfriendly and socially disinhibited and increased rates of overactivity, inattention, excess anxiety, specific preoccupations and hyperacusis. Although there are general cognitive deficits in Williams syndrome so that affected individuals have low scores on tests of general intelligence, the conversational aspects of linguistic abilities are spared (Bellugi *et al.* 1999). An excellent review of this condition was published by Metcalfe (1999).

Smith–Magenis syndrome

Smith–Magenis syndrome is associated with moderate to severe mental retardation and is caused by a microdeletion of 17p11.2. Affected individuals have striking behavioural abnormalities and indulge in self-abusive behaviours such as onychotillomania (pulling off finger-nails) and polyembolokoilomania (insertion of objects into body orifices). Sleep disturbance and a tendency to 'self-hugging' are also common (Dykens and Smith 1998).

Microdeletions of chromosome 15q11–13

The region of chromosome 15q11–13 is subject to the phenomenon known as genomic imprinting. Within this region genetic information of maternal origin is expressed in a different way to that of paternal origin, but information from both parental genomes is required for normal development. Individuals who have a microdeletion of 15q11–13 manifest different characteristics depending upon whether the deletion is of maternal or paternal origin. A maternal deletion will give rise to the symptoms of Angelman syndrome (Angelman 1965) which include ataxia, absent speech, severe mental retardation, seizures and a happy, sociable disposition. A paternal deletion gives rise to features of the Prader–Willi syndrome (Cassidy 1997) which include neonatal hypotonia and feeding problems, hyperphagia (over eating) from early childhood with the development of obesity, hypogonadism and mild to moderate mental retardation. The hands and feet are often small and the facies are distinctive, with bitemporal narrowing and almond-shaped palpebral fissures. Tantrums, skin picking and obsessional behaviour are frequently observed (Einfeld et al. 1999). The genes responsible for the two conditions are in fact different, but lie in close proximity on chromosome 15q11–13. Although 60 per cent of patients with Prader–Willi syndrome and 70 per cent of Angelman patients have deletions, both conditions may also be caused by other genetic abnormalities which interfere with expression of genes in the 15q11–13 region. These conditions can also arise from maternal and paternal uniparental disomy for chromosome 15 leading to Prader–Willi syndrome and Angelman sydrome respectively.

Sex chromosome abnormalities

Sex chromosomal anomalies are most commonly due to chromosomal non-disjunction, the risk of which increases with maternal age. Most affected individuals, apart from those with Turner's syndrome, do not come to medical attention and remain undetected in the community.

Turner's syndrome

Turner's syndrome, which is characterized by the complete or partial absence of one X chromosome (45 XO), affects between 1 in 2,000 and 1 in 5,000 females. The clinical features of this syndrome include short stature, a webbed neck and increased carrying angle of the elbow and the failure to develop secondary sexual characteristics. Until recently no specific genes or critical regions of the X chromosome associated with the clinical features of Turner's syndrome had been identified. However, recent reports suggest that SHOX (short stature homeobox-containing gene), which has been cloned and located in the pseudoautosomal region, may play a role in the growth failure of Turner's syndrome (Rao et al. 1997).

Although not associated with mental retardation, a recent study of Turner's syndrome has suggested that the X chromosome might also play an important role for other sorts of cognitive skills which faciliate social interaction. Girls tend to show superior levels of skills such as the ability to respond to cues in the behaviour of others, to inhibit distractions, and to develop strategies of action compared with boys. Skuse et al. (1997) showed that girls affected by Turner's syndrome whose X chromosome was of maternal origin showed poorer social cognitive social skills than those who possessed a paternally derived X chromosome. The authors suggested that social cognitive skills might be influenced by an imprinted X-linked locus, which would also account for the lower social cognitive scores of boys whose X chromosome comes from their mother. More recently further work has been conducted to identify a critical region associated with the neurocognitive deficits characteristic of Turner's syndrome (most typically affecting visuospatial/perceptual skills). Initial findings from one study of 34 females with Turner's syndrome who showed partial deletions of chromosome Xp (Ross et al. 2000), have suggested the involvement of a critical region within the pseudoautosomal region of the X chromosome.

Klinefelter's syndrome

Klinefelter's syndrome affects approximately 1 in 1000 males and is characterized by the karyotype 47 XXY. In about two-thirds of cases the anomaly is due to maternal non disjunction and is non familial (Sanger et al. 1977). It was originally thought that affected individuals showed increased rates of mental retardation, psychiatric disorder and criminality. However early studies of the XXY behavioural phenotype were based on highly selected samples of those who had been institutionalized (Hook 1973), and it is now recognized that the majority who possess this karyotype are of normal intelligence and show no increased rates of behavioural or psychiatric difficulties (Stewart et al. 1982; Mandoki et al. 1991).

XYY syndrome

The rate of XYY syndrome in the general population has been estimated at between 1 and 2 per 2000. Affected males are characteristically tall and have lower than average IQ scores. Early studies based on samples from institutions suggested an increased rate of criminality and psychiatric disorder (Jacobs et al. 1968; Hook 1973). However, research based on an unselected sample suggests that contrary to earlier beliefs, XYY is not associated with excessive aggression. In this study, (Witkin et al. 1976) which is the largest to date, an unselected sample of over 4000 non-institutionalized males with height greater than 1.84 metres, was screened for the XYY karyotype. Twelve affected males were identified, and although 42 per cent of these had a criminal record compared with 9 per cent of normal males, the crimes were relatively minor in nature and there was no evidence of an association with violent criminality (see also chapter 9).

Conclusions

The progress in mapping single gene disorders is undoubtedly of importance, given that so many of these defects are associated with mental retardation. Although such conditions and chromosomal anomalies are relatively rare and thus might be considered less important at a population-wide level, collectively they represent a significant health burden. Moreover research in this area might have an impact in a broader sense. Given the increased interest in studying behavioural phenotypes associated with some of these conditions (Flint 1999), it is hoped that the localization of specific genes and the identification of genes within critical chromosomal regions might pave the way for understanding more about the genetic basis of behaviour and psychiatric disorders. Novel methods of identifying previously undetected submicroscopic chromosomal anomalies are also exciting, and highlight the fact that where mental retardation is currently thought to be idiopathic, specific aetiologies may yet be discovered. Nevertheless for many, particularly those with mild mental retardation, it is likely that no specific defects will be identified. The challenge in understanding more about aetiological factors for this group has much in common with the goal of unravelling the complex genetic basis of psychiatric disorders. Clearly properly designed twin and adoption studies are an essential step in determining the extent to which genes and environment contribute, co-act and interact. Molecular genetic studies of normal intelligence (see chapter 4) are also likely to have an impact and may also help pave the way towards understanding more about risk mechanisms.

References

Amir R. E., Van den Veyver I. B., Wan M. *et al.* (1999) Rett syndrome is caused by mutations in X-linked MECP2, encoding methyl-CpG-binding protein 2. *Nature Genetics* **23**:185–8.

Angelman H. (1965) Puppet children. *Developmental Medicine and Child Neurology* **7**:681–8.

Bellugi U., Adolphs R., Cassady C. *et al.* (1999) Towards the neural basis for hypersociability in a genetic syndrome. *Neuroreport* **10**:1653–7.

Börjeson M., Forssman H., Lehmann O. (1962) An X-linked recessively inherited syndrome characterised by grave mental deficiency, epilepsy and endocrine disorder. *Acta Medical Scand* **171**:12–21.

Brown W. T., Jenkins E., Neri G. *et al.* (1991) Conference report: Fourth International Workshop on the fragile X and X-linked mental retardation. *Am J Med Genet* **38**:158–72.

Bundey S., Thake A., Todd J. (1989) The recurrence risks for mild idiopathic mental retardation. *J Med Genet* **26**:260–6.

Cassidy S. B. (1997) Prader–Willi Syndrome. *J Med Genet* **34**:917–23.

Chiurazzi P., Hamel B. C. J., Neri G. (2001) XLMR genes: update 2000. *European Journal of Human Genetics* **9**:71–81.

Cohen M. M. Jr, Hall B. D., Smith D. W. (1973) A new syndrome with hypotonia, obesity, mental deficiency and facial, oral and limb anomalies. *J Pediatr* **83**:280–4.

Cornish K. M. (1996) Verbal-performance discrepancies in a family with Noonan syndrome. *Am J Med Genet* (letter) **66**:235–6.

Costeff H. and Weller L. (1987). The risk of having a second retarded child. *Am J Med Genet* **27**:753–66.

Crow Y. J. and Tolmie J. L. (1998) Recurrence risks in mental retardation. *J Med Genet* **35**:177–82.

De Fries J. C. and Fulker D. W. (1985) Multiple regression analysis of twin data. *Behaviour Genetics* **15**:467–73.

De Fries J. C. and Fulker D. W. (1988) Multiple regression analysis of twin data: Etiology of deviant scores versus individual differences. *Acta Geneticae Medicae et Gemellologicae* **37**:205–16.

Delabar J-M., Theophile D., Rahmani Z. *et al.* (1993) Molecular mapping of twenty-four features of Down syndrome on chromosome 21. *Eur J Hum Genet* **1**:114–24.

De Vries B. A., Halley D. J. J., Oostra B. A. *et al.* (1998) The fragile X syndrome. *J Med Genet* **35**:579–89.

Dykens E. M. and Smith A. C. (1998) Distinctiveness and correlates of maladaptive behaviour in children and adolescents with Smith-Magenis syndrome. *J Intellect Disabil Res* **42**:481–9.

Einfeld S. L., Smith A., Durvasula S. *et al.* (1999) Behaviour and emotional disturbance in Prader–Willi syndrome. *Am J Med Genet* **82**:123–7.

Epstein C. J., Korenberg J. R., Anneren G. *et al.* (1991) Protocols to establish genotype–phenotype correlations in Down syndrome. *Am J Hum Genet* **49**:201–7.

Fisch G. S. (1992) Is autism associated with the fragile X syndrome? *Am J Hum Genet* **43**:47–55.

Flint J., Wilkie A. O. M., Buckle V. J. *et al.* (1995) The detection of subtelomeric chromosomal rearrangements in idiopathic mental retardation. *Nature Genetics* **9**:132–9.

Flint J. (1999). The genetic basis of cognition. *Brain* **122**:2015–31.

Gagliardi A. R. T., Tajara E. H., Varella-Garcia M. *et al.* (1988) Trisomy 8 mosaicism. *J Med Genet* **15**:70–2.

Ghaziuddin M., Bolyard B., Alessi N. (1994) Autistic disorder in Noonan syndrome. *J Intellect Disabil Res* **38**:67–72.

Gillberg I. C., Gillberg C., Ahlsen G. (1994) Autistic behaviour and attention deficit in tuberous sclerosis: a population based study. *Dev Med Child Neurol* **36**: 603–14.

Glass I. (1991) X linked mental retardation. *J Med Genet* 361–71.

Goate A., Chartier-Harlin M-C., Mullan M. *et al.* (1991) Segregation of a missense mutation in the amyloid precursor protein gene with familial Alzheimer's disease. *Nature* **349**:704–6.

Gostason R., Wahlstrom J., Johannisson T. *et al.* (1991) Chromosomal aberrations in the mildly mentally retarded. *Journal of Mental Deficiency Research* **35**:240–6.

Gottesman I. I. (1971) An introduction to behavioural genetics of mental retardation, in R. M. Allen *et al.* (eds) *Role of Genetics in Mental Retardation*. Coral Gables: University of Miami.

Hagberg G., Lewerth A. *et al.* (1981) Mild mental retardation in Swedish schoolchildren II. Etiologic and pathogenetic aspects. *Acta Paediatrica Scandinavica* **70**:445–52.

Hagerman R. J. and Sobesky W. E. (1989) Psychopathology in fragile X syndrome. *American Journal of Orthopsychiatry* **59**:142–52.

Harrison J. E., O'Callaghan F. J., Hancock E. *et al.* (1999) Cognitive deficits in normally intelligent patients with tuberous sclerosis. *Am J Med Genet* **88**:642–6.

Hook E. B. (1973) Behavioural implications of the human XYY genotype. *Science* **179**:139–79.

Jacobs P. A., Price W. H., Cower-Brown W. M. *et al.* (1968) Chromosome studies on men in maximum security hospitals. *Annals of Human Genetics* **31**:339–58.

Kivitie-Kallio S., Larsen A., Kajasto K. *et al.* (1999) Neurological and psychological findings in patients with Cohen syndrome: a study of 18 patients aged 11 months to 57 years. *Neuropediatrics* **30**:181–9.

Knight S. J. L., Regan R., Nicod A. *et al.* (1999) Subtle chromosomal rearrangements in children with unexplained mental retardation. *The Lancet* **354**:1676–81.

Lehrke R. (1972) Theory of X-linkage of major intellectual traits. *Am J of Ment Defic* **76**:611–9.

Lyon M. F. (1993) Epigenetic inheritance in mammals. *Trends in Genetics* 9:123–8.

Mandoki M. W., Sumner G. S., Hoffman R. P. *et al.* (1991) A review of Klinefelter's syndrome in children and adolescents. *Journal of the American Academy of Child and Adolescent Psychology* **30**:167–72.

Martin J. P. and Bell J. (1943) A pedigree of mental defect showing sex-linkage. *J Neurol Psych* **6**:154–7.

Metcalfe K. (1999) Williams syndrome: an update on clinical and molecular aspects. *Arch Dis Child* **81**:198–200.

Murphy K. C., Jones L. A., Owen M. J. (1999) High rates of schizophrenia in adults with velo-cardio-facial syndrome. *Archives of General Psychiatry* **56**:940–5.

Oberle I., Rousseau F., Heitz D. *et al.* (1991) Instability of a 550 base pair DNA segment and abnormal methylation in fragile X syndrome. *Science* **252**:1097–102.

Penrose L. S. (1938) *A Clinical and Genetic Study of 1,280 Cases of Mental Defect.* Special Report Series 229. London: Medical Research Council.

Petril S. A., Saudino K. J., Cherny S. S. *et al.* (1997) Exploring the genetic etiology of low general cognitive ability from 14 to 36 months. *Developmental Psychology* **33**:544–8.

Rao E., Weiss B., Fukami M. *et al.* (1997) Pseudoautosomal deletions encompassing a novel homeobox gene cause growth failure in idiopathic short stature and Turner syndrome. *Nature Genetics* **16**(1):54–63.

Raynham H., Gibbons R., Flint J. *et al.* (1996) The genetic basis for mental retardation. *QJM* **89**:169–75.

Roeleveld N., Zielhuis G. A., Gabreels F. (1997) The prevalence of mental retardation: a critical review of recent literature. *Dev Med Child Neurol* **39**:125–32.

Rosanoff A. J., Handy L. M., Plesset I. R. (1937) The etiology of mental deficiency with special reference to its occurrence in twins. *Psychological Monographs* **216**:1–137.

Ross J. L., Roeltgen D., Kushner H. *et al.* (2000) The Turner Syndrome – associated neurocognitive phenotype maps to distal Xp. *Am J Hum Genet* **67**:672–81.

Ryan A. K., Goodship J. A., Wilson D. *et al.* (1997) Spectrum of clinical features associated with interstitial chromosome 22q11 deletions: a European collaborative study. *J Med Genet* **34**:798–804.

Sanger R., Tippett P., Gavin J. *et al.* (1977) Xg groups and sex chromosome abnormalities in people of northern European ancestry: an addendum. *J Med Genet* **14**:210–13.

Saudino K. J., Plomin R., Pedersen N. L. *et al.* (1994) The etiology of high and low cognitive ability during the second half of the life span. *Intelligence* **19**:353–71.

Schinzel A. (1991) Syndrome of the month – tetrasomy 12p (Pallister-Killian syndrome). *J Med Genet* **28**:122–5.

Scriver C. R., Beaudet A. L., Sly W. S. *et al.* (1995) *Metabolic and Molecular Basis of Inherited Disease*, 7th edn. McGraw Hill.

Skuse D., James R. S., Bishop D. V. M. *et al.* (1997) Evidence from Turner's syndrome of an imprinted X-linked locus affecting cognitive function. *Nature* **387**:705–8.

Slavotinek A., Rosenberg M., Knight S. *et al.* (1999) Screening for submicroscopic chromosome rearrangements in children with idiopathic mental retardation using microsatellite markers for the chromosome telomeres. *J Med Genet* **36**(5):405–11.

Stewart D. A., Netley C. T., Park E. (1982) Summary of clinical findings of children with 47XXY, 47XYY and 47XXX karyotypes, in D. A. Stewart (ed.) *Birth Defects: Original Article Series.* New York: Alan R Liss, Inc.

Strachan T. and Read A. P. (1996) Genetic mapping, in *Human Molecular Genetics*. Bios Scientific Publishers.

Strelnikov V., Nemtsova M., Chesnokova G. *et al.* (1999) A simple multiplex FRAXA, FRAXE and FRAXF PCR assay convenient for wide screening programs. *Hum Mutation* **13**(2):166–9.

Touraine R. L., Delauney J. P., Nivalon-Chevallier A. *et al.* (2000) New insight in the phenotype of Coffin–Lowry syndrome/RSK2 mutant patients: lessons from 40 families. *Eur J Hum Genet* **8**(Supplement 1):63.

Trask B. J. (1991) Fluorescent in situ hybridisation: applications in cytogenetics and gene mapping. *Trends in Genetics* **7**(5):149–54.

Trivier E., De Cesare D., Jacquot S. *et al.* (1996) Mutations in the kinase Rsk-2 associated with Coffin–Lowry syndrome. *Nature* **384**:567–70.

Turner G., Collins E., Turner B. (1971) Recurrence risk of mental retardation in sibs. *Med J Aust* **1**:1165–7.

Verkerk A. J., Pieretti M., Sutcliffe J. S. *et al.* (1991) Identification of a gene (FMR-1) containing a CGG repeat coincident with a breakage point cluster region exhibiting length variation in fragile X syndrome. *Cell* **65**:905–14.

Webb T. and Thake A. (1991) Moderate and mild mental retardation in the Martin–Bell syndrome. *Journal of Mental Deficiency Research* **35**:521–8.

Winter R. and Baraitser M. (2000) *London Dysmorphology Database*. Oxford: Oxford University Press.

Witkin H. A., Mednick S. A., Schulsinger F. (1976) Criminality in XYY and XXY men. *Science* **193**:547–55.

Wood A., Massarano A., Super M. *et al.* (1995) Behavioural aspects and psychiatric findings in Noonan's syndrome. *Arch Dis Child* **72**:153–5.

Young I. D. (1988) Syndrome of the month: The Coffin–Lowry Syndrome. *J Med Genet* **25**:344–8.

Yu S., Pritchard M., Kremer E. *et al.* (1991). Fragile X genotype characterised by an unstable region of DNA. *Science* **252**:1179–81.

Chapter 6

Reading and language disorders

Julie Williams

Reading and language disorders

A variety of disorders reflect abnormalities in the processing of language. Genetic research has concentrated predominantly on specific language impairment (SLI) and specific reading disability (SRD). Current quantitative and molecular research in these areas is producing many interesting findings that show strong evidence for the presence of a number of genes contributing to these disorders. Identifying genes will help us understand how the disorders develop and may also provide a novel insight into the biology of language processing, a crucial component of human cognition.

Specific language impairment

Specific language impairment (SLI) can be defined broadly as a significant difficulty in the acquisition of spoken language that occurs in the context of normal hearing and non-verbal intelligence and that is not associated with other developmental disorders. It is important to distinguish it from acquired language impairments that arise after normal language ability has developed and that are a consequence of neurological damage or deterioration. The definition also differentiates SLI from language delay or disability that is part of a global deficit in cognitive ability, such as Down syndrome, and from developmental disorders such as autism, where language impairment is part of the pattern of symptoms defining the condition.

This definition is based on relative underachievement of language skills compared to that expected from general cognitive ability measures. The definition can be operationalized by the use of a regression procedure whereby a language measure is regressed onto non-verbal intelligence and the residual obtained for each child, that is, the difference between observed and expected language scores. This more rigorous approach to the identification of SLI requires that the regression relationship be known for the general population. This knowledge may not be available, and an alternative is to identify SLI as a language age that is below chronological age but where the child has a non-verbal IQ above 85. There is then an issue of how much impairment needs to be present to diagnose SLI and in practice this varies between studies, but most take a cut off of between one and two standard deviations below age expectations (Tomblin *et al.* 1992).

Epidemiology

Epidemiological studies of SLI have produced a variety of prevalence estimates ranging from 0.57 per cent to 7.4 per cent (Dale *et al.* 1998; Fundudis *et al.* 1979; Randall *et al.* 1974; Stevenson and Richman 1976; Tomblin 1996). The lower estimates of prevalence were observed in studies with either more stringent definitions of SLI, such as language skills at least two standard deviations below those predicted (Randall *et al.* 1974), or where the definition was limited to specific forms of language impairment, for example, severe specific expressive language impairment (Stevenson and Richman 1976). Tomblin (1996) and Rice (1997) reported a prevalence rate of 7.4 per cent in boys and 6 per cent in girls. Their sample of 849 kindergarten children was screened for a broad range of language skills, including vocabulary, grammar, narrative, receptive and expressive language, which were used to produce a criterion for SLI that aimed to reflect clinical diagnosis. Tomblin also observed that 80 per cent of the children with deficits displayed both expressive and receptive language problems, and the majority showed evidence of selective deficit in grammar development, but few showed deficits in speech impairment.

Family studies

Most studies of familiality have focused on single families with multiple cases of language impairment (Arnold 1961; Borges-Osorio and Salzano 1985; Hurst *et al.* 1990; McReady 1926; Samples and Lane 1985). However, it is likely that these studies may have limited generalizability, as many cases of SLI do not show such strong within-family transmission. Another strategy to test familiality has been to estimate the frequency of language impairment among the relatives of language-impaired probands. Earlier studies generated estimates of language impairment rates in relatives ranging from 24 per cent to 63 per cent (Bishop and Edmundson 1986; Byrne *et al.* 1974; Hier and Rosenberger 1980; Ingram 1959; Luchsinger 1970) but suffered from a lack of matched control families ascertained via probands unaffected by SLI. More recently, family studies have included matched control families in their designs and have also reported high rates of language impairment in first degree relatives that ranged between 17 per cent and 43 per cent and represented a 2 to7-fold increase over controls (Lewis 1992; Neils and Aram 1986; Rice 1997; Spitz *et al.* 1997; Tallal *et al.* 1989; Tomblin 1989, 1996; Van der Lely and Stollwerck 1996).

Van der Lely and Stollwerck studied the familiality of SLI characterized by persistent and disproportionate impairment of grammatical comprehension and expressive language. They observed that 77.8 per cent of children with SLI had a positive family history of language impairment compared with 28.5 per cent of control children. Lewis (1992) studied the first-degree relatives of probands with phonological disorders and found rates of language and/or speech disorders ranging between 12 per cent and 31 per cent. In a prospective study of language development (between the ages of 16–26 months), Spitz and colleagues (1997) found that 50 per cent of children with a positive family history demonstrated language delay, whereas none of the children with a negative family history experienced delay. Tomblin (1996) estimated the risk of SLI in first-degree relatives to be 21 per cent, and in same-sex family members to be 28 per cent, which rose to 33 per cent in same-sex dizygotic twins, and to 71 per cent in monozygotic twins. More recently, Rice (1997) studied the rates of SLI in a wide range of relatives of SLI probands. She observed that the percentage of those affected was 26 per cent in brothers and 29 per cent in sisters, in contrast to controls, who displayed rates of 3 per cent and 4 per cent respectively. Fathers showed rates of 29 per cent, mothers 7 per cent, paternal aunts 21 per cent, paternal uncles 27 per cent, paternal grandparents 10 per cent, maternal aunts 22 per cent, maternal uncles 15 per cent, and maternal grandparents between 0 and 7 per cent.

In summary, SLI appears to be highly familial, although the rates depend on the definition used. As in other complex disorders, familiality suggests a genetic contribution but cannot be taken as definitive proof. Family similarities can stem from genes or common environmental influences, and twin studies have attempted explore the importance of each of these.

Twin studies

There have been few large twin studies of SLI, although several twin studies of global language impairments have shown good evidence of moderate heritability (Aoki and Asaka 1993; DeFries and Plomin 1983; Hardy-Brown 1983; Hardy-Brown and Plomin 1985; Hardy-Brown et al. 1981; Hay et al. 1987; Locke and Mather 1989; Matheny and Bruggemann 1973; Mather and Black 1984; Munsinger and Douglass 1976; Osborne et al. 1968; Plomin 1986; Plomin et al. 1988; Scarr and Carter-Saltzman 1983). Tomblin and Buckwalter (1998) studied 43 twin pairs in which at least one twin had SLI. They found a probandwise concordance rate for history of speech and/or language therapy of 0.86 in monozygotic (MZ) twins and 0.48 in dizygotic (DZ) twins. Elsewhere, Bishop and colleagues (1995) studied 90 twin pairs where at least one twin met strict criteria for the development of speech and language disorder. Using DSM-IIIR criteria and operational definitions, they found that only 8 (30 per cent) of the 27 DZ pairs were concordant for developmental articulation disorder, whereas 34 (54 per cent) of the 63 MZ twins were concordant. They reported a probandwise concordance rate of 46 per cent for DZ twins and 70 per cent for MZ twins for male twin pairs. However, they observed little difference between the small groups of female twins (MZ = 43 per cent, DZ = 44 per cent). Stromswold (1994) performed a meta-analysis of 188 MZ twin pairs and 94 DZ pairs and found a pairwise concordance rate of 72.9 per cent for MZ and 35.1 per cent for DZ twins. From these data, he calculated a heritability of 0.76.

Recently, Dale and colleagues (1998) performed a twin study focusing on the first signs of language problems in children aged two years. They performed an extremes analysis based on the lowest 5 per cent of a sample of 6,862 two-year-old twins from a potential sample of 15,512 assessed for productive vocabulary as measured by the MacArthur Communicative Development Inventory (Fenson *et al.* 1994). From a final sample of 3,000 twins, they found a group difference heritability of 73 per cent for low productive vocabulary scores, with a shared environment accounting for 18 per cent of the variance. However, almost the opposite pattern was observed for productive vocabulary in the sample as a whole, with heritability estimates in the region of 25 per cent and shared environment accounting for the majority of the variance (69 per cent). This is an interesting result, because it strongly suggests that the genetic aetiology of productive vocabulary performance is different at the extreme lower end of the dimension. It thus appears that genes play a significant role in the development of severe productive vocabulary impairment but play a smaller role in the normal variation of this trait. This argues against a continuum in the phenotypic–genotypic relationship and supports a more categorical perspective.

It has also been observed (Bishop *et al.* 1995; Rutter *et al.* 1993) that twinning itself is a risk factor for language delay, in that twins tend to show a deficit compared to normal children. However, this cannot explain the difference between MZ and DZ twins, which still supports a genetic component to SLI. We can therefore conclude that SLI has a significant genetic basis and that its familiality results in large part from the action of genes.

Mode of transmission

In a segregation analysis of speech and language disorders, Lewis *et al.* (1993) were unable to distinguish between a major gene model and a multifactorial transmission model for SLI. However, Van der Lely and Stollwerck (1996) found support for an autosomal-dominant mode of inheritance. The high rates of affection in first-degree relatives of probands with SLI have also been considered to be consistent with a common autosomal recessive allele. However, with a complex disorder such as SLI, it is likely that various modes of transmission may exist, e.g. some of the large multi-generational pedigrees may well show autosomal dominant inheritance, whereas more common forms of SLI may be contributed to by a number of genes. This type of genetic heterogeneity is difficult to pin down using purely statistical methods, but evidence in support of this has recently emerged form molecular studies.

Molecular genetic studies

Molecular genetic research into SLI has already identified the first gene contributing to language impairment (Lai *et al.* 2001). In 1998 Fisher and colleagues identified a region on chromosome 7 that cosegregated with speech and language disorder in a large three-generation family (which they named KE), where half the members were affected (i.e. 15 of the 30 relatives). Their findings resulted from a full genome scan, which initially identified a 27.4 cM region of linkage on chromosome 7 between genetic markers D7S527 and D7S530. Fine mapping of this region using additional markers showed a peak multipoint LOD score of 6.62 in a 3.8 cM interval between D7S2425 and CFTR (the gene coding for cystic fibrosis). A LOD score of [3]3.3 indicates significant linkage in complex disorders (Terwilleger and Ott 1994).

However, it should be noted that the language disorder segregating in the KE family is not typical of SLI and is characterized by deficient grammatical ability and expressive language, and grossly defective articulation manifest as a severe speech dyspraxia (Vargha-Khadem and Passingham 1990) and orofacial dyspraxia (Vargha-Khadem et al. 1995) in addition to language delay. Gopnik (1990) and Gopnik and Crago (1991) have suggested that the disorder is an impairment of word generation, specifically of word endings in accordance with grammatical rules, and have cited the linkage as evidence supporting genes related to specific aspects of grammar processing (Maynard Smith and Szathmary 1995). However, there is also evidence of extensive cognitive impairment in affected family members, with nine individuals having IQs below 82 (Vargha-Khadem et al. 1995). Brain imaging studies of this family (Watkins et al. 1999) suggest functional abnormalities in areas of the frontal lobe related to motor activity and also observe anatomical abnormalities in several brain regions, including the neostriatum.

Subsequently, Lai and colleagues (2001) published strong evidence that a mutation in a forkhead-domain gene underlies the development of SLI in this family. They identified an unrelated individual who showed the same speech and language impairment as that observed within the family and who had a chromosomal translocation (Lai et al. 2000) involving the linked region. The translocation was balanced and reciprocal between chromosomes 5 and 7 (q22;q31.2). The group went on to identify a gene within the region of the breakpoint which showed significant homology with the DNA-binding domain of the forkhead/winged-helix (FOX) family of transcription factors which they designated FOXP2. They further localized the translocation breakpoint to a 200-bp region in the intron between exons 3b and 4 in the translocated individual. However, a G-2-A neucleoteid transition in Exon 14 was found cosegregate with SLI in the KE family. The DNA mutation results in an arginine-to-histidine substitution in the forkhead DNA-binding domain of FOXP2 and may therefore disrupt its DNA-binding or transactivation properties. The FOXP2 gene is expressed in the brain and members of the forkhead family are known to regulate embryogenesis, which implies that the observed mutation in the FOXP2 gene may lead to abnormal development of neural structures important for speech and language.

The relationship between SLI and specific reading disability (SRD)

The acquisition of spoken language is a common human feature and is likely to have been subject to the direct pressures of evolution. In contrast processing written language, a more recent feature, is unlikely to have been the subject of selection pressure. However, there is considerable evidence that speaking and reading difficulties are related, which may reflect the action of common biological processes. For example, there is evidence that children with pre-school SLI or who have diagnosed language disorders when reading tuition begins, are at greater risk of developing reading disability (Catz 1993; Scarborough and Dobrich 1990; Silva *et al.* 1985).

Specific reading disability (SRD)

Understanding how language is processed in the brain is one of the major challenges for neuroscience research today. Identifying genes that contribute to the processes underlying the way written language is manipulated cognitively may provide important insights into how language is processed in general. Specific reading disability (SRD), or developmental dyslexia, is characterized as a gross difficulty in reading and writing that is not attributable to a general intellectual impairment or to a lack of exposure to an appropriate educational environment. Specific reading disability is usually characterized as a deficit of at least two years in reading age, compared to that predicted from chronological age. However, as with SLI, measures of reading ability differ considerably between studies, and, as yet, there are no universal operational definitions of reading disability to allow accurate generalizability.

A variety of cognitive components have been implicated in the development of SRD. These include deficits of visual processing (Stein 1993), deficits in the language processing system (Shankweiler *et al.* 1979), and deficits in temporal processing (Stein and Walsh 1997). However, most evidence suggests that deficits in phonological processing are central to the development of SRD. The functional unit of phonological processing is the phoneme, the smallest discernible segment of speech. Phonological processing includes phoneme awareness, decoding, storage and retrieval. Another component of the reading process is the visual appearance or shape (orthographic content) of a written word (Olson *et al.* 1994). The speed at which language-based information is processed may also be of importance (Wolf and Bowers 1999). Indeed our own work (Robinson *et al.* unpublished) has identified three factors, that describe severe SRD (children with at least two and a half years deficit in reading age compared to chronological age; N = 250) as comprising phonologic, orthographic and rapid-naming dimensions. Many of the genetic studies of SRD have attempted to characterize components of reading disability alongside measures of general reading ability (e.g. Grigorenko *et al.* 1997, 2000; Fisher *et al.* 1997, 1999; Gayan *et al.* 1999).

Family studies

From the very first systematic studies of SRD (Hinshelwood 1907) the tendency for SRD to reoccur in families was noted. Numerous studies since have supported this observation (see Hallgren 1950; Owens *et al.* 1971; Yule and Rutter 1975; Walker and Cole 1965). In an early family study of SRD, Rutter and colleagues (1970) observed that 34 per cent of children with specific reading retardation had a parent or sibling with a reading problem compared to 9 per cent of control children. Four of the major family studies undertaken have reported consistently high sibling recurrence risks of 40.8 per cent (N = 174, Hallgren, 1950), 42.5 per cent (N = 40, Finucci *et al.* 1976), 43 per cent (N = 168, Vogler *et al.* 1985) and 38.5 per cent (N = 52, Gilger *et al.* 1991).

Gilger and colleagues (1991) studied the probability of a mother or father being affected, given an affected or control male or female offspring, in three separate population samples from Iowa and Colorado, USA. They found that for male probands the risk in fathers ranged between 30 per cent and 35 per cent and between 12 per cent and 15 per cent for mothers, whereas the risk for female probands to fathers ranged from 17 per cent to 41 per cent and from 30 per cent to 42 per cent for mothers. The risk of SRD in fathers or mothers of a normal control proband was 4 per cent and 3 per cent respectively, which is slightly lower than but roughly in keeping with estimates of the population frequency of SRD, which range between 5 per cent and 10 per cent (Pennington 1990). Thus the relative risk (λs) for SRD in parents is between 4 to 10.

Twin studies

Early twin studies of SRD showed significantly greater MZ than DZ concordance for specific reading disability but suffered from methodological problems, especially ascertainment bias and inconsistent definitions of SRD (Bakwin 1973; Zerbin-Rudin 1967). The first compelling evidence that the high familiality of SRD was due to genetic, rather than shared, environmental factors came in the 1980s. During this time, two important twin studies were published: The Colorado Twin Reading Study (DeFries and Fulker 1987; DeFries *et al.* 1987) and the London Twin Study (Stevenson *et al.* 1987). Both provided strong evidence for the role of genes in SRD.

These studies differed in some aspects of method. The Colorado study used twins where one member of each twin pair was reading disabled, whereas the London study selected twins from the normal population who represented the full range of reading ability. Nevertheless, there was convergent evidence from the two studies about the heritability of reading and spelling. The twins in the Colorado study were obtained from schools, where at least one member of the twin pair was suspected by their teachers of having a reading disability. The reading ability of both twins was then assessed by a standard battery of tests that had been shown previously to discriminate between reading-disabled and normal readers.

The London study (Stevenson *et al.* 1987) was of 285 twin pairs identified from the general population on the basis of birth records and from the registers of schools in the London area. Although the number of reading-disabled twins was far fewer than was available in the Colorado study, the London study was able to investigate genetic and environmental influences on the full range of reading and spelling ability.

The Colorado study found substantial heritable components to reading (44 per cent) and spelling (62 per cent) and highly heritable deficits in phonological processing as indicated by nonword reading (75 per cent), but did not detect a significant heritable component for orthographic processing (31 per cent). Common family environment was also shown to have a significant influence (48 per cent). The London Twin Study, in contrast, found no evidence supporting heritability of reading per se. However, it produced strong evidence for some aspects of reading disability. Phonological coding (non-word reading) showed heritability of 82 per cent where probands had evidence of specific reading disability, and homophone recognition showed heritability of 67 per cent in the normal range, which was not significant when probands showed evidence of SRD. Stevenson (1991) also observed significant heritability for impaired spelling (62 per cent) and a greater pattern of heritability for deficits in phonological processing than for deficits in orthographic coding.

Olson and colleagues extended their analysis of phonological and orthographic dimensions within the Colorado Twin Study (Gayan et al. 1994, 1997; Olson et al. 1994). They observed significant heritability for orthographic coding (56 per cent), which was approximately the same as that observed for phonological coding (59 per cent). Similarly, Castles and colleagues (1999) studied orthographic and phonological subgroups that were adjusted to be independent of overall reading ability to provide a clearer measure of associated skills. Their study involved 592 pairs of twins—272 MZ and 320 DZ—in which at least one twin fell within the SRD category. Children whose subtype scores fell in the top third of the distribution (i.e., their standardized scores on the phonological processing measure were higher than their scores on the orthographic processing measure) were allocated to the orthographic group. Children whose subtype scores fell in the bottom third of the distribution were allocated to the phonological dyslexia group. They observed a highly significant heritable component for word recognition (67 per cent) for the phonological subgroup that became more significant for the extreme form of the subtype (lower fifth of the distribution, heritability = 78 per cent). They also found a significant heritable component for word recognition deficit in the orthographic group (heritability of 31 per cent) although, interestingly, a greater environmental component to word recognition deficit was observed in the orthographic group.

Other factors have also been shown to affect heritability of specific reading disability. In a further analysis the Colorado dataset, DeFries, Alarcon and Olson (1997) showed that reading had a higher heritability in younger than older children, but that the heritability of spelling was maintained across all ages and indeed, there was evidence that it actually increased with age. Second, again using the Colorado twin study, Wadsworth, Olson, Pennington and DeFries (1998) found there was a significant difference ($p < 0.03$) between the heritabilities for children with higher IQs (0.72) than children with below average intelligence (0.43), indicating that environmental factors play a greater role in reading difficulties experienced by children with a lower IQ.

Sex has also been shown to moderate genetic effects. Knopnik, Alacron and DeFries (1998) found that the genetic correlation between the sexes was significantly less than 1 (0.59), which can be taken to indicate that the genes that influence SRD in males are not identical to those operating in females. Finally, in attempting to analyse the complex patterns of genetic influences on reading Hohnen and Stevenson (1999) were able to show that the genes influencing general intelligence and also general language ability were also related to individual differences in the early stages of reading ability. Although they did not find a direct genetic relationship between phonological ability and literacy, they found that these two aspects of cognitive development were associated via shared genetic influences on general language and via a specific shared environmental influence.

Mode of transmission

Pennington and colleagues (1991) have performed the most extensive complex segregation analysis of SRD. They examined the relatives of SRD probands in four independent samples from the states of Colorado (rural and urban), Washington, and Iowa, producing a sample of 204 families and 1,698 individuals. Their results supported a major locus transmission in three of the four samples and a polygenic transmission in the fourth. In the first three samples the estimates of penetrance of the AA, Aa and aa genotypes (where A is the abnormal allele) were, respectively, 1, 1 and 0.001 to 0.039 in males and 0.56 to 1, 0.55 to 0.897, and 0 in females. The estimated gene frequency of the major locus was between 3 per cent and 5 per cent; thus sex influenced, additive or dominant transmission was observed in three out of the four families. One of these family sets were ascertained for linkage and are therefore likely to show bias towards autosomal dominant-like pedigree structures. Furthermore, as outlined in Chapter 2, segregation analysis is far from infallible in complex traits and complex diseases are likely to involve multiple genetic and environmental effects. However, the existence of at least one family, discussed earlier, showing major gene transmission of speech disorder suggests that we should be open to the existence of major genes in reading disorder.

There was also evidence of sex differences in penetrance rates in the analyses of Pennington *et al.* (1991), with females showing lower estimates of penetrance in the autosomal-dominant family. It is possible that this might go some way to explain the slight excess of males observed in SRD (i.e., sex ratio of 1.5:1 male:female, Shaywitz *et al.* 1990; Wadsworth *et al.* 1992), although the higher ratio of males to females could also be attributed to an artefact of clinical ascertainment (Shaywitz *et al.* 1990). Pennington and colleagues (1991) have also shown that the slight excess of males is unlikely to be due to classic X-linked transmission. There was no robust evidence of transmission from fathers to sons, or for mitochondrial transmission, as transmission rates from each parental sex were essentially equal (0.34 for fathers and mothers, Colorado sample), or for imprinting, for the above reasons. Finally there was no observed effect of parental sex on the severity of SRD in the offspring.

Molecular genetics

Studies of SRD have produced strong evidence of linkage to chromosome 6p (Cardon *et al.* 1994, 1995; Fisher *et al.* 1999; Gayan *et al.* 1999; Grigorenko *et al.* 1997, 2000), significant evidence of linkage to chromosome 15 (Grigorenko *et al.* 1997; Schulte-Korne *et al.* 1998; Smith *et al.* 1983), significant linkage to chromosome 2 (Fagerheim *et al.* 1999; Petryshen *et al.* 2000b), and suggestive evidence of linkage to chromosome 1 (Rabin *et al.* 1993). In addition, studies using linkage disequilibrium mapping have fine mapped the region on chromosomes 6 and 15 to areas of approximately 1–6 cM (Morris *et al.* 2000; Turic *et al.* in press), which show good evidence of containing the putative genes.

Cardon and colleagues (1994) were the first to report linkage between SRD and chromosome 6. They detected significant linkage (p = 0.0002) implicating a region at 6p21.3 in a sample of 114 sibling pairs from 19 families and a sample of DZ twins taken from the Colorado Twin Register (50 pairs). Their measure of SRD was based on a composite score of reading ability, which they treated as a quantitative variable to test for linkage. Although the results of the original study were later amended to take account of MZ twins included inadvertently, the linkage remained significant (twins p = 0.009; siblings p = 0.0003: Cardon *et al.* 1995).

Grigorenko and colleagues (1997, 2000) used samples of large multiplex families (six, later expanded to eight) and found significant evidence of linkage to 6p21.3 in a region spanned by markers D6S464 and D6S273. In the 1997 study, they appeared to show strongest evidence of linkage to a phenotype characterized as phonemic awareness, although significant linkage was also observed with phonological decoding and single-word reading. However, their extended study (Grigorenko *et al.* 2000) produced a less clear relationship showing linkage to single-word reading, vocabulary, and spelling with phonemic awareness and phonological decoding showing little evidence of a relationship. However, it must be noted that the evidence for relationships with specific SRD phenotypes or dimensions may be influenced by the methods of ascertainment, phenotypic definition, and the frequency of the phenotypes in the sample studied. For example, in one Grigorenko study, 38 per cent of individuals were categorized as having a phonemic-awareness deficiency while only 18 per cent had a vocabulary deficit. Furthermore, as most of these phenotypes show some correlation with each other, which results in many individuals having numerous phenotypes, clear relationships may be difficult to establish. In addition, studies that have attempted to explore relationships with subphenotypes or phenotypic dimensions have varied in their approaches both to the phenotypic definitions and in the assumptions underlying their analysis, namely whether the phenotype is viewed as a categorical variable, (as is the case in the Grigorenko studies) or whether the phenotype is viewed as a continuous dimension (e.g. by Cardon *et al.* 1994).

Fisher and colleagues (1999) studied a sample of 181 sibling pairs in which one member of each pair was selected for SRD. They observed evidence of linkage with phonological (non-word reading, p = 0.0035) and orthographic processing (irregular-word reading, p = 0.0035), with a peak around markers D6S276 and D6S105. Similarly, Gayan and colleagues (1999) tested 79 sibships including 180 individuals not included in previous studies in which at least one sibling had evidence of SRD. They observed evidence for linkage to orthographic (LOD = 3.1) and phonological (LOD = 2.42) deficits that peaked at marker D6S461 for the former and 2–3 cM proximal to this for the latter. However, Field and Kaplan (1998) and Petryshen *et al.* (2000b) found no evidence of linkage with SRD with markers in the region 6p23–6p21.3 using either qualitative or quantitative measures of the SRD phenotype. The design of their study differed from previous approaches. They used families containing at least two individuals with severe phonological reading deficits, and have speculated that their families may have a different balance of SRD subtypes within them that may not be strongly linked to the region studied. Turic and colleagues have mapped the region of putative linkage (D6S109–D6S1538) for linkage disequilibrium and have identified a region between D6S109 and D6S1260 that showed significant evidence of haplotype association with SRD in two independent samples of parent–proband trios (Turic *et al.* in press). However, moderate evidence of association was also observed around locus D6S258.

Smith *et al.* (1983) were the first to report some evidence of linkage between SRD and chromosome 15. Their study used a limited number of chromosomal heteromorphisms and produced a LOD score of 3.2. Smith and colleagues (1986, 1990) showed that the positive linkage result was due to 20 per cent of the families studied. However, Bisgaard and colleagues (1987) failed to show linkage between SRD and chromosome 15 heteromorphisms. However, the assessment method for SRD was far less rigorous and involved no direct testing, only questionnaires and interviews. In 1997, Grigorenko and colleagues reported significant evidence of linkage (LOD score = 3.15) with marker D15S143 and single-word reading in six extended families, each of which contained four individuals with significant reading disability. No significant linkage was observed with phonological phenotypes, although phonological awareness was found to be linked to chromosome 6 in the same families. Schulte-Korne *et al.* (1998) also reported suggestive evidence of linkage (maximum LOD score = 1.78: D15S132–D15S143) to chromosome 15 for spelling disability in seven multiplex families.

More recently, Morris and colleagues (2000) observed an association with SRD and haplotypes resulting from the combination of microsatellite markers D15S146/ D15S214/ D15S994 using a two-stage, family-based, linkage disequilibrium mapping design, observing a significant haplotypic association (p < 0.001) in both stage one (101 parent-RD proband trios) and stage 2 samples (77 trios). Twenty genes have been identified within this region of association, of which three are involved in regulating neurotransmission and one is a member of the *phosphoinosotinic* gene family with another family member mapping to the candidate region for SRD on chromosome 6.

Other chromosomal regions have been identified as possibly containing genes contributing to SRD. In 1993, Rabin and his colleagues reported suggestive evidence of linkage (Z_{max} = 21.95, 1p34–p36) in a series of families, a proportion of whom had been linked previously to chromosome 15 (Smith *et al.* 1983). Coincidentally, Froster *et al.* (1993) identified a family in whom SRD appeared to cosegregate with a balanced translocation [t(1;2) (p22:q31)], suggesting linkage to a gene on the distal region of 1p or 2q. It is interesting, therefore, that Fagerheim *et al.* (1999) reported linkage in a large Norwegian family to a region with maximum LOD scores of 3.54, 2.92, and 4.32 for three diagnostic models. These findings were the result of a genome-wide search for linkage, using a 20 cM marker map in 36 members of a family of more than 80 identified members and implicated a region on chromosome 2p15–16 that appears to show evidence of replication in an independent sample (Petryshen *et al.* 2000a). However, this region is different from that implicated by the results of Froster and colleagues.

Finally, recent genome scans show preliminary evidence of loci on chromosomes 2, 4, 6, 9, 13 and 18 for components of reading disability (Fisher and Smith 2001).

Conclusions

There is little doubt that genes play a significant role in the development of language disorders, such as specific language impairment and specific reading disability. Indeed, recent findings have identified the first gene which results in a form of language delay, as well as other chromosomal regions which show good evidence of susceptibility genes for reading disabililty. As the pace of this research increases, there is great optimism that more genes will soon be identified and our understanding of the origins of these disorders will improve dramatically.

References

Aoki S. and Asaka A. (1993) Genetic analysis of motor development, language development and some behavior characteristics in infancy. Paper presented at the annual Behavior Genetics Association Meeting, Sydney, Australia.

Arnold G. E. (1961) The genetic background of developmental language disorders. *Folia Phoniatrica* **13**:246–54.

Bakwin H. (1973) Reading disability in twins. *Developmental Medicine and Child Neurology* **15**:184–7.

Bisgaard M. L., Eiberg H., Moller N. *et al.* (1987) Dyslexia and chromosome 15 heteromorphism— negative lod in a Danish material. *Clinical Genetics* **32**:118–19.

Bishop D. V. M. and Edmundson A. (1986) Is otitis media a major cause of specific developmental language disorders? *British Journal of Disorders of Communication* **21**:321–38.

Bishop D. V. M. *et al.* (1995) Genetic basis of specific language impairment: evidence from a twin study. *Developmental Medicine and Child Neurology* **37**:56–71.

Borges-Osorio M. R. and Salzano F. M. (1985) Language disabilities in three twin pairs and their relatives. *Acta Geneticae Medicae et Gemellologiae (Roma)* **34**:95–100.

Byrne B., Willerman L., and Ashmore L. (1974) Severe and moderate language impairment: Evidence for distinctive etiologies. *Behavioral Genetics* **4**:331–45.

Cardon L. R., Smith S. D., Fulker D. W. *et al.* (1994) Quantitative trait locus for reading disability on chromosome 6. *Science* **266**:276–9.

Cardon L. R., Smith S. D., Fulker D. W. *et al.* (1995) Quantitative trait locus for reading disability. *Science* **268**:1553.

Castles A., Datta H., Gayan J. *et al.* (1999) Varieties of developmental reading disorder: Genetic and environmental influences. *Journal of Experimental Child Psychology* **72**:73–94.

Catz H. W. (1993) The relationship between speech-language impairments and reading disabilities. *Journal of Speech and Hearing Research* **36**:948–58.

Dale P. S., Simonoff E., Bishop D. V. M. *et al.* (1998) Genetic influence on language delay in two-year-old children. *Nature Neuroscience* **1**(4):324–8.

De Fries J. C., Alarcon M., and Olson R. C. (1997) Genetic aetiologies of reading and spelling deficits: Developmental differences. In C. Hulme & M. Snowling (eds), *Dyslexia: Biology, cognition and intervention*. (pp. 20–37). London: Whurr.

De Fries J. C. and Plomin R. (1983) Adoption designs for the study of complex behavioural characteristics, in C. L. Ludlow and J. A. Cooper (eds) *Genetic Aspects of Speech and Language Disorders*. New York, Academic Press.

De Fries J. C., Fulker D. W., and LaBuda M. C. (1987) Evidence for a genetic aetiology in reading disability of twins. *Nature* **329**:537–9.

Fagerheim T., Raeymaekers P., Tonnessen F. E. *et al.* (1999) A new gene (DYX3) for dyslexia is located on chromosome 2. *Journal of Medical Genetics* **36**:664–9.

Fenson L., Dale P. S., Reznick J. S. *et al.* (1994) Variability in early communicative development. *Monogr Soc Res Child Development* **59**:1–173.

Field L. L. and Kaplan B. J. (1998) Absence of linkage of phonological coding dyslexia to chromosome 6p23-p21.3 in a large family data set. *Am J Human Genet* **63**:1448–56.

Finucci J. M., Guthrie J. T., Childs A. L. *et al.* (1976) The genetics of specific reading disability. *Annual Review of Human Genetics* **40**:1–23.

Fisher S. E., Vargha-Khadem F., Watkins K. E. *et al.* (1998) Localisation of a gene implicated in a severe speech and language disorder. *Nature Genetics* **18**:168–70.

Fisher S. E., Marlow A. J., Lamb J. *et al.* (1999) A quantitative-trait locus on chromosome 6p influences aspects of developmental dyslexia. *Am J Human Genet* **64**(1):146–56.

Fisher S. E. and Smith S. D. (2001) *Progress Towards the Identification of Genes Influencing Developmental Dyslexia*. London: Whurr Publishers.

Froster U., Schulte Korne G., Hebebrand J. *et al.* (1993) Cosegregation of balanced translocation (1;2) with retarded speech development and dyslexia. *Lancet* **8864**:178–9.

Fundudis T., Kolvin I., Garside R. (1979) *Speech Retarded and Deaf Children: Their Psychological Development*. New York: Academic Press.

Gayan J., Forsberg H., Olson R. K. (1994) Genetic influences on subtypes of dyslexia. *Behavioral Genetics* **24**:513.

Gayan J., Datta H. E., Castles A. E. *et al.* (1997) The aetiology of group deficits in word decoding across levels of phonological decoding and orthographic decoding. Paper presented at the Annual Meeting of the Society for the Scientific Study of Reading, Chicago, 22–24 March.

Gayan J., Smith S. D., Cherny S. S. *et al.* (1999) Quantitative-trait for specific language and reading deficits on chromosome 6p. *Am J Human Genet* **64**:157–64.

Gilger J. W., Pennington B. F., De Fries J. C. (1991) Risk for reading disability as a function of parental history in three family studies. *Reading and Writing: An Inter Disciplinary Journal* **3**:205–17.

Gopnik M. (1990) Feature-blind grammar and dysphasia. *Nature* 344:715.

Gopnik M. and Crago M. B. (1991) Familial aggregation of a developmental language disorder. *Cognition* **39**:1–50.

Grigorenko E. L., Wood F. B., Meyer M. S. *et al.* (1997) Susceptibility loci for distinct components of developmental dyslexia on chromosomes 6 and 15. *Am J Human Genet* **60**:27–39.

Grigorenko E. L., Wood F. B., Meyer M. S. *et al.* (2000) Chromosome 6p influences on different dyslexia-related cognitive processes: further confirmation. *Am J Human Genet* **66**:715–23.

Hallgren B. (1950) Specific dyslexia: a clinical and genetic study. *Acta Psychiatria et Neurologica Scandinavia* (Suppl. 65).

Hardy-Brown K. (1983) Universals and individual differences: Disentangling two approaches to the study of language acquisition. *Developmental Psychology* **19**:610–24.

Hardy-Brown K. and Plomin R. (1985) Infant communicative development: Evidence from adoptive and biological families for genetic and environmental influences on rate differences. *Developmental Psychology* **21**:378–85.

Hardy-Brown K., Plomin R., and De Fries J. C. (1981) Genetic and environmental influences on the rate of communicative development in the first year of life. *Developmental Psychology* **17**:704–17.

Hay D. A., Prior M., Collett S. *et al.* (1987) Speech and language development on twins. *Acta Geneticae Medicae et Gemellogiae* **36**:213–23.

Hier D. B. and Rosenberger P. B. (1980) Focal left temporal lobe lesions and delayed speech acquisition. *Journal of Developmental and Behavioral Pediatrics* **1**:54–7.

Hinshelwood J. (1907) Four cases of congenital word-blindness occurring in the same family. *British Medical Journal* **1**:608–9.

Hohnen B. and Stevenson J. (1999) The structure of genetic influences on general cognitive, language, phonological and reading abilities. Devel Psychol, **35**:590–603.

Hurst J. A., Baraitser M., Auger E. *et al.* (1990) An extended family with a dominantly inherited speech disorder. *Developmental Medicine and Child Neurology* **32**:352–5.

Ingram T. T. S. (1959) Specific developmental disorders of speech in childhood. *Brain* **82**:450–4.

Knopik V. S., Alarcon M. and De Fries J. C. (1998) Common and specific gender influences on individual differences in reading performance: a twin study. *Personality and Individual Differences* **25**:269–77.

Lai C. S. L. *et al.* (2000) The SPCH1 region on human 7q31: genomic characterization of the critical interval and localization of translocation associated with speech and language disorder. *Am J Hum Genet* **67**:357–68.

Lai C. S. L. *et al.* (2001) A forkhead-domain gene is mutated in a severe speech and language disorder. *Nature* **413**:519–22.

Lewis B. A. (1992) Pedigree analysis of children with phonology disorders. *Journal of Learning Disability* **25**:586–97.

Lewis B. A., Cox N. J., and Byard P. J. (1993) Segregation analysis of speech and language disorders. *Behavior Genetics* **23**:291–7.

Locke J. L. and Mather P. L. (1989) Genetic factors in the ontogeny of spoken language: Evidence from monozygotic and dizygotic twins. *Journal of Child Language* **16**:553–9.

Luchsinger R. (1970) Inheritance of speech deficits. *Folia Phoniatrica* **22**:216–30.

Matheny A. P. and Bruggemann C. E. (1973) Children's speech: Heredity components and sex differences. *Folia Phoniatrica* **25**:442–9.

Mather P. L. and Black K. N. (1984) Heredity and environmental influences on preschool twins' language skills. *Developmental Psychology* **20**:303–8.

Maynard-Smith J. and Szathmary E. (1995) *The Major Transitions in Evolution*. W. H. Freeman: Oxford.

McReady E. B. (1926) Defects in the zone of language (word-deafness and word-blindness) and their influence in education and behavior. *American Journal of Psychiatry*, **6**:267.

Morris D. W., Robinson L., Turic D. *et al.* (2000) Family-based association mapping provides evidence for a gene for reading disability on chromosome 15q. *Human Molecular Genetics* **9**(5):855–60.

Munsinger H. and Douglass A. (1976) The synactic abilities of identical twins, fraternal twins, and their siblings. *Child Development* **47**:40–50.

Neils J. and Aram D. (1986) Family history of children with developmental language disorders. *Perception and Motor Skills* **63**:655–8.

Olson R. K., Forsberg H., and Wise B. (1994) Genes, environment, and the development of orthographic skills, in V. W. Berninger (ed.) *The Varieties of Orthographic Knowledge. 1: Theoretical Developmental Issues.* Dordrecht: Kluwer Academic Publishers.

Osborne R. T., Gregor A. J., and Miele F. (1968) Heritability of factor V: verbal comprehension. *Perceptual and Motor Skills* **26**:191–202.

Owen F., Adams P., Forrest T. *et al.* (1971) Learning disorders in children: Sibling studies. Monographs of the Society for Research in Child Development. *Child Development* **36**(4):.

Pennington B. F. (1990) Annotation: the genetics of dyslexia. *Journal of Child Psychiatry and Psychology* **31**(2):193–201.

Pennington B. F., Gilger J. W., Pauls D. *et al.* (1991) Evidence for major gene transmission of developmental dyslexia. *Journal of the American Medical Association* **266**(11):1527–34.

Petryshen T. L., Kaplan B. L., Hughes M. L. *et al.* (2000a) Evidence for the chromosome 2p15-p16 dyslexia susceptibility locus (DYX3) in a large Canadian data set. *Am J Med Genet (Neuropsychiatric Genetics)* **96**(4):473.

Petryshen T. L., Kaplan B. J., Liu M. F. *et al.* (2000b) Absence of significant linkage between phonological coding dyslexia and chromosome 6p23–21.3, as determined by use of quantitative-trait methods: confirmation of qualitative analyses. *Am J Human Genet* **66**:708–14.

Plomin R. (1986) *Development, Genetics, and Psychology.* Hillsdale, NJ: Lawrence Erlbaum Associates.

Plomin R., DeFries J. C., and Fulker D. W. (1988) *Nature and Nurture During Infancy and Early Childhood.* New York: Cambridge University Press.

Rabin M., Wen X. L., Hepburn M. *et al.* (1993) Suggestive linkage of developmental dyslexia to chromosome 1p34–p36. *Lancet* **342**:178.

Randall D., Reynell J., and Curwen M. (1974) A study of language development in a sample of 3 year-old children. *British Journal of Disorders of Communication* **9**:3–16.

Rice M. L. (1997) Specific language impairments: in search of diagnostic markers and genetic contributions. *Mental Retardation and Developmental Disabilities Research Reviews* **3**(4):350–7.

Rutter M. E., Simonoff E., and Silberg J. (1993) How informative are twin studies of child psychopathology? In T. J. Bouchard and P. Propping (eds) *Twins as Tools of Behavioural Genetics.* Chichester, England: Wiley, 179–94.

Rutter M., Tizard J., and Whitmore K. (1970) *Education, Health and Behaviour.* London: Longmans.

Samples J. M. and Lane V. W. (1985) Genetic possibilities in six siblings with specific language learning disorders. *ASHA* **27**:27–32.

Scarr S. and Carter-Saltzman L. (1983) Genetics and intelligence, in J. Fuller and E. Simmel (eds) *Behavior Genetics: Principles and Applications.* Hillsdale, NJ: Lawrence Erlbaum Associates.

Scarborough H. S. and Dobrich W. (1990) Development of children with early language delay. *Journal of Speech and Hearing Research* **33**:70–83.

Schulte-Korne G., Grimm T., Nothen N. M. *et al.* (1998) Evidence for linkage of spelling disability to chromosome 15. *Am J Human Genet* **63**:279–82.

Shankweiler D., Liberman I. Y., Mark L. S. *et al.* (1979) The speech code and learning to read. *Journal of Experimental Psychology: Human Learning and Memory* **5**:531–45.

Shaywitz S. E., Shaywitz B. A., Fletcher J. M. *et al.* (1990) Prevalence of reading disability in boys and girls. *Journal of the American Medical Association* **264**:998–1002.

Silva P. A., McGee R., and Williams S. (1985) Some characteristics of 9-year-old boys with general reading backwardness or specific reading retardation. *Journal of Child Psychology and Psychiatry* **26**:407–21.

Smith S. D., Kimberling W. J., Pennington B. F. *et al.* (1983) Specific reading disability—identification of an inherited form through linkage analysis. *Science* **219**:1345.

Smith S. D., Pennington B. F., Kimberling W. J. *et al.* (1986) Genetic heterogeneity in specific reading disabilities. *Am J Human Genet* **39**:169a.

Smith S. D., Pennington B. F., Kimberling W. J. *et al.* (1990) Genetic linkage analysis with specific dyslexia: use of multiple markers to include and exclude possible loci, in G. T. Pavlidis (ed.) *Perspectives on Dyslexia*. Vol. 1. *Neurology, Neuropsychology and Genetics.* Chichester, West Sussex: John Wiley and Sons.

Spitz R. V., Tallal P., Flax J. *et al.* (1997) Look who's talking: A prospective study of familial transmission of language impairments. *Journal of Speech and Hearing Research* **40**(5):990–1001.

Stein J. F. (1993) Visuospatial perception in disabled readers, in D.M Willows, R. S. Kruk, E. Corcos (eds) *Visual Processes in Reading and Reading Disabilities.* Hillsdale, NJ: Lawrence Erlbaum.

Stein J. and Walsh V. (1997) To see but not to read: the magnocellular theory of dyslexia. *Trends in Neuroscience* **20**:147–52.

Stevenson J. (1991) Which aspects of processing text mediate genetic effects? *Reading and Writing* **3**:249–69.

Stevenson J. and Richman N. (1976) The prevalence of language delay in a population of 3 year-old children and its association with general retardation. *Developmental Medicine and Child Neurology* **18**:431–41.

Stevenson J., Graham P., Fredman G. *et al.* (1987) A twin study of genetic influences on reading and spelling ability and disability. *Journal of Child Psychiatry and Psychology* **28**: 229–47.

Stromswold K. (1994) The nature of children's early grammar: evidence from inversion errors. Paper presented at the LSA, Boston, MA, in M. L. Rice (ed.) *Toward a Genetics of Language.* Mahwah, NJ: Lawrence Erlbaum Associates.

Tallal P., Ross R., and Curtiss S. (1989) Familial aggregation in specific language impairment. *Journal of Speech and Hearing Disorders* **54**:167–73.

Terwilliger J. D. and Ott J. (1994) *Handbook of Human Genetic Language.* Baltimore, Maryland: The Johns Hopkins University Press.

Tomblin B. (1989) Familial concentration of developmental language impairment. *Journal of Speech and Hearing Disorders* **54**:287–95.

Tomblin B. J. (1996) Genetic and environmental contributions to the risk for specific language impairment, in M. L. Rice (ed.) *Towards a Genetics of Language.* Mahwah, NJ: Lawrence Erlbaum Associates.

Tomblin J. B. and Buckwalter P. R. (1998) Heritability of poor language achievement among twins. *Journal of Speech and Hearing Research* **41**:188–99.

Tomblin J. B., Freese P. R., and Records N. L. (1992) Diagnosing specific language impairment in adults for the purpose of pedigree analysis. *Journal of Speech and Hearing Research* **35**:832–43.

Turic D., Robinson L., Duke M. *et al.* (2002) Linkage disequilibrium mapping provides further evidence for a gene for reading disability on chromosome 6p21.3–22. *Molecular Psychiatry.* In press.

Van der Lely H. K. and Stollwerck L. (1996) A grammatical specific language impairment in children: An autosomal dominant inheritance? *Brain and Language* **52**:484–504.

Vargha-Khadem F. and Passingham R. E. (1990) Speech and language defects. *Nature* **346**:226.

Vargha-Khadem F., Watkins K., Alcock K. *et al.* (1995) Praxic and nonverbal cognitive deficits in a large family with a genetically transmitted speech and language disorder. *Proceedings of the National Academy of Science USA* **92**:930–3.

Vogler G. P., De Fries J. C., and Decker S. N. (1985) Family history as an indicator of risk for reading disability. *Journal of Learning Disabilities* **18**:419–21.

Wadsworth S. J., De Fries J. C., Stevenson J. *et al.* (1992) Gender ratios among reading-disabled children and their siblings. *Journal of Child Psychiatry and Psychology* **33**:1229–39.

Wadsworth S. J., Olson, R. K., Pennington, B. F. *et al.* (1998) Differential genetic etiology of reading disability as a function of IQ. *Behavior Genetics* **28**:483–4.

Walker L. and Cole E. (1965) Familial patterns of expression of specific reading disability in a population sample. *Bulletin of the Orton Society* **15**:12–24.

Watkins K. E., Gadian D. G., and Vargha-Khadem F. (1999) Functional and structural brain abnormalities associated with a genetic disorder of speech and language. *Am J Human Genet* **65**:1215–21.

Wolf M. and Bowers P. G. (1999) The double-deficit hypothesis for the developmental dyslexias. *Journal of Educational Psychology* **91**(3):415–38.

Yule W. and Rutter M. (1975) The concept of specific reading retardation. *Journal of Child Psychiatry and Psychology* **16**:181–97.

Zerbin-Rudin E. (1967) Congenital word-blindness. *Bulletin of the Orton Society* **17**:47–54.

Chapter 7

Childhood disorders

Anita Thapar and Jane Scourfield

Introduction

As recently as a decade ago, it might have been considered reasonable to make the accusation that child psychiatry, unlike other medical specialties, was slow in embracing genetics research. This is clearly not the case now. There has been a rapidly escalating interest in conducting genetic studies of childhood psychopathology, and this has undoubtedly been accompanied by a marked increase in our knowledge base. There is now a wealth of evidence from family and twin studies, not only demonstrating that many child psychiatric disorders run in families but that genes play an important role in contributing to this familial transmission. For conditions such as autism and attention deficit hyperactivity disorder (ADHD) that have been shown to be strongly influenced by genetic factors, the search for susceptibility genes has attracted many researchers. Molecular genetics research in child psychiatry is undoubtedly an exciting development, and the identification of susceptibility genes will represent a major stepping stone towards unravelling risk mechanisms, identifying other important aetiological factors (both risk and protective) and developing new methods of treatment. Nevertheless, although identifying and locating specific susceptibility genes is one goal, it must be appreciated that genetic research strategies provide a means of examining aetiology at a much broader level. There is now an increasing recognition that genetic research designs can and must be utilized to examine the contribution of and interaction with other types of risk factors. New research findings on childhood depression and conduct disorder for example, have been particularly exciting in providing empirical evidence that although environmental risk factors matter, their impact can be moderated by genes (gene–environment interaction [Rutter and Silberg 2002]) and they may often go hand in hand with genetic risk for psychopathology (gene–environment correlation; Ge et al. 1996; O'Connor et al. 1998) thereby leading to a 'double dose' risk effect.

Findings from genetic studies are also having an impact on clinical understanding of many disorders. For example at one time, before the neurobiological basis of autism was accepted, it was thought that autism might have been the consequence of 'refrigerator parents' and a culture of blaming parent's caregiving for the child's psychopathology prevailed. Genetic study findings have provided clear evidence that such an attitude, apart from being unhelpful, is also incorrect.

In this chapter we review current findings from genetic studies of the major psychiatric disorders of childhood and adolescence.

Autism

Autism is a disorder characterized by abnormalities in communication, reciprocal social interaction and restricted, repetitive patterns of behaviour that has onset in early childhood. 'Infantile autism' was originally described in 1943 by Kanner, an American child psychiatrist who initially considered that the disorder was constitutional. However, subsequently for many years it was presumed that genetic influences were of little importance (Hanson and Gottesman 1976). Modern genetics research has dispelled this belief and there is now a wealth of evidence showing the importance of genetic factors for autism. More recent research into the genetic basis of autism has undoubtedly been strengthened by the availability of standardized methods of assessment, for example instruments such as the Autism Diagnostic Interview (ADI-R) (Lord *et al.* 1994) and the Autism Diagnostic Observation Schedule (ADOS) that have been used across different studies. In contrast to many other childhood psychiatric disorders, classic autism is a condition that nearly always results in affected children presenting to clinical services rather than remaining undetected in the community. This has meant that family studies of referred children are more likely to provide representative samples. Moreover twin studies of autism are unusual within the field of child and adolescent psychiatry in having been based on clinical cases rather than on non-referred populations, as has been the rule with most other diagnoses.

Family studies

Family studies have shown that between 2 per cent and 6 per cent of siblings of those with autism are similarly affected (Smalley *et al.* 1988; Bailey *et al.* 1996). Reported prevalence rates of autism in the general population have varied widely from 0.7 to 21.1/10,000 with a recent review of 23 epidemiological surveys of autism yielding a median prevalence estimate of 5.2/10,000 and demonstrating that prevalence rates have increased in more recent surveys (Fombonne 1999). Nevertheless overall given that the population prevalence for autism is relatively low (although with each new survey, reported rates seem to be higher), the relative increased risk to siblings of those with classic autism is between 30 and 120 times.

Family studies have also consistently shown that relatives of affected individuals display increased communication and social difficulties (e.g. Bolton *et al.* 1994; Piven *et al.* 1997; Bailey *et al.* 1998), and it appears that those types of deficits that are milder manifestations of narrowly defined autism have a true familial relationship to autism. That is, they represent an extended broader phenotype of autism. There have also been reports of elevated rates of depression and anxiety as well as cognitive impairments, including reading and spelling difficulties and cognitive dysfunction assessed using specific tasks, in relatives (e.g. Bailey *et al*. 1998; Delong and Nohria 1994; Piven *et al*. 1997; Bolton *et al*. 1998; Murphy *et al*. 1998; Hughes *et al*. 1999). However family study findings for these phenotypes are conflicting, and at present there is insufficient support for these characteristics to be considered as manifestations of a broader autistic phenotype in themselves (Bailey *et al*. 1998; Szatmari *et al*. 1998; Fombonne *et al*. 1997).

Twin studies

There have been three published population-based twin studies of autism, which as mentioned earlier have the merit of having been based on twins who fulfilled clinical diagnostic criteria for autism (Folstein and Rutter 1978; Steffenburg *et al*. 1989; Bailey *et al*. 1995). The original Folstein and Rutter (1978) series was subsequently enlarged and studied in greater detail by Bailey *et al*. (1995). These studies (see Table 7.1) have shown that the concordance rates for autism are very much higher in MZ twins (36 per cent to 91 per cent) than amongst DZ twins (0 per cent) with heritability estimates of above 90 per cent. One other twin study (Ritvo *et al*. 1985) found an MZ concordance rate of 95 per cent and a DZ concordance rate of 23 per cent, but has methodological problems in that it was based on a non-systematically ascertained sample of twins. This provides convincing evidence that autism is not only familial but is one of the most heritable psychiatric disorders. Two of the twin studies additionally provided further support for the concept of a broader autism phenotype, in that concordance rates for communication and social difficulties were found to be much higher for MZ twin pairs than for DZ twin pairs (Folstein and Rutter 1978; Bailey *et al*. 1995). These results, together with family study findings, suggest that the autism phenotype, at least when considered in terms of genetic liability, extends beyond traditional diagnostic criteria to include milder manifestations of the same core features as full-blown autism. Genetic studies which focus on probands with the broader phenotype and other types of pervasive developmental disorder, for example Asperger's syndrome, are awaited.

Table 7.1 Epidemiologically based twin studies of autism

Author	Population	Sample size	Diagnostic criteria	MZ pairwise concordance rate	DZ pairwise concordance rate
Folstein & Rutter 1977	UK	11 MZ, 10 DZ	Kanner and Rutter criteria defined autism	36%	0%
Steffenburg *et al*. 1989	Scandinavia	11 MZ, 10 DZ	DSM-III-R autistic disorder	91%	0%
Bailey *et al*. 1995*	UK	17 MZ, 11 DZ	ICD-10 autism	69%	0%

* New sample only

Heterogeneity

An important question to consider when embarking on molecular genetic studies is whether the autism phenotype can be dissected out in genetically meaningful ways. First, it is clearly of importance to identify recognizable genetic anomalies that are associated with autism. These are of additional interest in the hope that the identification of such anomalies might shed light on potential candidate regions and genes. Although there are some single gene defects associated with autism, most notably tuberous sclerosis and fragile X syndrome, these occur in less than 5 per cent of those with autism (Bailey *et al.* 1993; Smalley 1998; Fombonne 1999). There has also been interest in associated chromosomal abnormalities, but no consistent pattern has emerged and there have been individual reports of anomalies associated with autism from virtually every chromosome (Lauritsen *et al.* 1999). The one exception is chromosome 15. There have been several reports of chromosome 15 anomalies, in particular, partial duplications of chromosome 15 (Baker *et al.* 1994; Flejter *et al.* 1996; Gillberg *et al.* 1991; Hotopf and Bolton 1995) and a replicated association between autism and a maternally derived chromosome 15q duplication (Cook *et al.* 1997; Schroer *et al.* 1998). This appears a promising lead, although it has not yet led to the identification of a specific susceptibility gene in the region of interest using candidate gene association studies.

In the majority of affected individuals who have no specific recognized genetic defects, it has been difficult to demonstrate whether autism can be subdivided in a meaningful way. Early findings suggested that there might be an increased familial loading in those with severe mental retardation (August *et al.* 1981). However this has not been supported by more recent work, which instead suggests that familial loading may differ according to whether or not there is a lack of useful language (Starr *et al.* 2000; Pickles *et al.* 2000).

The difficulty in delineating genetic heterogeneity at a phenotypic level is illustrated by the finding from twin data that there appears to be as much variability in symptom expression within MZ twin pairs as between MZ pairs (LeCouteur *et al.* 1996). This suggests that non-genetic influences play an important part in determining the pattern of phenotypic expression. Thus using phenotypic data to establish clinical subtypes that are meaningful at a genotypic level has not, so far, proved possible.

Molecular genetic studies

There is now considerable international effort focused on the search for susceptibility genes for autism, and preliminary results are emerging. The published results of an initial full genome scan from the International Molecular Genetic Study of Autism Consortium (1998) showed a maximum lod score of 2.53 in a region on chromosome 7q in 87 affected sib-pair families. The 7q finding was replicated in another genome-wide scan based on 51 families from the Paris Autism Research International Sibpair Study (MLS = 0.83; Philippe *et al.* 1999) with 3 other regions (of the total 11 regions yielding nominal p values of 0.05 or over) overlapping with regions identified from the first study (2q, 16p, and 19p). The Collaborative Linkage Study of Autism (1999) showed a lod score of 2.2 in this region although genome screens undertaken by two different groups based on 90 families (Risch *et al.* 1999) and 95 families (Buxbaum *et al.* 2001) failed to show similar findings.

The statistically most significant linkage findings from these genome scans are summarized in Table 7.2.

Identifying one susceptibility locus will be particularly important in facilitating the detection of other loci, given that genetic epidemiological evidence suggests that autism is likely to be influenced by the epistatic effect of at least several (Pickles *et al.* 1995) if not many (Risch *et al.* 1999) different genes. There have also been a few reported candidate gene studies. So far most attention has focused on candidate genes in the chromosome 15q region mentioned earlier, but there has also been interest in other candidate genes such as the promoter region of the 5HT transporter gene and HOXA1 and HOX1B. However findings of association, linkage and linkage disequilibrium for polymorphisms within these candidate genes have so far been conflicting and inconclusive or remain unreplicated (e.g. Cook *et al.* 1997, 1998; Klauck *et al.* 1997; Salmon *et al.* 1999; Maestrini *et al.* 1999; Ingram *et al.* 2000).

In summary, there is robust evidence that autism is a highly heritable condition and emerging linkage findings on chromosome 7q look promising, although not entirely consistent. Recent linkage disequilibrium mapping of the chromosome 7q locus undertaken by the International Molecular Genetic Study of Autism Consortium (2001) has suggested two regions of association. It remains to be seen whether these findings are further replicated by other groups and how quickly such findings can lead to the identification of a specific susceptibility locus.

Table 7.2 Most significant linkage findings from full genome screens of autism MLS value

Chromosome	IMGSAC 1998	MLS value Philippe *et al.* 1999	Risch *et al.* 1999	CLSA 1999	Buxbaum *et al.* 2001
7q	2.53*	0.83	0.93	2.2	
6q		2.23*			
1p			2.15*		
13q				3.0*	
2q					1.96*[a]

IMGSAC International Molecular Genetic Study of Autism Consortium

CLSA Collaborative Linkage Study of Autism

* = Most significant linkage region from each genome screen

[a] = Maximal multipoint HLOD score of 2.99 for subgroup of those with autism plus delayed onset of phrase speech

Childhood schizophrenia

Compared with the literature on adult onset schizophrenia, which is reviewed in Chapter 10, there is a relative dearth of evidence regarding childhood schizophrenia. Indeed, it was not until the 1970s that the diagnosis was recognized as being separate from infantile autism and other early onset disorders which today would probably not be regarded as including psychotic symptoms (Kolvin 1971). DSMIII and ICD9 were the first diagnostic systems to include a separate category of infantile autism and adopt the current practice of using the same criteria to diagnose schizophrenia in children and adults.

Family and twin studies

Early onset of schizophrenia might result from an increased genetic load or a greater environmental insult. An increase in genetic loading might therefore be expected to result in increased familiality. This has, indeed, been observed in schizophrenic males with onset of psychosis before 17 (Pulver *et al.* 1990) and a large Swedish family study of schizophrenia showed that earlier age of onset in the proband was associated with an increased risk of schizophrenia in the relatives (Sham *et al.* 1994). The small number of family studies of childhood schizophrenia have so far not demonstrated higher rates of disorder among relatives than is found for adult probands. Kallman and Roth (1956) found rates of schizophrenia of 8.8 per cent (12.5 per cent age corrected) among the parents of their sample of childhood schizophrenics, and of 12.2 per cent among siblings. A more recent study (Gordon *et al.* 1994) using structured interviews and best estimate diagnostic methods has reported an overall rate of 13 per cent for non-affective psychotic illness among the first degree relatives of subjects with childhood-onset schizophrenia. These rates are comparable to those found among relatives of adult schizophrenics. However, when a subgroup of childhood onset schizophrenics with premorbid speech and language impairments were examined, the rate of non-affective psychotic disorder among relatives rose to 21 per cent (Nicolson *et al.* 2000). To date, then, there is not conclusive support for an increased risk to relatives of childhood-onset as opposed to adult-onset schizophrenics although those with premorbid speech and language impairment show greater familial loading.

We are not aware of any adoption studies in the area of childhood onset schizophrenia.

There has been one twin study of childhood schizophrenia which warrants discussion. Kallman and Roth (1956) recruited a sample of 52 twins and 50 singletons under age 15 from admissions to a hospital in the State of New York. Diagnoses were made by a single clinician. Pairwise concordance rates for monozygotic and dizygotic twins were 70.6 per cent and 17.1 per cent respectively, rates which are comparable to those found in adults, although the MZ concordance rate is slightly higher, possibly a result of the clinical sampling frame which can inflate the genetic effect. There was also an unexpectedly large proportion of same sex DZ pairs which might have been due to the mistaken identification of MZ pairs as DZ. This would have increased DZ and lowered MZ concordances, tending to reduce estimates of heritability.

Molecular genetic studies

A recent report (Gordon *et al.* 1994; Yan *et al.* 2000) described a balanced translocation between chromosomes 1 and 7 in a boy with childhood onset schizophrenia. The breakpoints were at p22 on chromosome 1 and q21 on chromosome 7. The report is interesting given a previous case of chromosomal rearrangement involving chromosomes 1, 7, and 21 in a 6-year-old autistic boy where the break point on chromosome 1 was also 1p22 (Lopreiato *et al.* 1992). A small number of case reports have described early autistic symptoms followed by development of schizophrenia in adolescence and early adulthood. Whilst the full syndromes of autism and schizophrenia appear to be distinct, as Gordon *et al.* (1994) suggest, it is possible that a subgroup of those with childhood schizophrenia and those with autism share a similar genetic abnormality. This observation makes the break points of chromosomes 1 and 7 likely sites for further molecular genetic studies.

Another report from the same group (Usiskin *et al.* 1999) has identified three cases of childhood-onset schizophrenia with a deletion of chromosome 22q11 (velocardiofacial syndrome) among a series of 47 cases. This is a rate higher than than in the general population and showed a trend towards being higher than in adult schizophrenics. However, the sample size is small and the result warrants replication in a larger sample. Additionally, the rate of cytogenetic abnormalities among childhood-onset schizophrenics has been found to be 10.6 per cent, significantly higher than in the general population (Nicholson *et al.* 1999) and 4 (6.1 per cent) of 66 patients with childhood onset schizophrenia or DSMIIIR psychotic disorder not otherwise specified were found to have sex chromosome anomalies, significantly more than expected from general population rates (Kumra *et al.* 1998).

An association study from Canada (Maziade *et al.* 1977) is the only one we are aware of which includes data from a sample of childhood onset schizophrenics. This group tested for an allelic association between schizophrenia and a marker at the dopamine D3 receptor locus in a sample from Eastern Quebec. The sample was divided into cases with onset before and after the age of 17 and was compared with controls. Positive association was found only among the adult-onset group and not among those with childhood-onset, supporting the authors' hypothesis that extreme age of onset may identify a homogeneous subgroup of schizophrenia. The finding of an association between D3 and adult schizophrenia had earlier been reported by groups in Wales and France, with homozygosity at this locus approximately doubling the risk of schizophrenia although findings have been mixed (see Chapter 10).

In conclusion, such evidence as exists suggests that childhood-onset schizophrenia is subject to genetic influence, but the nature of this influence and whether it is different from that in the adult onset disorder is unclear.

Hyperkinetic Disorder/Attention Deficit Hyperactivity Disorder

Attention Deficit Hyperactivity Disorder (ADHD) (the DSM-IV diagnostic category) affects between 2 per cent and 6 per cent of children, although prevalence figures vary according to the diagnostic criteria used (Costello *et al.* 1996). The prevalence rate for the more narrowly defined ICD-10 Hyperkinetic Disorder is 1 in 200 (Taylor *et al.* 1991). Phenotypic definition is a particularly important issue when considering the genetic basis of ADHD. Apart from transatlantic differences in diagnostic practice, it remains uncertain as to whether ADHD is best conceptualized as a dimensional measure or as a category and whether it is preferable to rely on parental ratings of the child's symptoms, teacher reports or both. Despite this debate on how the phenotype should be defined, there has been much scientific interest in the genetic basis of ADHD.

Family studies

Family studies have shown an increased risk of ADHD in the families of children with ADHD (whether defined using DSM-III or DSM-IIIR diagnostic criteria) with reported relative risks (λ) of between 4 and 5.4 for first degree relatives (Biederman *et al.* 1992; Faraone *et al.* 2000). Relatives of children with ADHD, in a referred US population at least, also show increased rates of other psychiatric disorders, in particular conduct disorder, depression, learning difficulties and bipolar disorder (Faraone *et al.* 1992; Faraone *et al.* 2000). Although family study findings suggest a comparatively modest recurrence risk of ADHD in first degree relatives, there is some evidence to suggest that the relative risk of ADHD may be greater for relatives of probands with ADHD that persists into adolescence and adult life ($\lambda = 17.2$) and those with comorbid conduct disorder and bipolar disorder ($\lambda = 26.2$) (Faraone *et al.* 2000).

Twin studies

There have been at least twelve published twin studies of ADHD related traits and categories based on non-clinical samples from the UK, US, Australia, and Norway (Goodman and Stevenson 1989; Thapar *et al.* 1995; Edelbrock *et al.* 1995; Schmitz *et al.* 1995; Gjone *et al.* 1996; Levy *et al.* 1997; Sherman *et al.* 1997; Eaves *et al.* 1997; Neuman *et al.* 1997; Hudziak *et al.* 2000; Thapar *et al.* 2000; Kuntsi and Stevenson 2001). These have consistently shown the importance of genetic influences on ADHD symptom scores, with reported heritability estimates of between 60 per cent and 88 per cent for parent ratings and between 39 per cent and 72 per cent for teacher ratings (Thapar *et al.* 1999). Although there is consistent evidence of a substantial genetic contribution to normal variation in ADHD symptom scores, a critical issue is whether the genetic aetiology differs at the extreme end of the dimension. However it appears that even extreme scores are highly heritable, with reported heritability estimates ranging between 75 per cent and 91 per cent (Stevenson 1992; Gjone *et al.* 1996; Levy *et al.* 1997). There is also consistent evidence of a genetic contribution when ADHD is defined categorically (Goodman and Stevenson 1989; Sherman *et al.* 1997; Thapar *et al.*

2000) although two of these studies (Goodman and Stevenson 1989; Thapar *et al.* 2000) generated categories by using cut-offs on questionnaire measures.

A puzzling finding is that many twin studies (Goodman and Stevenson 1989; Thapar *et al.* 1995; Eaves *et al.* 1997), have shown that the DZ twin correlation for ADHD symptom scores is extremely low and in some instances even negative. This could be explained by twin competition effects that would make them more dissimilar, or by rater effects. There is evidence to suggest that these findings are likely to be be due to maternal rating contrast effects whereby mothers exaggerate differences between their twins (Simonoff *et al.* 1997). Non-additive genetic influences such as genetic domin-ance, epistasis and gene–environment interaction could also be contributory factors although these would be unlikely to account for negative DZ twin correlations. However low DZ twin correlations have not been reported in all twin studies (e.g. Levy *et al.* 1997; Thapar *et al.* 2000) and thus rater contrast effects may vary depending on the instrument used (Thapar *et al.* 2000a).

Adoption studies

Although published adoption studies of ADHD are much less recent than the twin studies and have some methodological drawbacks such as small sample size, non-systematic ascertainment or the failure to use standardized measures or diagnostic criteria, overall the findings have been consistent in showing the importance of genetic factors. Biological parents of hyperactive children appear to show higher rates of hyperactivity and ADHD (Morrison and Stewart 1973; Cantwell 1975; Faraone *et al.* 2000) and poorer performance on cognitive measures of attention (Alberts-Corush *et al.* 1986) than adoptive relatives. Similarly in a study of separately fostered siblings, in accordance with expectations of a genetic aetiology, hyperactive children showed greater concordance with their biological siblings than their half siblings (Safer 1973).

Comorbidity

The overlap of ADHD and Conduct Disorder is a particularly important one, given that these disorders frequently coexist in both clinical and non clinical populations. Indeed ICD-10 recognizes a separate subtype of Hyperkinetic Conduct Disorder. Family study findings have shown that ADHD and Conduct Disorder are co-transmitted in families, and thus appear to share a common familial aetiology which could of course be genetic or environmental in origin (Faraone *et al.* 1991; Biederman *et al.* 1991; Faraone *et al.* 1998). Three twin studies have further shown that ADHD and con-duct problems, when defined as symptom scores (Silberg *et al.* 1996; Kuntsi and Stevenson, 2001; Thapar *et al.* 2001) and broad categories (Thapar *et al.* 2001), are influenced by a common genetic aetiology.

Family data additionally suggest that children who show both ADHD and conduct problems may represent a somewhat distinct subcategory which is either quantitatively or qualitatively different from pure ADHD (Faraone *et al.* 1991; Biederman *et al.* 1992;

Faraone *et al.* 1998) and is associated with increased familial loading (Faraone *et al.* 2000). Recent twin evidence suggests that the subgroup with both ADHD behaviours and conduct problems may represent a category with greater genetic loading (Thapar *et al.* 2001).

The overlap of reading disability (RD) and ADHD behaviours has also been examined (see Chapter_). Twin studies suggest that this comorbidity is mainly explained by a shared genetic aetiology, particularly for RD and symptoms of inattention (Wilcutt *et al.* 2000).

Molecular genetic studies

Molecular genetic findings for ADHD are now beginning to emerge. So far most research groups have adopted a candidate gene association approach and on the basis that at least 70 per cent of children with ADHD/Hyperkinetic Disorder show a rapid symptomatic improvement when given psychostimulants e.g. methylphenidate, there has been particular interest in genes involved in dopaminergic pathways.

To date most attention has focused on the dopamine transporter gene DAT1 and the dopamine receptor DRD4 gene. There is a theoretical appeal in considering DAT1 as a candidate gene, given that methylphenidate inhibits the dopamine transporter mechanism (Amara and Kuhar 1993) and DAT1 knockout mice show motor overactivity (Caron 1996). The initial positive findings of an association of ADHD with DAT1(Cook *et al.* 1995) have so far (by February 2001) been independently replicated by three groups (Gill *et al.* 1997; Daly *et al.* 1999 (extension of Gill *et al.*'s sample); Waldman *et al.* 1998; Curran *et al.* 2001). However negative findings have also been reported by at least three other groups (Swanson *et al.* 1998; Holmes *et al.* 2000; Palmer *et al.* 1999). Heterogeneity appears to be one likely explanation, although differences in methods of ascertainment and measurement are also likely to be important. A meta-analysis of published and unpublished data is currently underway.

The findings for DRD4 have been more consistent (see Table 7.2) in that so far there have been seven published reports of positive findings from family-based and case control studies (LaHoste *et al.* 1996; Swanson *et al.* 1998; Smalley *et al.* 1998; Faraone *et al.* 1998; Muglia *et al.* 2000; Tahir *et al.* 2000; Sunohara *et al.* 2000). However three groups have reported positive case control findings yet negative TDT results (Rowe *et al.* 1998; Holmes *et al.* 2000; Mill *et al.* 2001) and there have also been some negative reports (Eisenberg *et al.* 2000; Hawi *et al.* 2000; Castellanos *et al.* 1998; Kotler *et al.* 2000). A meta-analysis of data from groups participating in the International ADHD Molecular Genetics Network—which includes unpublished work—suggests that the DRD4 7 repeat allele confers susceptiblity (although small) to ADHD with an overall estimated odds ratio of 1.9 (1.5–2.2) from case control studies and 1.4 (CI; 1.1–1.6) from family-based studies (Faraone *et al.* 2001).

Negative and some positive association findings for polymorphisms from other candidate genes involved in dopamine, serotonin, adrenergic, and degradation

Table 7.3 Molecular genetic studies of DAT1[a] and ADHD (data published or known to be in press before 1.2.01)

Authors	Sample (n) and origin	Design	Association/ linkage finding	Chi Square	odds ratio/rel risk (where reported)	p value
Cook et al. 1995	n = 49 USA	family-based (HHRR)	positive	7.29		0.007
Gill et al. 1997 extended sample-	n = 40 Ireland n = 118 (includes	family-based (HRR) family-based (HHRR)	positive	6.07		0.014
Daly et al. 1999	original 40)			7.5	1.29	0.006
Waldman et al. 1999	n = 117 USA	family-based (TDT) quantitative TDT	positive positive	6.53[b] t = 1.87[c]	2.75	<0.05 0.032
Palmer et al. 1999	n = 209 USA	family-based (TDT)	negative		not reported	0.80
Holmes et al. 2000	n = 133 UK n = 108	case control family-based (TDT)	negative negative	— —	0.96 0.89	

[a] = candidate allele-3' region VNTR 480bp repeat/allele 10

[b] = for 'high severity ADHD'

[c] = hyperactive-impulsive symptoms (not inattentive)

Table 7.4 Molecular genetic studies of DRD4[a] and ADHD (data published or known to be in press before 1.2.01)

Authors	Sample (n) and origin	Design	Association Linkage finding
La Hoste, et al. 1996	n = 39 USA	case control	positive
Castellanos, et al. 1998	n = 41 USA	case control	negative
Swanson et al. 1998	n = 52 USA	family-based (HRR)	positive
Rowe et al. 1998	n = 70 USA	case control family-based (TDT)	positive negative
Smalley et al. 1998	n = 133 USA	family-based (TDT)	positive
Faraone et al. 1999	n = 27 adults USA	family-based (TDT)	positive
Tahir et al. 2000	n = 111 Turkey	family-based (TDT)	positive
Holmes et al. 2000	n = 129 UK n = 110	case control family-based (TDT)	positive negative
Hawi et al. 2000	n = 99 Eire	family-based (HHRR)	negative
Eisenberg et al. 2000	n = 46 Israel	family-based (HRR)	negative
Muglia et al. 2000	n = 66 adults Canada n = 44	case control family-based (TDT)	positive negative (positive trend)
Kotler et al. 2000	n = 49 Israel	family-based (HRR)	negative
Mill et al. 2001	n = 132 UK n = 85	case control family-based (TDT)	positive negative
Sunohara et al. 2000	n = 88 Canada n = 59 USA-new cases* plus n = 52 from Swanson et al. study 1998 USA	family-based (TDT)	positive (negative* new USA cases alone)

[a] = candidate allele: exon 4 VNTR 48bp, − 7 repeat allele

[b] = see text for pooled odds ratios for DRD4 replicated with permission from Faraone et al. 2001
  sample characteristics obtained from original published papers

pathways are being published at a rapid rate, although none of these findings have been consistently replicated at the time of this book going to press.

In summary, there is consistent and robust evidence that ADHD, whether defined in terms of symptom scores or as a broadly defined category, is highly heritable. Molecular genetic studies are underway and initial results for DRD4 have been very promising. There is clearly a need for further work on DRD4 examining other polymorphisms and exploring phenotype issues more carefully. Findings that consistently emerge for other candidate loci are also awaited.

Conduct/oppositional behaviour

There has been increasing awareness in recent years that genetic influences may have a role in the development of conduct problems. In common with most of the disorders considered in this book, it is unlikely that there will turn out to be 'genes for' conduct problems. Rather, the path from genes to behaviour is likely to be a complex one involving environmental stresses as well as individual differences in liability. Genetic and environmental influences need not necessarily be seen as competing explanations; distal and proximal influences on behaviour can be different in kind and the factors which influence the origins of behavioural inclinations may be quite different from those which precipitate or maintain actual behaviour (Goldsmith and Gottesman 1996). There may be basic tendencies of behaviour such as impulsiveness, sensation seeking or cognitive style which are partially under genetic influence and make the probability of certain behaviours higher in some individuals, given a particular set of environmental circumstances.

Defining phenotypes for genetic analysis

The boundaries are blurred between conduct disorder, oppositional defiant disorder, delinquency, criminality, aggression, and antisocial personality disorder in adults. Although there is overlap between conduct disorder and delinquency the two are differently defined, one being a clinical syndrome and the other a legal definition. Much of the literature uses general measures of antisocial behaviour rather than categorical definitions, but it is likely that either approach would subsume several different phenotypes and that antisocial behaviour in general represents a heterogeneous group. For example, Moffitt (1993) makes a distinction between adolescence-limited and life course persistent antisocial behaviour, suggesting that the adolescence-limited variety is a reflection of secular trends while the life course persistent group is characterized by more neuropsychological problems and environmental risks. This is supported by reviews of the genetic literature concluding that adult criminality and juvenile delinquency have somewhat different roots (McGuffin and Gottesmann 1985; Dillalla and Gottesman 1989). The comorbidity between conduct disorder and hyperactivity may represent another grouping (Farrington 1995), with early-onset hyperactivity being a step on the developmental path to conduct disorder which is different from some other pathways. A predominance of aggressive symptoms represents another possible distinction, with some evidence suggesting that aggressive antisocial behaviour is more heritable than non aggressive antisocial behaviour (Edelbrock et al. 1995; Eley et al. 1999).

In any investigation of behaviour or disorder in children, information will need to be obtained from carers and preferably teachers too in order to assess the child in the different domains of home and school. For conduct problems, as in other types of behaviour, there is evidence that different informants bring different biases and only modest agreement (Simonoff et al. 1995). This is relevant to genetic studies, as estimates of heritability can vary with informant. Using composite scores incorporating data from several informants is one approach to phenotypic measurement, although clearly the reasons for informant differences need to be better understood.

Family and adoption studies

The demonstration that a trait or disorder clusters in families is a prerequisite for any genetic research. However, such an observation cannot distinguish between environmental and genetic causes of similarity among relatives but leads on to adoption and twin studies, which can. There is evidence from both community and clinical samples that conduct problems cluster in families (Szatmari et al. 1993; Faraone et al. 1997).

Adoption studies suggest that both genes and environment contribute to the clustering of antisocial behaviours (Mednick et al. 1984; Van den Oord et al. 1994) and have also highlighted that antisocial behaviour is a phenotype which shows clear evidence of gene–environment interaction and correlation. A recent adoption study by Cadoret et al. (1995) examined 287 adoptees and, using symptom scores, demonstrated that biological offspring of parents with antisocial personality disorder showed significantly increased aggression and conduct disorder when placed in an adverse adoptive home environment, compared with similar offspring who were not placed in adverse adoptive environments. Another adoption study has shown that adoptees at risk of antisocial behaviour based on their biological mothers' reports of their own antisocial behaviour were consistently more likely to evoke more negative parenting from their adoptive parents than other adoptees (O'Connor et al. 1998). This suggests that genes and environment are interacting in a non-additive way in the development of conduct disorder.

Twin studies

To date there have only been two published twin studies of conduct disorder as a diagnostic category (Eaves et al. 1997; Slutske et al. 1997) and only one of oppositional defiant disorder per se (Eaves et al. 1997). Another has used symptoms of DSM-III conduct disorder (Lyons et al. 1995). Most investigators have used the alternative approach of symptom checklists which give scores for conduct problems/antisocial symptoms. From a large number of studies, estimates of heritability have varied from virtually zero (Thapar and McGuffin 1996) to 0.94 (Ghodsian-Carpey and Baker 1987) when the phenotype was aggression in younger children. There should not be too much emphasis on the actual values of the different parameters, but the consensus from this literature is that there is considerable (although not unanimous) support for genetic influences on conduct problems in children and adolescents, both as a disorder and as a continuous dimension of behaviour. The other important finding is the frequency with which shared environmental influences have been found to be significant. Conduct problems show the largest effect of shared environment of any behaviour disorder (Plomin et al. 1997).

Issues complicating genetic studies of conduct/antisocial behaviours

There are several features of this type of behaviour which complicate genetic studies.

1 Gene x environment interaction: as mentioned above, there is evidence from family studies (Cadoret *et al.* 1995) that genes and environment interact (not merely co-act) in the genesis of conduct problems, a phenomenon which is worthy of note because standard twin analysis assumes no significant effect from this type of non additive effect.

2 Assortative mating: this refers to non-random selection of a mate so that parents are more alike than expected for a particular trait. There is evidence from population samples that assortative mating occurs for antisocial behaviour (Krueger *et al.* 1998), an effect which might tend to inflate estimates of shared environmental influence and bias any genetic effects downwards in the basic twin model. Nevertheless it has been suggested that such biases in twin studies are likely to be small (Maes *et al.* 1998).

3 Twin imitation: Rowe (1985), in his study of adolescent twins, found that twins tended to imitate each other's antisocial behaviour and to be 'partners in crime'. This would tend to make them more similar and would appear as a shared environmental effect. It is possible to take account of this type of effect using 'sibling interaction' models (Carey 1986) but this is an issue that has been largely ignored so far in the conduct disorder genetic literature.

4 Gene–environment correlation: genes and environment may be correlated in as much as genes influence our behaviour, and our behaviour affects how we interact with the environment. With regard to conduct problems, there is evidence from adoption studies that the children of parents with antisocial personality disorder display hostile and aggressive behaviour which evokes more harsh/inconsistent parenting from their adoptive family, an example of gene–environment correlation (Ge *et al.* 1996). The importance of gene–environment correlations is thought to increase as the individual develops (Scarr and McCartney 1983), with related individuals becoming more similar over time and their environment being selected by behaviour which is genetically influenced. This might appear as an increasing genetic influence with age, a phenomenon which has been seen in genetic studies of conduct problems and is outlined in the following section.

5 Changing influences with development: there is some evidence that genetic influences on conduct problems may increase as children develop and grow up. In a recent large study of an American sample of adult male twins, Lyons *et al.* (1995) found a considerable genetic influence on antisocial symptoms in adulthood but no significant genetic influence on retrospectively reported childhood symptoms of conduct disorder. This was in keeping with the earlier review by McGuffin and Gottesman in which they concluded that adult criminality showed genetic influences whilst juvenile delinquency

showed largely environmental influences (McGuffin and Gottesmann 1985). Additionally, findings from the Cardiff Twin Study suggest that whilst conduct problems in younger children are probably due entirely to environmental factors, similar problems in adolescents have a significant genetic influence (Thapar and McGuffin 1996; McGuffin and Thapar 1997).

6 Sex differences: epidemiological studies in child psychiatry have consistently found higher rates of conduct disorder in boys. There is some evidence of different genetic influences on this type of behaviour in boys and girls, with a report from the Virginia Twin Study of Adolescent Behavioural Development finding that in younger children, genetic influences appeared greater in boys whilst shared and non-shared environmental influences were more important in girls (Silberg *et al.* 1994).

Conclusion

Existing genetic studies of conduct problems using family, adoption, and twin methods point fairly consistently to a genetic influence on this type of behaviour. However, the use of different measures and samples means that the size of genetic and environmental influences vary between studies. A complex phenotype such as this is likely to involve genes and environment in its development, with complex interplay between the two. There will not be 'genes for' bad behaviour, but it is possible that our genetic make up influences our levels of activity or our interpretation of the world around us which, when combined with certain environmental stresses, makes conduct problems more likely.

Tourette's syndrome

Classic Tourette's syndrome (TS) is a rare disorder with a prevalence of 4–5 per 10,000 (Apter *et al.* 1993). The condition most commonly begins in childhood or adolescence and is characterized by multiple motor and vocal tics. However simple tics are much commoner, while ADHD, obsessive-compulsive behaviours and conduct problems are common comorbid findings and the extent to which these different symptoms should be considered as part of the Tourette's syndrome phenotype is controversial. Family and twin data provide useful information on the diagnostic boundaries of this condition. Genetic epidemiological studies have suggested that chronic tics and obsessive-compulsive symptoms should be included in its definition, which would make it a more common disorder, tics and OCD having a prevalence of around 2 per cent (Rapoport *et al.* 1994).

Family studies

It is well documented that relatives of patients with TS show increased rates of the disorder (Nee *et al.* 1980; Kidd *et al.* 1980; Pauls *et al.* 1981; Pauls *et al.* 1984; Pauls *et al.* 1991). In addition to this finding, there is evidence that chronic tics and obsessive-compulsive disorder are increased in the relatives of TS probands with females tending to express OCD and males to express chronic tics or TS (Pauls *et al.* 1991; Santangelo *et al.* 1996).

This suggests that symptoms of OCD and chronic tics should be included in the TS phenotype for genetic studies. To complicate matters further, there have been reports of increased rates of attention deficit disorder in the relatives of TS probands (Pauls *et al.* 1993; Knell and Comings 1993; Comings and Comings 1990a), and one group advocates the inclusion of a wide range of disorders within the extended TS phenotype including affective disorders (Comings and Comings 1990b) drug and alcohol abuse (Comings and Comings 1990c) and agoraphobia with panic attacks (Comings and Comings 1987). However the evidence is probably not strong enough to support such a considerable broadening of the TS concept.

Twin studies

Case report of pairs of twins concordant for TS have appeared in the literature (Jenkins and Ashby 1983; Vieregge, *et al.* 1988) but the most compelling evidence of genetic influence on the syndrome comes from a study of 43 pairs of same sex twins, in which at least one member of the pair had TS (Price *et al.* 1985). Concordance rates for TS were 53 per cent for MZ twins and 8 per cent for DZ using narrow diagnostic criteria. When diagnostic boundaries were broadened to include any tics in the co-twin, the concordance rates rose to 77 per cent and 23 per cent for MZ and DZ twins. These results suggest a considerable genetic effect for TS, although some caution is required since the twins were ascertained through membership of the Tourette's Association, a non-systematic sampling strategy.

Molecular genetics

Several groups have carried out segregation analyses to attempt to determine the mode of inheritance. Findings have differed between studies with some suggesting the likelihood of an autosomal dominant gene, others finding evidence for a mixed model of inheritance, i.e. a major locus against a multifactorial background (Walkup *et al.* 1996) and others ruling out the possibility of Mendelian transmission (Seuchter *et al.* 2000).

Although some segregation analyses findings are consistent with a major locus effect, the results of molecular genetic studies have so far been disappointing (Barr and Sandor 1998). Whole genome screens for linkage in multiplex families and affected sib pairs have not yet demonstrated definite replicated findings (Barr *et al.* 1999; The Tourette Syndrome Association International Consortium for Genetics 1999). There has also been interest in a variety of candidate genes, particularly those involved in dopamine pathways (e.g. Barr *et al.* 1999; Devor *et al.* 1998). However there have been no consistent replications of findings of association or linkage. Genetic heterogeneity and difficulties in defining the phenotype are likely to be important contributory factors. Given that linkage strategies have so far been unsuccessful, it may be that future whole genome linkage disequilibrium and association studies will prove more fruitful if susceptibility loci of smaller effect size are to be successfully detected.

Anxiety

The literature on the role of genetic influences on anxiety in children is relatively sparse. Family studies have shown that there is a degree of familial aggregation for anxiety disorders in children (Rutter *et al.* 1999), for example, the relatives of children with anxiety disorders show significantly higher rates of these disorders than do children of controls (Last *et al.* 1991). The children of parents with depression and an anxiety disorder show increased rates of both disorders, particularly if the parental anxiety disorder is panic disorder (Weissman *et al.* 1984), suggesting that the type of anxiety disorder in a parent may influence disorder in the child. However, it is not clear that there is familial aggregation for specific subtypes of anxiety, the likelihood of psychopathology among relatives of anxious children extending to other types of psychopathology such as depression and alcohol abuse (Bell-Dolan *et al.* 1990).

Twin studies

Parent-reported anxiety symptoms in children and adolescents appear to be moderately genetically influenced. In the Cardiff twin study (Thapar and McGuffin 1995), based on parental questionnaire data, the heritability of manifest anxiety was 59 per cent. These results are consistent with those of the later and larger Virginia Twin Study (VTSABD)which found similar estimates of heritability based on parental questionnaire (Eaves *et al.* 1997).

However twin study findings for self reports of anxiety are more mixed. In the Cardiff twin study, when the adolescents were the informants no significant genetic effects were detected, with shared environment accounting for similarity between twins. Shared environmental influences were also found to be significant on child self-reports of anxiety (in boys only) in the Virginia twin study (Eaves *et al.* 1997). In two studies based on the State-Trait Anxiety Inventory for Children, the results were conflicting. Shared environmental influences were found to contribute to trait anxiety in one study (Eley and Stevenson 1999) but only state, not trait anxiety, in the other study (Legrand *et al.* 1999). However the three latter twin studies are consistent in showing small-modest genetic influences on self-reported manifest and trait anxiety. In the Virginia twin study, the genetic influences on manifest anxiety were stronger in girls than boys and also showed age effects. Genetic effects accounted for 37 per cent of the variance in child-reported symptoms of overanxious disorder in boys and girls, but did not have a significant effect on symptoms of separation anxiety (Topolski *et al.* 1997).

Twin studies also suggest that depressive and anxiety symptoms share a common genetic aetiology, and this is considered further in the section on childhood depression.

Conclusion

The evidence regarding genetic influences on anxiety in childhood is mixed. The results of family studies suggest that these disorders do aggregate within families, but the extent to which this is due to genetic similarity is unclear. Twin studies suggest genetic influences on some measures of anxiety, although parameter estimates vary with informant, and it seems that the genetic liability to anxiety overlaps with that for depression.

Depression

Depressive disorder affects approximately 0.5 per cent of children and 2–6 per cent of adolescents, with follow-up studies showing that there are strong continuities between childhood depression and depressive disorder in later adult life (Harrington *et al.* 1991; Lewinsohn *et al.* 1999; Weissman *et al.* 1999). Although it might be supposed that early-onset disorder represents a more heritable variant, there have been no twin or adoption studies of clinical depressive disorder in childhood and adolescence. Thus remarkably little is known about the genetic basis of depression in a younger age group. Many genetic studies, particularly twin and adoption studies, have adopted a dimensional approach to defining depression. There is empirical evidence to support this approach in so far as high symptom scores are often associated with impairment and predict later disorder (Gotlib *et al.* 1995; Harrington and Clark 1998) although clearly high scores cannot be equated with categorically defined clinical disorder. Another important issue that distinguishes research of depression in younger people is that it is not clear whether it is preferable to rely on parent or child ratings of depressive symptoms and how to interpret findings for these different raters.

Family studies

There have been two main approaches that have been used to examine the familial transmission of childhood depression. First there are those studies in which the offspring of depressed parents have been examined ('top-down' studies) and second those that have focused on the adult relatives of depressed children ('bottom-up' studies). Although some of the published family studies have suffered from methodological drawbacks, most notably, non-blind assessments of relatives and the use of non-psychiatric control groups, findings have been remarkably consistent across studies (Rice *et al.* 2002).

There is good evidence that the children of depressed parents show higher rates of depression (e.g. Weissman *et al.* 1997; Radke Yarrow *et al.* 1992; Hammen *et al.* 1990; Mufson *et al.* 1992; Warner *et al.* 1995) even when compared to the offspring of parents with other psychiatric disorders and when followed up over time (Weissman *et al.* 1997). Children of depressed parents also appear to show increased rates of anxiety, particularly when there is high familial loading for depressive disorder (Warner *et al.* 1999). Indeed follow up studies suggest that childhood anxiety disorder, at least in families where depression is clustering, indexes high familial liability (that could be common familial adversity or genetic in origin) and may represent a developmental precursor to depression (Rende *et al.* 1999).

Studies of adult relatives of depressed children and adolescents have similarly shown increased rates of depressive disorder in first degree relatives (e.g. Mitchell *et al.* 1989; Puig Antich *et al.* 1989; Harrington *et al.* 1993; Goodyer *et al.* 1993; Williamson *et al.* 1995; Kovacs *et al.* 1997; Neuman *et al.* 1997). A recent review of family studies of depression in childhood and adolescence (Rice *et al.* 2002) showed a pooled odds ratio of 3.98 for depression in relatives of affected children and in the offspring of depressed parents compared to normal controls and an overall odds ratio of 1.7 to 2.3 when psychiatric controls were used (Rice *et al.* 2002).

Findings from some, but not all, family studies have suggested that early-onset depression (onset <20 years old, Weissman *et al.* 1992; child vs. adult probands; Neuman *et al.* 1997) represents a more familial variant. However the estimates of familial risk based on family studies of childhood depression (Rice *et al.* 2002) are very similar to findings from a recent meta-analysis of family study data for adult depression (odds ratio = 2.84; Sullivan *et al.* 2000 (see Chapter 9). Further examination of childhood (or prepubertal) onset and adolescent (or postpubertal) onset depression has yielded mixed findings. Two studies have showed no difference in the rate of major depressive disorder in the relatives of children with prepubertal depression than among relatives of those with postpubertal depression (Harrington *et al.* 1997; Wickramaratne *et al.* 2000). However findings from a follow-up study (Weissman *et al.* 1999) showed an increased prevalence of mood disorders among relatives of children with recurrent prepubertal depression compared with relatives of those with non-recurrent prepubertal depression. This led to the suggestion that prepubertal recurrent depression may represent a familial variant, although this clearly needs to be further examined.

Finally in one family study, lower rates of depression were noted in the relatives of depressed children with conduct disorder and suicidality (Puig-Antich *et al.* 1989). The London follow-up study (Harrington *et al.* 1997) also found that lower rates of continuity with adult depression amongst the group of depressed children with comorbid conduct disorder. These findings suggest that children with major depression and comorbid conduct disorder may represent a distinct variant of depression, but this needs to be further examined.

Overall it can be concluded that depressive disorder in childhood and adolescence is highly familial, and that there are strong familial links between depression in early and adult life. Findings on the further distinction between prepubertal and postpubertal onset depression are mixed and so far inconclusive. Although it has been suggested that childhood onset depressive disorders may represent subforms with greater genetic loading, family study findings do not allow for such a conclusion (Wickramaratne *et al.* 2000). Twin studies are needed to examine to what extent familial transmission is accounted for by genes and shared environmental factors.

Twin and blended family studies

Published twin studies of childhood depression have differed from family studies, in that the twins have been ascertained from non-clinical populations and depression has been conceptualized as a dimensional measure rather than as a diagnostic category. To date there have been no twin studies of clinical depressive disorder in children and adolescents.

Overall, most twin studies in which depression has been measured using depression specific questionnaires or interviews have shown that depression symptom scores are genetically influenced. Heritability estimates have varied widely from between 15 per cent to 80 per cent for self-rated symptom scores and 30–80 per cent for maternal ratings (Wierzbicki *et al.* 1987; Rende *et al.* 1993; Thapar and McGuffin 1994; Edelbrock *et al.* 1995; Schmitz *et al.* 1995; Eley 1997; Eaves *et al.* 1997; Gjone and Stevenson 1997). Most of these studies have also suggested an important contribution of non-shared environmental influences.

Twin study findings have highlighted additional issues which need to be further considered. First, in the large population based Virginia Twin Study of Adolescent Behaviour and Development, heritability estimates were considerably higher for parent reports than self reports, although this has not been a consistent finding (Thapar and McGuffin 1994). The reasons for this are not clear, although it has to be considered that parental biases in completing depression questionnaires could potentially result in inflated heritability estimates.

Second, the results of three different studies suggest that the contribution of genetic factors varies with age, in that depressive symptoms appear to be more heritable in adolescence than in childhood (Thapar and McGuffin 1994, 1996; Silberg *et al.* 1999; Eley *et al.* 1999). However in one of the twin studies this increased heritability was only detected in adolescent females (Silberg *et al.* 1999), and in another study high heritability estimates were only found for adolescent males (Eley *et al.* 1999). This type of age effect has been shown for other phenotypes e.g. antisocial behaviour (see earlier in this chapter), IQ (Chapter 4) and could be due to developmental changes associated with puberty that result in for example, the switching on of genes. Another more likely explanation that seems increasingly plausible is that gene–environment correlation (e.g. the selection of high risk environments) increases through adolescence and this manifests as increased heritability estimates (see later). However findings on age effects have been mixed. Two studies found genetic influences seemed to be less important in older children (Gjone and Stevenson 1997; O'Connor *et al.* 1998). These studies differed in that the study by Gjone and Stevenson 1997 examined internalizing symptom scores rather than pure depression and O'Connor *et al.* utilized a longitudinal mixed family and twin study design.

Third, as mentioned earlier, there have been no twin studies of clinical disorder and although normal variation in depressive symptom scores may be influenced by genetic factors, the aetiology of clinical depression may differ. This issue has been partly addressed by those studies in which the DeFries and Fulker method of analysis has been used to examine the genetic aetiology of extreme symptom scores. The results of two studies based on self rated CDI (Children's Depression Inventory) scores suggest that genetic influences are less important for extreme scores in children and adolescents. Heritability estimates were considerably lower for extreme scores on the CDI (20–23 per cent) than for normal variation in symptom scores (34–48 per cent) and the contribution of shared environment was substantially higher (26–44 per cent vs. 4–10 per cent) (Rende *et al.* 1993; Eley 1997). Similarly in a combined twin and step-family, although internalizing rather than depressive symptoms were examined, once again, heritability estimates for extreme scores were found to be lower (Deater Deckard 1997). Although high symptom scores cannot be equated with clinical disorder, most of these findings rather argue against the suggestion that early-onset depression represents a more heritable subform. However results have not been consistent. In a Norwegian twin study, higher levels of internalizing symptoms were found to be more heritable (Gjone *et al.* 1996). Overall the results are puzzling given that adult studies of depression have convincingly shown the importance of genetic factors (Chapter 9), and that family and follow up studies show clear links between depression in children and adolescents and in adults. It may be that high depressive symptom scores are too non-specific (for example will include children with other disorders such as conduct disorder) to draw conclusions about the genetic aetiology of disorder and that closer attention needs to be given to whether the sample includes children or adolescents, the persistence and recurrence of symptoms, and patterns of comorbidity. Given these uncertainties there is clearly a need for a twin study of adolescent depressive disorder.

Finally twin studies have been used to further examine the overlap of depressive symptoms with other types of symptoms. We discussed earlier how family studies of childhood depression have suggested a common familial influence on anxiety and depression. Twin studies which have further investigated the genetic basis to the comorbidity of depression and anxiety symptoms have shown that common genetic influences appear to influence both parent-rated (Thapar and McGuffin 1997) and self-rated depressive and anxiety symptoms (Eley and Stevenson 1999). These results are strikingly similar to those from studies of adult twins (Chapter 9). The overlap of externalizing/conduct or antisocial symptoms and depression has also been examined. In keeping with family study findings that suggested lower familiality for depression comorbid with conduct disorder, in the study by Gjone and Stevenson (1997), genetic influences were found to be less important for comorbid internalizing and externalizing symptoms (extreme scorers) than for pure internalizing problems. Findings on covariation of depression and antisocial symptom scores are mixed, however, with one study showing a common genetic influence accounting for most of the covariation (O'Connor *et al.* 1998) and another finding a common shared environmental factor primarily contributing to the overlap (Gjone and Stevenson 1997).

Adoption studies

There have been no adoption studies of clinical depression in children and adolescents. However adoption study designs have been used to examine the genetic influence on depression symptom scores in non-clinical populations (Van den Oord *et al.* 1994; Eley *et al.* 1998). The reported findings were striking, in that there was no evidence of a genetic contribution to depression symptom scores for either self-reported or parent-reported symptoms and nearly all of the variation in one study was explained by non-shared environmental factors (Eley *et al.* 1998) and by shared environmental influences in the other study (Van den Oord *et al.* 1994). It is not clear at present what accounts for the discrepancy between twin and adoption study findings. Two possibilities have to be considered. First it may be that the higher heritability estimates from twin studies reflect gene–environment correlation and/or gene–environment interaction where the environmental risk is contingent upon the genotype of the biological parent. An alternative explanation is that adoption study findings reported have been based on an age group for whom genetic factors may be less important and are based on selected samples. Clearly, further adoption evidence is greatly needed to further examine this.

Genetic findings on environmental adversity and depression

The importance of environmental adversity for childhood depression has been widely researched and more recently examined within genetically sensitive study designs. There is likely to be increasing interest in examining the action of measured environmental risk factors and the co-action and interaction with genes. So far most interest has focused on two types of risk factors, parent–child relationships and life events.

Twin and adoption study evidence suggests that many factors such as parental warmth, negativity and parent–child conflict are in part influenced by genetic factors (Plomin 1994; Kendler *et al.* 1997; Deater Deckard *et al.* 1999). Twin and blended family study analyses that have been used to further examine the relationship between parental conflict–negativity and depressive symptoms suggests that the association is partly mediated by genes (Neiderhiser *et al.* 1999). To some extent this may be due to the influence of a common perception of both relationships and symptoms (Neiderhiser *et al.* 1998; O'Connor *et al.* 1999) but also because of gene–environment correlation, where the child's genotype influences parental response.

Genetic factors also appear to influence some types of life events in childhood and adolescence (Thapar and McGuffin 1996; Silberg *et al.* 1999). Again there is some evidence to suggest that the association between life events and depressive symptoms in childhood, apart from independent life events, is partly mediated by a common genetic influence (Thapar *et al.* 1998; Silberg *et al.* 1999).

Findings from the Virginia twin study have additionally shown that the magnitude of genetic influences on life events in children increases with age (Silberg *et al.* 1999). These results would support the hypothesis that an increased heritability of depressive symptom scores in adolescence may be accounted for by increased gene–environmental correlation.

As noted in chapter 2, gene–environment interactions are difficult to detect in twin studies of human behaviour and psychopathology, partly because so many environmental risk factors are likely to be influenced by gene–environment correlation which reduces the power to detect interactions (Rutter *et al.* 1999; Andrieu and Goldstein

1998). Nevertheless, in a recent analysis of twin data, it was possible to show that genetic factors moderated the impact of those life events that could be considered as independent of genotype (Silberg *et al.* 2001).

In conclusion, there is good evidence that childhood depression is familial, although insufficient evidence to conclude that either childhood or adolescent depression represents a more strongly familial variant than adult depression. Most twin study findings suggest that depressive symptom scores are heritable. However there are insufficient data to draw conclusions about the role of genetic influences on clinical depression, and further work is needed to examine age effects. Twin and adoption study findings also point to the likely importance of gene–environment correlation and interaction for depression in childhood and adolescence.

Conclusion

In our introduction we highlighted the rapid growth of genetics research and have reviewed some of the major findings that are emerging, There is now strong evidence that neurodevelopmental disorders such as autism and ADHD are highly heritable. Initial linkage (and association) findings for autism and ADHD look promising, although the history of psychiatric genetics research suggests that caution is warranted until findings have been consistently reported. Genetic epidemiological findings for conduct disorder suggest that both genes and environment are important influential factors and illustrate the complexity of the relationship between genetic and environmental influences. Findings on depression and anxiety in childhood are much more mixed, and clearly further genetic epidemiological work is needed to be able to draw firmer conclusions. Overall it is clear that genetics research of child psychiatric disorders is now flourishing and yielding important and interesting findings. It is hoped that increasingly, genetic designs will not be viewed as the sole domain of geneticists but rather as useful tools that can assist us in understanding more about childhood psychopathology.

References

Alberts-Corush J., Firestone P., Goodman, J. T. (1986) Attention and impulsivity characteristics of the biological and adoptive parents of hyperactive and normal control children. *Am J Orthopsychiatry* **56**(3):413–23.

Amara S. G. and Kuhar M. J. (1993) Neurotransmitter transporters: recent progress. *Annu Rev Neurosci* **16**:73–93.

Andrieu N. and Goldstein M. (1998) Epidemiological and genetic approaches in the study of gene-environment interaction: an overview of available methods. *Epidemiologic Reviews* **20**(2):137–47.

Apter A., Pauls D. L., Bleich A. *et al.* (1993) An epidemiologic study of Gilles de la Tourette's syndrome in Israel. *Arch Gen Psychiatry* **50**(9):734–8.

August G. J., Stewart M. A., Tsai, L. (1981) The incidence of cognitive disabilities in the siblings of autistic children. *Br J Psychiatry* **138**:416–22.

Bailey A., Bolton P., Butler L. *et al.* (1993) Prevalence of the fragile X anomaly amongst autistic twins and singletons. *J Child Psychol Psychiatry* **34**(5):673–88.

Bailey A., Le Couteur A., Gottesman I. *et al.* (1995) Autism as a strongly genetic disorder: evidence from a British twin study. *Psychol Med* **25**(1):63–77.

Bailey A., Palferman S., Heavey L. *et al.* (1998) Autism: the phenotype in relatives. *J Autism Dev Disord* **28**(5):369–92.

Bailey A., Phillips W., Rutter M. (1996) Autism: towards an integration of clinical, genetic, neuropsychological, and neurobiological perspectives. *J Child Psychol Psychiatry* **37**(1):89–126.

Baker P., Piven J., Schwartz S. *et al.* (1994) Brief report: duplication of chromosome 15q11–13 in two individuals with autistic disorder. *Journal of Autism and Developmental Disorders* **24**:529–35.

Barr C. L. and Sandor P. (1998) Current status of genetic studies of Gilles de la Tourette Syndrome. *Canadian Journal of Psychiatry* **43**:351–7.

Barr C. L., Wigg K. G., Pakstis A. J. *et al.* (1999) Genome scan for linkage to Gilles de la Tourette Syndrome. *Am J Med Genet (Neuropsychiatric Genetics)* **88**:437–45.

Bell-Dolan D. J., Last C. G., Strauss C. C. (1990) Symptoms of anxiety disorders in normal children. *J Am Acad Child Adolesc Psychiatry* **29**(5):759–65.

Biederman J., Faraone S. V., Keenan K. *et al.* (1992) Further evidence for family-genetic risk factors in attention deficit hyperactivity disorder. Patterns of comorbidity in probands and relatives psychiatrically and pediatrically referred samples. *Arch Gen Psychiatry* **49**(9):728–38.

Biederman J., Rosenbaum J. F., Bolduc E. *et al.* (1991) A high risk study of young children of parents with panic disorder and agoraphobia with and without comorbid major depression. *Psychiatry Research* **37**:333–48.

Bolton P. F., Pickles A., Murphy M. *et al.* (1998) Autism, affective and other psychiatric disorders: patterns of familial aggregation. *Psychol Med* **28**(2):385–95.

Bolton P., Macdonald H., Pickles A. *et al.* (1994) A case-control family history study of autism. *J Child Psychol Psychiatry* **35**(5):877–900.

Buxbaum J. D., Silverman J. M., Smith C. J. *et al.* (2001) Evidence for a susceptibility gene for autism on chromosome 2 and for genetic heterogeneity. *Am J Hum Genet* **68**:1514–20.

Cadoret R. J., Yates W. R., Troughton E. *et al.* (1995) Adoption study demonstrating two genetic pathways to drug abuse. *Arch Gen Psychiatry* **52**(1):42–52.

Cadoret R. J., Yates W. R., Troughton E. *et al.* (1995) Genetic-environmental interaction in the genesis of aggressivity and conduct disorders. *Arch Gen Psychiatry* **52**(11):916–24.

Cantwell D. P. (1975) Genetics of hyperactivity. *Journal of Child Psychology and Psychiatry* **16**:261–4.

Carey G. (1986) Sibling imitation and contrast effects. *Behavior Genetics* **16**:319–41.

Caron M. G. (1996) A dopamine transporter mouse knockout. *American Journal of Psychiatry* **153**(12):1515.

Castellanos F. X., Lau E., Tayebi N. *et al.* (1998) Lack of an association between a dopamine-4 receptor polymorphism and attention-deficit hyperactivity disorder: genetic and brain morphometric analyses. *Molecular Psychiatry* **3**:431–4.

Collaborative Linkage Study of Autism (CLSA) (1999) An autosomal genomic screen for autism. *American Journal of Medical Genetics (Neuropsychiatric Genetics)* **88**:609–15.

Comings D. E. and Comings B. G. (1987) A controlled study of Tourette Syndrome—phobias and panic attacks. *Am J Hum Genet* **41**(5):761–81.

Comings D. E. and Comings B. G. (1990a) A controlled family history study of Tourette's syndrome, I: Attention-deficit hyperactivity disorder and learning disorders. *J Clin Psychiatry* **51**(7):275–80.

Comings D. E. and Comings B. G. (1990b) A controlled family history study of Tourette's syndrome, II: Alcoholism, drug abuse, and obesity. *J Clin Psychiatry* **51**(7):281–7.

Comings D. E. and Comings B. G. (1990c) A controlled family history study of Tourette's syndrome, III: Affective and other disorders. *J Clin Psychiatry* **51**(7):288–91.

Cook E. H. Jr, Lindgren V., Leventhal B. L. *et al.* (1997) Autism or atypical autism in maternally but not paternally derived proximal 15q duplication. *Am J Hum Genet* **60**:928–34.

Cook E. H. Jr, Stein M. A., Krasowski M. D. *et al.* (1995) Association of attention-deficit disorder and the dopamine transporter gene. *Am J Hum Genet* **56**:993–8.

Cook E. H. Jr, Courchesne R. Y., Cox N. J. *et al.* (1998) Linkage-disequilibrium mapping of autistic disorder, with 15q11–13 markers. *Am J Hum Genet* **62**(5):1077–83.

Costello E. J., Angold A., Burns B. J. *et al.* (1996) The Great Smoky Mountains Study of Youth. Goals, design, methods, and the prevalence of DSM-III-R disorders. *Archives of General Psychiatry* **53**(12):1129–36.

Curran S., Mill J., Tahir E. *et al.* (2001) Association study of a dopamine transporter polymorphism and attention deficit hyperactivity disorder in UK and Turkish samples. *Am J Med Genet* **105**(4): 387–93.

Daly G., Hawi Z., Fitzgerald M. *et al.* (1999) Mapping susceptibility loci in attention deficit hyperactivity disorder: preferential transmission of parental alleles at DAT1, DBH and DRD5 to affected children. *Molecular Psychiatry* **4**:192–6.

Deater-Deckard K., Fulker D. W., Plomin R. (1999) A genetic study of the family environment in the transition to early adolescence. *Journal of Child Psychology and Psychiatry and Allied Disciplines* **40**(5):769–75.

Deater-Deckard K., Reiss D., Hetherington E. M. *et al.* (1997) Dimensions and disorders of adolescent adjustment: a quantitative genetic analysis of unselected samples and selected extremes. *Journal of Child Psychology and Psychiatry* **38**(5):515–25.

DeLong R. and Nohria C. (1994) Psychiatric family history and neurological disease in autistic spectrum disorders. *Dev Med Child Neurol* **36**(5):441–8.

Devor E. J., Dill-Devor R. M., Magee H. J. (1998) The Bal I and Msp I polymorphisms in the dopamine D3 receptor gene display, linkage disequilibrium with each other but no association with Tourette syndrome. *Psychiatr Genet* **8**(2):49–52.

Dillalla L. A. and Gottesman I. I. (1989) Heterogeneity of causes for delinquency and criminality: lifespan perspectives. *Development and Psychopathology* **1**:339–49.

Eaves L. J., Silberg J. L., Meyer J. M. *et al.* (1997) Genetics and developmental psychopathology: 2. The main effects of genes and environment on behavioral problems in the Virginia Twin Study of Adolescent Behavioral Development. *J Child Psychol Psychiatry* **38**(8):965–80.

Edelbrock C., Rende R., Plomin R. *et al.* (1995) A twin study of competence and problem behavior in childhood and early adolescence. *J Child Psychol Psychiatry* **36**(5):775–85.

Eisenberg J., Zohar A., Mei-Tal G. *et al.* (2000) A haplotype relative risk study of the dopamine D4 receptor (DRD4) exon III repeat polymorphism and attention deficit hyperactivity disorder (ADHD). *Am J Med Genet (Neuropsychiatric Genetics)* **96**:258–61.

Eley T. C. and Stevenson J. (1999) Exploring the covariation between anxiety and depression symptoms: a genetics analysis of the effects of age and sex. *Journal of Child Psychology and Psychiatry* **40**(8):1273–82.

Eley, T. C. (1997) Depressive symptoms in children and adolescents: etiological links between normality and abnormality: a research note. *Journal of Child Psychology and Psychiatry* **38**(7):861–5.

Eley T. C., Deater-Deckard K., Fombonne E. *et al.* (1998) An adoption study of depressive symptoms in middle childhood. *Journal of Child Psychology and Psychiatry* **39**(3):337–45.

Eley T. C., Lichtenstein P., Stevenson J. (1999) Sex differences in the etiology of aggressive and nonaggressive antisocial behavior: results from two twin studies. *Child Development* **70**(1):155–68.

Faraone S. V., Biederman J., Monuteaux M. C. (2000) Toward guidelines for pedigree selection in genetic studies of attention deficit hyperactivity disorder. *Genet Epidemiol* **18**(1):1–16.

Faraone S. V., Biederman J., Chen W. J. *et al.* (1992) Segregation analysis of attention deficit hyperactivity disorder. *Psychiatric Genetics* 2:257–75.

Faraone S. V., Biederman J., Keenan K. *et al.* (1991) Separation of DSM-III attention deficit disorder and conduct disorder: evidence from a family-genetic study of American child psychiatric patients. *Psychol Med* 21(1):109–21.

Faraone S. V., Biederman J., Mennin D. *et al.* (1998) Familial subtypes of attention deficit hyperactivity disorder: a 4-year follow-up study of children from antisocial-ADHD families. *J Child Psychol Psychiatry* 39(7):1045–53.

Faraone S. V., Doyle A. E., Mick E. *et al.* J. (2001) Meta-analysis of the association bewteen the dopamine D4 gene 7 repeat allele and attention deficit hyperactivity disorder. *American Journal of Psychiatry* 158(7):1052–7.

Faraone S., Biederman J., Jetton J. *et al.* (1997) Attention deficit disorder and conduct disorder, longitudinal evidence for a familial subtype. *Psychological Medicine* 27:291–300.

Farrington D. P. (1995) The development of offending and antisocial behaviour from childhood, key findings from the Cambridge study in delinquent development. *Journal of Child Psychology and Psychiatry* 360(6):929–64.

Flejter W. L., Bennett-Baker P. E., Ghaziuddin M. *et al.* (1996) Cytogenetic and molecular analysis of inv dup(15) chromosomes observed in two patients with autistic disorder and mental retardation. *Am J Med Genet* 61(2):182–7.

Folstein S. E. and Rutter M. (1977) Infantile autism: a genetic study of 21 twin pairs. *Journal of Child Psychology and Psychiatry* 18:297–321.

Fombonne E. (1999) The epidemiology of autism: a review. *Psychological Medicine* 29:769–86.

Fombonne E., Bolton P., Prior J. *et al.* (1997) A family study of autism: cognitive patterns and levels in parents and siblings. *J Child Psychol Psychiatry* 38(6):667–83.

Ge X., Rand D., Cadoret R. *et al.* (1996) The developmental interface between nature and nurture, a mutual influence model of child antisocial behavior and parent behaviors. *Developmental Psychology* 32(4):574–89.

Ghodsion-Carpey J. and Baker L. A. (1987) Genetic and environmental influences on aggression in 4 to 7 year old twins. *Aggressive Behavior* 13:173–86.

Gill M., Daly G., Heron S. *et al.* (1997) Confirmation of association between attention deficit hyperactivity disorder and a dopamine transporter polymorphism. *Molecular Psychiatry* 2:311–13.

Gillberg C., Steffenburg S., Wahlstrom J. *et al.* (1991) Autism associated with marker chromosome. *J Am Acad Child Adolesc Psychiatry* 30(3):489–94.

Gjone H. and Stevenson J. (1997) A longitudinal twin study of temperament and behavior problems: common genetic or environmental influences? *J Am Acad Child Adolesc Psychiatry* 36(10):1448–56.

Gjone H., Stevenson J., Sundet J. M. (1996) Genetic influence on parent-reported attention-related problems in a Norwegian general population twin sample. *J Am Acad Child Adolesc Psychiatry* 35(5):588–96.

Goldsmith H. H. and Gottesman I. (1996) Heritable variability and variable heritability in developmental psychopathology, in M. Lenzenweger and J. Haugaard (eds) *Frontiers of Developmental Psychopathology*. Oxford, Oxford University Press.

Goodman R. and Stevenson J. (1989) A twin study of hyperactivity–II. The aetiological role of genes, family relationships and perinatal adversity. *J Child Psychol Psychiatry* 30(5):691–709.

Goodyer I. M., Cooper P. J., Vize C. M. *et al.* (1993) Depression in 11–16-year-old girls: the role of past parental psychopathology and exposure to recent life events. *J Child Psychol Psychiatry* 34(7): 1103–15.

Gordon C. T., Frazier J. A., McKenna K. *et al.* (1994) Childhood-onset schizophrenia: an NIMH study in progress. *Schizophr Bull* **20**(4):697–712.

Gotlib I. H., Lewinsohn P. M., Seeley J. R. (1995) Symptoms versus a diagnosis of depression: differences in psychosocial functioning. *J Consult Clin Psychol* **63**(1):90–100.

Hammen C., Burge D., Burney E. *et al.* (1990) Longitudinal study of diagnoses in children of women with unipolar and bipolar affective disorder. *Arch Gen Psychiatry* **47**(12):1112–17.

Hanson D. R. and Gottesman I. I. (1976) The genetics, if any, of infantile autism and childhood schizophrenia. *J Autism Child Schizophr* **6**(3):209–34.

Harrington R. and Clark A. (1998) Prevention and early intervention for depression in adolescence and early adult life. *Eur Arch Psychiatry Clin Neurosci* **248**(1):32–45.

Harrington R. C., Fudge H., Rutter M. L. *et al.* (1993) Child and adult depession: A test of continuities with data from a family study. *British Journal of Psychiatry* **162**:627–33.

Harrington R., Fudge H., Rutter M. *et al.* (1991) Adult outcomes of childhood and adolescent depression: II. Links with antisocial disorders. *J Am Acad Child Adolesc Psychiatry* **30**(3):434–9.

Harrington R., Rutter M., Weissman M. *et al.* (1997) Psychiatric disorders in the relatives of depressed probands. I. Comparison of prepubertal, adolescent and early adult onset cases. *J Affect Disord* **42**(1):9–22.

Hawi Z., McCarron M., Kirley A. *et al.* (2000) No association of the dopamine DRD4 receptor (DRD4) gene polymorphism with attention deficit hyperactivity disorder (ADHD) in the Irish population. *Am J Med Genet (Neuropsychiatric Genetics)* **96**:268–72.

Holmes J., Payton A., Barrett J. H. *et al.* (2000) A family-based and case-control association study of the dopamine D4 receptor gene and dopamine transporter gene in attention deficit hyperactivity disorder. *Molecular Psychiatry* **5**(523):530.

Hotopf M. and Bolton P. (1995) A case of autism associated with partial tetrasomy 15. *J Autism Dev Disord* **25**(1):41–9.

Hudziak J. J., Rudiger L. P., Neale M. C. *et al.* (2000) A twin study of inattentive, aggressive, and anxious/depressed behaviors. *J Am Acad Child Adolesc Psychiatry* **39**(4):469–76.

Hughes C., Plumet M.-H., Leboyer M. (1999) Towards a cognitive phenotype for autism: increased prevalence of executive dysfunction and superior spatial span amongst siblings of children with autism. *Journal of Child Psychology and Psychiatry* **40**(5):705–18.

International Molecular Genetic Study of Autism Consortium (1998) A full genome screen for autism with evidence for linkage to a region on chromosome 7q. *Human Molecular Genetics* **7**(3):571–8.

Ingram J. L., Stodgell C. J., Hyman S. L. *et al.* (2000) Discovery of allelic variants of HOXA1 and HOXB1: genetic susceptibility to autism spectrum disorders. *Teratology* **62**:393–405.

Jenkins R. L. and Ashby H. B. (1983) Gilles de la Tourette's syndrome in identical twins. *Arch Neurol* **40**(4):249–51.

Kallman F. J. and Roth B. (1956) Genetic aspects of preadolescent schizophrenia. *American Journal of Psychiatry* **112**:599–606.

Kanner L. (1943) Autistic disturbances of affective contact. *Nervous Child* **2**:217–50.

Kendler K. S., Sham P. C., MacLean C. J. (1997) The determinants of parenting: an epidemiological, multi-informant, retrospective study. *Psychological Medicine* **27**:549–63.

Kidd K. K., Prusoff B. A., Cohen D. J. (1980) Familial pattern of Gilles de la Tourette syndrome. *Arch Gen Psychiatry* **37**(12):1336–9.

Klauck S. M., Poustka F., Benner A., Lesch K. P., Poustka A. (1997) Serotonin transporter (5-HTT) gene variants associated with autism? *Hum Mol Genet* **6**(13):2233–8.

Knell E. R. and Comings D. E. (1993) Tourette's syndrome and attention-deficit hyperactivity disorder: evidence for a genetic relationship. *J Clin Psychiatry* **54**(9):331–7.

Kolvin I. (1971) Studies in the childhood psychoses. I. Diagnostic criteria and classification. *Br J Psychiatry* **118**(545):381–4.

Kotler M., Manor I., Sever Y. *et al.* (2000) Failure to replicate an excess of the long dopamine D4 exon III repeat polymorphism in ADHD in a family-based study. *Am J Med Genet (Neuropsychiatric Genetics)* **96**:278–81.

Kovacs M., Devlin B., Pollock M. *et al.* (1997) A controlled family history study of childhood-onset depressive disorder. *Archives of General Psychiatry* **54**:613–23.

Krueger R., Moffitt T., Caspi A. *et al.* (1998) Assortative mating for antisocial behavior, developmental and methodological implications. *Behavior Genetics*, **28**(3):173–86.

Kumra S., Wiggs E., Krasnewich D. *et al.* (1998) Association of sex chromosome anomalies with childhood-onset psychotic disorders. *Journal of the American Academy of Child and Adolescent Psychiatry* **37**(3):292–6.

Kuntsi J. and Stevenson J. (2001) Psychological mechanisms in hyperactivity: II. The role of genetic factors. *J Child Psychol Psychiatry* **42**(2):211–19.

LaHoste G. J., Swanson J. M., Wigal S. B. *et al.* (1996) Dopamine D4 receptor polymorphism is associated with attention deficit hyperactivity disorder. *Molecular Psychiatry* **1**:121–4.

Last C. G., Hersen M., Kazdin A. *et al.* (1991) Anxiety disorders in children and their families. *Archives of General Psychiatry* **48**(10):928–34.

Lauritsen M., Mors O., Mortensen P. B. *et al.* (1999) Infantile autism and associated autosomal chromosome abnormalities: a register-based study and a literature survey. *J Child Psychol Psychiatry* **40**(3):335–45.

Le Couteur A., Bailey A., Goode S. *et al.* (1996) A broader phenotype of autism: the clinical spectrum in twins. *J Child Psychol Psychiatry* **37**(7):785–801.

Legrand L. N., McGue M., Iacono W. G. (1999) A twin study of state and trait anxiety in childhood and adolescence. *J Child Psychol Psychiatry* **40**(6):953–8.

Levy F., Hay D. A., McStephen M. *et al.* (1997) Attention-deficit hyperactivity disorder: a category or a continuum? Genetic analysis of a large-scale twin study. *J Am Acad Child Adolesc Psychiatry* **36**(6):737–44.

Lewinsohn P. M., Rohde P., Klein D. N. *et al.* (1999) Natural course of adolescent major depressive disorder: I. Continuity into young adulthood. *J Am Acad Child Adolesc Psychiatry* **38**(1):56–63.

Lopreiato J. O. and Wulfsberg E. A. (1992) A complex chromosome rearrangement in a boy with autism. *Journal of Developmental and Behavioral Pediatrics* **13**:281–3.

Lord C., Rutter M., Le Couteur A. (1994) Autism Diagnostic Interview-Revised: a revised version of a diagnostic interview for care givers of individuals with possible pervasive developmental disorders. *J Autism Dev Disord* **24**:659–85.

Lyons M. J., True W. R., Eisen S. A. *et al.* (1995) Differential heritability of adult and juvenile antisocial traits. *Arch Gen Psychiatry* **52**(11):906–15.

Maes H., Neale M., Kendler K. *et al.* (1998) Assortive mating for major psychiatric diagnoses in two population-based samples. *Psychological Medicine* **28**(1389):1401.

Maestrini E., Lai C., Marlow A. *et al.* (1999) Serotonin transporter (5-HTT) and gamma-aminobutyric acid receptor subunit beta3 (GABRB3) gene polymorphisms are not associated with autism in the IMGSA families. The International Molecular Genetic Study of Autism Consortium. *Am J Med Genet* **88**(5):492–6.

Maziade M., Matinez M., Rodrigue C. (1977) Childhood-early adolescence onset and adult onset schizophrenia: heterogeneity at the dopamine D3 receptor gene. *British Journal of Psychiatry*.

McGuffin P. and Gottesman I. (1985) Genetic influences on normal and abnormal development, in M. Rutter and L. Hersov (eds) *Child and Adolescent Psychiatry, Modern Approaches*. Boston, MA, Blackwell Scientific Publications Inc.

McGuffin P. and Thapar A. (1997) Genetic basis of bad behaviour in adolescents. *Lancet* **350**(9075):411–12.

Mednick S. A., Gabrielli W. F., Jr, Hutchings B. (1984) Genetic influences in criminal convictions: evidence from an adoption cohort. *Science* **224**(4651):891–4.

Mill J., Curran S., Kent L. *et al.* (2001) Attention deficit hyperactivity disoder (ADHD) and the dopamine D4 receptor gene: evidence of association but no linkage in a UK sample. *Molecular Psychiatry* **6**:440–4.

Mitchell J., McCauley E., Burke P. *et al.* (1989) Psychopathology in parents of depressed children and adolescents. *Journal of the American Academy of Child and Adolescent Psychiatry* **28**(3):352–7.

Moffitt T. (1993) Adolescence-limited and life-course-persistent antisocial behavior, a developmental taxonomy. *Psychological Review* **100**:674–701.

Morrison J. R. and Stewart M. A. (1973) The psychiatric status of the legal families of adopted hyperactive children. *Arch Gen Psychiatry* **28**(6):888–91.

Mufson L., Weissman M. M., Warner, V. (1992) Depression and anxiety in parents and children: A direct interview study. *Journal of Anxiety Disorders* **6**(1):1–13.

Muglia P., Jain U., Macciardi F. *et al.* (2000) Adult attention deficit hyperactivity disorder and the dopamine D4 receptor gene. *Am J Med Genet (Neuropsychiatric Genetics)* **96**:273–7.

Murphy M., Bolton P., Pickles A. *et al.* (1998) Personality traits of the relatives of autistic probands. *Psychological Medicine* **30**:1411–24.

Nee L. E., Caine E. D., Polinsky R. J. *et al.* (1980) Gilles de la Tourette syndrome: clinical and family study of 50 cases. *Ann Neurol* **7**(1):41–9.

Neiderhiser J. M., Pike A., Hetherington E. M. *et al.* (1998) Adolescent perceptions as mediators of parenting: genetic and environmental contributions. *Dev Psychol* **34**(6):1459–69.

Neiderhiser J. M., Reiss D., Hetherington E. M. *et al.* (1999) Relationships between parenting and adolescent adjustment over time: genetic and environmental contributions. *Dev Psychol* **35**(3):680–92.

Neuman R. J., Geller B., Rice J. P. *et al.* (1997) Increased prevalence and earlier onset of mood disorders among relatives of prepubertal versus adult probands. *J Am Acad Child Adolesc Psychiatry* **36**(4):466–73.

Nicolson R., Giedd J. N., Lenane M. *et al.* (1999) Clinical and neurobiological correlates of cytogenetic abnormalities in childhood-onset schizophrenia. *Am J Psychiatry* **156**(10):1575–9.

Nicolson R., Lenane M., Singaracharlu S. *et al.* (2000) Premorbid speech and language impairments in childhood-onset schizophrenia: association with risk factors. *Am J Psychiatry* **157**(5):794–800.

O'Connor T. G., Deater-Deckard K., Fulker D. *et al.* (1998) Genotype-environment correlations in late childhood and early adolescence: antisocial behavioral problems and coercive parenting. *Dev Psychol* **34**(5):970–81.

Palmer C. G. S., Bailey J. N., Ramsey, C. *et al.* (1999) No evidence of linkage or linkage disequilibrium between DAT1 and attention deficit hyperactivity disorder in a large sample. *Psychiatric Genetics* **9**:157–60.

Pauls D. L., Cohen D. J., Heimbuch R. *et al.* (1981) Familial pattern and transmission of Gilles de la Tourette syndrome and multiple tics. *Arch Gen Psychiatry* **38**(10):1091–3.

Pauls D. L., Kruger S. D., Leckman J. F. *et al.* (1984) The risk of Tourette's syndrome and chronic multiple tics among relatives of Tourette's syndrome patients obtained by direct interview. *J Am Acad Child Psychiatry* **23**(2):134–7.

Pauls D. L., Leckman J. F. Cohen D. J. (1993) Familial relationship between Gilles de la Tourette's syndrome, attention deficit disorder, learning disabilities, speech disorders, and stuttering. *J Am Acad Child Adolesc Psychiatry* **32**(5):1044–50.

Pauls D. L., Raymond C. L., Stevenson J. M. *et al.* (1991) A family study of Gilles de la Tourette syndrome. *Am J Hum Genet* **48**(1):154–63.

Phillippe A., Martinez M., Guilloud-Bataille M. *et al.* (1999) genome-wide scan for autism susceptibility genes. *Human Molecular Genetics* **8**(5):805–12.

Pickles A., Bolton P., Macdonald H. *et al.* (1995) Latent-class analysis of recurrence risks for complex phenotypes with selection and measurement error: a twin and family history study of autism. *Am J Hum Genet* **57**(3):717–26.

Pickles A., Starr E., Kazak S. *et al.* (2000) Variable expression of the autism broader phenotype: findings from extended pedigrees. *Journal of Child Psychology and Psychiatry* **41**(4):491–502.

Piven J., Palmer P., Landa R. *et al.* (1997) Personality and language characteristics in parents from multiple-incidence autism families. *Am J Med Genet* **74**(4):398–411.

Plomin R., De Fries J. C., McClearn G. E. *et al.* (1997) *Behavioral Genetics*. London, St Martins Press.

Plomin R. (1994) The Emanuel Miller Memorial Lecture 1993. Genetic research and identification of environmental influences. *J Child Psychol Psychiatry* **35**(5):817–34.

Price R. A., Kidd K. K., Cohen D. J. *et al.* (1985) A twin study of Tourette syndrome. *Arch Gen Psychiatry* **42**(8):815–20.

Puig-Antich J., Goetz D., Davies M. *et al.* (1989) A controlled family history study of prepubertal major depressive disorder. *Archives of General Psychiatry* **46**:406–17.

Pulver A. E., Brown C. H., Wolyniec P. *et al.* (1990) Schizophrenia: age at onset, gender and familial risk. *Acta Psychiatr Scand* **82**(5):344–51.

Radke-Yarrow M., Nottelmann E., Martinez P. *et al.* (1992) Young children of affectively ill parents: a longitudinal study of psychosocial development. *J Am Acad Child Adolesc Psychiatry* **31**(1):68–77.

Rapoport J. L., Swado S., Leonard, H. (1994) Obsessive-compulsive disorder, in *Child and Adolescent Psychiatry*, Oxford, Blackwell Scientific.

Rende R. D., Plomin R., Reiss D. *et al.* (1993) Genetic and environmental influences on depressive symptomatology in adolescence: individual differences and extreme scores. *J Child Psychol Psychiatry* **34**(8):1387–98.

Rende R., Warner V., Wickramarante P. *et al.* (1999) Sibling aggregation for psychiatric disorders in offspring at high and low risk for depression: 10-year follow-up. *Psychol Med* **29**(6):1291–8.

Rice F., Harold G., Thapar A. (2002) The aetiology of childhood depression: a review of genetic influences. *Journal of Child Psychology and Psychiatry* **43**(1):65–79.

Risch N., Spiker D., Lotspeich L. *et al.* (1999) A genomic screen of autism: evidence for a multilocus etiology. *Am J Hum Genet* **65**:493–507.

Ritvo E. R., Freeman B. J., Mason-Brothers A. *et al.* (1985) Concordance for the syndrome of autism in 40 pairs of afflicted twins. *Am J Psychiatry* **142**(1):74–7.

Rowe D. (1985) Sibling interaction and self-reported delinquent behavior, a study of 265 twin pairs. *Criminology* **23**:223–40.

Rowe D. C., Stever C., Giedinghagen L. N. *et al.* (1998) Dopamine DRD4 receptor polymorphism and attention deficit hyperactivity disorder. *Molecular Psychiatry* **3**:419–426.

Rutter M., Silberg J., O'Connor T. *et al.* (1999) Genetics and child psychiatry: I. Advances in quantitative and molecular genetics. *J Child Psychol Psychiatry* **40**(1):3–18.

Rutter M. and Silberg J. (2002) Gene-environment interplay in relation to emotional and behavioral disturbance. *Annual Review of Psychology* **54**:463–90.

Safer D. J. (1973) A familial factor in minimal brain dysfunction. *Behav Genet* 3(2):175–86.

Salmon B., Hallmayer J., Rogers T. *et al.* (1999) Absence of linkage and linkage disequilibrium to chromosome 15q11-q13 markers in 139 multiplex families with autism. *Am J Med Genet* 88(5):551–6.

Santangelo S. L., Pauls D. L., Lavori P. W. *et al.* (1996) Assessing risk for the Tourette spectrum of disorders among first-degree relatives of probands with Tourette syndrome. *Am J Med Genet* 67(1):107–16.

Scarr S. and McCartney K. (1983) How people make their own environments: a theory of genotype greater than environment effects. *Child Development* 54(2):424–35.

Schmitz S., Fulker D. W., Mrazek D. A. (1995) Problem behavior in early and middle childhood: an initial behavior genetic analysis. *Journal of Child Psychology and Psychiatry* 36(8):1443–58.

Schroer R. J., Phelan M. C., Michaelis R. C. *et al.* (1998) Autism and maternally derived aberrations of chromosome 15q. *Am J Med Genet* 76:327–36.

Seuchter S. A., Hebebrand J., Klug B. *et al.* (2000) Complex segregation analysis of families ascertained through Gilles de la Tourette syndrome. *Genet Epidemiol* 18(1):33–47.

Sham P. C., MacLean C. J., Kendler K. S. (1994) A typological model of schizophrenia based on age at onset, sex and familial morbidity. *Acta Psychiatr Scand* 89(2):135–41.

Sherman D. K., McGue M. K., Iacono W. G. (1997) Twin concordance for attention deficit hyperactivity disorder: a comparison of teachers' and mothers' reports. *Am J Psychiatry* 154(4):532–5.

Silberg J., Erickson M., Meyer J. *et al.* (1994) The application of structural equation modelling to maternal ratings of twins' behavioral and emotional problems. *Journal of Consulting and Clinical Psychology* 62:510–21.

Silberg J., Rutter M., Meyer J. *et al.* (1996) Genetic and environmental influences on the covariation between hyperactivity and conduct disturbance in juvenile twins. *J Child Psychol Psychiatry* 37(7):803–16.

Silberg J., Rutter M., Neale M. *et al.* (2001) Genetic moderation of environmental risk for depression and anxiety in adolescent girls. *The British Journal of Psychiatry* 179:116–21.

Simonoff E., Pickles A., Hewitt J. *et al.* (1995) Multiple raters of disruptive child behavior: using a genetic strategy to examine shared views and bias. *Behav Genet* 25(4):311–26.

Simonoff E., Pickles A., Meyer J. M. *et al.* (1997) The Virginia Twin Study of Adolescent Behavioral Development. Influences of age, sex, and impairment on rates of disorder. *Arch Gen Psychiatry* 54(9):801–8.

Slutske W. S., Heath A. C., Dinwiddie S. H. *et al.* (1997) Modeling genetic and environmental influences in the etiology of conduct disorder: a study of 2,682 adult twin pairs. *J Abnorm Psychol* 106(2):266–79.

Smalley S. L., Asarnow R. F., Spence M. A. (1988) Autism and genetics: a decade of research. *Archives of General Psychiatry* 45:953–61.

Smalley S. L. (1998) Autism and tuberous sclerosis. *J Autism Dev Disord* 28(5):407–14.

Smalley S. L., Bailey J. N., Palmer C. G. *et al.* (1998) Evidence that the dopamine D4 receptor is a susceptibility gene in attention deficit hyperactivity disorder. *Mol Psychiatry* 3(5):427–30.

Starr E., Berument S. K., Pickles A. *et al.* (2001) Family genetic-study of autism associated with profound mental retardation. *Journal of Autism and Developmental Disorders* 31:89–96.

Steffenburg S., Gillberg C., Hellgren L. *et al.* (1989) A twin study of autism in Denmark, Finland, Iceland, Norway and Sweden. *Journal of Child Psychology and Psychiatry* 30(3):405–16.

Stevenson J. (1992) Evidence for a genetic etiology in hyperactivity in children. *Behav Genet* **22**(3):337–44.

Sullivan P. F., Neale M. C., Kendler K. S. (2000) Genetic epidemiology of major depression: review and meta-analysis. *American Journal of Psychiatry* **157**:1552–62.

Sunohara G. A., Roberts W., Malone M. *et al.* (2000) Linkage of the dopamine D4 receptor gene and attention-deficit/hyperactivity disorder. *Journal of the American Academy of Child and Adolescent Psychiatry* **39**(12):1537–42.

Swanson J. M., Sunohara G. A., Kennedy J. L. *et al.* (1998) Association of the dopamine receptor D4 (DRD4) gene with a refined phenotype of attention deficit hyperactivity disorder (ADHD): a family-based approach. *Molecular Psychiatry* **3**:38–41.

Szatmari P., Boyle M. H., Offord, D. R. (1993) Familial aggregation of emotional and behavioral problems of childhood in the general population. *Am J Psychiatry* **150**(9):1398–403.

Szatmari P., Jones M. B., Zwaigenbaum L. *et al.* (1998) Genetics of autism: overview and new directions. *Journal of Autism and Developmental Disorders* **28**(5):351–68.

Tahir E., Yazgan Y., Cirakoglu B. *et al.* (2000) Association and linkage of DRD4 and DRD5 with attention deficit hyperactivity disorder (ADHD) in a sample of Turkish children. *Molecular Psychiatry* **5**(4):396–404.

Taylor L. D., Krizman D. B., Jankovic J. *et al.* (1991) 9p monosomy in a patient with Gilles de la Tourette's syndrome. *Neurology* **41**(9):1513–15.

Thapar A. and McGuffin P. (1994) A twin study of depressive symptoms in childhood. *Br J Psychiatry* **165**(2):259–65.

Thapar A. and McGuffin P. (1995) Are anxiety symptoms in childhood heritable? *J Child Psychol Psychiatry* **36**(3):439–47.

Thapar A. and McGuffin P. (1996) A twin study of antisocial and neurotic symptoms in childhood *Psychol Med* **26**(6):1111–18.

Thapar A. and McGuffin P. (1996) Genetic influences on life events in childhood. *Psychol Med* **26**(4):813–20.

Thapar A. and McGuffin P. (1997) Anxiety and depressive symptoms in childhood—a genetic study of comorbidity. *J Child Psychol Psychiatry* **38**(6):651–6.

Thapar A., Harold G., McGuffin P. (1998) Life events and depressive symptoms in childhood–shared genes or shared adversity? A research note. *J Child Psychol Psychiatry* **39**(8):1153–8.

Thapar A., Harrington R., McGuffin P. (2001) Examining the comorbidity of ADHD related behaviours and conduct problems using a twin study design. *British Journal of Psychiatry* **179**:224–9.

Thapar A., Harrington R., Ross K., and McGuffin P. (2000) Does the definition of ADHD affect heritability? *J Am Acad Child Adolesc Psychiatry* **39**(12):1528–36.

Thapar A., Hervas A., McGuffin P. (1995) Childhood hyperactivity scores are highly heritable and show sibling competition effects: twin study evidence. *Behav Genet* **25**(6):537–44.

Thapar A., Hervas A., and McGuffin P. (1995) Childhood hyperactivity scores are highly heritable and show sibling competition effects: twin study evidence. *Behav Genet* **25**(6):537–44.

Thapar A., Holmes J., Poulton K. *et al.* (1999) Genetic basis of attention deficit and hyperactivity. *British Journal of Psychiatry* **174**:105–11.

The Tourette Syndrome Association International Consortium in Genetics (1999) A complete genome screen in sib pairs affected by Gilles de la Tourette syndrome. *Am J Hum Genet* **65**(5):1428–36.

Topolski T. D., Hewitt J. K., Eaves L. J. *et al.* (1997) Genetic and environmental influences on child reports of manifest anxiety and symptoms of separation anxiety and overanxious disorders: a community-based twin study. *Behav Genet* **27**(1):15–28.

Usiskin S. I., Nicolson R., Krasnewich D. M. *et al.* (1999) Velocardiofacial syndrome in childhood-onset schizophrenia. *J Am Acad Child Adolesc Psychiatry* **38**(12):1536–43.

Van den Oord E. J., Boomsma D. I., Verhulst F. C. (1994) A study of problem behaviors in 10- to 15-year-old biologically related and unrelated international adoptees. *Behav Genet* **24**(3):193–205.

Vieregge P., Schafer C., Jorg J. (1988) Concordant Gilles de la Tourette's syndrome in monozygotic twins: a clinical, neurophysiological and CT study. *J Neurol* **235**(6):366–7.

Waldman I. D., Rowe D. C., Abramowitz A. *et al.* (1998) Association and linkage of the dopamine transporter gene and attention-deficit hyperactivity disorder in children: heterogeneity owing to diagnostic subtype and severity. *Am J Hum Genet* **63**:1767–76.

Walkup J. T., LaBuda M. C., Singer H. S. *et al.* (1996) Family study and segregation analysis of Tourette syndrome: evidence for a mixed model of inheritance. *Am J Hum Genet* **59**(3):684–693.

Warner V., Mufson L., Weissman M. M. (1995) Offspring at high and low risk for depression and anxiety: mechanisms of psychiatric disorder. *J Am Acad Child Adolesc Psychiatry* **34**(6):786–97.

Warner V., Weissman M. M., Mufson L. *et al.* (1999) Grandparents, parents, and grandchildren at high risk for depression: a three-generation study. *J Am Acad Child Adolesc Psychiatry* **38**(3):289–96.

Weissman M. M., Fendrich M., Warner V. *et al.* (1992) Incidence of psychiatric disorder in offspring at high and low risk for depression. *J Am Acad Child Adolesc Psychiatry* **31**(4):640–8.

Weissman M. M., Leckman J. F., Merikangas K. R. *et al.* (1984) Depression and anxiety disorders in parents and children. Results from the Yale family study. *Arch Gen Psychiatry* **41**(9):845–52.

Weissman M. M., Warner V., Wickramaratne P. *et al.* (1997) Offspring of depressed parents. 10 years later. *Arch Gen Psychiatry* **54**(10):932–40.

Weissman M. M., Wolk S., Goldstein R. B. *et al.* (1999) Depressed adolescents grown up. *JAMA* **281**(18):1707–13.

Wickramaratne P. J., Greenwald S., Weissman M. M. (2000) Psychiatric disorders in the relatives of probands with prepubertal-onset or adolescent-onset major depression. *J Am Acad Child Adolesc Psychiatry* **39**(11):1396–405.

Wierzbicki M. (1987) Similarity of monozygotic and dizygotic child twins in level and liability of subclinically depressed mood. *American Journal of Orthopsychiatry* **57**(1):33–40.

Willcutt E. G., Pennington B. F., De Fries J. C. (2000) Twin study of etiology of comorbidity between reading disability and attention-deficit/hyperactivity disorder. *Am J Med Genet* **96**:293–301.

Williamson D. E., Neal B. A., Ryan M. D. *et al.* (1995) A Case-Control Family History Study of Depression in Adolescents. *J Am Acad Child Adolesc Psychiatry* **34, 12**:1596–607.

Yan W. L., Guan X. Y., Green E. D. *et al.* (2000) Childhood-onset schizophrenia/autistic disorder and t(1;7) reciprocal translocation: identification of a BAC contig spanning the translocation breakpoint at 7q21. *Am J Med Genet* **96**(6):749–53.

Part 3

The genetics of abnormal behaviour in adult life

Chapter 8

Personality disorders

Peter McGuffin, Terrie Moffitt, and Anita Thapar

What is personality disorder?

Personality can be defined as the individual's entire repertoire of relatively stable and enduring behaviours (genetic aspects of normal personality are considered in Chapter 4). By the same token, personality disorders are usually seen as reflecting enduring traits, and not as transient or alterable states that may change greatly over time. Hence they are considered to be different from those forms of mental illness that have a definable onset and may later show remissions and relapses. The difference between personality disorders and other types of psychiatric disorder is recognized in the Diagnostic and Statistical Manual 4th edition (DSM-IV) (American Psychiatric Association 1993) whereby axis I diagnoses consist of clinical disorders and 'other conditions that may be a focus of clinical attention' and disorders of personality are included separately under axis II. Although the other main psychiatric classification scheme contained in the International Classification of Diseases 10th edition (ICD-10) does not include axial separation, the difference between personality disorder and other types of clinical condition is implicitly recognized. In most other respects the DSM-IV and ICD-10 definition and classification of personality disorders is very similar. Both systems are entirely categorical rather than dimensional and both use very similar wording in defining personality disorder as 'an enduring pattern of inner experience and behaviour that deviates markedly from the expectations of the individual's culture'.

Deviations in four domains can occur. These are

1 Cognition, the way that the individual perceives and interprets their world and forms attitudes towards their own self or others;

2 Affectivity, the range and appropriateness of the individual's emotional responses;

3 Interpersonal functioning, the way the individual relates to others and manages interactions with others;

4 Impulse control, the degree to which the individual can control impulses or delayed gratification of needs.

Both DSM and ICD definitions require that there are deviations in two or more of these areas, that the pattern is inflexible and pervasive across a broad range of situations, and the pattern of behaviour leads to *distress or impairment*. Finally both systems require that the pattern of behaviour is stable and can be traced back to adolescence or early adult life and that the behaviour is not simply accounted for by another physical or mental disorder or drug use.

DSM-IV divides personality disorder into three clusters, as shown in Table 8.1. ICD-10 does not, but contains a very similar set of categories, with a few noticeable differences. The first is that schizotypal disorder is classified as an axis II condition in DSM-IV whereas ICD-10 recognizes it as the equivalent of an axis I disorder, classifying it along with schizophrenia and delusional disorders. Also ICD-10 does not recognize narcissistic personality disorders but does contain a category called emotionally unstable personality disorder, impulsive type that is not found in DSM-IV. By contrast DSM-IV contains a rather broad concept of borderline personality disorder which ICD-10 classifies as a particular subtype of emotionally unstable disorder. Elsewhere the differences are largely terminological, so that for example ICD-10 uses the term anankastic to describe obsessive-compulsive personality disorder.

In this chapter we will discuss the genetic evidence concerning personality disorders, broadly following the main DSM-IV clusters and, where necessary, attempt to 'map' other terminologies on to these. However, we should first note two other complications in the definition of personality disorder. The first is that, in contrast to the DSM or ICD categorical approach, personality psychologists usually consider personality traits, even 'abnormal' ones, as dimensions and thus tend to use trait definitions to apply to the entire population. Traits such as antisocial behaviour or aggression are not seen as present/absent attributes but as continuous, whereby some individuals will

Table 8.1 A summary of the DSM-IV and ICD-10 classification of personality disorders (adapted from Farmer et al. 2002)

Cluster	Characteristics	DSM-IV personality disorders	ICD-10 equivalent personality disorder
A	Paranoid: persecutory beliefs, self reference, socially aloof or tending to mistake fantasy for reality	Paranoid Schizoid Schizotypal	Paranoid Schizoid
B	Antisocial: disruptive, aggressive, compulsive and histrionic behaviours	Antisocial Borderline Histrionic Narcissistic	Dissocial Emotionally unstable Borderline type Impulsive type Histrionic
C	Anxious/avoidant: high levels of anxiety, fearfulness, dependency or obsessionality	Avoidant Dependent Obsessive-compulsive	Anxious Dependent Anankastic

have high levels, some will have low levels, but the majority have scores that are somewhere in between.

The other complication is that some forms of personality disorder, particularly antisocial personality are also of interest to the legal system and to criminologists. Although defining a disorder in terms of criminal convictions is likely to be unsatisfactory and mixes cultural, legal, psychological and medical concepts, a conviction for crime is a clearly definable end point which has certain attractions.

The best single source for an overview of the epidemiology of personality disorders, including an historical perspective, is provided by Weissman (1993); a timely updating is needed to take account of post DSM-III and ICD-9 criteria in a rapidly changing world.

Cluster A—schizotypal, schizoid and paranoid personality disorders

Although DSM-IV guidelines define schizotypal, paranoid and schizoid personality disorders separately, they have tended to be considered as one group in genetic studies. The term schizotypy was used by Meehl (1962, 1990), following S. Rado (1960), to describe characteristic symptoms in non-schizophrenic individuals who are predisposed to schizophrenia. Meehl also introduced 'schizotaxia' to denote an endophenotype, closer to the implicated genotype in a causal chain (Lenzenweger 1998). Faraone *et al.* (2001) have extended Meehl's concept of schizotaxia by reviewing the literature on the relatives of schizophrenics who do not themselves show psychotic or schizotypal features (cf. Kendler and Gardner 1997). They conclude that such relatives show increased rates of psychosocial dysfunction, various forms of neuropsychological impairment, and attenuated negative symptoms of schizophrenia.

Other important early influences on current concepts of Cluster A personality disorders came from adoption studies of schizophrenia. Heston (1970) was impressed by the range of deviant personalities that he found in the biological relatives of schizophrenics versus the relatives of control adoptees, and put forward the notion of 'schizoid disease' (Essen-Moller 1946). This was particularly broad in that it not only encompassed schizoid and paranoid features but was also broadened to include a range of biological and psychopathological indicators (Shields *et al.* 1975). The Danish adoption studies (Kety *et al.* 1971; Ingraham and Kety 2000) introduced the concept of 'schizophrenia spectrum disorders' that, along with overt schizophrenia, was found to be increased among the biological relatives of schizophrenics who had been adopted away from their families early in life. The rather broad concept of schizophrenia spectrum disorders (a condition that affected more than 10 per cent of the relatives of *control* adoptees) was narrowed and operationalized by Spitzer *et al.* (1979) for incorporation into DSM-III (American Psychiatric Association 1980) as schizotypal personality disorder. Subsequently nearly all genetic studies have been based upon a definition of schizotypal personality disorder that is identical to or closely akin to the DSM-III definition.

Most family studies where the families have been ascertained via probands affected by schizophrenia have shown increased rates of both schizotypal personality disorder and schizophrenia among first degree relatives (Baron *et al.* 1983, 1985; Kendler 2000). Only one study of families ascertained via schizophrenic probands failed to show this relationship (Coryell and Zimmerman 1989). Family studies also suggest a familiar overlap between schizotypal personality, paranoid personality disorder and schizophrenia (Baron *et al.* 1985; Kendler 2000; Siever *et al.* 1990). More recently a study of the relatives of probands with childhood-onset schizophrenia (Asarnow *et al.* 2001; Gottesman *et al.* 1982) found an increased risk of both schizophrenia and schizotypal personality disorder in the parents compared with the parents of probands with attention deficit hyperactivity disorder or no disorder. Interestingly there was also an increased risk of avoidant personality disorder (a cluster C personality disorder) in the parents of the childhood-onset schizophrenics compared with the parents of controls.

Family studies that have ascertained families via probands not with schizophrenia but with schizotypal or other cluster A personality types have yielded less clear-cut findings. Several such studies (Baron 1985b; Schulz *et al.* 1986; Soloffa and Millward 1983) failed to find an increased risk of schizophrenia among first degree relatives of probands with schizotypal personality disorder. Such negative results could be explained by lack of power, since the expected rate of schizophrenia may be lower than among the relatives of schizophrenics. This would certainly be true if we were to postulate a liability–threshold model (Gottesman and Shields 1967; Reich *et al.* 1972, see also Chapter 2) where schizotypy represents a milder more common variant lying on the same continuum of liability as schizophrenia, which would then be considered as a narrow more severe disorder with a higher threshold for being affected.

Despite this some studies have found increased rates of schizophrenia among the relatives of schizotypal probands. Battaglia *et al.* (1991) found an increased risk of schizophrenia among the relatives of 'pure' schizotypal personality disorder probands but not among the relatives of probands with a mixed borderline/schizotypal personality disorder. Elsewhere an increased risk of schizophrenia was also found among the relatives of probands who had schizotypal and/or paranoid personality disorder (Thacker *et al.* 1993). Part of the problem of interpreting the results of family studies of schizotypal disorder which take as their starting point index cases with personality disorder rather than schizophrenia is that a schizotypal personality disorder commonly coexists with other personality disorder such as borderline personality disorder (see below). Family studies suggest that comorbidity may be an important issue. For example it seems that 'schizophrenia-related personality disorder traits' (schizotypal personality disorder/paranoid personality disorder not comorbid with borderline personality disorder) have a more specific familiar association with schizophrenia (Silverman *et al.* 1993).

Some workers have attempted to clarify the relationship between schizophrenia and schizotypal disorder by taking dimensional approaches and either using factor analysis to define subscales or by considering schizotypy as a single continuum. Mata *et al.* (2000)

investigated the relatives of 90 consecutively admitted schizophrenic patients. They carried out detailed examinations of the psychopathology presented by these probands, and interviewed their mothers regarding childhood personality, including premorbid schizotypal traits and social adjustment. Three schizotypy scales were derived by factor analysis from schizotypy questionnaires completed by the first-degree relatives. There was a correlation between all three schizotypy scales in relatives and positive symptoms (such as hallucinations and delusions) in the probands as well as a correlation between schizotypy in the relatives and pre-morbid schizoid–schizotypal traits in the probands. In a somewhat different design Bergman *et al.* (2000) compared the patterns of schizotypal symptoms in the first-degree relatives of two groups of probands, the first having schizophrenia and the second having personality disorders. Again using a factor analysis derived set of subscales these authors found that the structure of schizotypal symptoms in the relatives of schizophrenics was similar to the three factor model that has often been found in overt schizophrenia consisting of cognitive/perceptual, interpersonal and disorganization factors. However, the model did not fit with the structure of schizotypal symptoms that was found in the relatives of personality disorder patients.

Finally Jones *et al.* (2000) measured schizotypy in the non-schizophrenic relatives of families containing at least two members affected by schizophrenia. Somewhat suprisingly they found no increase in the schizotypy scores, but speculated that this may have been due to 'volunteer effect', since the study set out to examine just one well relative per family and possibly this resulted in a bias in favour of the least schizotypal members.

Twin and adoption studies of schizotypy

Twin data on schizotypal personality disorder are sparse. In a re-analysis of the twin data on schizophrenia of Gottesman and Shields (1972), Farmer *et al.* (1987) found a higher MZ/DZ concordance ratio when both schizophrenia and schizotypal personality disorder were included in the definition of concordance. Although there is a need for caution in using the MZ/DZ ratio as an index of the degree of genetic contribution, the results support a genetic relationship between schizotypal personality disorder and schizophrenia, and suggest that a slightly broadened concept of schizophrenia may constitute a more 'genetically valid' phenotype. A subsequent study by Onstad and Colleagues (1991) found an essentially similar result with the addition of schizotypal personality disorder to schizophrenia increasing the MZ/DZ concordance ratio.

A study concentrating on schizotypy indicators alone and taking a dimensional approach adds another perspective. MacDonald *et al.* (2001) used a set of self-report questionnaires to investigate schizotypy in a sample of normal young adult twins consisting of 98 monozygotic and 59 same sex dizygotic pairs. Although there was evidence of a familial effect, the analysis could not clearly distinguish whether a common schizotypy factor received its most important influence from genes or shared environment, and concluded that contrary to their expectations 'positive' schizotypal symptoms were not strongly genetically influenced in their community-based sample. This was in contrast with some earlier studies using a measure of schizotypy devised by Claridge which was found to be heritable in community-based twin samples (Claridge and Hewitt 1987; Kendler and Hewitt 1992). The discrepancy maybe explained by the fact that these earlier studies found that some components of schizotypy such as anhedonia (loss of the ability to experience pleasure) that are akin to negative schizophrenic symptoms, appear to be more heritable, although lacking specificity for schizotypy per se.

Current concepts of schizotypy owe much to the Danish adoption studies of schizophrenia. The original rates of schizophrenia and schizophrenia related personality disorder were 20 per cent among the biologic relatives of schizophrenics and 6 per cent among the adopted relatives of controls. When the same data were re-examined using stricter DSM-III criteria, the rates of schizophrenia and schizotypal personality disorder rose to 22 per cent among biologic relatives and dropped to 2 per cent among the adopted relatives and controls (Kendler *et al.* 1981; Kendler and Gruenberg 1984), which suggests that the narrower DSM concept of schizotypal personality disorder is more meaningful at a genetic level than the previously broader concept of spectrum disorder. Similarly, higher rates of paranoid personality disorder were found among the biological relatives of schizophrenics, although this was less common than schizotypal personality disorder.

In conclusion the family, twin and adoption data overall on schizotypal and paranoid personality disorders indicate a genetic component and a genetic relationship with schizophrenia (Kendler 2000; Prescott and Gottesman 1993). The results are most compelling at the severe end of the spectrum in studies where the ascertainment has been via probands with schizophrenia and the results are less clear and more difficult to interpret in studies where ascertainment has been via broader groups of probands, including those with schizotypal personality disorder, and in studies that have conceptualized schizotypy as a dimension that is continuously distributed in the population. There is also at least a hint that the components of schizotypy that are similar to schizophrenic positive symptoms may be less heritable than negative symptom-like components.

Cluster B personality disorders

Antisocial personality disorder

Most of the early studies of the genetics of antisocial behaviour were based upon probands who had been convicted of crimes, often including minor ones, rather than probands who met clinical criteria for antisocial personality disorder. As have we have already noted, few would accept that recorded criminal activity and antisocial personality disorder are the same, but criminal conviction is at least a definable marker. One of the earliest twin studies was by Lange (1931), the results of which were published in a provocatively titled book '*Crime as Destiny*'. Ten of 13 MZ pairs were found to be concordant for criminality compared with 2 of 17 DZ pairs. Although, on the face of it, this suggested a genetic component, the study was open to criticism both for its small sample size and its lack of systematic ascertainment that may have led to bias in the direction of an over estimate of concordant MZ pairs.

Subsequent studies using criminality as an indicator of antisocial behaviour have taken some trouble to overcome ascertainment biases and, considering the imprecision of the definition, have produced consistent results. The results of twin studies of adult criminality for North America, Japan and Europe combined gave a pairwise concordance of 52 per cent in a total of 229 MZ pairs compared with 23 per cent in 316 DZ pairs (Goldsmith and Gottesman 1996). Results from the single largest study to date of criminality in adult twins are summarized in Table 8.2. In this study, Cloninger and Gottesman (1987), drawing on the monumental work of the late Karl Otto Christiansen in Denmark, attempted to tackle one of the problems that needs to be confronted in applying a quantitative genetic model to criminality. As all societies contain many more criminal men than women, Cloninger and Gottesman adopted a threshold model of criminality where it is assumed that liability to criminal behaviour is contributed by multiple genetic and environmental factors such that the distribution of liability in the general population will tend to be normally distributed; only those individuals whose liability exceeds a threshold become classified as 'affected'. This is the general model first put forward by Falconer (1965) to explain the inheritance of common familial disorders (see Chapter 2 for a more detailed discussion of liability–threshold models) and first applied to psychopathology by Gottesman and Shields (1967). Cloninger and Gottesman applied an extension of the Falconer model to include two thresholds, one for a broad or common form of the disorder and the other for the narrow or more severe form (Reich *et al.* 1972). In the most straightforward form of such a model it is predicted that the relatives of narrow form probands will be more often affected than the relatives of broad form cases. This is because a narrow form disorder occupies a more extreme position on a liability continuum and thus includes cases with greater 'loading' of genetic and/or environmental risk factors. As applied here, the broad form was male criminality and the narrow form was female criminality even though, as we see in the table, the cross-sex correlations (male–female) were somewhat lower than the within sex (male–male and female–female), Cloninger and Gottesman found that all of these correlations could be constrained to be equal whilst still giving a satisfactory fit to the data. They therefore concluded that criminality is equally heritable in men and women with the heritability estimate of approximately 54 per cent. Furthermore, in contrast with a great majority of normal personality traits (see Chapter 4) antisocial behaviour, as indexed by criminality, showed a substantial shared environmental effect accounting for about 20 per cent of the variance.

Evidence for genetic contribution to antisocial personality disorder or criminal behaviour also comes from adoption studies. As discussed in Chapter 2, there are three types of adoption studies. Studies of adoptees themselves compare the rates of criminality in the

Table 8.2 Concordance and correlation for registered criminality in a Danish Twin Sample (data from Cloninger and Gottesman 1987)

Zygosity	Pairing: proband–twin	N of pairs	N of affected	N of concordant pairs	Probandwise rates		Tetrachoric correlation
					Freq./n	(%)	
MZ	Male–male	365	73	25	50/98	51.0	0.74 ± 0.07
MZ	Female–female	347	15	3	6/18	33.3	0.74 ± 0.12
DZ	Male–male	700	146	26	52/172	30.2	0.47 ± 0.06
DZ	Female–male	2,073	30	7	7/30	23.3	0.23 ± 0.10
DZ	Male–female	2,073	198	7	7/198	3.5	0.23 ± 0.10
DZ	Female–female	690	28	2	4/30	13.3	0.46 ± 0.11

adopted away offspring of criminal parents with the rate in control adoptees who do not have a criminal biological parent. Studies of adoptees' families compares the rates of criminality or antisocial personality among the biological and adopted families who are adopted away early in life and have been convicted as criminals. Finally, cross-fostering studies compare the rates of criminality in the adopted away offspring of criminal biological parents raised by non-criminal adoptive parents, with the rates in the offspring of non-criminal biological parents raised by criminal adopting parents. Studies of the first type have shown higher rates of antisocial personality as well as significantly more convictions, arrests and imprisonments among this group compared with controls (Crowe 1972, 1974).

In a study of psychopathy in adoptees' families in Denmark, the rate of disorder was significantly higher among the biological relatives of adoptees who were psychopathic compared with adoptive relatives and controls (Schulsinger 1972). Similarly, a genetic influence on antisocial behaviour and antisocial personality was found in a series of studies of adoptees families in the United States (Cadoret 1978; Cadoret and Cain 1980; Cadoret *et al.* 1985). On the other hand, the initial results of a Swedish adoption study suggested that genetic factors had little or no influence on antisocial behaviour (Bohman 1978), but later analysis, allowing for the confounding effects of alcohol abuse, showed that genetic influences were important for recurrent petty crime. Violent repetitive crime appeared to be largely related to alcoholism (Bohman *et al.* 1982; Cloninger *et al.* 1982). Both this study and a subsequent Danish study (Mednick *et al.* 1984) suggested that criminality, although influenced by genes, is heterogenous with genetic factors being more influential for property crimes and petty recidivism but less important for violent crimes against persons.

Just as twin studies provide important evidence that liabiltity to criminality and antisocial personality are influenced by environmental factors of some sort, adoptive studies have pointed to some of those environmental factors. In Swedish studies (Sigvardsson *et al.* 1982) men who had been in multiple temporary placements and whose adoptive homes were of low socio-economic status were at higher risk for criminality, whereas in women the important environmental risk factors were prolonged institutional care and being reared in an urban environment.

Using the Danish National Criminal register Medick *et al.* (1984), were able to carry out a cross-fostering analysis investigating the effects of genetic and environmental factors simultaneously. It was found that where neither the biological nor the adopting parents had a criminal record 13.5 per cent of adoptees had a history of conviction, compared with a rate of 14.7 per cent of adoptees who had ever been convicted where the *adoptive* fathers were 'known to the police' but there was no known history of criminal activity in the biological parents. This small difference was not significant. However, the rate for adopted away offspring of biological criminal fathers reared by non-criminal adoptive parents was elevated at 20 per cent, and the highest rate of criminality, at 24.5 per cent, was found in the adoptees whose biological and adoptive fathers had a police record. These results point to both genetic influences and influences of the environment of rearing and hint at, but do not provide clear evidence for, non additive gene–environment interaction. It is of interest that the base rate in the non-adoptee population at the time was 9 per cent for felony offending; the difference from 13.5 per cent as the base rate in adoptees *per se* implicates some aspect of parents whose offspring appear in an adoption pool (Loehlin *et al.* 1985).

The question of interaction between genes and environment was explored in a subsequent study in the United States (Cadoret *et al.* 1995). A total of 95 men and 102 women who were separated at birth from their biological parents were studied, and a comparison was made between those adoptees whose biological parents had a documented history in prison or hospital records of antisocial behaviour or alcohol abuse and those whose biological parents had no known history of such traits. As well as assessing antisocial behaviours in adulthood the investigators enquired about adolescent aggression and conduct disorder in the adoptees. The main findings were that a biological background of antisocial personality disorder predicted increased adolescent aggression, conduct disorder and antisocial behaviour. Adverse adoptive home environment, which included marital or legal problems, psychiatric disorder or substance abuse also predicted increased adult antisocial behaviours. Furthermore there was a significant interaction between adverse adoptive home environment and having a history of antisocial behaviour in a biological parent, in increasing the rate of aggression and conduct disorder in adoptees. In the absence of a biological background of antisocial behaviour an adverse adoptive home environment did not appear to increase the risk of aggression or conduct disorder in the adoptees.

Despite this fairly large amount of consistent evidence favouring a genetic contribution to antisocial behaviour as indexed by transgressions of the law, we would agree with Raine's (1993) conclusion that none of these studies included a compelling diagnosis of psychopathy, and this remains the case even in those studies published since his review. The alternative therefore is to take a dimensional approach and attempt to estimate the contributions of genes and environmental effects to measures of self reported antisocial behaviour, aggression or violence in population-based samples of twins. Studies carried out over the past two decades are summarized in Table 8.3. The majority suggest moderate heritability of around 50 per cent or less with just two studies focusing on aggression or hostility giving heritability estimates of around 70 per cent (Cates *et al.* 1993; Rushton *et al.* 1986). Interestingly, and in contrast with the studies just reviewed taking a categorical approach toward criminality or antisocial behaviour, the majority of dimensional studies of antisocial behaviour find no evidence of shared environmental effects with the exception of Livesley *et al.* (1993) on conduct problems; they found no additive genetic effects with over half of the variance accounted for by shared environment.

Symptoms of antisocial personality disorder, or scales designed to measure aggression or psychopathy, have also been explored in twins reared apart. The main results are summarized in Table 8.4, the majority of estimates of heritability are somewhat higher

Table 8.3 Estimates of genetic and environmental influences on population variation in antisocial behaviour from twin studies

Heritability (%)	Environment (%)		Measure	Data source	N. of pairs	Age	Nation/sample	Authors	Year
	Common	Non-shared (and error)							
34	1	65	MMPI psychopathy	Self	133	20–25	USA, Indiana	Pogue-Geile and Rose in M&C	1985
70	No report	No report	Buss–Durkee, verbal, indirect, anger scales	Self	98	40–45	USA, midwest	Cates et al.	1993
58	0	42	MPQ aggression	Self	331	19–41	USA, Minnesota	Tellegen et al.	1988
0	53	47	Conduct problems	Self	175	16–71	Canada, Vancouver	Livesley et al.	1993
28–47	Nil	53–72	Buss–Durkee aggress. scales	Self	300	36–54	USA, Vietman Era Study	Coccaro et al.	1997
43	5	52	Antisocial personality disorder	Self	3226	36–55	USA, Vietnam Era Study	Lyons et al.	1995
74	0	26	Aggression	Self	136	19–64	UK, London	Rushton et al.	1986
55	0	45	Violence	Self	274	19–64	UK, London	Rushton	1996
35–39	0	61–65	MPQ aggression	Self	1257	27–64	USA, Minnesota	Finkel and McGue	1997
54	0	46	Composite of 18 aggression questionnaires	Self	247	Adult	Ohio and British Columbia	Vernon et al.	1999

Table 8.4 Twins reared apart

Heritability (%)	Environment (%)		Measure	Data source	N. of pairs	Age	Nation/sample	Authors	Year
	Common	Non-shared (and error)							
28	No report	No report	ASPD symptoms	Self	32	16–68	USA-UK	Grove et al.	1990
80	0	20	MPQ aggression	Self	71	19–68	USA, Minnesota	Tellegen et al.	1988
28	25	47	Socialization	Self	71	19–68	USA, Minnesota	Bouchard and McGue	1990
60	40	0	MMPI psychopathy	Self	76	19–68	USA, Minnesota	Gottesman et al.	1984
61	0	39	MMPI psychopathy	Self	119	18–77	USA, Minnesota	DiLalla et al.	1996

than in population-based studies of twins reared together, but we must note that four out of the five reports summarized in the table are based on the well known Minnesota series of twins reared apart and therefore the subjects in each of these reports overlap markedly or completely. Nevertheless we can conclude that the reared-apart twin studies and the population-based studies provide generally consistent support for a moderate genetic contribution to individual differences in antisocial behaviours. The value of a quantitative approach that attempts to grade antisocial behaviour along a severity continuum is also supported by adoption data. We earlier considered the Danish adoption data of Mednick *et al.* (1984) and the results favouring genetic effects when family members were classified as convicted/not convicted. However when the number of convictions in an individual's lifetime from age 15 to 50 was considered, stronger heritability estimates emerged for individuals who had multiple convictions.

Changes over time

Several reviews comparing twin study results for juvenile delinquency and adult criminality (e.g. McGuffin and Gottesman 1985; DiLalla and Gottesman 1989) have suggested that the genetic influence on delinquency and adolescent samples is low in contrast to the stronger genetic influence in crime in adult samples. It has been suggested that the transient involvement in delinquency is common among adolescents, and that this high prevalence of delinquency among ordinary adolescents having little heritable liability might make it difficult to uncover the genetic predisposition to antisocial behaviour, if it exists, in samples of young people.

A large scale attempt to take a longitudinal perspective and investigate both adolescent symptoms and behaviours during adult life in the same sample was carried out in 3226 pairs of male twins identified via a register of men who had served in the armed forces of the United States of America during the Vietnam era (May 1965–August 1975) (Lyons *et al.* 1995). All subjects were interviewed by telephone with a structured interview and genetic models were fitted. A summary of the main findings in adult life is shown in Table 8.5. Eight of the adult symptoms were significantly heritable and only one ('no regard for the truth') was significantly influenced by shared environment. However, the patterns of findings in the same subjects before the age of 15 years was different. Although significant heritabilities were found for some fairly prevalent items (e.g. truancy, fighting and cruelty to animals) other self-reported items such as lying, stealing and damaging property received stronger influences from shared environmental effects. Lyons *et al.* (1995) went on to carry out bivariate model fitting to examine the continuities between juvenile and adult antisocial behaviours. Although additive genetic factors explained about six times more variance overall in adult than juvenile traits, the juvenile genetic determinants overlapped completely with the genetic influences on adult traits. This would be compatible with a view that adult or antisocial behaviours in adolescents in general shows important environmental influences including substantial shared environmental effects, but as adolescents emerge into adulthood, those who show persistent antisocial behaviour are more likely to be those with a genetic predisposition towards such behaviour.

Table 8.5 Antisocial symptoms in adults (data from Lyons *et al.* 1995)

Symptoms	Prevalence (%)	rmz	rdz	A	C	E
Inconsistent work	16.1	0.34	0.15	34	—	66
Fails to conform to social norms	20.5	0.49	0.32	52	—	48
Aggressive	38.5	0.5	0.27	50	—	50
Fails to honour financial obligations	5.1	0.39	0.2	38	—	62
Impulsive	7.0	0.41	0.23	41	—	59
No regard for truth	2.7	0.15	0.28	—	77	23
Reckless	47.8	0.47	0.31	48	—	52
Irresponsible parent	1.2	0.22	—	—	—	—
Never monogamous	4.2	0.3	0.19	31	—	69
Lacks remorse	4.0	0.22	0.14	22	—	78

rmz and rdz are the tetrachoric correlations for MZ and DZ twins respectively. A, C and E are the proportions of phenotypic variance explained by additive genes, common environment and non shared environment respectively under the best fitting model

It is again worthwhile to consider composite or continuous measures of antisocial behaviour in adolescents. The lowest estimates of heritability tend to emerge from observational or self report measures which sample only a narrow type of behaviour (e.g. arguing) over brief spans of time. The next lowest estimates emerge from measures of official offending in juveniles which again necessarily sample a narrow range of behaviour, indeed one that is seldom detected (it has been estimated that fewer than half of juveniles who offend are arrested and of these 75 per cent are arrested only once or twice (Farrington *et al.* 1986). Medium estimates of heritability tend to emerge from measures of self-reported offending aggregated across a moderately wide sample of behaviours and longer periods of time. The largest estimates of heritability tend to emerge from studies using measures along a continuum of non antisocial to severely and persistently antisocial behaviours. This suggests that the way forward in researching antisocial behaviour in adolescents, as in adults, is to employ composite measures. Thus far studies in adolescents taking this general approach have been remarkably consistent. In contrast with the studies based on adjudicated juvenile delinquency (McGuffin and Gottesman 1985; Dilalla and Gottesman 1989) studies based on composite measures of behaviours have found heritabilities in the region of 50–60 per cent (Simonoff *et al.*1995; O'Connor *et al.* 1998; Burt *et al.* 2001). Taylor *et al.* (2000) were able to show differential effects of genetic factors and of peer influence for early versus late (after age 11) starters for childhood antisocial behaviours in twins followed until age 17. The findings provide an excellent start on resolving the issues raised earlier (DiLalla and Gottesman 1989; Moffitt 1993).

Borderline personality disorder

The concept of borderline personality or 'borderline states' originates primarily from psychoanalytic practice. Borderline states are said to be characteristic of unstable personality types that are associated with poor impulse control, self mutilation, suicide attempts and, often, promiscuity or petty criminal acts such as shoplifting. Borderline states and personality types have been well described by Gunderson and Singer (1975) and the general concept was modified and incorporated into DSM-III and then DSM-IV as borderline personality disorder. Because of the prominence of impulsivity and irresponsible or minor criminal behaviour, the authors of DSM-III placed this type of personality disorder in the same cluster as antisocial personality. However, the term borderline personality remains somewhat confusing since some authors have emphasised periods of 'micro-psychosis' and others have used the term 'borderline schizophrenia' to describe schizophrenia related disorders. In general borderline personality disorders are characterized more by depressive than by schizophrenic features (Stone 1981). Family studies of borderline probands also suggest that type of personality disorder has a familial relationship with affective disorder, with several studies showing higher than expected rates of affective disorder among relatives (Baron *et al.* 1985b; Gasperini *et al.* 1991; Loranger *et al.* 1982). However family studies that have sought to find borderline personality disorder among the relatives of depressed probands have, by and large, been negative (Coryell and Zimmerman 1989; Maier *et al.* 1992). Riso (2000) attempted to clarify the situation by comparing the rates of psychiatric disorders in the relatives of 119 outpatients with mood disorders, the relatives of 11 patients with borderline personality disorder and those of 45 never psychiatrically ill controls. Although the sample was rather small, the authors interpreted their result of increased rates of both mood disorders and personality disorders in the relatives of borderline probands as showing that the familial aggregation pattern was generally similar between the borderline personality and the mood disorder group. They postulated that there may be common aetiological factors between borderline personality disorders and mood disorders. It has also been reported that other cluster B personality disorders, histrionic and antisocial personality disorder are increased among the relatives of borderline personality disorder probands (Pope *et al.* 1983).

There have only been two published twin studies of borderline personality disorder, both from Norway. Torgersen (1984) first studied 25 pairs where at least one twin had a diagnosis of borderline personality disorder, but found no evidence for genetic effect. Torgersen and colleagues (2000) expanded their work to 92 monozygotic and 129 dizygotic twin pairs ascertained via twin and patient registers. All pairs were interviewed using a structured interview and DSM-IIIR personality disorder diagnoses were made. There was considerable overlap between personality disorders with many subjects fulfilling the criteria for two or more disorders. Interestingly, in this enlarged series the heritability for borderline personality was estimated at 69 per cent. This presumably was not detected by the earlier study because the small sample lacked power.

Hysterical/histrionic personality disorder

There has been so little agreement over the concept of hysteria that it is not surprising that the results of genetic studies of hysterical personality have also been inconsistent. In the only twin study of classical hysterical conversion or dissociative states (Slater 1961), the concordance rates of both 12 MZ and 12 DZ pairs were zero. In another early twin study where hysteria was measured as a personality trait using the Minnesota Multiphasic Personality Inventory (MMPI) (Gottesman 1963) there was again no evidence of significant genetic effects. These results contrast with the findings of two later twin studies taking a dimensional approach to hysterical personality traits. A study using the Middlesex Hospital Questionnaire (MHQ) found that hysterical personality traits were heritable (Young *et al.* 1971) but the measure used in this study may have been 'contaminated' by items measuring extraversion, which as has been discussed in Chapter 4, has consistently been found to be heritable. Using an hysterical dimension derived from factor analysis, Torgersen (1980) found higher correlations among MZ twins compared with DZ twins, especially in female pairs. Subsequently Torgersen *et al.* (2000) estimated that the heritability of histrionic personality disorder as defined by DSM-IIIR was 67 per cent. A Canadian study looking at dissociative experiences in normal volunteer twins found heritabilities of 48 per cent and 55 per cent for two measures derived from an instrument called the Dissociative Experiences Scale (Jang *et al.* 1998).

A somewhat different concept of hysteria, where the presentation involves multiple somatic complaints affecting multiple systems, has been described as Briquet Syndrome after the nineteenth century French physician who first described it; but it has also been called 'St Louis Hysteria' after the Washington University St Louis group who first revived interest in this condition (Guze *et al.* 1967). The St Louis group observed that there were high rates of Briquet Syndrome among the female relatives of men who were criminal or had antisocial personality disorder (Cloninger and Guze 1973). Briquet Syndrome corresponds closely to what is now classified in DSM-IV as somatization disorder characterized by multiple physical symptoms with no identifiable organic cause. Although somatization disorder is classed as an Axis I disorder, it is in many ways more akin to an Axis II or personality disorder because of its enduring nature and because of the strong familial overlap with antisocial personality disorder (Cloninger *et al.* 1975).

Cluster C anxious and avoidant personalities

Although, as reviewed in Chapter 12, there is now fairly good evidence that anxiety disorders are influenced by genetic factors, comparatively little is known about the genetics of anxious personality disorders specifically. This may in part be due to the difficulty in reliably distinguishing between anxiety states and more enduring personality disorders or traits. Several studies have focused on 'normal' phobias or fears, for example Torgersen's (1979) study based on a Norwegian twin register found that MZ twins were more alike than DZ twins on questionnaire-based measures fears of animals, social fears, mutilation fears (e.g. medical procedures, injury and blood), separation fears and 'nature' fears (e.g. heights). In all cases except for separation fears the MZ/DZ differences were significant. Genetic factors also appeared to be important for a wide range of common fears in a study of college age twins (Rose *et al.* 1981) and in a study of adolescents (Stevenson *et al.* 1992). It has been pointed out (Marks 1986) that the common cues for fears tend to consist of hazards that would be appropriate for our ancestors (e.g. snakes or spiders) rather than more common modern hazards such as weapons or traffic, and it has been suggested that the capacity to react with anxiety to certain 'prepotent' stimuli may in earlier times have conferred an evolutionary advantage. The most recent twin data on avoidant personality disorder from Torgersen and colleagues (2000) found it to be one of the less genetically influenced personality types with a heritability of 28 per cent, but the accuracy of the estimate is limited by the small sample size.

Regarding obsessional personality disorder the distinction between obsessional traits and obsessive-compulsive disorder is again not always easily made. As outlined in Chapter 12, the data on clinically defined obsessive-compulsive disorder are fragmentary and somewhat inconsistent but nevertheless overall suggest a genetic contribution. Torgersen *et al.* (2000) in their twin study found that obsessional personality disorder was one of the most highly heritable personality disorder types. As noted in Chapter 12, a dimensional approach was adopted by Clifford *et al.* (1984) who used a questionnaire, the Leyton Obsessional Inventory, to assess the genetic influences on obsessional traits and symptoms in volunteer twins. They found moderate heritabilities for both (47 per cent for obsessional symptoms and 40 per cent for obsessional traits) and in addition found a highly significant correlation between obsessional symptom scores and neuroticism. They suggested that genetic factors could influence obsessive-compulsive disorder by contributing both to neuroticism and obsessional personality traits, a hypothesis that is yet to be tested using current multivariate approaches to structural equation modelling.

The biology of personality disorder

The evidence reviewed so far from quantitative genetic studies suggests that despite problems in definition, overlap between disorders and gaps in the information, personality disorder is substantially influenced by genes. A crucial question, therefore, is can we progress further and discover mutations or genetic variations that predispose individuals to antisocial behaviour? The mode of transmission of the personality disorders seems highly likely to involve multiple genes in combination with environmental effects and although some single gene defects, for example a mutation in the monoamine oxidase A (MAO A) gene (Brunner *et al.* 1993), are associated with aggression in a single extended pedigree, such defects are extremely rare and may not be relevant to common forms of antisocial behaviour.

There has also been much debate on the relationship between chromosomal anomalies and antisocial behaviour. In particular a form of aneuploidy, the XYY syndrome has a putative association with antisocial behaviour and criminality. Males who have the karyotype 47,XYY are now thought to have no characteristic physical abnormalities other than their markedly above average height. The majority probably have IQs in the dull–normal range, although most of the first described cases were institutionalized with mental handicap. The relationship of the syndrome to antisocial behaviour has been much debated. One of the earlier studies (Jacobs *et al.* 1968) found that XYY males accounted for about 3 per cent of the inmates of Carstairs Special Hospital in Scotland, an Institution for mentally abnormal offenders. Subsequently other studies in institutions found a consistent slight excess of males with the 47,XYY karoytype. However, the instance of the abnormality among new born males is in the region of 1–2 per 2000, and since the syndrome has not been shown to shorten life it is likely that the majority of such infants reach adulthood. Therefore, it is clear that only a minority of XYY males are so conspicuously abnormal as to be hospitalized or incarcerated in special institutions. A survey of over 4000 ostensibly normal men of more than 1.84 m in height (Witkin *et al.* 1976) called up for universal military service in Denmark found 12 individuals (0.3 per cent of the total population) who were XYY. Five of these 12 (42 per cent) had criminal records compared with only 9 per cent of XY tall males. However, the offences committed by the XYY men included relatively minor crimes and were not predominantly acts of aggression. Thus there appears to be some evidence of deviant tendencies among non-institutionalized XYY men, but seriously psychopathic or criminal individuals account for only a small proportion of those with the syndrome. One might therefore speculate that the excess of XYY individuals in special hospitals is as much related to the response of the judiciary to convicted individuals who are tall and below average intelligence as to a specific association between criminality and the 47,XYY karyotype. Some support for this comes from the population-based study that identified 17,XYY men from over 34,000 British males who were screened at birth (Gotz *et al.* 1999). Compared with XY controls and men with the 47,XXY karyotype (Klinefelter's syndrome), XYY men showed a significantly higher rate of antisocial behaviour in adolescence and adulthood and of criminal convictions. However, multiple regression analysis showed that this was mainly mediated through lower intelligence.

Perhaps because of the difficulties of studying families in which personality disorders segregate, there has been much less molecular research on personality disorders than on many other forms of psychiatric disease. With the exception of MAO A mutation family (Brunner *et al.* 1993) there have been no modern linkage studies on personality disorder, however, two alternative types of study that are beginning to provide some leads on the molecular basis of aggressive, antisocial and other disordered forms of behaviour are animal studies and studies involving candidate genes in humans.

Animal studies

Animal models have been pursued particularly with regard to the trait of 'fearfulness', which might be considered as a model for Cluster C personality disorders (see Chapter 12) and in aggression, which, in humans, forms an important part of the symptoms of Cluster B disorders, particularly antisocial personality disorder. Genetic influences on aggressive behaviour in mice have been reviewed by Maxson (1999, 2000).

Breeding experiments show that aggressive behaviour and non-aggressive behaviour can be selected for. For example, so called short attack latency and long attack latency strains have been selected from wild type parental stock upon the basis of a resident/intruder paradigm, that is how quickly a resident mouse attacks when an intruder is introduced to its cage (Van Oortmerssen and Sluyter 1994). More generally, aggressive behaviour in mice has been described as consisting of four types: offence, defence, infanticide and predation. Clearly one of the problems about mouse models is that each of these, with the exception of infanticide, can be seen as biologically adaptive and therefore not necessarily a good model for maladaptive, violent, aggressive or antisocial behaviour in humans. Nevertheless, findings in mice could help narrow down chromosomal regions or highlight candidate genes which can subsequently be explored in humans.

As in humans male mice are more aggressive than females, and hence there has been considerable interest in the role of the Y chromosome. Although the results are controversial (Maxson 2000) the sex determining region of the Y chromosome (SRY) is a candidate gene for the affects of the male specific part of the Y chromosome on offensive behaviour. There is also evidence that one or more genes in the t region of chromosome 7 have effects on offence in male mice. The region is involved in normal testis development and is linked to the major histocompatability complex (MHC) that is homologous to the region in man that contains the genes for the HLA system.

An approach that has been widely used in recent years is to study behaviour in genetically engineered mice where both working copies of a particular gene are disrupted or 'knocked out' (Campbell and Gold 1996). Two types of knock-out mice that show high levels of aggressive behaviour have been created recently. In one such strain there is a lack of brain nitric oxide synthase (NOS) (Nelson *et al.* 1995) and in the other strain the gene encoding the serotonin receptor, 5-HT1b is absent (Saudau *et al.* 1994). In both cases the abnormal behaviour is seen in homozygous knock outs, that is, animals are first engineered that are effectively heterozygous with one working copy and one disrupted copy of the gene and the heterozygotes are then crossed to produce a homozygous offspring with no working copies. It can be argued that such engineered loss of function does not necessarily tell us much about the biological basis of naturally occurring aggression, and certainly it should not be assumed that either the nitric oxide synthase gene or the 5-HT1b receptor gene is 'the gene for' aggression. But it could be that spontaneously occurring variations in these genes have a role in aggressive behaviour in mice and, by extrapolation, they have some role in antisocial behaviour in humans. A transgenic mouse in which the MAO-A gene was fortuitously disrupted by an insertion mutation has also been found to show increased aggression (Cases *et al*. 1995) and, as already noted, the human mutation that disrupts the activity of this gene has been found to be associated with violent behaviour and mild mental retardation in one family.

It has recently been shown that the excessive aggressiveness and impulsiveness of NOS knockout mice is associated with a decrease in serotonin turnover in the brain and deficient 5-HT1B and 5HT1A receptor function. Therefore the findings on these 3 types of genetically engineered mice, the NOS and 5-HT1B knockouts and the mutated MAOA mouse would seem to converge and lead us to conclude that the serotonergic transmission plays an important role in aggressive behaviour (Chiavegatto *et al*. 2001). Although genetic data on higher mammals are lacking, pharmacological and neurochemical studies support a relationship between lowered serotonegic transmission and increased impulsivity or aggression. For example a study of male vervet monkeys that used a resident-intruder type paradigm found that impulsivity as indexed by a shorter latency in the resident approaching the intruder was inversely correlated with levels of the serotonin metabolite 5-HIAA in cerebrospinal fluid and that such impulsivity could be reduced by pretreatment with the anti-depressant fluoxetine which increases the availability serotonin synapses (Fairbanks *et al*. 2001).

Genes in other systems that have been implicated in the control of aggressive behaviour fall into seven main groups or systems plus a miscellaneous group. These include neurotransmitters, steroids, signalling proteins, growth factors, genes involved in neurodevelopment, transcription factors and an immune system gene (Maxson 2000; Nelson and Chiavagatto 2001; Sluyter *et al*. 2002). The mutant form, knockout or rare variant by no means always results in increased aggression and some knockout mice, such as those lacking the oxytocin gene or the histamine 1 receptor gene, show reduced aggression. Table 8.6 summarizes the main published findings and the types of genetic study involved, which include linkage studies and studies of transgenic animals as well as gene knockouts.

Most of the examples given in the table relate to studies on male mice and Maxson (2000) points out that there is a long-standing debate on whether the same genes affect aggression in males and females. For example some genes such as NOS appear to affect only male aggression. Clearly sex effects and the candidate genes involved in aggression in mice need to be explored further. An added difficulty is that there are, at least descriptively, four different types of aggression and few mutant strains have been tested across multiple types of experiment designed to measure aggression. Indeed only one of the genes listed in Table 8.6 (CamK2a) has been tested for its effect on both offence and defence, both of which are reduced in homozygous mutants.

Candidate genes in humans

Nearly all of the recent work in humans has focused on components or dimensions of personality disorder rather than specific categories as such. Thus, as with animal studies,

Table 8.6 Genes implicated in aggression in male mice (data from Maxson 1998, 2000; Nelson and Chiavegatto 2001; Sluyter *et al.* 2002)

System	Gene (symbol)	Method	Effect of mutation/variation on aggression
Steroid Metabolism	Androgen receptor (Ar)	Linkage	
	Aromatase (cyp19)	KO	Decrease
	Estrogen receptor A (Estra)	KO	Decrease
	Estrogen receptor B (Estrb)	KO	Increase [but depending on test day]
	Estrogen receptor A and B (Estra and Estrb)	KO	Decrease
	Steroid sulfatase	Linkage	
Neuro-transmission	Adenosine 2a receptor (Adora2a)	KO	Increase
	Alpha2-adrenoceptor (A$_{2C}$)	KO	Increase
	Brain creathine kinase (B-CK)	KO	Increase
	Catechol-O-methyltransferase (COMT)	KO	Increase [+/−; −/− = normal]
	Dopamine receptors	KO	Increase
	Enkephalin (Penk)	KO	Increase
	Glutamic-acid decarboxylase (GAD65)	KO	Decrease
	Histamine 1 receptor (H1r)	KO	Decrease
	5-HT1$_B$recepter (Htr 1b)	KO	Increase
	5-HT1$_A$recepter (Htr 1a)	KO	Possibly decrease
	Monoamine oxidase A (Maoa)	IM	Increase
	Neurokinin-1 (Nk1)	KO	Decrease
	Neutral endopeptidase (NEP)	KO	Increase
	Nitric oxide synthase (nNos1)	KO	Increase [but depending on background]
	Oxytocin (Oxt)	KO	Decrease/increase [depending on design]
	Vasopressin 1b (V1b)	KO	Decrease
	Vgf polypeptide (vgf)	KO	Decrease
Signalling proteins	α-Calcium/calmodulin kinase II (CamK2A)	KO	Increase/decrease [depending on gene dosage and type of aggression]
	Regulator of G protein signalling (RSG2)	KO	Decrease
	Breakpoint cluster region (BCR)	KO	Increase
Growth factors	Transforming growth factor α (Tgfa)	Transgenic	Increase
	Brain-derived neurotrophic factor (BDNF)	KO	Increase
Transcription factor Development	Pet-1	KO	Increase
	Neural cell adhesion molecule (Ncam)	KO	Increase
	Tailless (T1x)	KO	Increase
	Dishevelled (Dvl1)	KO	Decrease
	Arg Tyrosine kinase (Arg)	KO	Decrease
Immunological factors Miscellaneous	Interleukin-6 (IL-6)	KO	Increase
	Nitric oxide synthase (eNOS)	KO	Decrease
	Gastric-releasing peptide	KO	Increase, [but depending on paradigm]
	Transformation related protein (p73)	KO	Decrease

KO = knockout, IM = instertional mutation, transgenic = over expressed human transgene particular attention has been paid to exploring the molecular genetic basis of impulsivity and aggression. While some studies have implicated increased noradrenergic transmission in impulsive aggression, other studies have demonstrated the opposite so that the role of nonadrenalin in such behaviour remains unclear (Oquendo and Mann 2000). However, as in mice, there are consistent data implicating abnormalities of serotonergic transmission in aggression, impulsiveness and self harm (Dolan *et al.* 2001; Mann *et al.* 2001).

As outlined in Chapter 9, there has also been considerable interest in the association between mood disorders or predisposition to mood disorders as reflected in neuroticism and serotonin. A functional polymorphism in the serotonin transporter gene has been the subject of several investigations. This polymorphism has essentially two alleles, a 'short' and a 'long' form. The 'short' allele has been associated with depression or increased neuroticism in some but not all studies. There does, however, appear to be greater consistency in studies that are examined in association with impulsivity. For example the 'short' serotonin transporter allele was found to be associated with an increased risk of so called type 2 alcoholism (see also Chapter 11) where prominent features are antisocial and impulsive, violent behaviour (Hallikainen *et al.* 1999). The same allele has been associated with an increased risk of suicide attempts, particularly severe or repetitive self harm in the context of alcoholism (Gorwood *et al.* 2000) and the authors speculated that there might also be a more general association with aggressive or impulsive behaviour.

Conclusions

There are clearly large problems remaining in the classification and nosology of personality disorders, and the debate over the usefulness of a dialectical solution to the categorical versus a dimensional approach is far from resolved. In spite of this there is consistent evidence of at least a moderate effect of genes in predisposing to antisocial behaviour. The interplay of such genes with the environment probably involves both additive effects (co-action) and non-additive gene–environment interaction. There is also consistent evidence that schizotypal disorder is genetically influenced and that the liability to schizotypal personality disorder overlaps with liability to schizophrenia. The data are less clear and consistent where investigators have attempted to measure schizotypy using a continuous scale. Quantitive genetic data are comparatively sparse on the anxious/avoidant cluster of personality disorders, but such results as are available tend to suggest familial clustering that is influenced by genes. Furthermore subjects who attract a diagnosis in the anxious/avoidant cluster tend to fulfil criteria for more than one type of personality disorder when they are investigated using standardized interviews or questionnaire measures (Farmer *et al.* 2002).

Animal models of personality disorder (Flint 2002) are imperfect but give strong support for the hypothesis that traits such as anxiety, impulsivity and aggression are influenced by genes. Studies using linkage designs or genetically engineered animals have implicated at least 15 different genes in aggression in male mice. The most consistent pattern is of an association between aggressive or impulsive behaviour and genetic variants or mutants that result in reduced serotonergic transmission. This is also consistent with neurochemical and pharmacological studies in humans as well as with some allelic association studies. However, molecular genetic studies of personality disorders and the extremes of personality traits are few, and future studies will need to follow the leads resulting from animal models as well as systematic search strategies to find the genes involved in personality disorders.

Further reading

Benjamin J., Ebstein R. P., Belmaker R. H. (2002) *Molecular Genetics and the Human Personality*. Washington, DC: American Psychiatric Press.

Moffitt T. E. (1993) Adolescence-limited and life-course-persistent antisocial behavior: a developmental taxonomy. *Psychological Review* **100**:674–701.

Lenzenweger M. F. (1998) Schizotypy and schizotypic psychopathology: mapping an alternative expression of schizophrenia liability, in M. F. Lenzenweger and R. H. Dworkin (eds) *Origins and Development of Schizophrenia: Advances in Experimental Psychopathology*. Washington, DC: American Psychological Association.

Meehl P. E. (1990) Toward an integrated theory of schizotaxia, schizotypy, and schizophrenia. *Journal of Personality Disorders* **4**:1–99.

References

American Psychiatric Association (1980) *Diagnostic Criteria from DSM-III*. American Psychiatric Association, Washington DC.

American Psychiatric Association (1993) *Diagnostic Criteria from DSM-IV*. American Psychiatric Association, Washington DC.

Asarnow R. F., Nuechterlein K. H., Fogelson D. *et al.* (2001) Schizophrenia and schizophrenia-spectrum personality disorders in the first-degree relatives of children with schizophrenia: the UCLA family study. *Archives of General Psychiatry* 58(6):581–8.

Baron M., Gruen R., Asnis L. *et al.* (1983) Familial relatedness of schizophrenic and schizotypal states. *American Journal of Psychiatry* **140**:1437–42.

Baron M., Gruen R., Rainer J. D. *et al.* (1985) A family study of schizophrenic and normal control probands: Implications for the spectrum concept of schizophrenia. *American Journal of Psychiatry* **142**:447–55.

Baron M., Gruen R., Asnis L. *et al.* (1985b) Familial transmission of schizotypal and borderline personality disorders. *American Journal of Psychiatry* **142**:927–34.

Battaglia M., Gasperini M., Siuto G. *et al.* (1991) Psychiatric disorders in the families of schizotypal subjects. *Schizophrenia Bulletin* **17**:659–65.

Benjamin J., Ebstein R. P. and Belmaker R. H. (2002) Genes for human personality traits: endophenotypes of psychiatric disorders? In J. Benjamin, R. P. Ebstein and R. H. Belmaker (eds). *Molecular Genetics and the Human Personality*.

Bergman A. J., Silverman J. M., Harvey P. D. *et al.* (2000) Schizotypal symptoms in the relatives of schizophrenia patients: an empirical analysis of the factor structure *Schizophrenia Bulletin* **26**(3):577–86.

Bohman M. (1978) Some genetic aspects of alcoholism and criminality. A population of adoptees. *Archives of General Psychiatry* **35**:267–76.

Bohman M., Cloninger C. R., Sigvardsson S. *et al.* (1982) Predisposition to petty criminality in Swedish adoptees. I. Genetic and environmental heterogeneity. *Archives of General Psychiatry* **35**:267–76.

Bouchard T. J. and McGue M. (1990) Genetic and rearing environmental influences on adult personality: An analysis of adopted twins reared apart. *Journal of Personality* **58**:263–92.

Brunner H. G., Nelen M. R., Van Zandvoort P. *et al.* (1993) X-linked borderline mental retardation with prominent behavioural disturbance: Phenotype genetic localization and evidence for disturbed monoamine metabolism. *American Journal of Human Genetics* **52**:1032–9.

Cadoret R. J. (1978) Psychopathology in adopted-away offspring of biologic parents with antisocial behaviour. *Archives of General Psychiatry* **35**:176–84.

Cadoret R. J. and Cain C. (1980) Sex differences in predictors of antisocial behaviour adoptees. *Archives of General Psychiatry* **37**:1171–5.

Cadoret R. J., O'Gorman T. W., Troughton E. *et al.* (1985) Alcoholism and antisocial personality—interrelationships, genetics and environmental factors. *Archives of General Psychiatry* **42**:161–7.

Cadoret R. J., Yates W. R., Troughton E. *et al.* (1995) Genetic-environmental interaction in the genesis of aggressivity and conduct disorders. *Arch Gen Psychiaty* **52**:916–24.

Campbell I. L. and Gold L. H. (1996) Transgenic modelling of neuropsychiatric disorders. *Molecular Psychiatry* **1**:105–20.

Cases O., Seif I., Grimsby J. *et al.* (1995) Aggressive behaviour and altered amounts of brain serotonin and norepinephrine in mice lacking MAOA. *Science* **268**:1763–6.

Cates D. S., Houston B. K., Vavak C. R. *et al.* (1993) Heritability of hostility-related emotions, attitudes, and behaviors. *J of Behavioral Medicine* **16**:237–56.

Chiavegatto S., Dawson V. L., Mamounas L. A. *et al.* (2001) Brain serotonin dysfunction accounts for aggression in male lacking neuronal nitric oxide synthase. *Proc Natl Acad Sci* **3**:1277–81.

Claridge G. and Hewitt J. K. (1987) A biometrical study of schizotypy in a normal population. *Personality and Individual Differences* **8**:303–12.

Clifford C. A., Murray R. M., Fulker D. W. (1984) Genetic and environmental influences on obsessional traits and symptoms. *Psychological Medicine* **14**:791–800.

Cloninger C. R. and Gottesman I. I. (1987) Genetic and environmental factors in antisocial behaviour disorders, in S. A. Mednick, T. E. Moffitt, S. A. Stacks (eds) *Causes of Crime. New Biological Approaches*. Cambridge, UK: Cambridge University Press.

Cloninger C. R., and I. I., Guze S. B. (1973) Psychiatric illness in the families of female criminals. A study of 288 first-degree relatives. *British Journal of Psychiatry* **127**:697–703.

Cloninger C. R., and Reich T., Guze S. B. (1975) The multifactorial model of disease transmission III. Familial relationship between sociopathy and hysteria (Briquet's Syndrome) *British Journal of Psychiatry* **127**:23–32.

Cloninger C. R., and Sigvardsoon S., Bohman M. *et al.* (1982) Predisposition to petty criminality in Swedish adoptees II. Cross fostering analysis of gene-environment interaction. *Archives of General Psychiatry* **27**:600–3.

Coccaro E. F., Bergeman C. S., Kavoussi R. J. *et al.* (1997) Heritability of aggression and irritability: A twin study of the Buss-durkee Aggression Scales in Adult Male subjects. *Biol Psychiatry* **41**:273–84.

Coryell W. M. and Zimmerman M. (1989) Personality disorder in the families of depressed, schizophrenic and never ill probands. *American Journal of Psychiatry* **146**:496–502.

Crowe R. R. (1972) The adopted offspring of women criminal offenders—a study of their arrest records. *Archives of General Psychiatry* **27**:600–3.

Crowe R. R. (1974) An adoption study of antisocial personality. *Archives of General Psychiatry* **31**:785–91.

DiLalla L. F. and Gottesman I. I. (1989) Heterogeneity of causes for delinquency and criminality: lifespan perspectives. *Development and Psycopathology* **1**:339–49.

DiLalla D. L., Carey G., Gottesman, I. I. *et al.* (1996) Heritability of MMPI personality indicators of psychopathology in twins reared apart. *Journal of Abnormal Psychology* **105**:491–9.

Dolan M., Anderson I. M., Deakin J. F. (2001) Relationship between 5-HT function and impulsivity and agression in male offenders with personality disorders. *British Journal of Psychiatry* **178**:352–9.

Essen-Moller E. (1946) The concept of schizoidia. *Monatsschrift fur Psychiatrie und Neurologie* **112**:258–71.

Fairbanks L. A., Melega W. P., Jorgensen M. J. *et al.* (2001) Social impulsivity inversely associated with CSF 5-H1AA and fluoxetine exposure in vervet monkeys. *Neuropsychopharmacology* **4**:370–8.

Falconer D. S. (1965) The inheritance of liability to certain diseases, estimated from the incidence among relatives. *Annals of Human Genetics* **29**:51–76.

Faraone S. V., Green A. I., Seidman L. J. *et al.* (2001) 'Schizotaxia': clinical implications and new directions for research. *Schizophrenia Bulletin* **27**(1):1–18.

Farmer A. E., McGuffin P., Gottesman I. I. (1987) Twin concordance for DSM-III schizophrenia. Scrutinizing the validity of the definition. *Archives of General Psychiatry* **44**:634–41.

Farmer A., McGuffin P. and Williams J. (2002) *Measuring Psychopathology.* Oxford University Press Inc. New York.

Farrington D. P., Ohlin L. and Wilson J. Q. (1986) Understanding and controlling crime. New York: Springer-Verlag.

Finkle D. and McGue M. (1997) Sex differences and nonadditivity in the heritability on the Multidimensional Personality Questionnaire scales. *Journal of Personality and Social Psychology* **72**:929–38.

Flint J. (2002) Animal models of personality, in J. Benjamin, R. P. Ebstein and R. H. Belmaker (eds). *Molecular Genetics and the Human Personality.*

Gasperini M., Battaglia M., Sherillo P. *et al.* (1991) Morbidity risk for mood disorders in the families of borderline patients. *Journal of Affective Disorders* **21**:265–72.

Goldsmith H. H. and Gottesman I. I. (1996) Heritable variability and variable heritability in developmental psychopathology, in M. F. Lenzenweger and J. Haugaard (eds) *Frontiers in Developmental Psychopathology.* Oxford: Oxford University Press.

Gorwood P., Batel P., Ades J. *et al.* (2000) Serotonin transporter gene polymorphisms, alcoholism and suicidal behaviour. *Biological Psychiatry* **48**(4):259–64.

Gottesman I. I., Shields J., (1963) Heritability of personality: a demonstration. *Psychol Monogr* **77**:1–21.

Gottesman I. I. and Shields J., (1972) *Schizophrenia and Genetics: A Twin Vantage Point.* New York and London: Academic Press.

Gottesman I. I. and Hanson D. R. (1982) *Schizophrenia—the Epigenetic Puzzle.* New York: Cambridge University Press.

Gottesman I. I. and Hanson D. R. (1967) A polygenic theory of schizophrenia. *Proceedings of the National Academy of Sciences* **58**:199–205.

Gottesman I. I., Carey G. and Bouchard T. J. (1984) *The Minnesota Multiphasic Personality Inventory of identical twins raised apart.* Paper presented at the 15th annual meeting of the behavior Genetics Association, Bloomington, IN.

Gotz M. J., Johnstone E. C., Ratcliffe S. G. (1999) Criminality and antisocial behaviour in unselected men with sex chromosome abnormalities. *Psychological Medicine* **29**(4):953–62.

Grove W. M., Eckert E. D., Heston L. *et al.* (1990) Heritability of substance abuse and antisocial behavior: A study of monozygotic twins reared apart. *Biological Psychiatry* **27**:1293–304.

Gunderson J. G. and Singer M. T. (1975) Defining borderline patients: An overview. *American Journal of Psychiatry* **132**:1–10.

Guze S. B., Wolfgran E. D., McKinney J. K. *et al.* (1967) Psychiatric illness in the families of convicted criminals. A study of 519 first-degree relatives. *Diseases of the Nervous System* **28**:651–9.

Hallikainen T., Saito T., Lachman H. M. *et al.* (1999) Association between low activity serotonin transporter promoter geneotype and early onset alcoholism with habitual impulsive violent behaviour. *Mol Psychiatry* **4**(4):385–8.

Heston L. L. (1970) The genetics of schizophrenia and schizoid disease. *Science* **167**:249–56.

Ingraham L. J. and Kety S. S. (2000) Adoption studies of schizophrenia. *Am J Med Genet* **97**:18–22.

Jacobs P. A., Price W. H., Cower-Brown W. M. *et al.* (1968) Chromosome studies on men in maximum security hospitals. *Annals of Human Genetics* **31**:339–58.

Jang K. L., Paris J., Zweigh-Frank H. *et al.* (1998) Twin study of dissociative experience. *Journal of Nervous and Mental Diseases* **186**(6):345–51.

Jones L. A., Cardno A. G., Murphy K. C. *et al.* (2000) The Kings Schizotypy Questionnaire as a Quantitative Measure of Schizophrenia Liability. *Schizophrenia Research* **45**:213–21.

Kendler K. S. and Gruenberg A. M. (1984) An independent analysis of the Danish Adoption Study of Schizophrenia. VI. The relationship between psychiatric disorders as defined by DSM-III in the relatives and adoptees. *Archives of General Psychiatry* **41**:555–64.

Kendler K. S. (2000) Schizophrenia: genetics, in B. J. Sadock and V. A. Sadock (eds) *Kaplan and Sadocks's Comprehensive Textbook of Psychiatry, Vol. 1.* Philadelphia: Lippincott, Williams, and Wilkins.

Kendler K. S. and and Gardner C. O. (1997) The risk for psychiatric disorders in relatives of schizophrenic and control probands: a comparison of three independent studies. *Psychological Medicine* **27**:411–19.

Kendler K. S. and and Hewitt J. K. (1992) The structure of self-report schizotypy in twins. *Journal of Personality Disorders* **6**:1–17.

Kendler K. S., Gruenberg A. M., Strauss J. S. (1981) An independent analysis of the Copenhagen sample of the Danish adoption study of schizophrenia. II. The relationship between schizotypal personality disorders and schizophrenia. *Archives of General Psychiatry* **38**:982–4.

Kendler K. S., Masterson C. C., Ungaro R. *et al.* (1984) A family history study of schizophrenic related personality disorders. *American Journal of Psychiatry* **139**:1185–6.

Kety S. S., Rosenthal D., Wender P. H. *et al.* (1971) Mental illness in the biological and adoptive families of adopted schizophrenics. *American Journal of Psychiatry* **128**:302–6.

Lange J. (1931) *Crime as Destiny.* London: Allen and Unwin.

Livesley W. J., Jang K. L., Jackson D. N. *et al.* (1993) Genetic and environmental contributions to dimensions of personality disorder. *Am J Psychiatry* **150**:1826–31.

Loehlin J. C., Willerman L., Horn J. M. (1985) Personality resemblances in adoptive families when the children are late adoleslecent or adult. *Journal of Personality and Social Psychology* **48**:376–92.

Loranger A. W., Oldham J. M., Tulis E. H. (1982) Familial transmissions of DSM-III borderline personality disorders. *Archives of General Psychiatry* **39**:795–9.

Lyons M. J., Truen W. R., Eisen S. A. *et al.* (1995) Differential heritability of adult and juvenile antisocial traits. *Archives of General Psychiatry* **52**:906–15.

MacDonald A. W. III, Pogue-Geile M. F., Debski T. T. *et al.* (2001) Genetic and environmental influences on schizotypy: a community based twin study. *Schizophrenia Bulletin* **27**(1):47–58.

Maier W., Lichtermann D., Minges J. *et al.* (1992) The familial relationship of personality disorders (DSM-111-R) to unipolar major depressions. *Journal of Affective Disorders* **26**:151–6.

Mann J. J., Brent D. A., Arango V. (2001) The neurobiology and genetics of suicide and attempted suicide focus on the serotonergic system. *Neuropsychopharmacology* **24**(5):467–77.

Marks I. (1986) Genetics of fear and anxiety disorders. A review. *British Journal of Psychiatry* **149**:406–18.

Mata I., Sham P. C., Gilvarry C. M. *et al.* (2000) Childhood schizotypy and positive symptoms in schizophrenic patients predict schizotypy in relatives. *Schizophrenia Research* **44**(2):129–36.

Maxson S. C. (1999) Sexual selection and the Y chromosome. *Trends Ecol Evol* **14**:236.

Maxson S. C. (2000) Genetic influences on aggressive behaviour, in D. Pfaff, W. Berrettini, T. Joh *et al.* (eds) *Genetic Influences on Neural and Behavioral Functions*. CRC Press LLC.

McGuffin P. and Gottesman I. I. (1985) Genetic influences on normal and abnormal development, in M. Rutter and L. Hersov (eds) *Child Psychiatry: Modern Approaches*, 2nd edn. London: Blackwell.

Mednick S. A., Gabrielli W. F. Jr, Hutchings B. (1984) Genetic influences in criminal convictions. Evidence from an adoption cohort. *Science* **22**:891–4.

Meehl P. E. (1962) Schizotaxi, schizotypy schizophrenia, *American Psychologist* **17**:827–31.

Nelson R. J., Demas G. E., Huange P. L. *et al.* (1995) Behavioural abnormalities in male mice lacking neuronal oxide synthase. *Nature* **378**:383–96.

Nelson R. J. and Chiagevatto (2001) Molecular basis of aggression. *Trends in Neuroscience* **24**:713–19.

O'Connor T. G., Neiderhiser J. M., Reiss D. *et al.* (1998) Genetic contributions to continuity, change, and co-occurrence of antisocial and depressive symptoms in adolescence. *Journal of Child Psychology and Psychiatry and Allied Disciplines* **39**:323–36.

Onstad S., Skre I., Torgersen S. *et al.* (1991) Twin concordance for DSM-III-R schizophrenia. *Acta Psychiatrica Scandinavica* **83**:395–401.

Oquendo M. A. and Mann J. J. (2000) The biology of impulsivity and suicidality. *Psychiatr Clin North Am* **23**(1):11–25.

Prescott C. A. and Gottesman I. I. (1993) Genetically mediated vulnerability to schizophrenia. *Psychiatric Clinics of North America* **16**:245–68.

Pope H. G., Jonas J. M., Hudson J. I. *et al.* (1983) The validity of DSM-III borderline personality disorder: A phenomenologic, family history, treatment response and long term follow-up study. *Archives of General Psychiatry* **40**:23–30.

Rado S. (1960) Theory and therapy: the theory of schizotypal organization and its application to the treatment of decompensated schizotypal behavior, in S. C. Scher and H. R. Davis (eds) *The Outpatient Treatment of Schizophrenia*. New York: Grune and Stratton.

Raine A. (1993) *The Psychopathology of Crime*. New York: Academic Press.

Reich T., James J. W., Morris C. A. (1972) The use of multiple thresholds in determining the mode of transmission of semi continuous traits. *Annals of Human Genetics* **36**:163–84.

Riso (2000) A family study of outpatients with borderline personality disorder and no history of mood disorder. *J Personality Disorder* **14**(3):208–17.

Rose R. J., Miller J. Z., Pogue-Geile M. F. *et al.* (1981) Twin-family studies of common fears and phobias. In *Twin Research 33: Intelligence, Personality and Development*, pp. 169–174. Alan R. Liss, New York.

Rushton J. P., Fulker D. W., Neale M. C. *et al.* (1986) Altruism and aggression: The heritability of individual differences. *Personality of Social Psychology* **50**:1192–8.

Rushton J. P. (1996) Self-report delinquency and violence in adult twins. *Psychiatric Genetics* **6**:87–9.

Saudau F., Amara D. A., Dierich A. *et al.* (1994) Enhanced aggressive behaviour in mice lacking 5-HT1b receptor. *Science* **265**:1875–8.

Schulsinger F. (1972) Psychopathy, heredity and environment. *International Journal of Mental Health* **1**:190–206.

Schulz P. M., Schulz S. C., Goldberg S. C. *et al.* (1986) Diagnoses of the relatives of schizotypal outpatients. *Journal of Nervous and Mental Diseases* **174**:457–63.

Shields J., Gottesman I. I., Heston L. L. (1975) Schizophrenia and the schizoid: the problem for genetic analysis, in R. R. Fieve, D. Rosenthal, H. Brill (eds) *Genetic Resesearch in Psychiatry*. Baltimore, MD: Johns Hopkins University Press.

Siever L. J., Silverman K. M., Horvath T. B. *et al.* (1990) Increased morbid risk for schizophrenia related disorders in relatives. *Archives of General Psychiatry* **47**:634–40.

Sigvardsson S., Cloninger C. R., Bohman M. *et al.* (1982) Predisposition to petty criminality in Swedish adoptees: III. Sex differences and validations of the male typology. *Archives of General Psychiatry* **39**:1248–353.

Silverman K. M., Siever L. J., Horvath T. B. *et al.* (1993) Schizophrenia-related and affective personality disorder traits in relatives of probands with schizophrenia and personality disorders. *Archives of General Psychiatry* **150**:435–42.

Simonoff E., Pickles A., Hewitt J. *et al.* (1995) Multiple rates of disruptive child behaviour: using a genetic strategy to examine shared views and bias. *Behaviour Genetics* **25**:311–26.

Slater E. (1961) The thirty-fifth Maudsley lecture: 'Hysteria 311'. *J Mental Sc* **107**:359–81.

Sluyter F., De Geus E. J. C., Van Luijtelaar E. L. J. M. *et al.* (2002 In press) Behavior genetics, in Academic Press *Encylopedia of the Human Brain*. San Diego: Academic Press.

Soloff P. H. and Millward J. W. (1983) Psychiatric disorders in the families of borderline personality and borderline schizophrenia. *Archives of General Psychiatry* **40**:37–44.

Spitzer R. L., Endicott J., Gibbon M. (1979) Crossing the border into borderline personality and borderline schizophrenia. *Archives of General Psychiatry* **365**:17–24.

Stevenson J., Batten N., Cherner M. (1992) Fears and fearfulness in children and adolescents: a genetic analysis of twin data. *Journal of Child Psychology and Psychiatry* **33**:977–85.

Stone M. H. (1981) Psychiatrically ill relatives of borderline patients: a family study. *Psychiatric Q* **58**:71–83.

Taylor J, Iacono W. G., McGue M. (2000) Evidence for a genetic etiology of early-onset delinquency. *Journal of Abnormal Psychology* **109**:634–43.

Tellegen A., Lykken D. T., Bouchard T. J. *et al.* (1988) Personality similarity in twins reared apart and together. *Journal of Personality and Social Psychology* **6**:1031–9.

Thacker G., Adami H., Moran M. *et al.* (1993) Psychiatric illnesses in families of subjects with schizophrenia-spectrum personality disorders: high morbidity risks for unspecified functional psychoses and schizophrenia. *American Journal of Psychiatry* **150**:66–71.

Torgersen S. (1980) The old obsessive and hysterical personality syndrome. A study of heredity and environmental factors by means of the twin method. *Archives of General Pscyhiatry* **37**:1272–7.

Torgersen S. (1984) Genetic and non sociological aspects of schizotypal and borderline personality disorders. A twin study. *Archives of General Psychiatry* **41**:546–54.

Torgersen S. (1979) The nature and origin of common phobic fears. *British Journal of Psychiatry* **134**:343–51.

Torgersen S., Lygren S., Oien P. A. *et al.* (2000) A twin study of personality disorders. *Comprehensive Psychiatry* **41**(6):416–25.

Van Oortmerssen G. A. and Sluyter F. (1994) Studies on wild house mice. V. Aggression in lines selected for attack latency and their Y-chromosomal congenics. *Behav Genet* **24**:73–8.

Vernon P. A., McCarthy J. M., Johnson A. M. *et al.* (1999) Individual differences in multiple dimensions of aggression: A univariate and multivariate genetic analysis. *Twin Research* **2**:16–21.

Weissman M. M. (1993) The epidemiology of personality disorders: a 1990 update. *Journal of Personality Disorders* **7**(Special Suppl):44–62.

Witkin H. A., Mednick S. A., Schulsinger F. (1976) Criminality in XYY and XXY men. *Science* **193**:547–55.

World Health Organization (1993) *The ICD-10 Classification of Mental and Behavioural Disorders. Diagnostic criteria for research*. World Health Organization. Geneva, Switzerland.

Young J. P. R., Fenton G. W., Lader M. H. (1971) The inheritance of neurotic traits: a twin study of the Middlesex Hospital Questionnaire. *British Journal of Psychiatry* **119**:393–8.

Chapter 9

Genetics of affective disorders

Ian Jones, Lindsey Kent, and Nick Craddock

Introduction

It has long been noted that episodes of psychological distress tend to run in families, but recent advances in molecular genetics now provide the tools needed to identify genes influencing susceptibility. Although psychiatric and behavioural traits represent, perhaps, the greatest challenge to molecular investigation of complex genetic disorders, they also offer arguably the greatest potential reward. Identifying susceptibility genes for psychiatric disorders will pinpoint biochemical pathways involved in pathogenesis, facilitate development of more effective, better targeted treatments and offer opportunities for improving the validity of diagnosis and classification. This is likely to have important public health implications, given that the recent review by the World Bank identified mood disorder as second only to ischaemic heart disease as a cause of global health burden in 2020 (Murray and Lopez 1996).

The term 'affective disorder' includes a wide variety of conditions, from mild and common mood variations to some of the most severe episodes of psychotic illness seen in clinical practice. In this chapter we shall limit our discussion to the more severe end of the diagnostic spectrum, and mainly concern ourselves with bipolar disorder and unipolar major depressive illness. We will first discuss some of the methodological issues facing genetic studies of affective illness before moving on to consider both classical and molecular genetic approaches to these disorders that cause suffering to so many people.

Diagnostic issues

Affective disorders are complex genetic disorders in which the core feature is a pathological disturbance of mood, ranging from extreme elation or mania to severe depression. Other symptoms also found in these disorders include disturbances in thinking and behaviour, which may include psychotic symptoms, such as delusions and hallucinations. Historically, affective disorders have been classified in a number of ways, including distinctions between endogenous and reactive episodes, psychotic and neurotic symptomatology and affective disorders arising *de novo* (primary) and those episodes arising in the context of another disorder (secondary) (Kendell 1976; Farmer and McGuffin 1989). The main nosological division in modern classification systems such as ICD-10 (WHO 1993) or DSM-IV (APA 1994) is between the unipolar and bipolar forms of the condition. The DSM-IV diagnostic criteria for manic and major depressive episodes are summarized in Table 9.1. The diagnosis of bipolar disorder (also known as manic depressive illness) requires that an individual has suffered one or more episodes of mania with or without episodes of depression at other times during the life history. This requirement for the occurrence of an episode of mania at some time during the course of illness distinguishes bipolar disorder from unipolar disorder (also commonly known as unipolar major depression, or simply

Table 9.1 DSM-IV diagnostic criteria for (1) major depressive episode and (2) manic episode. Presence of all of A,B,C,D and E is required

Major depressive episode	Manic episode
	(A) Abnormally and persistently elevated, expansive or irritable mood lasting at least one week (any duration if hospitalized).
(A) Five (or more) of the following symptoms present during the same two week period and represent a change from normal functioning: at least one of (1) depressed mood or (2) loss of interest or pleasure.	(B) Three or more of the following symptoms (four if mood only irritable) present to a significant degree.
(1) Depressed mood (2) Loss of interest or pleasure (3) Change in appetite and or weight (4) Sleep change (5) Psychomotor agitation or retardation (6) Loss of energy (7) Feelings of worthlessness or excessive or inappropriate guilt (8) Diminished concentration (9) Recurrent thoughts of death/suicidal ideation	(1) Increased self-esteem or grandiosity (2) Decreased need for sleep (3) More talkative than usual or pressure to keep talking (4) Flight of ideas or racing thoughts (5) Distractibility (6) Increase in goal directed activity or psychomotor agitation (7) Excessive involvement in pleasurable activities that have a high potential for a painful consequences
(B) The symptoms do not meet the criteria for a mixed episode.	(C) The symptoms do not meet the criteria for a mixed episode.
(C) The symptoms cause clinically significant distress or impairment in social, occupational, or other important areas of functioning.	(D) The mood disturbance is sufficiently severe to cause marked impairment in occupational functioning or in usual social activities or relationships with others or to necessitate hospitalization to prevent harm to self or others, or there are psychotic features.
(D) The symptoms are not due to the direct physiological effects of a substance (e.g. a drug of abuse, a medication or other treatment) or a general medical condition (E) The symptoms are not better accounted for by bereavement.	(E) The symptoms are not due to the direct physiological effects of a substance (e.g. a drug of abuse, a medication or other treatment) or a general medical condition.

unipolar depression) in which individuals suffer one or more episodes of depression without ever experiencing episodes of pathologically elevated mood. Although bipolar and unipolar disorders are not completely distinct nosological entities their separation for the purposes of diagnosis and research is supported by evidence from outcome, treatment and genetic studies (Kendell 1987; Farmer and McGuffin 1989). In DSM-IV, Bipolar Disorder is subclassified into Bipolar I Disorder, in which episodes of clear-cut mania occur, and Bipolar II Disorder, in which only milder forms of mania (so-called 'hypomania') occur. The validity of this subclassification, however, awaits robust validation. The lifetime prevalence of narrowly defined bipolar disorder is in the region of 0.5–1.5 per cent with similar rates in males and females and a mean age of onset around the age of 21 years (Smith and Weissman 1992).

Unipolar disorder is substantially more common than bipolar illness but measured prevalence rates differ markedly according to the diagnostic criteria, methodology and sample employed. For example, the large US multi-site Epidemiological Catchment Area (ECA) study reported a lifetime population prevalence for DSM-III major depression of approximately 4.4 per cent (Weissman *et al.* 1988), whereas the US National Comorbidity Survey estimated the lifetime prevalence of DSM IIIR major depression to be 17.1 per cent with 10.3 per cent of the population experiencing a major depressive episode in the preceding 12 months (Kessler *et al.* 1994). In contrast to bipolar illness, the rate of unipolar disorder for women is about twice that for men—21.3 per cent and 12.7 per cent respectively in the US National Comorbidity Survey (Kessler *et al.* 1994), and this gender difference is a consistent finding, at least in studies in the developed world. Affective disorders are associated with high levels of service utilization and morbidity and often prove fatal, with up to 15 per cent of patients eventually committing suicide (Guze and Robins 1970). Reasonably effective treatments are available for both manic and depressive episodes (Daly 1997) but current treatments have undesirable side-effects, are not effective in all patients and the pathogenesis of affective disorders remain poorly understood. These facts act as a major motivation for genetic investigation of affective illness, with its promise of improved understanding of aetiology and more effective treatments.

Methodological issues

There are several methodological challenges facing genetic research in affective disorders but many, we will see, apply to all complex diseases.

Unknown diagnostic validity

In the absence of a clear understanding of the biology of psychiatric illnesses the most appropriate boundaries between the subtypes of affective disorders and between affective disorders and other psychiatric conditions remain unclear. At the psychotic end of the spectrum, there are a large number of patients who have illnesses with features both of schizophrenia and affective disorder (usually called 'Schizoaffective Disorder'). At the less severe end of the spectrum considerable comorbidity occurs between depressive disorders and anxiety states. Alcohol problems and substance abuse disorders are also over-represented in those with affective illness. Current diagnostic boundaries are based on best available evidence (Kendell 1987; Farmer and McGuffin 1989) but the extent to which they reflect genetic vulnerability will only become clear as susceptibility genes are identified.

Lifetime diagnosis

Although psychiatric diagnoses tend to remain stable over time, a change in diagnosis from one episode to another is occasionally observed. In psychiatric genetic studies a 'lifetime diagnosis' is made in an attempt to classify individuals on the basis of a presumed diathesis for illness. This requires that the sum of an individual's abnormal behaviour and experience over his/her lifetime be reduced to a small number of diagnostic categories, usually just one. Although this task is non-trivial, a relatively robust methodology of lifetime diagnosis has been developed which allows integration of information from different sources in an unbiased manner in order to produce acceptably reliable diagnoses (Leckman *et al.* 1982; Farmer *et al.* 1994).

Variable age at onset

In common with many other diseases, individuals can develop the first episode of affective disorder at any time of life. For this reason, unaffected individuals are much less useful in genetic studies than are affected individuals, because they provide less information about genetic risk.

Secular changes (e.g. birth cohort effect)

A change in the measured rate of mood disorder in successive birth cohorts has been described (Klerman 1988; Gershon *et al.* 1987). However, this effect may be an artefact of research methodology (Giuffra and Risch 1994) and the effect has been more clearly demonstrated in unipolar than in bipolar disorder. This effect complicates prevalence-dependent analysis of data that include differing birth cohorts, further reducing the usefulness of unaffected subjects.

These challenges have resulted in a substantial refinement of psychiatric diagnosis, and the evolving methodologies of complex disease genetics provide approaches that in large part address many of the issues outlined above. Currently the major problem is the unknown biological validity of current psychiatric classifications, and it is worth bearing in mind that advances in molecular genetics are likely to be instrumental in providing the first robust validation of our diagnostic schemata.

Mindful of these issues we shall now turn to a consideration of classical and molecular genetic studies of affective disorders. Although less common than unipolar illness, bipolar disorder is a more clear-cut and homogeneous clinical entity, it represents a relatively severe subset of affective disorder and, as we will see, is a more familial form of mood disorder. An NIMH Genetics Workshop recently identified bipolar disorder as a preferred target for large-scale genetic studies of major psychiatric disorder (Barondes 1999). For these reasons the major focus of research to date, and consequently greater emphasis in this chapter will be accorded to, the bipolar form of affective illness.

Family studies of bipolar disorder

Early studies of affective illness failed to distinguish between clinical subtypes, but almost all demonstrated familial aggregation of widely defined mood disorder and are reviewed in Tsuang and Faraone (1990). In more recent decades many studies employing a modern conception of bipolar disorder have been conducted, and Fig. 9.1 provides a graphical representation of their results (modified from Craddock and Jones 1999). It includes all published studies (a) which use the modern concept of bipolar disorder, (b) measure lifetime risk of bipolar disorder in first degree relatives of a bipolar proband, and (c) in which at least some of the relatives were interviewed directly (see Fig. 9.1 for references). There were 21 studies that met these criteria, of which eight included a sample of controls. Figure 9.1 shows the relative risk of narrowly defined bipolar disorder

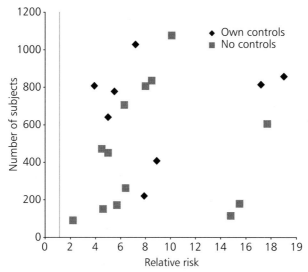

Fig. 9.1 Family Studies of bipolar disorder. All studies can be seen to give a relative risk of greater than one and therefore provide evidence of familial aggregation of bipolar disorder. (Studies included: Perris 1966; Mendlewicz and Rainer 1974; Helzer and Winokur 1974; Gershon *et al.* 1975; James and Chapman 1975; Smeraldi *et al.* 1977; Johnson and Leeman 1977; Petterson 1977; Taylor *et al.* 1980; Scharfetter and Nusperli 1980; Gershon *et al.* 1982; Winokur *et al.* 1982; Weissman *et al.* 1984a; Rice *et al.* 1987; Andreasen *et al.* 1987; Strober *et al.* 1988; Pauls *et al.* 1992; Maier *et al.* 1993; Heun and Maier 1993; Winkour *et al.* 1995; Kendler *et al.* 1998a.)

(equivalent to DSM-IV Bipolar I Disorder) in first degree relatives of bipolar probands as a function of the number of subjects included in the study. Relative risk is defined as the ratio of risk of bipolar disorder in first degree relatives of bipolar probands to the risk in first degree relatives of controls or, for studies that did not include controls, to an assumed general population baseline risk of 1 per cent. As can be seen, all of these studies showed an increased risk of bipolar disorder in the relatives of bipolar probands. Using the eight studies (Gershon *et al.* 1975; Gershon *et al.* 1982; Winokur *et al.* 1982; Weissman *et al.* 1984a; Pauls *et al.* 1992; Maier *et al.* 1993; Heun and Maier 1993; Kendler *et al.* 1998a) that included their own control groups, a meta-analysis (Woolf 1955), showed no evidence of heterogeneity between studies and provided an overall estimate of risk (as measured by odds ratios, OR, and 95 per cent confidence intervals, CI) in first degree relatives of bipolar I probands of 7 per cent (CI 5–10 per cent).

It is not only bipolar illness that is found in the families of bipolar probands. Other affective disorders also occur at increased rates compared with the population. The family studies outlined above demonstrate an increased risk of *unipolar* disorder in first degree relatives of bipolar probands and indeed the absolute risk of unipolar depression is higher than the risk of bipolar illness. However, the background population prevalence is much higher for unipolar illness than for bipolar disorder (of the order of 10 per cent versus 1 per cent) and therefore the *relative* increase in risk is much lower at approximately a doubling of risk (McGuffin and Katz 1989). According to one estimate two-thirds to three-quarters of cases of unipolar depression in the relatives of a bipolar individual can be considered to be 'genetically bipolar', that is they share a common genetic susceptibility with the bipolar form of affective illness (Blacker and Tsuang 1993).

Undoubtedly Bipolar II disorder is found with increased frequency in the families of Bipolar I probands compared with the general population (Gershon *et al.* 1982; Rice *et al.* 1987; Heun and Maier 1993). It has been suggested that this form of the disorder is the most common manifestation of the bipolar phenotype (Simpson *et al.* 1993), others have argued for its status as a separate disorder (Strober 1992) and at present this point remains unresolved. Although the occasional individual with schizophrenia is found in the family of a bipolar proband, there is no compelling evidence of a substantial elevation of risk of this disorder above population rates (Gershon *et al.* 1982; Rice *et al.* 1987; Maier *et al.* 1993; Kendler *et al.* 1998a). In contrast, schizoaffective disorder in which manic features occur has shown consistent familial aggregation with bipolar disorder (Gershon *et al.* 1982; Rice *et al.* 1987; Maier *et al.* 1993; Kendler *et al.* 1998a). The interpretation of these observations is that there may be some familial (probably genetic) factors shared across the functional psychoses but that there is substantial specificity for factors that contribute risk to the two extreme ends of the Kraepelinian dichotomy. This is in accord with findings coming from more recent twin analyses of functional psychosis and from the pattern of results emerging from systematic genome linkage studies of major psychiatric disorders.

We turn now to the issue of whether and which characteristics of proband and illness predict morbidity rates in relatives. It has been demonstrated that lifetime risk of affective disorder in family members is increased with early age of onset (Strober 1992), with number of affected relatives (Gershon *et al.* 1982) and it is possible that a vulnerability to puerperal triggering of episodes is a marker for a more familial form of bipolar illness (Dean *et al.* 1989; Jones and Craddock in press). Lifetime risk in relatives does not appear to vary with sex of the relative or indeed with the sex of the proband (Rice *et al.* 1987; Pauls *et al.* 1992; Heun and Maier 1993). Many studies have reported only pooled estimates of risk for all types of first degree relatives, but a number have provided data that can address the question of whether lifetime rates of illness in first degree relatives of bipolar probands vary according to type of relative (i.e. parent, offspring or sibling). Gershon *et al.* (1975) found the risk in siblings to be greater than that in parents and offspring, but in a later study Gershon and colleagues (1982) found the difference to be reversed. There are no adequate data to provide meaningful estimates of the rates of affective disorder in second degree relatives of bipolar probands, but available evidence suggests that the rates lie between those for first degree relatives and those for the general population.

Family studies of unipolar disorder

The earliest family studies of mood disorder, which did not make the BP/UP distinction, provided evidence for familial aggregation and are reviewed in Tsuang and Faraone (1990). The important BP/UP distinction was introduced into research in the 1960s, since when there have been a number of studies of affective disorders in the relatives of depressed probands. Earlier studies (e.g. Angst 1966; Perris 1966), although supporting the familial aggregation of major affective disorders, did not employ control groups. Controlled family studies of unipolar depression have consistently demonstrated a higher risk of unipolar depression in relatives of probands with unipolar depression compared to the relatives of non-depressed controls. There is no evidence, however, to suggest that first degree relatives of unipolar probands are at any increased risk of bipolar illness (McGuffin and Katz 1989).

Figure 9.2 demonstrates the relative risk of unipolar depression in first degree relatives of unipolar probands as a function of the number of subjects included in the study, for the controlled family studies of unipolar depression. The relative risk is defined as the ratio of risk of illness in first degree relatives of unipolar probands to the risk in first degree relatives of controls. These studies vary in their methodology, which may account for some of the variation in the results. For example, studies have employed either Feighner, DSM-III, or SADS-RDC diagnostic criteria, only some used age correction procedures and some employed screened (i.e. 'supernormal') controls.

The majority of studies report a relative risk of 1.5–3, except the earlier study of Gershon *et al.* (1975) which reports a dramatically higher relative risk of 20 and the recent study of Farmer *et al.* (2000) which found a figure of nearly 10. The Gershon

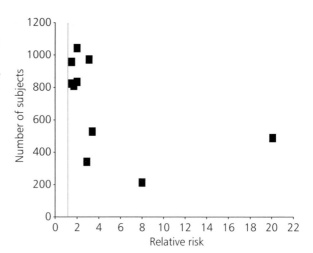

Fig. 9.2 Controlled Family Studies of unipolar disorder. All studies can be seen to give a relative risk of greater than one and therefore provide evidence of familial aggregation of unipolar disorder. (Studies included: Gershon *et al.* 1975; Tsuang *et al.* 1980; Gershon *et al.* 1982; Winokur *et al.* 1982; Weissman *et al.* 1984a; Maier *et al.* 1993; Heun and Maier 1993; Winokur *et al.* 1995; Kendler *et al.* 1998; Farmer *et al.* 2000.)

study consisted of only 16 unipolar probands and reported a prevalence rate of only 0.7 per cent for unipolar depression in the relatives of controls, which is extremely low and a later sample from the same group (Gershon *et al.* 1982) reported a prevalence rate of 5.8 per cent for depression in the relatives of controls, resulting in a much reduced relative risk of 2.9. The high relative risk in the Farmer study also reflects the very low rate of depression in the controls, which in turn probably reflects the ascertainment through siblings screened for health (i.e. a 'supernormal' control group).

A reasonably consistent finding in family studies of unipolar depression is that earlier onset is associated with increased familiality. Earlier studies that examined this issue are reviewed in Tsuang and Faraone (1990) and this has been supported by most, but not all, modern controlled family (and twin) studies which have examined the effect of age of onset in probands and the risk of unipolar depression in relatives (reviewed by Sullivan *et al.* 2000). Gershon *et al.* (1975) reported a trend, that did not reach significance, for an early age of onset to be associated with a greater risk of depression in relatives, which was later supported by Weissman *et al.* (1984b) whereas Weissman *et al.* (1982), reported no age effect. A recent controlled family history study demonstrated increased familiality in childhood-onset depression compared with adult-onset (Kovacs *et al.* 1997). However, the age relationship is not completely linear and there is evidence that pre-pubertal (i.e. very early) onset depression may have a distinct, perhaps *less* genetic aetiology (Harrington *et al.* 1997).

The majority of early family studies report an excess of females affected by depression (reviewed in McGuffin and Katz 1989). Several of the controlled family studies have also examined whether a sex effect is apparent in the risk for unipolar depression. Winokur *et al.* (1982) reported significantly more women than men with unipolar depression amongst the relatives of both unipolar probands and the relatives of the controls. Similarly Weissman *et al.* (1984a) reported a significant sex of relative effect for major depression with rates being higher in women. The reasons for this disparity remain controversial and subject to a heated debate between those advocating the primacy of biological factors and those who believe that social factors account for the excess of depression in women. The sex of the proband, however, does not appear to affect rates of affective illness amongst the relatives of probands with unipolar depression.

One interesting concept arising out of family studies of unipolar depression is that of depression spectrum and pure depressive disease (Winokur *et al.* 1971; Winokur 1997). However, the division of families into those with pure depressive disease (in which only depression is found) and depressive spectrum disease (characterized by a family history of alcoholism in males and depression in females) has not been supported by other family studies (Merikangas *et al.* 1985; Merikangas 1990) and remains an intriguing but controversial concept.

Twin studies of affective disorders

Family studies can demonstrate familial aggregation of a disorder. In order to determine the cause of this familiality we need to turn to the twin and adoption paradigms. Early twin studies did not distinguish between bipolar and unipolar illness but supported the involvement of genes in broadly defined mood disorders (reviewed by Tsuang and Faraone 1990). More modern studies have used this important distinction to examine BP and UP disorders separately, and these are reviewed below.

Bipolar disorder

Six studies have employed a modern concept of bipolar disorder and are consistent in showing an increased probandwise concordance rate in monozygotic (MZ) twins when compared with dizygotic (DZ) twins (Kringlen 1967; Allen *et al.* 1974; Bertelsen *et al.* 1977; Torgersen 1986; Kendler *et al.* 1993a; Cardno *et al.* 1999). The results of these studies are summarized in Table 9.2. Pooling the data from these studies provides an estimate of MZ concordance for narrowly defined bipolar disorder of 50 per cent (95 per cent confidence intervals 40–60 per cent). However, it is likely that three of the studies underestimate the concordance, so the true MZ concordance is probably closer to the 60 per cent found in the study of Bertelsen.

It is important to note that incomplete concordance in MZ twins provides the most robust evidence that non-genetic factors play an important role in determining susceptibility to bipolar disorder. Studies of genetically identical individuals illustrate the phenotypic spectrum that may be associated with bipolar susceptibility genes: in addition to bipolar disorder, unipolar disorder or absence of illness, MZ co-twins or co-triplets of a bipolar proband have been described who have a diagnosis of schizoaffective disorder (Bertelsen *et al.* 1977) or (very rarely) schizophrenia (McGuffin *et al.* 1982).

An interesting and illuminating approach to the study of bipolar twins is to follow the offspring of the unaffected members of discordant MZ pairs. Gottesman and

Table 9.2 Lifetime rates of affecting disorder in co-twins of bipolar twin probands

Authors	Sample	Lifetime rates of affective illness in co-twin of bipolar twin probands (probandwise concordance rate) (%)	Comment
Kringlen (1967)	Norway twin and psychosis register. 6MZ pairs with BP proband.	BP–BP: MZ 67	Small sample and no operationalized diagnostic criteria.
Allen et al. (1974)	USA Veteran twin register. 5MZ, 15DZ pairs (out of 15,909 twin pairs on register)	BP–BP: MZ: 20 DZ: 0	Inferior methodology. Note the very low rate of BP disorder detected in the twin sample (0.07%)
Bertelsen et al. (1977)	Denmark twin and psychiatry registers. 34MZ, 37DZ pairs.	BP–BP: MZ: 62 DZ: 8 BP–BP/UP: MZ: 79 DZ: 19	Despite lack of operationalized diagnostic criteria this is a detailed study and is the best available.
Torgersen (1986)	Norway twin register. 4MZ, 6DZ pairs.	BP–BP: MZ: 75 DZ: 0 BP–BP/UP: MZ: 100 DZ: 0	Small sample which may overlap partly with that of Kringlen.
Kendler et al. (1993a)	Sweden twin and psychiatric registers. 13MZ, 22DZ pairs.	BP–BP: MZ: 39 DZ: 5 BP–BP/UP: MZ: 62 DZ: 14	This has the strength of being a large epidemiological twin sample but with the weakness of using questionnaire assessments. Likely to underestimate concordance.
Cardno et al. (1999)	UK psychiatric hospital twin register. 22MZ, 27DZ pairs.	BP–BP: MZ: 36 DZ: 7	Diagnoses based on hospital notes. Likely to underestimate concordance.

MZ: monozygotic; DZ: dizygotic. Sample size refers to number of twin pairs in which at least one twin suffered with bipolar disorder. BP–BP refers to twin pairs in which both have narrowly defined bipolar disorder. BP–BP/UP refers to twin pairs in which one has bipolar disorder and the other has broadly defined bipolar phenotype (including unipolar depression)

Bertelsen (1989) found an elevated risk of bipolar illness in this group indistinguishable from that in the offspring of individuals affected by bipolar disorder.

Unipolar disorder

Making comparisons between studies of unipolar disorder is not a trivial task, as ascertainment and diagnostic methodologies differ widely. Some general observations can, however, be made. Modern twin studies that make the distinction suggest that although unipolar major depression does not receive as strong a genetic contribution as does bipolar disorder (reviewed by McGuffin and Katz 1989), most (Bertelsen *et al.* 1977; Torgersen 1986; McGuffin *et al.* 1996; Kendler *et al.* 1995a; Lyons *et al.* 1998; Kendler and Prescott 1999a; Bierut *et al.* 1999) but not all (Andrews *et al.* 1990) point to a substantial genetic effect.

Because unipolar disorder is more common than bipolar illness and large samples are more straightforward to assemble, a range of newer and more powerful analytic approaches have been used. The use of population-based twin registries has been one of the most significant developments in twin methodology in recent years. Although the strategy of ascertaining twins independent of their history of treatment may be inappropriate for relatively rare disorders such as schizophrenia and bipolar illness, for more common conditions such as unipolar depression it is likely to give a sample that is more typical of the condition in the community. Another important development has been the application of biometrical model fitting to twin data (Kendler 1993b). This powerful approach enables the relative contribution of genetic, familial and unique environmental factors to the aetiology of a condition to be evaluated. Studies of unipolar disorder (in common with many other psychiatric and behavioural traits) have demonstrated substantial effects of both genetic and unique environmental factors but suggest a much more modest or absent contribution from familial (i.e. shared) environment. In a meta-analysis of the data from five twin studies meeting their inclusion criteria, Sullivan and colleagues (2000) estimated the heritability of unipolar depression to be 37 per cent (95 per cent CI = 31–42 per cent) with a minimal contribution of environmental effects common to siblings (point estimate = 0 per cent, 95 per cent CI = 0–5 per cent) and substantial individual specific environmental effects (point estimate = 63 per cent, 95 per cent CI = 58–67 per cent). However, in a study of 177 probands with major depression from the Maudsley Hospital Twin Register (McGuffin *et al.* 1996) higher estimates of heritability were obtained (between 48 per cent and 75 per cent depending on the assumed population risk). One possibility to account for the higher heritability is that clinically ascertained samples consist of a more severe and familial form of the disorder. Alternatively, reliability of diagnoses may be higher in clinical than in community ascertained samples. In the standard twin model unique environment and unreliability of measurement are confounded. When the reliability of assessment in the community based Virginia twin registry was corrected for by multiple assessments of twins, estimates of the heritability of major depression increased substantially from around 40 per cent to approximately 70 per cent (Kendler *et al.* 1993d).

As we have discussed previously, there is significant comorbidity between affective disorders and other psychiatric conditions. Multivariate twin models can be used to decompose the observed correlation between two or more disorders into that due to additive genes, that due to familial environment and that due to unique environmental influences. This approach has been used to examine the genetic correlation between major depression and generalized anxiety disorder in female twins from the Virginia Twin Registry (Kendler *et al.* 1992). The genetic correlation did not differ significantly from one, which suggests that there are shared genetic factors contributing to susceptibility to both depression and anxiety in this sample. In contrast, the genetic correlation between major depression and phobias in the same twin sample was much smaller (Kendler *et al.* 1993c) and the genetic influences on alcoholism were largely disorder specific and distinct from those influencing susceptibility to depression (Kendler *et al.* 1995b). Thapar and McGuffin (1997) examined the comorbidity of maternally rated depressive and anxiety symptoms in a sample of twin pairs aged 8 to 16 and, consistent with the Virginia study, found evidence for a common set of genes. In addition to the shared genetic effects they found evidence for non-shared environmental factors accounting for part of the comorbidity of these symptoms and demonstrated the influence of specific genetic effects on depressive symptomatology.

By including both male/male and female/female pairs, three studies have examined the genetic risk factors for major depression separately in men and women and are consistent in suggesting that they are of equal importance (McGuffin *et al.* 1996; Kendler *et al.* 1995a; Kendler *et al.* 1999b). Two studies, by including opposite sex pairs, examined the issue of whether the genes themselves are the same in both sexes. In a study of a Swedish twin sample of moderate size and employing a questionnaire-based assessment, the best fitting model estimated the genetic risk factors to be the same in both sexes (Kendler *et al.* 1995a). A larger study, however, employing the direct interview of probands, suggested that men and women share some but not all of their genes for major depression (Kendler *et al.* 1999a).

Studies employing twin samples have also begun to address issues regarding clinical subtypes of depression. Kendler and colleagues (1996) employed latent class analysis in the identification of three clinically significant depressive syndromes (mild typical depression, atypical depression and severe typical depression) that were at least partially distinct from a clinical, longitudinal and familial genetic perspective. There is evidence that the melancholic or endogenous form of depressive illness identifies a subset of individuals with a particularly high familial liability to depressive illness (McGuffin *et al.* 1996; Kendler 1997). Multiple episodes may also characterize those forms of depression with a more marked degree of genetic determination (McGuffin *et al.* 1996) but in the Virginia Twin sample the best-fitting model indicated an inverted U-shaped function with greatest co-twin risk for major depression with seven to nine lifetime episodes (Kendler *et al.* 1999b).

As we have seen, the increasing sophistication of twin methodology has allowed a number of interesting questions to be addressed that go beyond the traditional concern with heritability. Other developments that have increased the scope of the twin approach include the inclusion of specified environmental risk factors and the assessment of other relatives in addition to the twin pair. It is likely that the future will also see an increased focus on particular subtypes of depressive symptomatology, for example in the puerperium (Treloar *et al.* 1999), premenstrually (Kendler *et al.* 1998b) or in relationship to the seasons (Madden *et al.* 1996).

Adoption studies

There have been few studies of affective disorder that have employed the adoption paradigm—but, with one exception, they provide support for an important genetic contribution to susceptibility to mood disorder. Mendlewicz and Rainer (1977) investigated the biological and adoptive parents of 29 bipolar and 22 normal adoptees and the biological parents of 31 bipolar non-adoptees and found significantly ($p < 0.05$) greater risk of affective disorder (bipolar, schizoaffective and unipolar) in the biological parents of bipolar adoptees (18 per cent risk) compared with the adoptive parents (7 per cent risk). This risk in biological relatives of bipolar adoptees was similar to that in the biological relatives of bipolar non-adoptees. The study of Cadoret (1978) dealt mainly with unipolar disorder and also found a significantly higher rate of affective disorder in adopted offspring of patients with affective illness. The study of Wender *et al.* (1986) from Denmark, again demonstrated the involvement of genetic factors in the aetiology of affective disorder. It found an increased risk of both UP depression and suicide in the biological relatives of adoptees with affective illness compared to both their adoptive relatives and the relatives of matched control adoptees. The Swedish study of von Knorring and colleagues (1983) investigated four groups, the biological and adoptive parents of probands with affective disorder and the biological and adoptive parents of matched control adoptees. In complete contrast to the findings of other studies, no significant differences in the rates of affective illness were found between any of the groups. However, the diagnoses of depression in this study were based on indirect sources, such as evidence of clinical treatment, which may underestimate the true prevalence.

Gene environment interplay

As we have established, the weight of evidence points to the role of both genetic and environmental factors in the aetiology of affective disorders. Increasingly, however, attention has been focused on how genes and environment interact to increase liability to depressive illness. It is unlikely that two subtypes of depression exist, one mainly genetic and the other mainly non-genetic (Andrew *et al.* 1998). Rather a model in which genes and environment interact in a complex fashion is more plausible. The common sense assumption has been that genetic and environmental factors interact *additively* to increase vulnerability—a certain genetic liability is inherited to which is added the liability which comes from environmental experiences. However, there are two other possibilities that must be considered. Kendler (1998c) has termed these 'genetic control of sensitivity to the environment' and 'genetic control of exposure to the environment' and we shall deal with each in turn.

In the case of the former, genetic factors can impact on the liability to depression by altering an individual's sensitivity to the depressogenic effects of stressful life events. There have been few studies that have examined this possibility but Kendler and colleagues in a sample of female twins from the Virginia Twin Registry provide data in support of the hypothesis (Kendler *et al.* 1995c). They found that, consistent with the model of genetic control of sensitivity to the environment, the increased risk for major depression given a severe life event was about twice as high in those at high genetic risk as in those at low genetic risk.

Genetic control of exposure to the environment has received more attention. Genes act in this model by influencing the probability that an individual will be exposed to a depressogenic environment. In an important study, McGuffin and colleagues (1988) found not only an increased rate of depression among relatives of depressed probands but also an increased reporting of life events. A number of subsequent twin or family studies have suggested that familial and/or genetic factors can influence an individual's risk for being exposed to severe life events (Kendler *et al.* 1993e; Breslau *et al.* 1991; Plomin *et al.* 1990; Moster 1991; Lyons *et al.* 1993). In a recently published sib pair study of depression, however, significant correlations between siblings were only found for life events that were shared between them such as death or illness of a parent (Farmer *et al.* 2000). Certainly, this is an area that will receive increasing attention and it is unlikely that a complete account of the aetiology of major affective disorders will be obtained without an understanding of the complex interaction of genetic and environmental factors.

Summary of classical genetic studies of mood disorder

Family studies provide consistent evidence of familial aggregation of both bipolar and unipolar disorder; twin and adoption studies point to genes as an important cause of this familial resemblance. In all, there is a consistent and impressive body of evidence that supports the existence of mood disorder susceptibility genes. These studies also demonstrate a graduation of risk of mood disorder between various classes of relatives with monozygotic co-twin showing highest risk, through first-degree relative to unrelated member of the general population showing the lowest risk. Producing exact figures for risk to different classes of relative is a difficult task but the figures given in Table 9.3 can be taken as a general guide to the order of magnitude of risk in different classes of relatives of a bipolar sufferer (Craddock 1995a), and may be used to give general information to patients and their relatives. The task of producing a similar meaningful table for unipolar disorder is even more challenging, and we have not presented a table in order to avoid the spurious impression of accuracy of knowledge.

Table 9.3 Approximate lifetime rates of mood disorder in various classes of relative of bipolar probands

Degree of relationship to bipolar proband	Risk of bipolar disorder (%)	(Additional) risk of unipolar depression (%)
Monozygotic co-twin	40–70	15–25
First degree relative	5–10	10–20
General population (ie. unrelated)	0.5–1.5	5–10

The lifetime risk of major mood disorder in a relative is obtained by *adding* the risk of bipolar disorder and the risk of unipolar depression. General population lifetime risk of unipolar depression is notoriously difficult to quantify, but the figures in the table are based on a definition comparable to that used in the genetic studies

Mode of inheritance

Having established that genes play a role in the aetiology of affective illness an important question then becomes how genes exert their effect—what is the *mode* of inheritance?

Bipolar disorder

Early linkage studies of bipolar disorder assumed, perhaps implausibly, single gene inheritance and large apparently autosomal dominant pedigrees were recruited. Although such pedigrees are rare, it is likely that some families exist in which single genes may play the major role in determining disease susceptibility. Segregation analyses on large pedigrees have produced mixed results with some studies consistent with single gene models (Rice *et al.* 1987; Pauls *et al.* 1992; Crowe and Smouse 1977; Spence *et al.* 1995) and others unable to demonstrate major locus transmission (Bucher *et al.* 1981; Goldin *et al.* 1983; Sham *et al.* 1991). However, caution is required in interpreting these results because of the limited power of the studies to distinguish between single gene and oligogenic models and because of the failure to take account of an important parameter, the recurrence risk in MZ co-twins of a bipolar proband (Craddock *et al.* 1997a). The observed very rapid decrease in recurrence risk from identical co-twins to first degree relatives and back to the general population (as shown in Table 9.3) is not consistent with single gene modes of inheritance (Craddock *et al.* 1995b). The recurrence risk data are consistent with epistatic interaction of multiple genes or with more complex genetic mechanisms. Several genetic mechanisms that are known to produce complex patterns of inheritance and which have been suggested as possible explanations for bipolar disorder are shown in Table 9.5.

X-linkage Several large pedigrees have been reported to show co-segregation of X-linked markers (e.g. colour-blindness or glucose-6-phosphatase deficiency) and bipolar disorder (Reich *et al.* 1969; Mendelwicz *et al.* 1972; Baron *et al.* 1987). These reports

Table 9.4 Some regions of interest from molecular genetic linkage studies of bipolar disorder

Chromosomal location	Study reference	Comment
4p16	Blackwood et al. (1996)	Single large UK pedigree with Max. Lod = 4.8
	Asherson et al. (1998)	Single moderate size UK pedigree (schizoaffective disorder) with Max. Lod = 1.9
12q23–q24	Craddock et al. (1994c)	Single UK pedigree in which Darier's disease [120, 121] and bipolar disorder co-segregate. Max. Lod = 2.1
	Dawson et al. (1995)	29 small-moderate UK pedigrees. Max. Lod = 2.0. No evidence in 16 German families
	Barden et al. (1996)	Single very large pedigree from an isolated French Canadian community. Max. Lod = 4.9
	Ewald et al. (1998)	2 moderate size Danish pedigrees. Max. Lod = 3.4
	Jones et al. (2002)	Single pedigree in which individuals affected by mood disorder show co-segregation with a haplotype in the region of the Darier's gene. Max. Lod = 3.58
18 centromeric	Berrettini et al. (1994)	22 moderate size US pedigrees. Max. non-parametric evidence $p < 0.0001$
	Stine et al. (1995)	Modest-strong evidence in 28 US pedigrees particularly those with paternal transmission. Max. non-parametric evidence $p < 0.0001$
18q22	Stine et al. (1995)	28 moderate size US pedigrees (particularly those showing paternal transmission). Max. sib pair evidence $p < 0.001$
18q22–q23	Freimer et al. (1996)	2 large Costa Rican pedigrees. Max. Lod for combined linkage-association = 4.06
21q22	Straub et al. (1994)	Single large US pedigree. Max. Lod = 3.4
	Detera-Wadleigh et al. (1996)	22 moderate size US pedigrees (particularly those showing maternal transmission). Max. sib pair evidence $p < 0.001$
	Detera-Wadleigh et al. (1997)	97 moderate size US pedigrees. Max. sib pair evidence $p < 0.001$
	Smyth et al. (1997)	23 UK and Icelandic pedigrees. Max. Lod = 2.2
	Aita et al. (1999)	40 US pedigrees. Max. heterogeneity Lod = 3.4
	Morisette et al. (1999)	One branch of single very large French-Canadian pedigree. Max. Lod = 1.6.

This table illustrates some current regions of interest. Readers seeking a systematic overview of findings including both negative and positive reports should consult Chromosome Workshop reports have received criticism on methodological grounds (Hebebrand 1992) and the question of X-linkage remains unresolved despite over half a century of debate. Certainly if X-linkage does occur, analyses suggest that it can only account for a minority of cases (Bucher *et al.* 1981; Risch *et al.* 1986).

Unipolar disorder

There have been fewer reports and claims of 'single gene' families for unipolar disorder than for bipolar disorder. Model fitting and segregation analyses of mood disorder have tended to focus either exclusively on bipolar disorder or have only included unipolar disorder as part of the presumed spectrum of bipolar illness. Consistent findings have not emerged from the earlier studies that have investigated the transmission of unipolar disorder *per se* (reviewed in Tsuang and Faraone [1990]). A recent complex segregation analysis on a dataset of 50 pedigrees ascertained through a proband with early onset recurrent unipolar major depression reported evidence consistent with a major transmissable effect (Marazita *et al.* 1997). It should, however, be recognized that this methodology has important limitations when applied to complex phenotypes such as mood disorders and, for datasets usually available is unable to distinguish between single major gene effects and more complex oligogenic epistatic models (Craddock *et al.* 1997a). All of the genetic mechanisms listed in Table 9.5 remain possibilities for unipolar disorder. However, most researchers have implicitly or explicitly adopted the assumption of multifactorial inheritance: the interaction and/or co-action of multiple genes and environmental factors. Because the parameters of such a model are loosely defined it is both intuitively appealing and relatively easily accommodates observed data. However, it is also difficult to test formally (because it is so poorly defined) in the absence of direct identification of the genes and environmental factors themselves. It is likely that only when this is achieved will it be possible to resolve important issues such as the relationship between susceptibility to bipolar disorder and unipolar disorder and the extent to which there are genetically determined gender specific risks.

Table 9.5 Complex genetic mechanisms that have been suggested in bipolar disorder

Genetic mechanism	Authors
Locus heterogeneity	Hodgkinson *et al.* (1987)
Allelic heterogeneity	Sandkuyl and Ott (1989)
Epistasis	Craddock *et al.* (1995)
Dynamic mutation	McInnis *et al.* (1993)
Imprinting	McMahon *et al.* (1995)
Mitochondrial inheritance	McMahon *et al.* (1995)

Genetic mechanisms causing complex patterns of inheritance and authors who have invoked the mechanisms to explain the inheritance of bipolar disorder

In summary, although affective disorders tend to aggregate in families, the pattern of inheritance in most pedigrees is not simple. Although it seems likely that occasional families exist in which a single gene plays a major role in determining susceptibility to illness, the evidence indicates that single gene transmission does not occur in most cases. This observation is consistent with the failure to identify genes of major effect in linkage studies predicated on single gene models.

Finding genes for affective disorder

Having established that genes play a role in influencing susceptibility to affective disorders and after discussing the possible mode of inheritance we now turn to the problem of identifying these susceptibility genes.

Chromosome studies

One possible pointer to the location of susceptibility genes for a disorder is the co-occurrence of the disorder and gross changes at the chromosomal level. Affective disorders are not consistently associated with chromosome abnormalities although a number of such reports have appeared in the literature (Craddock and Owen 1994a). Perhaps the most interesting observation is that individuals with trisomy 21 appear to be less susceptible to mania than are members of the general population (Craddock and Owen 1994b). This is consistent with the existence of a bipolar susceptibility gene on chromosome 21, a possibility that finds support from recent molecular genetic studies (see Table 9.4).

Molecular genetic studies

Molecular genetic approaches have achieved great success in discovering the mutations which lead to simple (Mendelian) genetic diseases. The challenge that remains is to develop methodologies that will uncover susceptibility genes for complex diseases. Conceptually, molecular genetic studies can be divided into positional and candidate gene approaches. The positional approach assumes no knowledge of disease pathophysiology but determines the broad chromosomal locations of susceptibility genes, usually by linkage studies. The candidate gene approach, however, involves the investigator making educated guesses at what genes may be involved in the pathophysiology of a condition and then testing the involvement of these genes by linkage or more commonly association studies. Of course the candidate gene approach is built on the assumption that the investigator has sufficient understanding of disease biology to be able to recognize suitable candidate genes. This point is certainly open to debate as far as affective disorders and indeed the majority of psychiatric conditions are concerned. In practice both positional and candidate approaches are often combined (Collins 1995; Craddock and Owen 1996). Our discussion of linkage and association studies of affective illness will primarily focus, as has the field, on bipolar disorder but we will turn our attention briefly to studies of unipolar major depression before concluding this section.

Linkage studies of bipolar disorder

Linkage studies employing very large pedigrees and based on the assumption of a single major gene are appropriate for simple Mendelian disorders, but when applied to complex disorders this approach is problematic. In the late 1980s two high profile claims for linkage appeared in the journal *Nature*: Baron *et al.* (1987) reported linkage to X chromosome markers in several Israeli pedigrees and Egeland *et al.* (1987) reported linkage to markers on chromosome 11p15 in a large pedigree of the Old Order Amish community. Other workers were unable to replicate these findings, and eventually in both cases the original groups published updated and extended analyses of their own data in which the significant evidence of linkage all but vanished (Kelsoe *et al.* 1989; Baron *et al.* 1993). Reasons for this dramatic change in findings included: (a) family members originally diagnosed as unaffected became ill for the first time during follow-up; (b) new family members were examined who did not show evidence for linkage, and (c) additional DNA markers were examined which reduced the evidence for linkage. More detailed discussion of this issue and the current status of linkage findings in the 11p15 chromosomal region can be found in the recent Chromosome 11 Workshop Report from the World Congress of Psychiatric Genetics in Bonn (Craddock and Lendon 1999).

Such a major setback was of course a disappointment, but the field has now moved forward with the development of methodologies more appropriate for the study of complex genetic traits, including a trend towards the use of smaller families (particularly affected sibling pairs) and analyses less sensitive to diagnostic changes (reviewed by Craddock and Owen 1996). Large molecular genetic linkage studies of bipolar disorder are underway in many centres around the world, and interesting findings are beginning to emerge. Systematic genome screens are being conducted on a variety of sample sets, ranging from large densely affected pedigrees in genetic isolates (Egeland *et al.* 1987) to large numbers of affected sib pairs (Bennett *et al.* 2002). The pattern of findings emerging from these studies is consistent with there being no gene of major effect to explain the majority of cases of bipolar disorder and with that expected in the search for genes for a complex disorder (Suarez *et al.* 1994; Lander and Schork 1994; Kruglyak and Lander 1995). Chromosomal regions of interest are typically broad (often < 20–30 cM). No finding replicates in all datasets and for individual positive findings levels of statistical significance and estimated effect sizes are usually modest. One trend that is becoming increasingly clear is that some chromosome regions are being implicated in linkage studies of more than one psychiatric phenotype, e.g. 15q11–q13 is implicated in studies of both bipolar disorder and schizophrenia (Craddock and Lendon 1999).

This is a rapidly moving field with new data becoming available regularly. While it is therefore difficult to provide a summary of the current state of the field Table 9.4 provides information about four regions that are currently attracting a great deal of interest (Craddock and Jones 1999). Of these 12q23–q24 and 21q22 are perhaps the most promising (as already mentioned, the reduced incidence of mania in trisomy 21 is consistent with a chromosome 21 locus [Craddock and Owen 1994b]). Chromosome 18 is the most confusing with positive reports scattered throughout the chromosome in a pattern that is not consistent with a single susceptibility gene (Van Broeckhoven and Verheyen 1999). Other regions of interest include 16p13 (Ewald *et al.* 1995), Xq24–q26 (Pekkarinen *et al.* 1995) and 15q11–q13 (reviewed in Craddock and Lendon 1999). Detailed reviews of linkage data for each chromosome can be found in the chromosome workshop reports of the World Congress of Psychiatric Genetics, published annually. As is the case for most complex genetic disorders, methodological differences between linkage studies make conventional meta-analysis all but impossible. However, new methods of meta-analysis are being developed in which the comparison of genome scan results is based on ranking of chromosomal regions according to linkage evidence rather than absolute values of linkage statistics. The application of such approaches to bipolar disorder will provide a more systematic and unified overview of the evidence.

Linkage studies of unipolar disorder

Unipolar major depression, when occurring in members of families identified through bipolar probands, has been included as part of the presumed spectrum of affected phenotype in most linkage studies of bipolar disorder. However, linkage analysis of unipolar disorder *per se* has been a relatively neglected area. In the pre-molecular era several small linkage studies were reported using classical markers such as blood groups or HLA type but there were no consistent positive findings (reviewed in Tsuang and Faraone 1990). Very few linkage studies have been reported using DNA markers. Samples have been relatively modest in size, markers have been confined to candidate regions or genes and positive findings have not emerged (e.g. Wang *et al.* 1992; Balciuniene *et al.* 1998; Neiswanger *et al.* 1998). There has been no large scale systematic genome scan reported, although appropriate samples are currently being recruited. Because the genetic effect size is smaller than for bipolar disorder, substantially larger samples are required to provide adequate power to detect linkage—an example being the 1200 sib pairs affected by recurrent major depression that is the target of a multi-centre international study funded by Glaxo Smith Kline Research and Development.

Association studies

Association studies represent an important paradigm for the investigation of complex genetic traits, both to follow up regions of interest from linkage studies (by systematic linkage disequilibrium mapping and positional candidate studies) and for pure positional candidate studies. The candidate gene approach is potentially very powerful, particularly when used within the context of a VAPSE (Variation Affecting Protein Sequence or Expression) paradigm (Sobell *et al.* 1992). This is a two stage process in which systematic polymorphism detection is first conducted across the gene of interest followed by association studies in disease comparison samples.

The advantages of the association approach include its relative robustness to genetic heterogeneity and the ability to detect much smaller effect sizes than are detectable using feasible sample sizes in linkage studies (Risch and Merikangas 1996). However, this approach is not without its problems, and spurious associations can arise from inadequate matching of cases and controls, especially when there is population stratification. Various family-based association methods have been developed to counter this problem and involve the construction of an artificial well-matched notional control sample from marker data from the family of each proband (Falk and Rubinstein 1987). A number of statistical approaches have been developed to be used in this design (Schaid and Sommer 1994) but the Transmission Disequilibrium Test (TDT) is perhaps the most popular (Spielman *et al.* 1993).

While problems of population stratification receive considerable attention, perhaps the major problem with candidate gene association studies involves the choice of candidates. Good candidate gene studies depend critically on the choice of good candidates—this inevitably depends on the current understanding of disease pathophysiology. To date, most candidate gene studies have focused on neurotransmitter systems that are influenced by medications used in clinical management of the disorders. Thus, as with other psychiatric phenotypes such as schizophrenia, studies of mood disorders have typically examined polymorphisms within genes encoding receptors or proteins and enzymes involved in metabolism, reuptake or action of Dopamine, Serotonin (5HT) and Noradrenalin (Jones and Craddock 2001a). Early studies often used anonymous markers in the hope of detecting linkage disequilibrium, but recently direct examination of polymorphisms of known or presumed functional relevance has become more usual. Most studies in the literature have been of the unrelated case-control design with samples rarely exceeding 200–300 subjects. No definitive positive findings have yet emerged in either unipolar or bipolar disorder, although there have been some interesting preliminary reports that await robust validation. We will illustrate the approach with some examples.

Association studies of bipolar disorder

Tyrosine hydroxylase

A plausible functional candidate that has been subjected to considerable scrutiny over the last decade is tyrosine hydroxylase (TH), the rate limiting enzyme that catalyses the first step in the synthesis of catecholamines. TH has been considered as an important candidate because catecholamine dysfunction has long been proposed as a possible biochemical mechanism in bipolar disorder. Moreover, the TH gene maps to 11p15, in the region implicated by the original Old Order Amish positive linkage report (Egeland *et al.* 1987). Leboyer and colleagues (1990) aroused interest when they reported significant association with alleles at 2 RFLPs at TH in a small French case-control sample (p < 0.01). However, several groups have examined these and other polymorphisms at

TH and a recent meta-analysis of 547 patients and 522 comparison individuals revealed no evidence that variation at this gene influences susceptibility to bipolar disorder (Turecki *et al.* 1997). However, it should be recognized that even in this meta-analysis the sample sizes have modest power to detect plausible effect sizes for susceptibility genes for complex diseases. Further, although a useful technique to enhance sample size, meta-analysis has limited utility in situations where there is substantial genetic heterogeneity between samples and is compromised in situations in which publication bias exists.

Serotonin transporter

An extremely plausible gene that has attracted recent interest is the human serotonin transporter (hSERT). This is undoubtedly an excellent functional candidate because it is the site of action of the selective serotonin reuptake inhibitors, a major class of anti-depressants which are affective in treatment of bipolar depression and which can induce mania in bipolar individuals (Prozac™ is a well known example of this class of drug). Several studies have produced modest evidence implicating this gene using both case-control (Collier *et al.* 1996; Rees *et al.* 1997; Kunugi *et al.* 1997) and family-based (Kirov *et al.* 1999) approaches. However, the picture is far from clear with other groups reporting negative findings with the same polymorphisms (Battersby *et al.* 1996; Stöber *et al.* 1996; Bellivier *et al.* 1997; Furlong *et al.* 1998; Hoehe *et al.* 1998). To make the situation more confusing, some groups have found association with one allele of a VNTR in intron 2, whereas others have found association with the short allele of a functional deletion/insertion polymorphism in the promoter and it does not appear that the different observations simply reflect linkage disequilbirium. hSERT is certainly not a gene of major effect. For example, in the studies reporting a statistically significant result, effect sizes, as measured by genotype odds ratios, are typically in the region of 1.5–2.5. The findings may simply reflect sampling variance or may result from a modest influence on disease susceptibility, perhaps in a specific subset of bipolar disorder. This latter possibility gains support from the recent report that variation at the VNTR influences susceptibility to bipolar affective puerperal psychosis (Coyle *et al.* 2000). Further studies of hSERT are required, and undoubtedly genes of modest effect size will require an accumulation of large independent studies to provide robust statistical evidence to support their involvement in disease. However, no matter how compelling the statistical significance, biological evidence from a cellular, tissue and ultimately animal model will be necessary to confirm the pathogenic relevance of any putative susceptibility allele.

Catechol-o-methyl transerfase (COMT) in rapid cycling

COMT is an enzyme involved in the degradation of monoamines and a common functional polymorphism occurs in the gene which influences enzyme activity. This polymorphism has been investigated in candidate studies of bipolar disorder and there is no evidence that it influences susceptibility to the illness (e.g. Craddock *et al.* 1997).

However, studies suggest that the low activity allele at this common polymorphism may be a *course modifier* associated with increased susceptibility to rapid cycling within bipolar patients (Lachman *et al.* 1996; Kirov *et al.* 1998; Papolos *et al.* 1998). This possibility has pleasing biological consistency with the observed tendency for antidepressants to induce rapid cycling in that both increase the availability of catecholamines at neuronal synapses. The data suggest that COMT makes only a modest contribution to rapid cycling observed in the bipolar population and confirmation is, of course, required in large independent samples.

Trinucleotide repeat expansion (dynamic mutation)

One intriguing possibility that has emerged in recent years is that the phenomenon of trinucleotide repeat expansion may play a role in the pathophysiology of bipolar disorder (Jones *et al.* 2002). The clinical phenomenon of anticipation in which progressively more severe disease and/or earlier age of onset is observed as the disease is transmitted through successive generations was first described in psychiatric disorders, but for many years it was thought it might merely be an artefact of ascertainment bias. Interest in this phenomenon has been rejuvenated by the discovery that anticipation in several neurological/neuropsychiatric conditions such as Huntington's disease, myotonic dystrophy and fragile X syndrome is caused by DNA sequences of trinucleotide repeats that expand as they are passed from parent to offspring. More recent studies that have examined the issue of anticipation have demonstrated its occurrence in bipolar disorder (reviewed by O'Donovan and Owen 1996a). Four independent association studies using the Repeat Expansion Detection (RED) method (Schalling *et al.* 1993) have shown trinucleotide repeat sequences to be significantly larger in bipolar patients compared with controls (Linbald *et al.* 1995; O'Donovan *et al.* 1995; O'Donovan *et al.* 1996b; Oruc *et al.* 1997) but other studies have failed to find this association (Vincent *et al.* 1996). One of the groups with the positive RED finding has reported that most of the large RED products may be explained by two specific loci, CTG18.1 and ERDA7 (Lindblad *et al.* 1998) and that CTG18.1 shows a modest association with bipolar disorder. However, this finding awaits confirmation.

Association studies of unipolar disorder

In contrast to the situation in linkage studies, many published association studies provide data directly examining samples of unipolar major depression. In most cases the same candidates have been studied in unipolar and bipolar samples. No consistent positive findings have emerged from studies to date although, as for bipolar disorder, there are tantalizing (e.g. Ogilvie *et al.* 1996; Collier *et al.* 1996b) but inconsistent (e.g. Collier *et al.* 1996a; Rees *et al.* 1997) reports suggesting that variation at the serotonin transporter may influence susceptibility to unipolar disorder. As for bipolar disorder, large, independent, representative samples are required to confirm findings. Given the assumption that the genetic effect sizes are smaller in unipolar than in bipolar disorder, it can be expected that even larger samples will be necessary to reliably detect susceptibility and course modifying genes in unipolar illness.

A deeper understanding of the pathogenesis of mood disorders will almost certainly extend to systems involved in signal transduction and modulation of gene expression. As the human genome project unfolds and genes involved in these systems are cloned, candidate gene studies will offer increasingly powerful approaches to explore their role in pathogenesis and modification of the course of bipolar and unipolar disorders.

Subtypes of affective disorders

One approach that may prove fruitful in the search for genes for complex diseases is to focus on specific phenotypic subtypes known or thought to represent more genetically homogenous forms of the disorder (Lander and Schork 1994). Researchers are beginning to take an interest in a number of such subtypes of affective disorder.

Rapid cycling

Rapid cycling is defined, perhaps arbitrarily, as the occurrence of four or more discrete episodes of mood disorder within a 12 month period. Although family studies have produced inconclusive results regarding the familiality of rapid cycling (Coryell *et al.* 1992; Nurnberger *et al.* 1988; Wehr *et al.* 1988), as we have mentioned above recent candidate gene association studies suggest that the low activity allele at a common polymorphism within the catechol-o-methyl transferase (COMT) gene may be associated with increased susceptibility to rapid cycling within bipolar individuals (Lachman *et al.* 1996; Kirov *et al.* 1998; Papolos *et al.* 1998).

Seasonal affective disorder

Seasonal Affective Disorder (SAD) describes mood disorder with a characteristic seasonal variation. Genes have been shown to influence seasonal variation in mood (Madden *et al.* 1996) and SAD has been shown to aggregate in families (Allen *et al.* 1993). Biological systems that have been implicated in the pathogenesis of SAD include systems involved in serotonergic neurotransmission and circadian and circannual clocks (Thompson *et al.* 1997). Recent candidate gene studies have supported the possible involvement in SAD of the serotonin transporter gene (hSERT) (Rosenthal *et al.* 1998; Sher *et al.* 1999) and the gene encoding the serotonin receptor 2A (5HT2A) (Sher *et al.* 1999) although the samples are small and statistical significance modest. As always, independent confirmations will be important.

Puerperal psychosis

Puerperal psychosis refers to severe (usually psychotic) psychiatric disorder occurring within a few weeks of parturition (Brockington 1996). Consistent and compelling evidence suggests the puerperium is a period of increased risk in bipolar women and points to genetic factors as influencing vulnerability to puerperal triggering of episodes (Jones and Craddock 2001b; Jones and Craddock in press). The exact nature of the puerperal trigger is unknown but the very high levels of estrogen in pregnancy and its precipitous fall following delivery has been implicated and has received the most attention to date (Jones *et al*. 2001a; Jones *et al*. 2000). Thus, neuronally relevant genes influenced by oestrogen are plausible candidates as susceptibility genes for bipolar affective puerperal psychosis. A recent case-control study of 96 bipolar women with puerperal episodes starting within two weeks of parturition, found that variation at the serotonin transporter gene (a major site of action of many antidepressants) exerted a significant ($p < 0.008$), substantial (odds ratio = 4) and important (population attributable fraction = 69 per cent) influence on susceptibility to bipolar affective puerperal episodes (Coyle *et al*. 2000). As with the other findings mentioned previously, it is important that this finding receives independent confirmation.

Lithium responsiveness

Lithium is widely used as a prophylaxis against recurrence of both manic and depressive episodes. Despite its chemical simplicity it has a complex pharmacology and although its therapeutic mode of action remains poorly understood, it is believed that its effect on the phosphatidyl inositol signal transduction pathway is important (Berridge *et al*. 1989). Up to 70 per cent of patients benefit from lithium prophylaxis, some remaining free from all episodes for prolonged periods. Lithium responsiveness appears to be familial (Alda *et al*. 1994; Grof *et al*. 1994) and samples are being used for molecular genetic studies (e.g. Turecki *et al*. 1999). Such samples are, however, difficult to recruit because accurate measurement of responsiveness to lithium (or any other prophylactic treatment) requires extended periods of observation of the course of illness both before and after initiation of the treatment. This can usually only be achieved satisfactorily in prospective designs.

Conclusions

In this chapter we have examined the evidence implicating genetic factors in the aetiology of affective disorders and have briefly covered the results of linkage and association studies to date. A number of groups around the world are currently assembling the large clinical samples that will be needed to identify susceptibility genes of relatively modest effect. A number of developments in the near future are likely to considerably facilitate this process, including the trend for genotyping methods to become increasingly automated with consequent dramatic increases in efficiency. It is likely that novel molecular and analytical approaches will supersede current linkage and association studies and all these developments will occur in the context of the completion of the Human Genome Project which will further facilitate the identification of genes of even small effect. A tantalizing prospect on the horizon is that the discovery of susceptibility genes will lead to the development of animal models of affective disorder. It should prove feasible to breed transgenic mice (or other species) that include the abnormal form of the gene, thereby providing a model in which the disease process can be studied *in vivo*.

Identification of genes conferring susceptibility to affective illness, while a major achievement in itself, will be only the first step on the path towards understanding the biological underpinnings of these disorders. The benefits of this research must not be underestimated (Jones *et al.* 2001b). Knowing a gene confers susceptibility will allow the identification of its product which in turn will lead to an understanding of the role of the individual protein in disease causation. This will lead to the development of treatments targeted specifically at the biochemical lesions involved in disease. It should also lead to the development of a more rational aetiologically based classification system which will provide a much better guide to treatment and prognosis than current systems. Importantly, identifying susceptibility genes will facilitate the identification of environmental factors that alter risk. Once these environmental factors are characterized, it may prove possible to provide helpful occupational, social and psychological advice to individuals at genetic risk of affective disorders. It is also likely that along this path we will learn much about the biological basis of normal affective responses and a recent twin study has demonstrated that normal happiness is in large a part under genetic influence (Lykken and Tellegen 1996).

In addition to the undoubted benefits discussed above potential costs must also be considered (Jones *et al.* 2001b). Major advances raise major ethical issues. Many of the issues are no different from those that arise in the context of other complex familial disorders, but the combination of genetics and mental illness raises particular concerns and has justifiably received close scrutiny of ethical and psychosocial issues (Nuffield Council on Bioethics 1998). It is important that we continue to address potential problems such as the availability of services, the right to information and the testing of individuals below the age of consent. The challenge that we face at the dawn of the new millennium is to translate advances in understanding of the complex aetiology of affective disorders into tangible improvements in clinical care.

References

Aita V. M., Liu J., Knowles J. A. *et al.* (1999) A comprehensive linkage analysis of chromosome 21q22 supports prior evidence for a putative bipolar affective disorder locus. *Am J Human Genet* **64**:217.

Alda M., Grof P., Grof E. *et al.* (1994) Mode of inheritance in families of patients with lithium-responsive affective disorders. *Acta Psychiatrica Scandinavica* **90**:304–10.

Allen J. M., Lam R. W., Remick R. A. *et al.* (1993) Depressive symptoms and family history in seasonal and nonseasonal mood disorders. *Am J Psychiatry* **150**:443–8.

Allen M. G., Cohen S., Pollin W. *et al.* (1974) Affective illness in veteran twins: a diagnostic review. *Am J Psychiatry* **131**:1234–9.

Andreasen N. C., Rice J., Endicott J. *et al.* (1987) Familial rates of affective disorder. *Archives of General Psychiatry* **44**:461–9.

Andrew M., McGuffin P., Katz R. (1998) Genetic and non-genetic subtypes of major depressive disorder. *British Journal of Psychiatry* **152**:775–82.

Andrews G., Stewart G., Allen R. *et al.* (1990) The genetics of six neurotic disorders: a twin study. *Journal of Affective Disorders* **19**:23–9.

Angst J. (1966) Zur atiologie und nosologie endogener depressive psychosen. *Monographen ans der Neurologie und Psychiatrie*, N112. Berlin: Springer

APA (1994) *Diagnostic and Statistical Manual for Mental Disorders*, 4th edn. Washington, DC, American Psychiatric Association.

Asherson P., Mant R., Williams N. *et al.* (1998) A study of chromosome 4p markers and dopamine D5 receptor gene in schizophrenia and bipolar disorder. *Mol Psychiatry* 3:310–20.

Balciuniene J., Yuan Q-P., Engström C. *et al.* (1998) Linkage analysis of candidate loci in families with recurrent major depression. *Molecular Psychiatr* 3:162–8.

Barden N., Plante M., Rochette D. *et al.* (1996) Genome-wide microsatellite marker linkage study of bipolar affective disorders in a very large pedigree derived from a homogeneous population in Quebec points to susceptibility locus on chromosome 12. *Psychiatr Genet* 6:145–6.

Baron M., Freimer N. F., Risch N. *et al.* (1993) Diminished support for linkage between manic-depressive illness and X-chromosome markers in three Israeli pedigrees. *Nat Genet* 3:49–55.

Baron M., Risch N., Hamburger R. *et al.* (1987) Genetic linkage between X-chromosome markers and bipolar affective illness. *Nature* 326:289–92.

Barondes S. H. (1999) An agenda for psychiatric genetics. *Archives of General Psychiatry* 56:549–52.

Battersby S., Ogilvie A. D., Smith C. A. D. *et al.* (1996) Structure of a variable number tandem repeat of the serotonin transporter gene and association with affective disorder. *Psychiatr Genet* 6:177–81.

Bellivier F., Laplanche J. L., Leboyer M. *et al.* (1997) Serotonin transporter gene and manic depressive illness: an association study. *Biological Psychiatry* 41:750–2.

Bennett P., Segurado R., Jones I. *et al.* (2002) The Wellcome Trust UK-Irish Bipolar Affective Disorder sibling-pair genome screen: first stage report. *Molecular Psychiatry* 7(2):189–200.

Berrettini W. H., Ferraro T. N., Goldin L. R. *et al.* (1994) Chromosome 18 DNA markers and manic-depressive illness: evidence for a susceptibility gene. *Proc Natl Acad Sci USA* 91:5918–21.

Berridge M. J., Downes C. P., Hanley M. R. (1989) Neural and developmental actions of lithium: a unifying hypothesis. *Cell* 59:411–19.

Bertelsen A., Harvald B., Hauge M. (1977) A Danish twin study of manic-depressive illness. *British Journal of Psychiatry* 130:330–51.

Bierut L. J., Heath A. C., Bucholz K. K. *et al.* (1999) Major depressive disorder in a community based twin sample. Are there different genetic and environmental contributions for men and women? *Archives of General Psychiatry* 56:557–63.

Blacker D. and Tsuang M. T. (1993) Unipolar relatives in bipolar pedigrees: are they bipolar? *Psychiatr Genet* 3:5–16.

Blackwood D. H. R., He L., Morris S. W. *et al.* (1996) A locus for bipolar affective disorder on chromosome 4p. *Nat Genet* 12:427–30.

Breslau N., Davis G. C., Andreski P. *et al.* (1991) Traumatic events and posttraumatic stress disorder in an urban population of young adults. *Archives of General Psychiatry* 48:216–22.

Brockington I. F. (1996) *Motherhood and Mental Health*. Oxford, Oxford University Press.

Bucher K. D., Elston R. C., Green R. *et al.* (1981) The transmission of manic depressive illness—II. Segregation analysis of three sets of family data. *J Psychiatr Res* 16:65–78.

Cadoret R. (1978) Evidence for genetic inheritance of primary affective disorder in adoptees. *Am J Psychiatry* 133:463–6.

Cardno A. G., Marshall E. J., Coid B. *et al.* (1999) Heritability estimates for psychotic disorders. *Archives of General Psychiatry* 56:162–8.

Collier D. A., Arranz M. J., Sham P. *et al.* (1996a) The serotonin transporter is a potential susceptibility factor for bipolar affective disorder. *Neuroreport* 7(10):1675–9.

Collier D. A., Stober G., Li T. *et al.* (1996b) A novel functional polymorphism within the promoter of the serotonin transporter gene: possible role in susceptibility to affective disorder. *Mol Psychiatry* 1:453–60.

Collins F. S. (1995) Positional cloning moves from periditional to traditional. *Nat Genet* 9:347–50.

Coryell W., Keller M., Endicott J. *et al.* (1989) Bipolar II illness: course and outcome over a five year period. *Psychol Med* 19:129–41.

Coyle N., Jones I., Robertson E. *et al.* (2000) Variation at the serotonin transporter gene influences susceptibility to bipolar affective puerperal psychosis. *Lancet* 356:1490–1.

Craddock N. and Owen M. (1994a) Chromosomal aberrations and bipolar affective disorder. *British Journal of Psychiatry* 164:507–12.

Craddock N. and Owen M. (1994b) Is there an inverse relationship between Down's syndrome and bipolar affective disorder? Literature review and genetic implications. *J Intellect Dis Res* 38:613–20.

Craddock N., Owen M. Burge S. *et al.* (1994c) Familial cosegregation of major affective disorder and Darier's disease (keratosis follicularis). *British Journal of Psychiatry* 164:355–8.

Craddock N. (1995a) Genetic linkage and association studies of Bipolar Disorder. Ph.D. thesis, University of Wales.

Craddock N., Khodel V., Van Eerdewegh P. *et al.* (1995b) Mathematical limits of multilocus models: the genetic transmission of bipolar disorder. *Am J Med Genet* 57:690–702.

Craddock N., Owen M. J. (1996) Modern molecular genetic approaches to psychiatric disease. *British Medical Bulletin* 52:434–452.

Craddock N., Van Eerdewegh P., Reich T. (1997a) Single major locus models for bipolar disorder are implausible. *Am J Med Genet* 74:18.

Craddock N., Spurlock G., McGuffin P. *et al.* (1997) No association between bipolar disorder and alleles at a functional polymorphism in the COMT gene. *British Journal of Psychiatry* 170:526–8.

Craddock N. and Lendon C. (1999) Chromosome Workshop: chromosomes 11, 14, and 15. *Am J Med Genet (Neuropsychiatr Genet)* 88:244–54.

Craddock N. and Jones I. (1999) Genetics of Bipolar Disorder. *Journal of Medical Genetics* 36(8):585–94.

Crowe R. R. and Smouse P. E. (1977) The genetic implications of age-dependent penetrance in manic-depressive illness. *J Psychiatr Res* 13:273.

Daly I. (1997) Mania. *Lancet* 349:1157–60.

Dawson E., Parfitt E., Robers Q. *et al.* (1995) Linkage studies of bipolar disorder in the region of the Darier's disease gene on chromosome 12q23-24.1. *American Journal of Medical Genetics (Neuropsychiatric Genetics)* 60:94–102.

Dean C., Williams R. J., Brockington I. F. (1989) Is puerperal psychosis the same as bipolar manic-depressive disorder? A family study. *Psychological Medicine* 19:637–47.

Detera-Wadleigh S. D., Badner J. A., Goldin L. R. *et al.* (1996) Affected sib-pair analyses reveal support of a prior evidence for a susceptibility locus for bipolar disorder, on 21q. *Am J Hum Genet* 58:1279–85.

Detera-Wadleigh S. D., Badner J. A., Yoshikawa T. *et al.* (1997) Initial genome scan of the NIMH genetics initiative bipolar pedigrees: chromosomes 4,7,9,18,19,20, and 21q. *Am J Med Genet* 74:254–62.

Egeland J. A., Gerhard D. S., Pauls D. L. *et al.* (1987) Bipolar affective disorders linked to DNA markers on chromosome 11. *Nature* 325:783–7.

Ewald H., Degn B., Mors O. *et al.* (1998) Significant linkage between bipolar affective disorder and chromosome 12q24. *Psychiatr Genet* **8**:131–40.

Ewald H., Mors O., Flint T. *et al.* (1995) A possible locus for manic depressive illness on chromosome 16p13. *Psychiatr Genet* **5**:71–81.

Falk C. T. and Rubinstein P. (1987) Haplotype relative risks: an easy and reliable way to construct a proper control sample for risk calculations. *Ann Hum Genet* **51**:227–33.

Farmer A. McGuffin P. (1989) The classification of the depressions contemporary confusion revisited. *British Journal of Psychiatry* **155**:437–43.

Farmer A. E., Williams J., Jones I. (1994) Phenotypic definitions of psychotic illness for molecular genetic research. *Am J Med Genet (Neuropsychiatr Genet)* **54**:365–71.

Farmer A. E., Harris T., Redman K. *et al.* (2000) Cardiff depression study: a sib-pair study of life events and familiality in major depression. *British Journal of Psychiatry* **176**:150–5.

Freimer N. B., Reus V. I., Escamilla M. A. *et al.* (1996) Genetic mapping using haplotype, association and linkage methods suggests a locus for severe bipolar disorder (BPI) at 18q22–123. *Nat Genet* **12**:436–41.

Furlong R. A., Ho L., Walsh C. *et al.* (1998) Analysis and meta-analysis of two serotonin transporter gene polymorphisms in bipolar and unipolar affective disorders. *Am J Med Genet (Neuropschiatr Genet)* **81**:58–63.

Gershon E. S., Mark A., Cohen N. *et al.* (1975) Transmitted factors in the morbid risk of affective disorders: a controlled study. *J Psychiatr Res* **12**:283–99.

Gershon E. S., Hamovit J., Guroff J. J. *et al.* (1982) A family study of schizoaffective, bipolar I, bipolar II, unipolar, and normal control probands. *Archives of General Psychiatry* **39**:1157–67.

Gershon E. S., Hamovit J., Guroff J. J. *et al.* (1987) Birth-cohort changes in manic and depressive disorders in relatives of bipolar and schizoaffective patients. *Archives of General Psychiatry* **44**:314–19.

Giuffra L. A. and Risch N. (1994) Diminished recall and the cohort effect of major depression: a simulation study. *Psychol Med* **24**:375–83.

Goldin L. R., Gershon E. S., Targum S. D. *et al.* (1983) Segregation and linkage analyses in families of patients with bipolar, unipolar, and schizoaffective mood disorders. *Am J Hum Genet* **45**:274–87.

Gottesman I. I. and Bertelsen A. (1989) Confirming unexpressed genotypes for schizophrenia. *Archives of General Psychiatry* **46**:867–72.

Grof P., Alda M., Grof E. *et al.* (1994) Lithium response and genetics of affective disorders. *J Affect Disord* **32**:85–95.

Guze S. B. and Robins E. (1970) Suicide and primary affective disorders. *British Journal of Psychiatry* **117**:437–8.

Harrington R. C., Rutter M., Weissman M. *et al.* (1997) Psychiatric disorders in the relatives of depressed probands I: comparison of pre-pubertal, adolescent and early adult onset cases. *J Affect Disord* **42**:9–22.

Hebebrand J. (1992) A critical appraisal of X-linked bipolar illness. Evidence for the assumed mode of inheritance is lacking. *British Journal of Psychiatry* **160**:7–11.

Helzer J. E. and Winokur G. (1974) Family interview study of male manic depressives. *Archives of General Psychiatry* **31**:73–7.

Heun R. and Maier W. (1993) The distinction of bipolar II disorder from bipolar I and recurrent unipolar depression: results of a controlled family study. *Acta Psychiatrica Scandinavica* **87**:279–84.

Hodgkinson S., Sherrington R., Gurling H. *et al.* (1987) Molecular genetic evidence for heterogeneity in manic depression. *Nature* **325**:805–6.

Hoehe M. R., Wendel B., Grunewald I. *et al.* (1998) Serotonin transporter (5-HTT) gene polymorphisms are not associated with susceptibility to mood disorders. *Am J Med Genet (Neuropsychiatr Genet)* 81:1–3.

James N. and Chapman C. J. (1975) A genetic study of bipolar affective disorder. *British Journal of Psychiatry* 126:449–56.

Johnson G. F. S. and Leeman M. M. (1977) Analysis of familial factors in bipolar affective illness. *Archives of General Psychiatry* 34:1074–83.

Jones I. and Craddock N. (In Press) Evidence that puerperal psychosis represents a more homogenous form of bipolar disorder. *Psychiatric Genetics.*

Jones I. Gordon-Smith K., Craddock N. (2002) Triplet repeats and bipolar disorder. *Current Psychiatry Reports* 4(2):134–40.

Jones I. Jacobsen N., Green E. *et al.* (2002) Evidence for familial cosegregation of major affective disorder and genetic markers flanking the gene for Darier's disease. *Molecular Psychiatry* 7(4):424–7.

Jones I., Lendon C., Coyle N. *et al.* (2001a) Molecular genetic approaches to puerperal psychosis. *Progress in Brain Research* 133:321–32.

Jones I., Kent L., Craddock N. (2001b) Clinical implications of psychiatric genetics in the new millennium—nightmare or nirvana? *Psychiatric Bulletin* 25:129–31.

Jones I. and Craddock N. (2001a) Candidate gene studies of bipolar disorder. *Annals of Medicine* 33:248–56.

Jones I. and Craddock N. (2001b) Familiality of the puerperal trigger in bipolar disorder: results of a family study. *American Journal of Psychiatry* 158:913–17.

Jones I., Middle F., McCandless F. *et al.* (2000) Molecular genetic studies of bipolar disorder and puerperal psychosis at two polymorphisms in the oestrogen receptor 1(a) gene. *Am J Med Genet (Neuropsychiatric Genetics)* 96(6):850–3.

Kelsoe J. R., Ginns E. I., Egeland J. A. *et al.* (1989) Re-evaluation of the linkage relationship between chromosome 11p loci and the gene for bipolar affective disorder in the Old Order Amish. *Nature* 342:238–43.

Kendell R. E. (1987) Diagnosis and classification of functional psychosis. *Br Med Bull* 43(3):499–513.

Kendell R. E. (1976) The classification of depressions: a review of contemporary confusion. *British Journal of Psychiatry* 129:15–28.

Kendell R. E. (1987) Diagnosis and classification of functional psychoses. *British Medical Bulletin* 43(3):499.

Kendler K. S., Neale M. C., Kessler R. C. *et al.* (1992) Major depression and generalized anxiety disorder: same genes, (partly) different environments? *Archives of General Psychiatry* 49:716–22.

Kendler K. S., Pedersen N., Johnson L. *et al.* (1993a) A pilot Swedish twin study of affective illness, including hospital- and population-ascertained subsamples. *Archives of General Psychiatry* 50:699–706.

Kendler K. S., (1993b) Twin studies of psychiatric illness. Current status and future directions. *Archives of General Psychiatry* 50:905–15.

Kendler K. S., Neale M. C., Kessler R. C. *et al.* (1993c) Major depression and phobias: the genetic and environmental sources of comorbidity. *Psychol Med* 23:361–71.

Kendler K. S., Neale M. C., Kessler R. *et al.* (1993d) The lifetime history of major depression in women. Reliability of diagnosis and heritability. *Archives of General Psychiatry* 50:863–70.

Kendler K. S., Neale M. C., Kessler R. *et al.* (1993e) A twin study of recent life events and difficulties. *Archives of General Psychiatry* 50:589–96.

Kendler K. S., Pedersen N. L., Neale M. C. *et al.* (1995a) A pilot Swedish twin study of affective illness including hospital and population ascertained subsamples: results of model fitting. *Behavior Genetics* 25:217–32.

Kendler K. S., Walters E. E., Neale M. C. *et al.* (1995b) The structure of the genetic and environmental risk factors for six major psychiatric disorders in women: phobia, generalized anxiety disorder, panic disorder, bulimia, major depression and alcoholism. *Archives of General Psychiatry* 52:374–83.

Kendler K. S., Kessler R. C., Walters E. E. *et al.* (1995c) Stressful life events, genetic liability and onset of major depression in women. *Am J Psychiatry* 152:833–42.

Kendler K. S., Eaves L. J., Walters M. S. *et al.* (1996) The identification and validation of distinct depressive syndromes in a population-based sample of female twins. *Archives of General Psychiatry* 53:391–9.

Kendler K. S., (1997) The diagnostic validity of melancholic major depression in a population-based sample of female twins. *Archives of General Psychiatry* 54:299–304.

Kendler K. S., Karkowski L. M., Walsh D. (1998a) The structure of psychosis. *Archives of General Psychiatry* 55:492–9.

Kendler K. S., Karkowski L. M., Corey L. A. *et al.* (1998b) Longitudinal population based twin study of retrospectively reported premenstrual symptoms and lifetime major depression. *Am J Psychiatry* 155:1234–40.

Kendler K. S. (1998c) Major depression and the environment: a psychiatric genetic perspective. *Pharmacopsychiat* 31:5–9.

Kendler K. S. and Prescott C. A. (1999a) A population based twin study of lifetime major depression in men and women. *Archives of General Psychiatry* 56:39–44.

Kendler K. S., Gardner C. O., Prescott C. A. (1999b) Clinical characteristics of major depression that predict risk of depression in relatives. *Archives of General Psychiatry* 56(4):322–7.

Kessler R. C., McGonagle K. A., Zhao S. *et al.* (1994) Lifetime and 12-month prevalence of DSM IIIR psychiatric disorders in the United States: results from the National Comorbidity Survey. *Archives of General Psychiatry* 51:8–19.

Kirov G., Murphy K. C., Arranz M. J. *et al.* (1998) Low activity allele of catechol-O-methyltransferase gene associated with rapid cycling bipolar disorder. *Mol Psychiatry* 3:342–5.

Kirov G., Rees M., Jones I. *et al.* (1999) Bipolar disorder and the serotonin transporter gene: a family-based association study. *Psychological Medicine* 29:1249–54.

Klerman G. L. (1988) The current age of youthful melancholia: evidence for increase in depression among adolescents and young adults. *British Journal of Psychiatry* 152:4–14.

Kovacs M., Devlin R., Pollock M. *et al.* (1997) A controlled family history study of childhood-onset depressive disorder. *Arch Gen Psychiatry* 54:13–623.

Kringlen E. (1967) *Heredity and Environment in the Functional Psychoses.* London: William Heinemann.

Kruglyak L. and Lander E. S. (1995) High-resolution genetic mapping of complex traits. *Am J Hum Genet* 56:1212–23.

Kunugi H., Hattori M., Kato T. *et al.* (1997) Serotonin transporter gene polymorphisms; ethnic difference and possible association with bipolar affective disorder. *Mol Psychiatry* 2:457–62.

Lachman H. M., Morrow B., Shprintzen R. *et al.* (1996) Association of codon 108/158 catechol-o-methyltransferase gene polymorphism with the psychiatric manifestations of velo-cardio-facial syndrome. *Am J Med Genet (Neuropsychiat Genet)* 67:468–72.

Lander E. S. and Schork N. J. (1994) Genetic dissection of complex traits. *Science* 265:2037–48.

Leboyer M., Malafosse A., Boularand S. *et al.* (1990) Tyrosine hydroxylase polymorphisms associated with manic-depressive illness. *Lancet* 335:1219.

Leckman J. F., Sholomskas D., Thompson D. *et al.* (1982) Best estimate of lifetime psychiatric diagnosis. *Archives of General Psychiatry* **39**:879–83.

Lindblad K., Nylander P.-O., De bruyn A. *et al.* (1995) Detection of expanded CAG repeats in bipolar affective disorder using the repeat expansion detection (RED) method. *Neurobiol Dis* **2**:55–62.

Lindblad K., Nylander P.-O., Zander C. *et al.* (1998) Two commonly expanded CAG/CTG repeat loci: involvement in affective disorders? *Mol Psychiatry* **3**:405–10.

Lykken D. and Tellegen A. (1996) Happiness is a stochastic phenomenon. *Psychol Science* **7**:186–9.

Lyons M. J., Goldberg J., Eisen E. A. *et al.* (1993) Do genes influence exposure to trauma? A twin study of combat. *Am J Med Genet* **48**:2–27.

Lyons M. J., Eisen E. A., Goldberg J. *et al.* (1998) A registry based twin study of depression in men. *Archives of General Psychiatry* **55**:468–72.

Madden P. A. F., Heath A. C., Rosenthal N. E. *et al.* (1996) Seasonal changes in mood and behavior—the role of genetic factors. *Archives of General Psychiatry* **53**:47–55.

Maier W., Lichtermann D., Minges J. *et al.* (1993) Continuity and discontinuity of affective disorders and schizophrenia. Results of a controlled family. *Archives of General Psychiatry* **50**(11):871–83.

Marazita M. L., Neiswanger K., Cooper M. *et al.* (1997) Genetic segregation analysis of early-onset recurrent unipolar depression. *Am J Hum Genet* **61**:1370–8.

Mendlewicz J. and Rainer J. D. (1977) Adoption study supporting genetic transmission in manic depressive illness. *Nature* **268**:327–9.

McGuffin P., Reveley A., Holland A. (1982) Identical triplets: non-identical psychosis? *British Journal of Psychiatry* **140**:1–6.

McGuffin P., Katz R., Bebbington P. (1988) The Camberwell Collaborative Study III. Depression and adversity in the relatives of depressed probands. *British Journal of Psychiatry* **152**:775–82.

McGuffin P., and Katz R. (1989) The genetics of depression and manic-depressive disorder. *British Journal of Psychiatry* **155**:294–304.

McGuffin P. Katz R. Watkins S. *et al.* (1996) A hospital based twin register of the heritability of DSMIV unipolar depression. *Archives of General Psychiatry* **53**:129–36.

McMahon F. J., Stine O. C., Meyers D. A. *et al.* (1995) Patterns of maternal transmission in bipolar affective disorder. *Am J Hum Genet* **56**:1277–86.

Mendlewicz J., Fleiss J. L., Fieve R. R. (1972) Evidence for X-linkage in the transmission of manic-depressive illness. *Journal of the American Medical Association* **222**:1624.

Mendlewicz J. and Rainer J. D. (1977) Adoption study supporting genetic transmission in manic-depressive illness. *Nature* **268**:327–9.

Mendlewicz J. and Rainer J. D. (1974) Morbidity risk and genetic transmission in manic-depressive illness. *Am J Hum Genet* **26**:692–701.

Merikangas K., Leckman J., Prusoff B. *et al.* (1985) Familial transmission of depression and alcoholism. *Archives of General Psychiatry* **42**:367–72.

Merikangas K. R. (1990) The genetic epidemiology of alcoholism. *Psychological Medicine* **20**:11–22.

Morissette J., Villeneuve A., Bordeleau L. *et al.* (1999). Genome-wide search for linkage of bipolar affective disorders in a very large pedigree deprived from a homogeneous population in Quebec points to a locus of major effect on chromosome 12q23–q24. *Am J Med Genet (Neuropsychiatric Genetics)* **88**:567–87.

Moster M. (1991) Stressful life events; genetic and environmental components and their relationship to affective symptomatology. Ph.D. thesis. University of Minnesota, St Paul, MN.

Murray C. J. L., Lopez A. D. (eds) (1996) *The Global Burden of Disease: A Comprehensive Assessment of Mortality, Injuries, and Risk Factors in 1990 and Projected to 2020.* Harvard School of Public Health and the World Health Organisation.

Neiswanger K., Zubenko G. S., Giles D. E. *et al.* (1998) Linkage and association analysis of chromosomal regions containing genes related to neuroendocrine or serotonin function in families with early-onset, recurrent major depression. *Am J Med Genet* 81(5):443–9.

Nuffield Council on Bioethics (1998) *Mental Disorders and Genetics: the Ethical Context.* Nuffield Council on Bioethics.

Nurnberger J., Guroff J. J., Hamovit J. *et al.* (1988) A family study of rapid-cycling bipolar illness. *J Affect Disord* 15:87–91.

O'Donovan M. C., Guy C., Craddock N. *et al.* (1995) Expanded CAG repeats in schizophrenia and bipolar disorder. *Nat Genet* 10:380–1.

O'Donovan M. C., and Owen M. J. (1996a) Dynamic mutations and psychiatric genetics. *Psychol Med* 26:1–6.

O'Donovan M. C., Guy C., Craddock N. *et al.* (1996b) Confirmation of association between expanded CAG/CTG repeats and both schizophrenia and bipolar disorder. *Psychol Med* 26:1145–53.

Ogilvie A. D., Battersby S., Bubb V. J. *et al.* (1996) Polymorphism in serotonin transporter gene associated with susceptibility to major depression. *Lancet* 347:731–3.

Oruc L., Lindblad K., Verheyen G. R. *et al.* (1997) CAG repeat expansions in bipolar and unipolar disorders. *Am J Hum Genet* 60:730–2.

Papolos D. F., Veit S., Faedda G. L. *et al.* (1998) Ultra-ultra rapid cycling bipolar disorder is associated with the low activity catecholamine-O-methyltransferase allele. *Mol Psychiatry* 3:346–9.

Pauls D. L., Morton L. A., Egeland J. A. (1992) Risks of affective illness among first-degree relatives of bipolar I Old-Order Amish probands. *Archives of General Psychiatry* 49:703–8.

Pekkarinen P. J., Bredbacka P.-E., Loonqvist J. *et al.* (1995) Evidence of a predisposing locus to bipolar disorder on Xq24–q27.1 in an extended Finnish pedigree. *Genome Res* 5:105115.

Perris C. (1966) A study of bipolar (manic-depressive) and unipolar recurrent depressive psychoses. *Acta Psychiatrica Scandinavica* 42(Sup):15–44.

Petterson U. (1977) Manic-depressive illness: a clinical, social and genetic study. *Acta Psych Scand* 269(Sup):1–93.

Plomin R., Lichtenstein P., Pederson N. *et al.* (1990) Genetic influences on life events during the last half of the life span. *Psychol Aging* 5:25–30.

Rees M., Norton N., Jones I. *et al.* (1997) Association studies of Bipolar Disorder at the human serotonin transporter gene (hSERT; 5HTT). *Mol Psychiatry* 2:398–402.

Reich T., Clayton P., Winokur G. (1969) Family history studies: V. The genetics of mania. *Am J Psychiatry* 125:1358–68.

Rice J., Reich T., Andreasen N. C. *et al.* (1987) The familial transmission of bipolar illness. *Archives of General Psychiatry* 44:441–7.

Risch N., Baron M., Mendlewicz J. (1986) Assessing the role of X-linked inheritance in bipolar-related major affective disorder. *J Psychiatr Res* 20:275–88.

Risch N. and Merikangas K. (1996) The future of genetic studies of complex human diseases. *Science* 273:1516–17.

Rosenthal N. E., Mazzanti C. M., Barnett R. L. *et al.* (1998) Role of serotonin transporter promoter repeat length polymorphism (5-HTTLPR) in seasonality and seasonal affective disorder. *Molecular Psychiatry* 3(2):175–7.

Sandkuyl L. A. and Ott J. (1989) Affective disorders: evaluation of a three-allele model accounting for clinical heterogeneity. *Genet Epidemiol* 6:265–9.

Schaid D. J. and Sommer S. S. (1994) Comparison of statistics for candidate-gene association studies using cases and parents. *Am J Hum Genet* 55:402–9.

Schalling M., Hudson T. J., Buetow K. H. *et al.* (1993) Direct detection of novel expanded trinucleotide repeats in the human genome. *Nat Genet* 4:135–9.

Scharfetter C. and Nusperli M. (1980) The group of schizophrenias, schizoaffective psychoses, and affective disorders. *Schizophr Bull* 6:586–91.

Sham P. C., Morton N. E., Rice J. P. (1991) Segregation analysis of the NIMH Collaborative study. Family data on bipolar disorder. *Psychiatr Genet* 2:175–84.

Sher L., Goldman D., Ozaki N. *et al.* (1999) The role of genetic factors in the etiology of seasonal affective disorder and seasonality. *Journal of Affective Disorders* 53(3):203–10.

Simpson S. G., Folstein S. E., Meyers D. A. *et al.* (1993) Bipolar II: the most common bipolar phenotype? *Am J Psychiatry* 150:901–3.

Smeraldi E., Negri F., Melica A. M. (1977) A genetic study of affective disorders. *Acta Psychiat Scand* 56:382–98.

Smith A. L., Weissman M. M. (1992) Epidemiology, in E. S. Paykel (ed.) *Handbook of Affective Disorders*. Edinburgh: Churchill Livingstone.

Smyth C., Kalsi G., Curtis D. *et al.* (1997) Two-locus admixture linkage analysis of bipolar and unipolar affective disorder supports the presence of susceptibility loci on chromosomes 11p15 and 21q22. *Genomics* 39:271–8.

Sobell J. L., Heston L. L., Sommer S. S. (1992) Delineation of genetic predisposition to multifactorial disease: a general approach on the threshold of feasibility. *Genomics* 12:1–6.

Spence M. A., Flodman P. L., Sadovnik A. D. *et al.* (1995) Bipolar disorder: evidence for a major locus. *Am J Med Genet (Neuropsychiatr Genet)* 60:370–6.

Spielman R. S., McGinnis R. E., Ewens W. J. (1993) Transmission test for linkage disequilibrium: the insulin gene region and insulin-dependent diabetes mellitus (IDDM). *Am J Hum Genet* 52:506–16.

Stöber G., Heils A., Lesch K. P. (1996) Serotonin transporter gene polymorphism and affective disorder. *Lancet* 347:1340–1.

Stine O. C., Xu J., Koskela R. *et al.* (1995) Evidence for linkage of bipolar disorder to chromosome 18 with a parent-of-origin effect. *Am J Hum Genet* 57:1384–94.

Straub R. E., Lehner T., Luo Y. *et al.* (1994) A possible vulnerability locus for bipolar affective disorder on chromosome 21q22.3. *Nat Genet* 8:291–6.

Strober M., Morrell W., Burroughs J. *et al.* (1988) A family study of Bipolar I disorder in adolescence. Early onset of symptoms linked to increased familial loading and lithium resistance. *J Affect Disord* 15:255–68.

Strober M. (1992) Relevance of early age-of-onset in genetic studies of bipolar affective disorder. *J Am Acad Child and Adolesc Psychiatry* 31:606–10.

Suarez B. K., Hample C. L., Van Eerdewegh P. (1994) *Problems of Replicating Linkage Claims in Psychiatry. Genetic Approaches to Mental Disorders*. Washington, DC: American Psychiatric Press Inc.

Sullivan P. F., Neale M. C., Kendler K. S. (2000) Genetic epidemiology of major depression: review and meta-analysis. *Am J Psychiatry* 157:1552–62.

Taylor M. A., Abrams R., Hayman M. A. (1980) The classification of affective disorders—a reassessment of the bipolar-unipolar dichotomy. A clinical, laboratory, and family study. *J Affect Disord* 2(2):95–109.

Thapar A. and McGuffin P. (1997) Anxiety and depressive symptoms in childhood—a genetic study of comorbidity. *J Child Psychol Psychiatry* 38(6):651–6.

Thompson C., Childs P. A., Martin N. J. *et al.* (1997) Effects of monitoring phototherapy on circadian markers in seasonal affective disorder. *British Journal of Psychiatry* 170:431–5.

Torgersen S. (1986) Genetic factors in moderately severe and mild affective disorders. *Archives of General Psychiatry* 43:222–6.

Tsuang M. T., Winokur G., Crowe R. R. (1980) Morbidity risks of schizophrenia and affective disorders among first degree relatives of patients with schizophrenia, mania, depression and surgical conditions. *Br J Psychiatr* **137**:497–504.

Tsuang M. T., and Faraone S. V. (1990) *The Genetics of Mood Disorders*. Baltimore: The Johns Hopkins University Press.

Treloar S. A., Martin N. G., Bucholz K. K. *et al.* (1999) Genetic influences on post-natal depressive symptoms: findings from an Australian twin sample. *Psychological Medicine* **29**:645–54.

Turecki G., Rouleau G. A., Mari J. *et al.* (1997) Lack of association between bipolar disorder and tyrosine hydroxylase: a meta-analysis. *Am J Med Genet (Neuropsychiatric Genet)* **74**:348–52.

Turecki G., Grof P., Cavazzoni P. *et al.* (1999) MAOA: association and linkage studies with lithium responsive bipolar disorder. *Psych Genet* **9**:13–16.

Van Broeckhoven D. and Verheyen G. (1999) Report of the chromosome 18 workshop. *Am J Med Genet (Neuropsychiatr Genet)* **88**:263–70.

Vincent J. B., Klempan T., Parikh S. S. *et al.* (1996) Frequency analysis of large CAG/CTG trinucleotide repeats in schizophrenia and bipolar affective disorder. *Mol Psychiatry* **1**:141–8.

Von Knorring A. L., Cloninger C. R., Bohman M. *et al.* (1983) An adoption study of depressive disorders and substance abuse. *Archives of General Psychiatry* **40**(9):943–50.

Wang Z., Crowe R. R., Tanna V. L. *et al.* (1992) Alpha 2 adrenergic receptor subtypes in depression: a candidate gene study. *J Affec Disord* **25**:191–6.

Wehr T. A., Sack D. A., Rosenthal N. E. *et al.* (1988) Rapid cycling affective disorder: contributing factors and treatment responses in 51 cases. *Am J Psychiatry* **145**:179–84.

Weissman M. M., Kidd K. K., Prusoff B. A. (1982) Variability in rates of affective disorders in relatives of depressed and normal probands. *Archives of General Psychiatry* **39**:1397–403.

Weissman M. M., Gershon E. S., Kidd K. K. *et al.* (1984a) Psychiatric disorders in the relatives of probands with affective disorders. *Archives of General Psychiatry* **41**:13–21.

Weissman M. M., Wickramaratne P., Merikangas K. R. *et al.* (1984b) Onset of major depression in early adulthood. *Archives of General Psychiatry* **41**:1136–43.

Weissman M. M., Leaf P. J., Tischler G. L. *et al.* (1988) Affective disorders in five United States communities. *Psychological Medicine* **18**:141–53.

Wender P. H., Kety S. S., Rosenthal D. *et al.* (1986) Psychiatric disorders in the biological and adoptive families of adopted individuals with affective disorders. *Archives of General Psychiatry* **43**:923–9.

WHO (1993) The ICD10 classification of mental and behavioural disorders.

Winokur G., (1997) All roads lead to depression: clinically homogeneous, etiologically heterogeneous. *J Affect Disord* **45**(1–2):97–108.

Winokur G., Coryell W., Keller M. *et al.* (1995) A family study of manic-depressive (bipolar i) disease. Is it a distinct illness separable from primary unipolar depression? *Archives of General Psychiatry* **52**:367–73.

Winokur G. Tsuang M. T., Crowe R. R. (1982) The Iowa 500—affective disorder in relatives of manic and depressed patients. *Am J Psychiatry* **139**:209–12.

Winokur G., Cadoret R., Dorzab J. *et al.* (1971) Depressive disease: a genetic study. *Archives of General Psychiatry* **24**:135–44.

Woolf B. (1955) On estimating the relationship between blood group and disease. *Ann Hum Genet* **19**:251–3.

Chapter 10

Schizophrenia

Michael J. Owen, Michael C. O'Donovan, and
Irving I. Gottesman

Introduction

Schizophrenia is a common, severe and disabling disorder that in a majority of instances
requires long-term medical and social care. The population incidence is 0.17–0.57 per
1000, and the point prevalence is 2.4–6.7 per 1000 reflecting the typical chronic course
(Jablensky 1999). The lifetime morbid risk in the general population is close to 1 per cent
(Gottesman 1991) for definite plus probable definitional standards, rising to higher values if spectrum cases are included (Farmer *et al*. 1987; Prescott and Gottesman 1993;
Kendler 2000). The incidence of schizophrenia is surprisingly uniform across a variety of
different populations with different cultures, and the evidence for pockets of particularly
high and low incidence and prevalence is not strong (Jablensky 1999), with the exception
of several population isolates with documented histories of 'genetic bottlenecks' (Hovatta
et al. 2000; Myles-Worsley *et al*. 1999). Most individuals with schizophrenia have an onset
in early adulthood, have long periods of illness, are unable to work at their pre-morbid
levels, and have difficulty in sustaining interpersonal relationships because of their illness. Schizophrenia has an enormous impact on family and national economics, owing
to such indirect costs as providing informal care at home and a high suicide rate
(Caldwell and Gottesman 1990; Wyatt *et al*. 1995).

Epidemiological studies have identified a number of putative risk factors associated
with an increased rate of schizophrenia. These include certain pregnancy and delivery
complications (Hultman *et al*. 1999; Jones *et al*. 1998), delayed developmental mile-
stones (Jones *et al*. 1994), maternal influenza or other viral exposures, winter births,
low IQ score (David *et al*. 1997), personality characteristics concerned with social rela-
tions (Malmberg *et al*. 1998), urban upbringing (Mortensen *et al*. 1999), immigration
(Hutchinson *et al*. 1996) and the use of illegal drugs, especially cannabis (Andreasson
et al. 1987; Dean *et al*. 2001).

Following the introduction of antipsychotic drugs in the 1950s, biological theories of
the origins of schizophrenia were dominated for many years by ideas derived from
neuropharmacological and psychopharmacological studies suggesting abnormalities in
monoamine neurotransmission, in particular of the dopaminergic and serotonergic systems. The predominant hypothesis for more than three decades was that schizophrenia

reflects excessive dopaminergic neurotransmission. This view is based on two observations. First, most conventional antipsychotic drugs block dopamine receptors to a degree that is proportionate to their efficacy. Second, dopaminergic drugs such as amphetamine are psychotomimetic. However, firm evidence for a primary dopaminergic abnormality in schizophrenia has not been established, and it remains possible that neurochemical abnormalities reflect downstream pathology, compensatory mechanisms, or even environmental influences. Moreover, the clinical efficacy of newer anti-psychotic drugs appears to be partly associated with actions on other neurotransmitter systems, and there is evidence to support hypotheses of serotonin, glutamate and GABA dysfunction in schizophrenia (Bray and Owen 2001).

In recent years, schizophrenia has increasingly been regarded as a neurodevelopmental disorder (Murray and Lewis 1987; Weinberger 1987). Views differ as to the timing and nature of the supposed developmental disturbance(s), but the view that schizophrenia has its origins in early, perhaps pre-natal, brain development is supported by epidemiological evidence of increased obstetric complications and childhood neuropsychological deficits in individuals who subsequently develop the disorder, together with a pattern of non-specific neuropathological anomalies (Weinberger 1995). Neuroimaging studies have shown that increased lateral ventricle size in schizophrenics is present at onset of symptoms (Degreef et al. 1992) and in currently unaffected adolescents who are of a high genetic risk for the disorder (Cannon et al. 1993). Volumetric studies have also shown a small but significant reduction in brain size, particularly within the temporal lobes (McCarley et al. 1999), for which there may be a significant genetic component (Lawrie et al. 1999). Histological studies, though prone to methodological problems, have provided evidence for subtle cytoarchitectural anomalies of putatively developmental origin within the frontal lobes and temporolimbic structures such as the hippocampus (Harrison 1999). However, it is not yet possible to base rational identification of candidate genes upon clear evidence for disturbances in specific neurodevelopmental processes (Bray and Owen 2001).

Genetic epidemiology

Familiality and genetic risk

The extensive literature on the genetic epidemiology of schizophrenia provides compelling support for vertically transmissible genetic factors in its etiology. Gottesman and Shields, based upon data from some 40 family and twin studies in western Europe spanning the twentieth century, were able to conclusively show that the risk of schizophrenia is increased in relatives of probands with the disorder (Gottesman 1991) (Fig. 10.1). The lifetime risk for schizophrenia in the general population worldwide is generally reported to be around 1 per cent. This contrasts with the averaged risk to siblings (9 per cent) and offspring of affected probands (13 per cent), and although the risk to parents is somewhat lower (6 per cent), this reflects the reproductive disadvantage conferred by schizophrenia

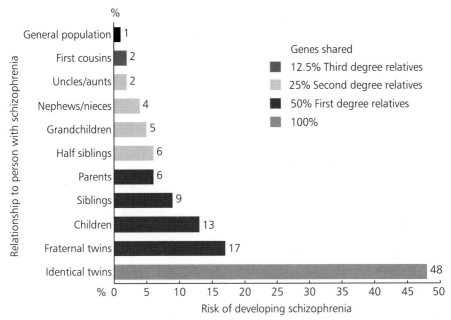

Fig. 10.1 Lifetime age-adjusted, averaged risks for the development of schizophrenia-related psychoses in classes of relatives differing in their degree of genetic relatedness. Compiled from family and twin studies in European populations between 1920 and 1987. Adapted with permission from Gottesman (1991).

(cf. McGuffin *et al.* 1994: 87). Many of the studies in this analysis predate contemporary methods, but in a review of 11 controlled studies using blinded, structured interviews, the finding of an approximately 10-fold increase in risk of schizophrenia in first-degree relatives remains (Kendler 2000). The offspring of so-called dual matings between parents both of whom suffer from schizophrenia have a 'super high-risk' for the disorder of 46 per cent. (Gottesman and Bertelsen 1989b, extracted from nine other studies). Although such unusual matings have not been studied in a fully systematic way, this remarkable finding provides further credence to the importance of familial factors in schizophrenia genesis.

Adoption studies using three study designs (adoptee studies, cross-fostering studies, and adoptee family studies) provide corroborative evidence that there is an increased risk of schizophrenia in biological first-degree relatives of probands, though not in non-biologically related adopted or adoptive family members who share the same environment as probands (Heston 1966; Rosenthal *et al.* 1971; Kety *et al.* 1994; Wender *et al.* 1974). The data from the Danish adoption study (Ingraham and Kety 2000) were re-analysed applying DSM-III criteria, with results that were consistent with a major genetic influence (Kendler *et al.* 1994): 7.9 per cent of first-degree biological relatives of schizophrenic adoptees had DSM-III schizophrenia, compared with 0.9 per cent of

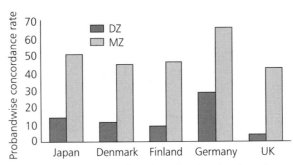

Fig. 10.2 Probandwise concordance rates for schizophrenia-related psychoses in 5 recent (1996–9) twin studies of major mental disorders (see Cardno and Gottesman 2000: 13 for references). Heritability estimates from these data range from 82 per cent (71–90) to 84 per cent (19–92) with 95 per cent confidence intervals given. From A. G. Cardno and I. I. Gottesman (personal communication, 27 July 1999) Copyright 1999 by A. G. Cardno and I. I. Gottesman. Adapted with permission.

first-degree relatives of control adoptees. The risk to relatives was raised even further to 15.8 per cent vs. 2.1 per cent using a broader spectrum of illness. Finally an ongoing Finnish study (Tienari *et al.* 2000) found a higher morbid risk of DSM- IIIR schizophrenia in the adopted-away offspring of mothers with schizophrenia compared with control offspring (8.1 per cent vs. 2.3 per cent).

The findings of adoption studies therefore suggest that shared genes rather than shared environments underlie the increased risk in relatives of schizophrenic probands, a conclusion that receives decisive support from twin studies (McGuffin *et al.* 1994). Based upon the five most recent systematically ascertained twin studies (none of which is in Fig. 9.1), the probandwise concordance rate for schizophrenia in monozygotic (MZ) twin pairs is 41–65 per cent compared with 0–28 per cent for dizygotic (DZ) twins-pairs, corresponding to heritability estimates of approximately 80–85 per cent (Cardno and Gottesman 2000) (Fig. 10.2).

Genetic epidemiology therefore points clearly to genetic aetiological mechanisms. Moreover, the relative risks associated with close genetic relatedness to an affected subject (Fig. 10.3) are considerably greater than those conferred by the putative environmental risk factors we have discussed above. Therefore if we are to understand schizophrenia, we must understand its genetics.

Mode of inheritance

Although it is clear that there is a genetic contribution to schizophrenia, the frequency with which MZ twins can be discordant for schizophrenia suggests that what is inherited is not the certainty of disease accompanying a particular genotype but rather a predisposition or liability to develop the disorder. This conclusion is supported by studies of the offspring of discordant MZ pairs showing comparable risks in the offspring of

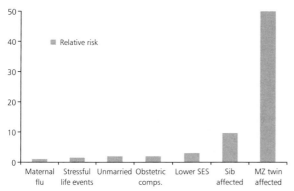

Fig. 10.3 Relative risks or odds ratios of developing schizophrenia, assuming a 1 per cent morbid risk in the general population, as a function of putative risk factors. Compiled from overviews of the field (Heinrichs 2001; Jablensky 2000; McGuffin and Gottesman 1999; Moises and Gottesman 2001).

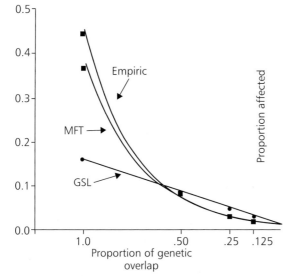

Fig. 10.4 Recurrence risk in various classes of relatives for developing schizophrenia under a generalized single locus (GSL) model and under a multifactorial threshold (MFT) model, both contrasted with the combined empirical observations. Adapted from McGue and Gottesman (1989).
Note: Proportion of genetic overlap is 1.0 for monozygotic twins, 0.50 for first-degree relatives. 0.25 for half-siblings and 0.125 for third-degree relatives.

both the affected and unaffected cotwins (Gottesman and Bertelsen 1989a; Kringlen and Cramer 1989). This indicates that specific genetic factors conferring susceptibility to schizophrenia are transmitted, but that they need not be expressed.

It is clear from studies on different classes of relative that the recurrence risk decreases too rapidly with increasing genetic distance from the proband for schizophrenia to be a single-gene disorder or collection of single gene disorders, even when incomplete penetrance is taken into account (Fig. 10.4). Rather, the mode of transmission like that of other common disorders is complex and non-Mendelian (Gottesman and Shields 1967; McGue and Gottesman 1989). The commonest mode of transmission is probably oligogenic, polygenic or a mixture of the two with a threshold effect (McGuffin *et al.* 1994). However the number of susceptibility loci, the disease risk conferred by each locus, and the degree of interaction among loci all remain unknown. Risch (1990) has calculated that the data for recurrence risks in the relatives of probands with schizophrenia are incompatible with the existence of a single locus conferring a relative risk (λs) of more than 3 and, unless extreme epistasis exists, models with two or three loci of λs < 2 are more plausible. It should be emphasized that these calculations are based upon the assumption of homogeneity. It is quite possible that alleles of larger effect are operating in some groups of patients, for example, families chosen for having a high density of illness. However, it has also been demonstrated that such families would be expected to occur even under polygenic inheritance, and their existence does not force the conclusion that alleles of large effect exist (McGue and Gottesman 1989).

Defining the phenotype

Ignorance of pathogenesis and the absence of a biological marker means that the disorder is effectively syndromic and diagnosis has to rely on a combination of hetero-phenomenology and clinical observation. Moreover, despite considerable effort, it has not been possible to delineate etiologically distinct subgroups. It therefore seems highly likely that the diagnosis of schizophrenia embraces a variety of overlapping conditions with different symptoms, courses, outcomes, responses to treatment and probably aetiologies. In spite of this, the use of structured and semi-structured interviews together with explicit operational diagnostic criteria means that it is possible to achieve high degrees of diagnostic reliability. Moreover, as we have seen, the syndrome so defined has a high heritability and it should in principle be possible to subject it to molecular genetic analysis. However, we also have to accept that the disorder, as defined by current diagnostic criteria, may well include a number of heterogeneous disease processes and that attempts at identifying genes and other aetiological factors would be greatly facilitated if it were possible to distinguish reliably between them using one or another refined taxonomy, phenotypic or endophenotypic (cf. Gottesman and Shields 1972).

The situation is further complicated by the fact that we are as yet unable to define the limits of the clinical phenotype to which genetic liability can lead. Family, twin and adoption studies have shown that the phenotype extends beyond the core diagnosis of schizophrenia to include a spectrum of disorders including mainly schizophrenic schizoaffective disorder, and schizotypal personality disorder (Farmer et al. 1987; Kendler et al. 1995). However, the limits of this spectrum and its relationship to other psychoses, affective and non-affective, and non-psychotic affective disorders remain uncertain (Kendler et al. 1998; Tienari et al. 2000; Cardno et al. 2002).

Finally, although it is clear that genotype confers liability to schizophrenia, we have no clear idea what form this 'liability' actually takes. That is to say, we do not understand the nature of the endophenotypes that predispose to schizophrenia. It follows then that unless it is manifested as disorder, we do not know who has inherited high liability, or more accurately, how liability should be measured and quantified.

However, in spite of the problems and uncertainties we have described, and the difficulties that ensue, schizophrenia is a compelling candidate for studies aimed at identifying disease genes, not unlike diabetes (Todd 1999), coronary artery disease (Sing et al. 1996;

Zerba *et al.* 2000) or Parkinson's disease (Scott *et al.* 2001). This is true not only because of its high heritability but also because, notwithstanding improvements in neuroimaging and gene-expression studies in autopsy material, the brain is still difficult to study directly, and the interpretation of cause and effect is problematic.

Molecular genetics

Linkage

The first wave of systematic molecular genetic studies of schizophrenia in the late 1980s and early 1990s focused upon linkage analysis of large pedigrees containing multiple affecteds. This approach had been highly effective for identifying genes of large effect in single-gene disorders. The proponents of this method were aware it was unlikely that such genes account for most of the genetic risk for schizophrenia. Instead, the hope was that under a mixed genetic model, the multiplex families, or at least a sufficient proportion of them, were the result of segregation of rare alleles of large effect.

This optimistic view proved to be well founded in a number of complex disorders, for example, Alzheimer's disease (Chapter 13). Early studies of large schizophrenia pedigrees also initially produced a strong positive finding (Sherrington *et al.* 1988). Unfortunately this could not be replicated. However, modest evidence for several linked loci has been reported (Fig. 10.5), some of which have received supportive evidence from international collaborative studies. These include chromosome 22q11–12, 6p24–22, 8p22–21 and 6q (Schizophrenia Linkage Collaborative 1996; Gill *et al.* 1996; Levinson *et al.* 2000). There are also a number of other promising areas of putative linkage. These include 13q14.1-q32 (Lin *et al.* 1995; Blouin *et al.* 1998; Brzustowicz *et al.* 1999; Camp *et al.* 2001), 5q21-q33 (Schwab *et al.* 2000; Straub *et al.* 1997; Camp *et al.* 2001; Gurling *et al.* 2001) and 10p15-p11 (Schwab *et al.* 1997; Faroane *et al.* 1998; Straub *et al.* 1998). One other region that is notable for the strength of evidence (maximum heterogeneity LOD score of 6.5) from a single study, rather than the more important convergence of evidence from different sources, is 1q21–q22 (Brzustowicz *et al.* 2000). The finding on 1q21–q22 more than meets standard criteria for claiming 'significant' linkage. Early experience in psychiatric genetics has, however, impressed the need for replication firmly upon the minds of most researchers, and although there have been other reports of linkage to markers on chromosome 1 (Hovatta *et al.* 2000; Ekelund *et al.* 2000; Gurling *et al.* 2001), these are distal to 1q21–q22 and it is not yet clear that these findings are consistent with the existence of a single susceptibility locus on 1q.

As more linkage data from genome scans of schizophrenia and bipolar disorder have emerged, putative linkages to both disorders have been reported in the same regions of the genome, for example 13q22 and 22q11–13 (Berrettini 2000). The observation of common linked loci may indicate at least a partial degree of overlap in the etiologies of these two disorders. However, there are several reasons for caution. First, none of the

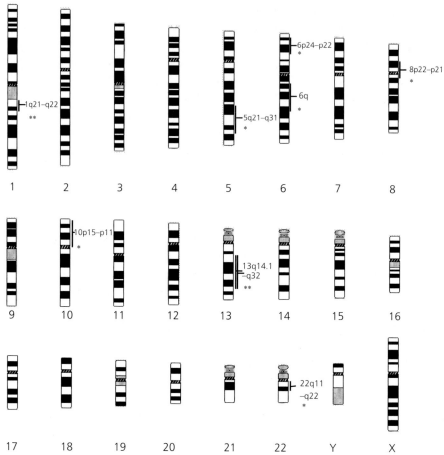

Fig. 10.5 Ideogram showing major chromosomal regions implicated by linkage studies of schizophrenia. Lines marked * indicate areas for which evidence of linkage has been found in more than one data set. Lines marked ** indicate regions where evidence of linkage has achieved genome-wide significance. From Bray and Owen (2001).

linkages are as yet definitive, and therefore the strength of the evidence for joint areas of linkage is not conclusive. Second, similarity is more noticeable than dissimilarity. Until the apparent 'clustering' has been subjected to rigorous statistical analysis, the pattern may be coincidence. Third, even if correct, the regions of linkage contain hundreds or even thousands of genes. It follows that common linked loci does not prove common genes. Fourth, it is possible that the joint linkages are to loci that modify the phenotypes in such a way as to influence measurement of affected status. The prognosis for both schizophrenia and bipolar disorder is generally worse in individuals who are subjected to chronic stress. Therefore loci that modulate the stress response might appear to be linked to both disorders simply because of an effect on chronicity, and, thereby, ease of ascertainment.

As an alternative to linkage based upon multiplex families, some groups have advocated the use of smaller families with two or more affecteds in a single sibship. This is in the belief that such families may be more suited for the analysis of complex traits, and being much more common, may also be more representative of disease in the general population. This approach has been successful in detecting linkage to several complex diseases including late-onset AD (Myers *et al.* 2000). Williams and colleagues (Williams *et al.* 1999) have recently reported the largest systematic search for linkage yet published using a sample of 196 affected sibling pairs (ASPs). This study was designed to have power > 0.95 to detect a susceptibility locus of $\lambda s = 3$ (the maximum effect size estimated by Risch (1990)) with genome-wide significance of 0.05. However, none of the findings approached genome-wide significance. Under the assumption of no dominance variance, Williams and colleagues were able to exclude susceptibility genes with an effect of $\lambda s = 3$ and 2 from 82.8 per cent and 48.7 per cent of the genome respectively, while genes with an effect size of $\lambda s = 1.5$ could only be excluded from 9.3 per cent of the genome.

At present, therefore, the linkage literature supports the predictions made by Risch (1990): it is highly unlikely that a commonly occurring locus of effect size $\lambda s > 3$ exists. However, there is suggestive evidence implicating a number of regions which is consistent with the existence of some susceptibility alleles of moderate effect (λs 1.5–3), and possibly uncommon loci of larger effect that can be identified in specific samples of large multiply affected families.

The findings from linkage studies of schizophrenia to date demonstrate several features that are to be expected in the search for genes for complex traits (Suarez *et al.* 1994; Lander and Kruglyak 1995; Roberts *et al.* 1999). First, no finding replicates in all data sets. Second, levels of statistical significance are unconvincing and estimated effect sizes are usually modest. Third, chromosomal regions of interest are typically broad (often > 20–30 cM).

While the linkage data for schizophrenia are not definitive, it seems likely that one or more of the above loci are true positives. However, it is difficult to know whether the statistical evidence for linkage is sufficiently strong to warrant large-scale and expensive efforts aimed at cloning putative linked loci. Of course, the proof that a positive linkage is correct comes when the disease gene is identified, and a number of the linked regions are currently being subjected to detailed analysis. Another way of resolving uncertainty will be to study much larger samples, of say 600–800 nuclear families, which should be sufficient to detect susceptibility genes of moderate effect size; collborative studies of this sort should now become a priority (Owen *et al.* 2000)

Association studies

Screening the whole genome by association presents difficulties because of the large, and as yet unknown, number of markers required to achieve comprehensive coverage. To date therefore most researchers have chosen to undertake positional and functional candidate gene studies. Because of doubts concerning the robustness of linkage findings in schizophrenia, most studies to date have focused upon functional candidate genes even in the absence of a consensus as to what constitutes such genes.

There are a number of potential problems with association studies that have been reviewed in Chapter 3. It is worth emphasizing that, for disorders of unknown etiology such as schizophrenia, the potential choice of candidate genes is limited largely by the number of genes in the human genome (probably around 30–40,000). This places a heavy burden of proof on positive results, because with very few exceptions each gene has an extremely low prior probability that it is involved in the disorder (Owen *et al.* 1997). It is also worth pointing out that most studies to date have not had sufficient power to detect or confirm the presence of alleles of small effect.

The most obvious candidate genes derive from the neuropharmacological literature. Thus, genes involved in dopaminergic and serotonergic neurotransmission have received a great deal of attention. With recent developments in genome analysis technology and the recent publication of a working draft of the human genome sequence (Lander *et al.* 2001) these data are being rapidly extended to other systems including the glutamatergic, GABAergic, and genes involved in neuromodulation and neuro-development (Williams *et al.* 2002). It is impossible for us to present the results of the hundreds, possibly thousands, of candidate gene analyses that have been published or reported at meetings. However, for the reasons outlined, the overwhelming number of negative reports cannot be considered as definitive exclusions of small genetic effects. Moreover, with perhaps the two exceptions we discuss below, no positive finding has received sufficient support from more than one group to suggest a true association with even a modest degree of probability.

Serotonin 5HT2a receptor

The serotonergic system is a therapeutic target for several antipsychotic drugs. The first genetic evidence for its involvement in schizophrenia was a report of association with a T > C polymorphism at nucleotide 102 in the HTR2A gene which encodes the 5HT2a receptor (Inayama *et al.* 1996). This association was based upon a small sample of Japanese subjects. Firmer evidence for association subsequently emerged from a large multi-centre European consortium (Williams *et al.* 1996) and a meta-analysis of all the data available in 1997 based upon more than 3,000 subjects (Williams *et al.* 1997).

Since the meta-analysis was undertaken, a few further negative reports have followed, none with the sample sizes required. If we assume homogeneity and that the association is correct, the odds ratio (OR) for the C allele is approximately 1.2. Sample sizes of 1000 cases and 1000 controls are required for 80 per cent power to detect an effect of this size even at a $p = 0.05$. However, while the negative studies do not refute the putative association, the evidence presented even in the meta-analysis ($p = 0.0009$) is well short of genome wide significance (estimated at around $p = 5 \times 10^{-8}$ by Risch and Merikangas (1996)). Admittedly, demanding genome-wide significance is inappropriate for such a strong candidate gene because the prior probability that it is involved in schizophrenia is greater than a gene selected at random. However, as there is no way of quantifying how much greater the prior probability is, we have no compelling alternative to genome-wide significance. Consequently, we believe that the evidence tends to indicate association between HTR2A and schizophrenia, but that the burden of proof has not yet been met. Another reason to be sceptical about the putative association is that there are no variants yet detected in the gene that clearly alter receptor function or expression. However, there is some indirect evidence for the existence of sequence variation else-where that alters HTR2A expression in some regions of the brain (Bunzel *et al.* 1998).

Dopamine receptor genes

The dominant neurochemical hypothesis of schizophrenia involves dysregulation of the dopaminergic system, with disordered transmission at the dopamine D2 receptor widely favoured. However, association analyses of DRD2 to date have essentially been negative, although recently, three groups, two Japanese (Arinami *et al.* 1997; Ohara *et al.* 1998) and one Swedish (Jonsson *et al.* 1999) have implicated a polymorphism in the promoter which alters DRD2 expression *in vitro*. Unfortunately, the largest single study to date found no evidence for association (Li *et al.* 1998a). Furthermore, when the UK data from this study were combined with a large dataset from Scotland, the allele that was associated with schizophrenia in the other three studies was actually significantly less common in the patient group (Breen *et al.* 1999). This finding has been interpreted as suggestive that the polymorphism tested in these studies is simply a marker for the true susceptibility variant and that the association is due to linkage disequilibrium (LD). While it is reasonable to attribute reversal of allelic association to LD differences between samples of different ethnic origins, the explanation would be more convincing if the same allele was associated in both UK and Swedish samples.

The dopamine hypothesis has received support from studies of DRD3, which encodes the dopamine D3 receptor. Association has been reported between schizo-phrenia and homozygosity for a Ser9Gly polymorphism in exon 1 of this gene (Crocq *et al.* 1992). As expected, positive and negative data have emerged, but unlike DRD2, sufficient data are available for meta-analysis. Based upon all available data from over 5000 individuals, Williams and colleagues (Williams *et al.* 1998) concluded in favour of association although the effect size was small (O.R. = 1.23) but nominally significant (p = 0.0002).

Two recent findings have however, begun to erode the DRD3 association. First, while the association is still significant, both the degree of statistical significance for associa-tion (p = 0.02) and the estimated O.R. (1.09) has been considerably diminished as around 3000 new subjects have been included in the meta-analysis (Johnsson *et al.* personal communication). Second, in the largest single homogeneous sample studied to date (and which has not been included in the meta-analysis) our own group did not find any evidence for association. In fact we found the converse, with extremely modest evidence for an excess of heterozygotes in the patient population compared both with controls and with Hardy-Weinberg Equilibrium. Given that in the extensive DRD3 literature, there is no support for a finding in this direction, this result is likely to be due to chance rather than to true association. Notwithstanding the fact that the sample is the largest in which the DRD3 polymorphism has been analysed to date, it is still small compared with that required to have high power for replication. However, it is surprising that we did not at least find a trend towards homozygosity, and the observation that the limit of the 95 per cent confidence interval (0.55–0.99) for an excess of homozygotes is less than 1. These findings, combined with the weakening of results in meta-analysis, now cast considerable doubt about the homozygosity hypothesis.

Anticipation and trinucleotide repeats

Anticipation is the phenomenon whereby a disease becomes more severe or has an earlier age at onset in successive generations. Numerous studies have now suggested that the pattern of illness in multiplex pedigrees segregating schizophrenia and other psychotic illnesses is at least consistent with anticipation, although ascertainment biases offer an alternative explanation (O'Donovan et al. 1996).

True genetic anticipation is a hallmark of a particular class of mutation called expanded trinucleotide repeats. The observations in families have therefore been taken as suggestive that such mutations might be involved in the etiology of psychosis (Petronis and Kennedy 1996). This hypothesis has received support from a number of studies reporting that large but unidentified CAG/CTG trinucleotide repeats are more common in patients with schizophrenia and other psychoses than in unaffected controls (e.g. O'Donovan et al. 1995; Lindblad et al. 1995; Morris et al. 1995; O'Donovan et al. 1996).

Unfortunately, these early findings have been followed by numerous failures of replication. Also, while two autosomal loci, one at 18q21.1 and the other at 17q21.3, account for around 50 per cent of large CAG/CTG repeats detected by RED, neither is associated with psychosis (e.g. Vincent et al. 2000; Bowen et al. 2000). These data have tempered much of the enthusiasm for the CAG/CTG hypothesis, but it is still premature to discard it. Indeed two independent groups have recently detected the presence of proteins in a small number of schizophrenics that react with an antibody against moderate-large polyglutamine sequences (see Ross 1999). This is of relevance here because polyglutamine sequences are often encoded in DNA by CAG repeats. It remains to be seen whether these proteins are involved in the disease rather than simply chance findings.

Cytogenetic abnormalities

A third approach by which investigators have sought to locate susceptibility genes for schizophrenia has been to identify chromosomal abnormalities in affected individuals.

Cytogenetic anomalies, such as translocations and deletions, may be pathogenic through several mechanisms; direct disruption of a gene or genes, the formation of a new gene comprised of a fusion of two genes that are normally spatially separated, by indirect disruption of the function of neighboring genes by a so-called position effect or an alteration of gene dosage in the case of deletions, duplications, and unbalanced translocations (see Chapter 1). It is also possible for an abnormality merely to be linked to a susceptibility variant in a particular family. Owing to the high prevalence of schizophrenia, a single incidence of a cytogenetic abnormality is insufficient to suggest causality. In order to warrant further investigation, a cytogenetic abnormality should either be shown to exist in greater frequencies in affected individuals, to disrupt a region already implicated by genetic analyses, or to show cosegregation with the condition in affected families.

There have been numerous reports of associations between schizophrenia and chromosomal abnormalities (Bassett *et al.* 2000; Baron 2001), but with two exceptions none has as yet provided convincing evidence to support the location of a gene conferring risk to schizophrenia. The first finding of interest is a (1;11)(q42;q14.3) balanced reciprocal translocation found to cosegregate with schizophrenia and other psychiatric disorders in a large Scottish family (St Clair *et al.* 1990; Blackwood *et al.* 2001). This translocation generates a LOD score of 3.6 when the disease phenotype is restricted to schizophrenia, of 4.5 when the disease phenotype is restricted to affective disorders, of 7.1 when relatives with recurrent major depression, with bipolar disorder, or with schizophrenia are all classed as affected (Blackwood *et al.* 2001). Recently the translocation has been reported to directly disrupt three genes of unknown function on chromosome 1 (Millar *et al.* 2000). These are located close to the chromosome 1 markers implicated in the two Finnish linkage studies mentioned previously (Hovatta *et al.* 2000; Ekelund *et al.* 2000). However, until mutations have been identified in other families or convincing biological evidence implicating this locus has been obtained, the mechanism by which the translocation confers risk to mental illness remains obscure and we cannot be sure whether genes or mechanisms of wider relevance have been identified.

The second finding of interest is the association between velocardiofacial syndrome and schizophrenia. velocardiofacial syndrome (VCFS), also known as DiGeorge or Shprintzen syndrome, is associated with small interstitial deletions of chromosome 22q11. The phenotype of VCFS is variable, but in addition to characteristic core features of dysmorphology and congenital heart disease, there is strong evidence that individuals with VCFS have a dramatic increase in the risk of psychosis, especially schizophrenia (Shprintzen *et al.* 1992; Pulver *et al.* 1994; Papolos *et al.* 1996; Murphy *et al.* 1999).

Clearly, with an estimated prevalence of 1 in 4000 live births, one can estimate that VCFS cannot be responsible for more than a small fraction (~1 per cent) of cases of schizophrenia and this estimate is in keeping with empirical data (Karayiorgou *et al.* 1995). From the practical perspective, clinicians should be vigilant for VCFS, especially when psychosis occurs in the presence of other features suggestive of the syndrome such as dysmorphology, mild learning disability or a history of cleft palate or congenital heart disease (Gothelf *et al.* 1997; Bassett *et al.* 1998). However, from the perspective of the genetic researcher, the most pressing question is whether the high rate of psychosis in VCFS provides a shortcut to a gene within the deleted region that is involved in susceptibility to schizophrenia in cases without a deletion. There are some reasons to believe that this might indeed be the case. As we have already seen, some linkage studies suggest the presence of a general schizophrenia susceptibility locus on 22q. While most suggest that this maps outside the VCFS region, linkage mapping in complex diseases is imprecise and modest evidence for linkage (Blouin *et al.* 1998; Lasseter *et al.* 1995; Shaw *et al.* 1998) and for linkage disequilibrium (Owen and O'Donovan, unpublished data) within the VCFS region has also been observed. Finally, mice that are heterozygously deleted for a subset of the genes that are deleted in VCFS show sensorimotor gating and memory impairments, both of which have been implicated as endophenotypes in schizophrenia (Paylor *et al.* 2001). Further studies of models of this sort may well allow the genetic basis for the behavioural and psychiatric phenotypes in VCFS to be determined, which in turn might point to genetic pathways relevant to schizophrenia.

Future directions

The linkage studies to date have shown that genes of major effect are not common causes of schizophrenia and allied psychoses. They have also provided information concerning the *possible* location of some genes of moderate effect. Association studies have provided further mapping information, with VCFS providing fairly strong evidence for a susceptibility locus in 22q11. To a lesser degree, with the proviso that the associations in HTR2A and DRD3 are weak and still provisional, they also suggest that the dominant neurochemical hypotheses are at least partially correct. It is clear however that the findings from molecular genetics are as fragile as those from genetic epidemiology are robust. How are we to make progress?

As in other common diseases, it is hoped that advances in the genetics of schizophrenia will come through the application of detailed information on the anatomy and sequence of the human genome combined with new methods of genotyping and statistical analysis. To take advantage of these developments, collection of large, well-characterized samples is a priority (Owen *et al.* 2000). Samples for genetic studies of schizophrenia are generally ascertained in differing and unsystematic ways, and this may be one of the explanations of failures to replicate. For example, if a positive association is detected in one patient group that, by virtue of the method of ascertainment, has been inadvertently selected for chronicity, the putative 'risk allele' might actually be an allele that modifies treatment response rather than predisposing to the disorder itself. Unless this is recognized, replication will be difficult except in samples with a bias towards chronicity.

It is possible that new genes and pathways might be implicated by emerging transcriptomic and proteomic approaches, and preliminary findings using both approaches have already been reported (Edgar *et al.* 1999; Hakak *et al.* 2001). The problem here is that human studies will be hampered by the many confounding variables associated with post mortem studies of brain, while the application of these methods in animal studies suffer from the difficulties inherent in extrapolating from animal behavior to a complex human psychiatric disorder such as schizophrenia.

Finally, it is worth reminding ourselves that the effectiveness of molecular genetic studies depends upon the genetic validity of the phenotypes studied. We therefore need to focus research on the development and refinement of phenotypic measures and biological markers, which might simplify the task of finding genes. It is likely that genetic validity will be improved by focusing upon quantitative aspects of clinical variation such as symptom profiles (e.g. Cardno *et al.* 1999; Pulver *et al.* 2000) or by identifying biological markers that define more homogeneous subgroups or those that predict degree of genetic risk; so-called endophenotypes. However, it seems unlikely that these will provide a rapid solution to the problem. First, we will need measures that can be practicably applied to a sufficient number of families or unrelated patients and second, we will need to ensure that the traits identified are highly heritable, which will itself require a return to classical genetic epidemiology and to model building and fitting.

References

Andreasson S., Allebeck A., Engstrom A. *et al.* (1987) Cannabis and schizophrenia: a longitudinal study of Swedish conscripts. *Lancet* 2:1483–6.

Arinami T., Gao M., Hamaguchi H. *et al.* (1997) A functional polymorphism in the promoter region of the dopamine D2 receptor gene is associated with schizophrenia. *Hum Mol Genet* 6:577–82.

Baron M. (2001) Genetics of schizophrenia and the new millennium: progress and pitfalls. *Am J Hum Genet* 68:299–312.

Bassett A., Hodgkinson K., Chow E. *et al.* (1998) 22q11 deletion syndrome in adults with schizophrenia. *Am J Med Genet* 81:328–37.

Bassett A. S., Chow E. W. C., Weksberg R. (2000) Chromosomal abnormalities and schizophrenia. *Am J Med Genet (Seminars in Medical Genetics)* 97:45–51.

Berrettini W. H. (2000) Susceptibility loci for bipolar disorder: overlap with inherited vulnerability to schizophrenia. *Biological Psychiatry* 47:245–51.

Blackwood D. H., Fordyce A., Walker M. T. *et al.* (2001) Schizophrenia and affective disorders— cosegregation with a translocation at chromosome 1q42 that directly disrupts brain-expressed genes: clinical and P300 findings in a family. *Am J Hum Genet* 69:428–33.

Blouin J. L., Dombroski B. A., Nath S. K. *et al.* (1998) Schizophrenia susceptibility loci on chromosomes 12q32 and 8p21. *Nat Genet* 20:70–3.

Bowen T., Guy C. A., Cardno A. G. *et al.* (2000) Repeat Sizes at CAG/CTG Loci CTG18.1, ERDA1 and TGC13–7a in schizophrenia. *Psychiatr Genet* 10:33–7.

Bray N. J. and Owen M. J. (2001) Searching for schizophrenia genes. *Trends Mol Med* 7:169–75.

Breen G., Brown J., Maude S. *et al.* (1999) -141 C Del/Ins polymorphism of the dopamine receptor 2 gene is associatedwith schizophrenia in a British population. *Am J Med Genet (Neuropsychiatr Genet)* 88:407–10.

Brzustowicz L. M., Honer W. G., Chow E. W. C. *et al.* (1999) Linkage of familial schizophrenia to chromosome 13q32. *Am J Hum Genet* 65:1096–103.

Brzustowicz L. M., Hodgkinson K. A., Chow E. W. C. *et al.* (2000) Location of a major susceptibility locus for familial schizophrenia on chromosome 1q21–22. *Science* **288**:678–82.

Bunzel R., Blumucke I., Cichon S. *et al.* (1998) Polymorphic imprinting of the serotonin-2A (5-HT2A) receptor gene in human adult brain. *Molec Brain Res* **59**:90–2.

Caldwell, C. B. and Gottesman, I. I. (1990) Schizophrenics kill themselves too: a review of risk factors for suicide. *Schizophr Bull* **16**:571–89.

Camp N. J., Neuhausen S. L., Tiobech J. *et al.* (2001) Genomewide multipoint linkage analysis of seven extended palauan pedigrees with schizophrenia, by a Markov-Chain Monte Carlo Method. *Am J Hum Genet* **69**:1278–89.

Cannon T. D. *et al.* (1993). Developmental brain abnormalities in the offspring of schizophrenic mothers. I. Contributions of genetic and perinatal factors. *Archives of General Psychiatry* **49**:531–64.

Cardno A. G., Jones L. A., Murphy K. C. *et al.* (1999) Dimensions of psychosis in affected sibling pairs. *Schizophr Bull* **25**:841–50.

Cardno A. G., Gottesman I. I. (2000) Twin studies of schizophrenia: from bow-and-arrow concordances to star wars Mx and functional genomics. *Am J Med Genet* **97**:12–17.

Cardno A. G., Rijsdijk F. V., Sham P. C. *et al.* (2002). A twin study of genetic relationships between psychotic symptoms. *Am J Psychiat* **159**:539–45.

Crocq M. A., Mant R., Asherson P. *et al.* (1992). Association between schizophrenia and homozygosity at the dopamine-d3 receptor gene. *J Med Genet* **29**:858–60.

David A. S., Malmberg A., Brandt L. *et al.* (1997) IQ and risk for schizophrenia: a population-based cohort study. *Psychol Med* **27**:1311–23.

Dean B., Sundram S., Bradbury R. *et al.* (2001) Studies on [3H]CP-55940 binding in the human central nervous system: regional specific changes in the density of cannabinoid-1 receptors associated with schizophrenia and cannabis use. *Neuroscience* **103**:9–15.

Degreef G. *et al.* (1992). Volumes of ventricular system subdivisions measured from magnetic resonance images in first-episode schizophrenia patients. *Archives of General Psychiatry* **49**:531–7.

Edgar P. F., Schonberger S. J., Dean B. *et al.* (1999) A comparative proteome analysis of hippocampal tissue from schizophrenic and Alzheimer's disease individuals. *Mole Psychiatry* **4**:173–8.

Ekelund J., Lichtermann D., Hovatta I. *et al.* (2000) Genome-wide scan for schizophrenia in the Finnish population: evidence for a locus on chromosome 7q22. *Hum Mol Genet* **9**:1049–57.

Faraone S. V., Matise T., Svrakic D. *et al.* (1998). Genome scan of European-American schizophrenia pedigrees: results of the nimh genetics initiative and millennium consortium. *Am J Med Genet* **81**:290–5.

Farmer A. E., McGuffin P., Gottesman I. I. (1987) Twin concordance for DSM-III schizophrenia. Scrutinizing the validity of the definition. *Archives of General Psychiatry* **44**:634–41.

Gill M., Vallada H., Collier D. *et al.* (1996) A combined analysis of D22s278 marker alleles in affected sib-pairs—support for a susceptibility locus for schizophrenia at chromosome 22Q12. *Am J Med Genet* **67**:40–5.

Gothelf D., Frisch A., Munitz H. *et al.* (1997) Velocardiofacial manifestations and microdeletions in schizophrenic patients. *Am J Med Genet* **72**:455–61.

Gottesman I. I. and Shields J. (1967) A polygenic theory of schizophrenia. *Proc Natl Acad Sci* **58**:199–205.

Gottesman I. I. and Shields J. (1972) *Schizophrenia and Genetics: a Twin Study Vantage Point.* New York: Academic Press.

Gottesman I. I., Bertelsen A. (1989) Confirming unexpressed genotypes for schizophrenia. Risks in the offspring of Fischer's Danish identical and fraternal discordant twins. *Archives of General Psychiatry* **46**:867–72.

Gottesman I. I. and Bertelsen A. (1989b) Dual mating studies in psychiatry—offspring of inpatients with examples from reactive (psychogenic) psychoses. *Int Rev Psychiatry* **1**:287–96.

Gottesman I. I. (1991) *Schizophrenia Genesis: The Origins of Madness.* Freeman: New York.

Gurling H. M., Kalsi G., Brynjolfson J. *et al.* (2001) Genomewide genetic linkage analysis confirms the presence of susceptibility loci for schizophrenia, on chromosomes 1q32.2 : 5q33.2, and 8p21–22 and provides support for linkage to schizophrenia, on chromosomes 11q23.3–24 and 20q12.1–11.23. *Am J Hum Genet* **68**:661–73.

Hakak Y., Walker J. R., Li C. *et al.* (2001) Genome-wide expression analysis reveals dysregulation of myelination-related genes in chronic schizophrenia. *Proc Natl Acad Sci* **98**:4746–51.

Harrison P. J. (1999) The neuropathology of schizophrenia. A critical review of the data and their interpretation. *Brain* **122**:593–624.

Heinrichs R. W. (2001) *In Search of Madness: Schizophrenia and Neuroscience.* New York: Oxford University Press.

Heston L. L. (1966) Psychiatric disorders in foster home reared children of schizophrenic mothers. *British Journal of Psychiatry* **112**:819–25.

Hovatta I., Varilo T., Suvisaari J. *et al.* (2000) Screen for schizophrenia genes in an isolated Finnish subpopulation, suggesting multiple susceptibility loci. *Am J Hum Genet* **65**:1114–25.

Hultman C. M., Sparen P., Takei N. *et al.* (1999) Prenatal and perinatal risk factors for schizophrenia, affective psychosis and reactive psychosis: case-control study. *BMJ* **318**:421–6.

Hutchinson G., Takei N., Fahy T. *et al.* (1996) Morbid risk of psychotic illness in first degree relatives of white and African-Caribbean patients with psychosis. *British Journal of Psychiatry* **169**:776–80.

Inayama Y., Yoneda H., Sakai T. *et al.* (1996) Positive association between a DNA sequence variant in the serotonin 2A receptor gene and schizophrenia. *Am J Med Genet* **67**:103–5.

Ingraham L. J. and Kety S. S. (2000) Adoption studies of schizophrenia. *Am J Med Genet (Seminars in Medical Genetics)* **97**:18–22.

Jablensky A. (1999) The 100-year epidemiology of schizophrenia. *Schizophr Res* **28**:111–25.

Jablensky A. (2000) Epidemiology of schizophrenia: The global burden of disease and disability. *European Archives of Psychiatry and Clinical neuroscience* **250**:274–85.

Jones P., Rodgers B., Murray R. *et al.* (1994) Child development risk factors for adult schizophrenia in the British 1946 birth cohort. *Lancet* **344**:1398–402.

Jones P., Rantakallio P., Hartikainen A.-L. *et al.* (1998) Schizophrenia as long-term outcome of pregnancy, delivery and perinatal complications: a 28-year follow-up of the 1966 North Finland general population birth cohort. *Am J Psychiatry* **155**:355–64.

Jonsson E. G., Nothen M. M., Neidt H. *et al.* (1999) Association between a promoter polymorphism in the dopamine D2 receptor gene and schizophrenia. *Schizophr Res* **40**:31–6.

Karayiorgou M., Morris M. A., Morrow B. *et al.* (1995) Schizophrenia susceptibility associated with interstitial deletions of chromosome 22q11. *Pr Natl Acad Sci USA* **92**:7612–6.

Kendler K. S., Gruenberg A. M., Kinney D. K. (1994) Independent diagnoses of adoptees and relatives as defined by DSM-III in the provincial and national samples of the Danish Adoption Study of Schizophrenia. *Archives of General Psychiatry* **51**:456–68.

Kendler K. S., Neale M. C., Walsh D. (1995) Evaluating the spectrum concept of schizophrenia in the Roscommon Family Study. *Am J Psychiatry* **152**:749–54.

Kendler K. S., Karkowski L. M., Walsh D. *et al.* (1998) The structure of psychosis: Latent class analysis of probands from the Roscommon family study. *Archives of General Psychiatry* **55**:492–509.

Kendler K. S. (2000) Schizophrenia genetics, in B. J. Sadock and V. A. Sadock (eds) *Kaplan and Sadock's Comprehensive Textbook of Psychiatry*, Vol. 1. Philadelphia: Lippincott, Williams and Wilkins.

Kety S. S., Wender P. H., Jacobsen B. *et al.* Mental illness in the biological and adoptive relatives of schizophrenic adoptees: replication of the Copenhagen study in the rest of Denmark. *Archives of General Psychiatry* **51**:442–55.

Kringlen E. and Cramer G. (1989) Offspring of monozygotic twins discordant for schizophrenia. *Archives of General Psychiatry* **46**:873–7.

Lander E. and Kruglyak L. (1995) Genetic dissection of complex traits: guidelines for interpreting and reporting linkage results. *Nat Genet* **11**:241–7.

Lander E. S. *et al.* (2001) Initial sequencing and analysis of the human genome. *Nature* **409**:860–921.

Lasseter V. K., Pulver A. E., Wolyniec P. S. *et al.* (1995) Follow-up report of potential linkage for schizophrenia on chromosome 22q.3. *Am J Med Genet* **60**:172–3.

Lawrie S. M., Whalley H., Kestelman J. N. *et al.* (1999) Magnetic resonance imaging of brain in people at high risk of developing schizophrenia. *Lancet* **353**:30–3.

Levinson D. F., Holmans P., Straub R. E. *et al.* (2000) Multicenter linkage study of schizophrenia candidate regions on chromosomes 5q, 6q, 10p, and 13q: Schizophrenia Linkage Collaborative Group III. *Am J Hum Genet* **67**:652–63.

Li T., Hu X., Chandy K. G. *et al.* (1998a) Transmission disequilibrium analysis of a triplet repeat within the HkCa3 gene using family trios with schizophrenia. *Biochem Biophys Res Comms* **251**:662–5.

Lin M. W., Curtis D., Williams N. *et al.* (1995) Suggestive evidence for linkage of schizophrenia to markers on chromosome 13q14.1–q32. *Psychiatr Genet* **5**:117–26.

Lindblad K., Nylander P.-O., De bruyn A. *et al.* (1995) Detection of expanded CAG repeats in bipolar affective disorder using the repeat expansion detection (RED) method. *Neurobiol Dis* **2**:55–62.

Malmberg A., Lewis G., David A. S. *et al.* (1998) Premorbid adjustment and personality in schizophrenia. *British Journal of Psychiatry* **172**:308–13.

McCarley R. W. *et al.* (1999) MRI anatomy of schizophrenia. *Biological Psychiatry* **45**:1099–119.

McGue M. and Gottesman I. I. (1989) A single dominant gene still cannot account for the transmission of schizophrenia. *Archives of General Psychiatry* **46**:478–9.

McGuffin P., Owen M. J., O'Donovan M. C. *et al.* (1994) *Seminars in Psychiatric Genetics*. London: Gaskell Press.

McGuffin P. and Gottesman I. I. (1999) Risk factors for schizophrenia. [Letter] *New England Journal of Medicine* 341:370–1.

Millar J. K., Wilson-Annan J. C., Anderson S. *et al.* (2000) Disruption of two novel genes by a translocation co-segregating with schizophrenia. *Hum Mol Genet* **9**:1415–23.

Moises H. W. and Gottesman I. I. (2001) Genetics, risk factors, and personality factors, in F. Henn, H. Helmchen, H. Lauter *et al.* (eds) *Contemporary Psychiatry*. Heidelberg: Springer Verlag.

Morris A. G., Gaitonde E., Mckenna P. J. *et al.* (1995) CAG repeat expansions and schizophrenia: association with disease in females and with early age at onset. *Hum Mol Genet* **4**:1957–61.

Mortensen P. B., Pedersen C. B., Westergaard T. *et al.* (1999) Effects of family history and place and season of birth on the risk of schizophrenia. *N Engl J Med* **340**:603–8.

Murphy K. C., Jones L. A., Owen M. J. (1999) High rates of schizophrenia in adults with velo-cardio-facial syndrome. *Archives of General Psychiatry* **56**:940–5.

Murray R. M. and Lewis S. W (1987) Is schizophrenia a neurodevelopmental disorder? *Br Med J* **295**:681–2.

Myers A., Holmans P., Marshall H. *et al.* (2000) Susceptibility locus for Alzheimer's disease on chromosome 10. *Science* **290**:2304–5.

Myles-Worsley M., Coon H., Tiobech J. *et al.* (1999) Genetic epidemiological study of schizophrenia in Palau, Micronesia: prevalence and familiality. *Am J Med Genet* **88**:4–10.

O'Donovan M. C., Guy C., Craddock N. *et al.* (1995) Expanded CAG repeats in schizophrenia and bipolar disorder. *Nat Genet* **10**:380–1.

O'Donovan M. C., Guy C., Craddock N. *et al.* (1996) Confirmation of association between expanded CAG/CTG repeats and both schizophrenia and bipolar disorder. *Psychol Med* **26**:1145–53.

Ohara K., Nagai M., Tani K. *et al.* (1998) Functional polymorhism of −141c Ins/Del in the dopamine D2 receptor gene promoter and schizophrenia. *Psychiatry Res* **16**(81):117–23.

Owen M. J., Holmans P., McGuffin P. (1997) Association studies in psychiatric genetics. *Mol Psychiatry* **2**:270–3.

Owen M. J., Cardno A. G., O'Donovan M. C. (2000) Psychiatric genetics: back to the future. *Mol Psychiatry* **5**:22–31.

Papolos D. F., Faedda G. I., Veit S. *et al.* (1996) Bipolar spectrum disorders in patients diagnosed with velo-cardio-facial syndrome:does a hemizygous deletion of chromosome 22q11 result in bipolar affective disorder? *Am J Psychiatry* **153**:1541–47.

Paylor R., McIlwain K. L., McAninch R. *et al.* (2001) Mice deleted for the DiGeorge/velocardiofacial syndrome region show abnormal sensorimotor gating and learning and memory impairments. *Hum Mol Genet* **10**:2645–50.

Petronis A., Kennedy J. L. (1996) Unstable genes—unstable mind? *Am J Hum Genet* **152**:164–72.

Prescott C. A. and Gottesman I. I. (1993) Genetically mediated vulnerability to schizophrenia. *Psychiatr Clin N Am* **16**:245–68.

Pulver A. E., Nestadt G., Goldberg R. *et al.* (1994) Psychotic illness in patients diagnosed with velo-cardio-facial syndrome and their relatives. *Journal of Nervous and Mental Diseases* **182**:476–8.

Pulver A. E., Mulle J., Nestadt G. *et al.* (2000) Genetic heterogeneity in schizophrenia: stratification of genome scan data using co-segregating related phenotypes. *Mol Psychiatry* **5**:650–3.

Risch N. (1990). Linkage strategies for genetically complex traits. 2. The power of affected relative pairs. *Am J Hum Genet* **46**:229–41.

Risch N., Merikangas K. (1996). The future of genetic studies of complex human diseases. *Science* **273**:1516–17.

Roberts S. B., MacLean C. J., Neale M. C. *et al.* (1999) Replication of linkage studies of complex traits: an examination of variation in location estimates. *Am J Hum Genet* **65**:876–84.

Rosenthal D., Wender P. H., Kety S. S. *et al.* (1971) The adopted-away offspring of schizophrenics. *Am J Psychiatry* **128**:307–11.

Ross C. A. (1999). Schizophrenia genetics: expansion of knowledge? *Mol Psychiatry* **4**:4–5.

Schizophrenia Linkage Collaborative Group for Chromosomes 3:6, and 8 (1996) Additional support for schizophrenia linkage on chromosomes 6 and 8: a multicenter study. *Am J Med Genet* **67**:580–94.

Schwab S. G., Eckstein G. N., Hallmayer J. *et al.* (1997) Evidence suggestive of a locus on chromosome 5q31 contributing to susceptibility for schizophrenia in German and Israeli families by multipoint affected sib-pair linkage analysis. *Mol Psychiatry* **2**:156–60.

Schwab S. G., Hallmayer J., Albus M. *et al.* (2000) A genome-wide autosomal screen for schizophrenia susceptibility loci in 71 families with affected siblings: support for loci on chromosome 10 p and 6. *Mol Psychiatry* **5**:638–49.

Scott W. K., Nance M. A. *et al.* (2001) Complete genomic screen in Parkinson disease: evidence for multiple genes. *JAMA* **286**:2239–44.

Shaw S. H., Kelly M., Smith A. B. *et al.* (1998) A genome wide search for schizophrenia susceptibility genes. *Am J Med Genet* **81**:364–76.

Sherrington R., Brynjolfsson J., Petursson H. *et al.* (1988) Localization of a susceptibility locus for schizophrenia on chromosome-5. *Nature* **336**:164–7.

Shprintzen R. J., Goldberg R. B., Golding-Kushner K. J. (1992) Late-onset psychosis in the Velo-Cardio-Facial syndrome. *Am J Med Genet* **42**:141–2.

Sing C. F., Haviland M. B., Reilly S. L. (1996) Genetic architecture of common multifactorial diseases, in G. Cardew (ed.) *Variation in the Human Geneotype*. Chichester: John Wiley.

St Clair D., Blackwood D., Muir W. *et al.* (1990) Association within a family of a balanced autosomal translocation with major mental illness. *Lancet* **336**:13–16.

Straub R. E., Maclean C. J., Oneill F. A. *et al.* (1997) Support for a possible schizophrenia vulnerability locus in region 5q22–31 in Irish families. *Mol Psychiatry* **2**:148–55.

Straub R. E., Maclean C. J., Martin R. B. *et al.* (1998) A schizophrenia locus may be located in region 10p15–p11. *Am J Med Genet* **81**:296–301.

Suarez B. K., Hampe C. L., Van Eerdewegh P. (1994) Problems of replicating linkage claims in psychiatry, in E. S. Gershon and C. R. Cloninger (eds) *Genetic Approaches to Mental Disorders*. Washington, DC: American Psychiatric Press.

Tienari P., Wynne L. C., Moring J. *et al.* (2000) Finnish adoptive family study: sample selection and adoptee DSM-III-R diagnoses. *Acta Psychiatrica Scandinavica* **101**:433–43.

Todd J. A. (1999) Interpretation of results from genetic studies of multifactorial diseases. *Lancet* **354**(suppl. 1):15–16.

Vincent J. B., Paterson A. D., Strong E. *et al.* (2000) The unstable trinucleotide story of major psychosis. *Am J Med Genet (Seminars in Medical Genetics)* **97**:77–97.

Weinberger D. R. (1987). Implications of normal brain development for the pathogenesis of schizophrenia. *Arch Gen Psychiatry* **44**:660–9.

Weinberger D. R. (1995). From neuropathology to neurodevelopment. *Lancet* **346**:552–7.

Wender P. H., Rosenthal D., Kety S. S. *et al.* (1974) Crossfostering: a research strategy for clarifying the role of genetic and experiential factors in the etiology of schizophrenia. *Archives of General Psychiatry* **30**:121–8.

Williams J., Spurlock G., McGuffin P. *et al.* (1996) Association between schizophrenia and T102C polymorphism of the 5-Hydroxytryptamine type 2A-Receptor gene. *Lancet* **347**:1294–6.

Williams J., McGuffin P., Nothen M. *et al.* (1997) Meta-analysis of association between the 5-HT2a receptor T102C polymorphism and schizophrenia. *Lancet* **349**:1221.

Williams J., Spurlock G., Holmans P. *et al.* (1998) A meta-analysis and transmission disequilibrium study of association between the dopamine D3 receptor gene and schizophrenia. *Mol Psychiatry* **3**:141–9.

Williams N. M., Rees M. I., Holmans P. *et al.* (1999) A two-stage genome scan for schizophrenia susceptibility genes in 196 affected sibling pairs. *Hum Mol Genet* **8**:1729–39.

Williams N. M., Bowen T., Spurlock G. *et al.* (2002). Determination of the genomic structure and mutation screening in schizophrenic individuals for five subunits of the N-methyl-D-aspartate glutamate receptor. *Mol Psychiatry* **7**:508–514.

Wyatt R. J., Henter I., Leary M. C. *et al.* (1995) An economic evaluation of schizophrenia—1991. *Soc Psychiatry Psychiatr Epidemiol* **30**:196–205.

Zerba K. E., Ferrell R. E., Sing C. F. (2000) Complex adaptive systems and human health: the influence of common genotypes of apolipoprotein E (Apo E) gene polymorphism and age on the relational order within the filed of lipid metabolism traits. *Hum Genet* **107**:466–75.

Chapter 11

Substance misuse

David Ball and David Collier

Introduction

At present the genetic and environmental dissection of alcohol dependence is more advanced than that of drug dependence, and this probably results from several factors. For example, the use of alcohol is sanctioned in most societies and generally the high rate of misuse changes slowly over long periods of time. As a result there is a relatively high prevalence of alcohol dependence, with many of those affected seeking treatment, and it is possible to recruit families and individuals for genetic studies. In contrast the legal status and occasionally faddish use of illicit drugs affect presentation, recruitment and the cooperation of family members. There has been an acceleration of interest recently in the genetics of misuse of substances other than alcohol, following clear demonstrations of genetic influences from family and twin studies and promising beginnings in molecular genetic research.

In parallel with developments in molecular genetics there has been a rapid progression from small, poorly controlled association studies of candidate genes to systematic genome scans to identify linkage. The combination of these two approaches, exploiting high throughput techniques, has great potential. However we can expect the genetics of substance misuse to be complex, with different genes and environmental factors interacting at various stages in the development of dependence. It seems likely that some of the genes that predispose to dependence will contribute to addictive behaviour in general whilst others will be substance specific. Some will be in pathways with well-established links to addiction, while others will be novel and the mechanism of their contribution subsequently identified.

Alcohol

Although some have elevated alcohol to the status of a god, others have accused it of being hell's best friend (Kobler 1974). So while alcohol, in its multiple and diverse forms, represents a valuable addition to the lives of many, some 29 per cent of adult males and 17 per cent of adult females in the United Kingdom drink more than the safe recommended limits, as stipulated by the Royal Colleges (Walker *et al.* 2001; Edwards 1996). Alcohol is thought to contribute to 33,000 deaths each year and is recognized to be the leading cause of disability in developed countries and the fourth largest in developing countries (Lopez and Murray 1998; Godfrey and Maynard 1992). Furthermore it is estimated that some 5–10 per cent of males and 3–5 per cent of females develop the symptoms of alcohol dependence during their lifetime (Glass 1991).

Alcohol dependence is characterized by the compulsive use of alcohol, tolerance to its effects, characteristic withdrawal symptoms and drinking to relieve these symptoms (Edwards 1982; Edwards *et al.* 1997). The development of alcohol dependence represents a complex interaction of multiple constitutional and environmental factors including availability and acceptability of alcohol, self-medication plus personality, psychological and hereditary factors (Ball and Strang 1999). During the development of alcohol dependence these multiple factors interact through various phases, including initiation and maintenance of alcohol consumption, through to the occurrence of psychological and physical dependence. For each individual the balance of factors will be different and vary between phases. The elucidation of these social, genetic and developmental factors that contribute to alcohol dependence will permit a better comprehension of the reasons why some individuals within society are unable to control their drinking and permit the development of more effective management approaches. Thus such findings will provide a better understanding of the functional involvement of genes in the development of alcohol dependence. Identification of the specific changes that underpin genetic predisposition to alcohol dependence will enable a better understanding of the biological mechanisms that are involved in the development and progression of this disorder. These will then provide a logical framework for the development of innovative treatment approaches. In addition they will be of value in making diagnoses and influencing treatment outcomes. For example treatments may be targeted to patients who demonstrate an enhanced response attributable to the presence of a particular genetic variant (Ball and Murray1994; Nuffield Council on Bioethics 1998). Such targeting may apply to all the treatment options, and their combination, across the therapeutic spectrum including spiritual, social, psychological and biological approaches.

Genetic predisposition to alcohol dependence

Convincing evidence for the operation of a genetic predisposition or vulnerability to alcohol dependence has been derived from family, twin and adoption studies.

It has long been recognized that alcohol dependence clusters in families, for example Plutarch (AD 45–125) has been quoted as stating *'ebrii gignunt ebrios'* ('drunks beget drunkards'), and this view has been confirmed by modern family studies. For example, Cotton (1979) reviewed 39 family studies, representing the families of 6251 alcoholics and 4083 non-alcoholics, and reported that an alcoholic was six times more likely than a non-alcoholic to report parental alcoholism. More generally the rates of alcoholism were significantly higher in relatives of alcoholic probands (15.3 per cent) than in non-alcoholic controls (8.7 per cent) (Guze *et al.* 1986). More recently, the odds of developing alcohol dependence were calculated from a study of 23,152 drinkers. This study demonstrated an increased risk of developing alcoholism that was dependent on the number and proximity of relatives affected. Thus the risk was increased by 167 per cent in individuals with both a first- and second-degree relative affected, 86 per cent in those an affected first-degree relative, and 45 per cent among those with a second- or third-degree affected relative (Dawson *et al.* 1992).

The results of twin studies are by and large consistent with the view that genes contribute to familiality and that alcoholism has a significant genetic component (see Table 11.1 for a summary). Three basic approaches have been taken; proband ascertainment via clinics with co-twin follow-up, the use of archival records with population-based twin registers and clinical assessment using population-based twin registers (Prescott 2001). Estimates of heritability using these approaches range widely from 0.0 to 0.98 with typical figures of 0.5 for males and 0.25 for females (Kaij 1960; Gurling *et al.* 1981; Hrubec and Omenn 1981; Gurling *et al.* 1984; Koskenvuo *et al.* 1984; Allgulander *et al.* 1991; Caldwell and Gottesman 1991; Pickens and Svikis 1991; Romanov *et al.* 1991; Kendler *et al.* 1992; McGue *et al.* 1992; True *et al.* 1996a; Reed *et al.* 1996; Heath *et al.* 1997; Kendler *et al.* 1997; Prescott *et al.* 1999).

Adoption studies provide the opportunity to study individuals reared by unrelated parents (see Table 11.2 for a summary). After a small negative study by Roe and Burks (1945) the subsequent three major adoption series have provided strong evidence for the action of genetic factors in males with weaker support in females (Goodwin *et al.* 1973, 1974, 1977a; Bohman 1978, 1981; Cadoret and Gath 1978; Cloninger *et al.* 1981; Bohman *et al.* 1985; Cadoret *et al.* 1987; Yates *et al.* 1996; Sigvardsson *et al.* 1996). In addition, there was no correlation between drinking behaviour in the adoptees and alcoholism in the adoptive parents and no protective effect conferred by being raised away from the biological parent (Goodwin *et al.* 1974, 1977b). Thus, the increased rate of alcoholism in adoptees with an alcoholic biological parent provides strong evidence for a significant genetic contribution to the development of alcoholism.

Molecular genetic studies

Both linkage and association studies have been employed in alcoholism. The pattern of inheritance suggests that alcohol dependence is the result of multiple genes interacting with environmental factors. Segregation analysis of alcoholism in 35 multigenerational families suggested that liability was in part controlled by a gene of major effect with additional multifactorial effects. The pattern could not be explained by a single genetic locus with strictly Mendelian transmission (Aston and Hill 1990; Yuan *et al.* 1996). If there are genes of reasonably large effect (as opposed to many genes of small effect) linkage strategies in alcohol dependence should be viable.

Table 11.1 Summary of twin studies

Source	Diagnosis	Proband ascertainment with proband follow-up				Components of variance		
		Gender	Concordance		Ratio	h^2	c^2	n^2
			MZ(%)	DZ(%)				
Kaij 1960	Chronic alcoholism	Male	71 (n = 14)	32 (n = 31)	2.2	0.98	0.01	0.01
Gurling et al. 1984	WHO alcohol dependence syndrome	Male	33 (n = 15)	30 (n = 20)	1.1			
	WHO alcohol dependence syndrome	Female	8 (n = 13)	13 (n = 8)	0.6			
	WHO alcohol dependence syndrome	Male/Female	21 (n = 28)	25 (n = 28)	0.8	0.00	0.44	0.56
Pickens et al. 1991	DSM-III Alcohol dependence	Male	59 (n = 39)	36 (n = 47)	1.6	0.60	0.17	0.23
	DSM-III Alcohol abuse and/or dependence	Male	76 (n = 50)	61 (n = 64)	1.2	0.36	0.51	0.13
	DSM-III Alcohol dependence	Female	25 (n = 24)	5 (n = 20)	5.0	0.42	0.00	0.58
	DSM-III Alcohol abuse and/or dependence	Female	36 (n = 31)	25 (n = 24)	1.4	0.26	0.29	0.45

Study	Criteria	Sex						
McGue et al. 1992	DSM-III Alcohol abuse and/or dependence	Male	77 (n = 85)	54 (n = 96)	1.4	0.54	0.33	0.13
	Alcohol abuse and/or dependence	Female	39 (n = 44)	42 (n = 43)	0.9	0.00	0.63	0.37
Caldwell and Gottesman 1991	DSM-III Alcohol dependence	Male	40 (n = 20)	13 (n = 15)	3.1	0.49	0.00	0.51
	DSM-III Alcohol abuse and/or dependence	Male	68 (n = 28)	46 (n = 26)	1.5	0.70	0.10	0.20
	DSM-III Alcohol dependence	Female	29 (n = 7)	25 (n = 12)	1.2	0.10	0.44	0.46
	DSM-III Alcohol abuse and/or dependence	Female	47 (n = 17)	42 (n = 24)	1.1	0.08	0.67	0.25
Archival records with population based twin registers								
Hrubec and Omenn 1981	ICD-8 Alcoholism	Male (prev. 3.0)	26 (n = 5932)	12 (n = 7554)	2.2	0.53	0.07	0.40
Allgulander et al. 1991	ICD-8 Alcoholism	Male (prev. 1.7)	40 (n = 2293)	36 (n = 3691)	1.1	0.08	0.32	0.60
		Female (prev. 0.5)	62 (n = 2736)	51 (n = 4164)	1.2	0.20	0.42	0.38
		Male/female				0.16	0.32	0.52

Table 11.1 (Cont.)

Source	Diagnosis	Proband ascertainment with proband follow-up						
		Gender	Concordance			Components of variance		
			MZ(%)	DZ(%)	Ratio	h^2	c^2	n^2
		Clinical assessment with population based twin registries						
Kendler et al. 1992	DSM-III-R Alcohol dependence	Female (prev. 9.0)	32 (n = 590)	24 (n = 440)	1.3	0.56	0.00	0.44
True et al. 1996	DSM-III-R Alcohol dependence	Male (prev. 38.3.)	20 (n = 1864)	17 (n = 1492)	1.2	0.55	0.00	0.45
Heath et al. 1997	DSM-III-R Alcohol dependence	Male (prev. 22.4)	56 (n = 396)	33 (n = 231)	1.7			
	DSM-III-R Alcohol dependence	Female (prev. 5.7)	30 (n = 932)	17 (n = 534)	1.8			
	DSM-III-R Alcohol dependence	Male/female	(n = 1328)	(n = 765)		0.64	0.01	0.35

This summary is derived from the original papers with additional information from Gottesman and Carey (1983) McGue (1994) and Prescott (2001); Strugstad and Sager (1998). Diagnosis describes the diagnostic criteria employed in the report. Gender identifies the sex studied, in some the analysis of heritability was completed on a combined male/female sample. 'Prev.' is the percentage prevalence rate of the diagnosis in the total twin sample from which the affected individuals are drawn. The concordance rates are expressed as a percentage and the ratio is that between monozygotic (MZ) and dizygotic (DZ) twin rates. The components of variance represent the proportion of alcoholism liability that is associated with genetic (h^2), shared environment (c^2) and non-shared environment (n^2)

Table 11.2 Summary of adoption studies

Source	Sample	Diagnosis	Birth cohort	Gender	History of parental alcoholism (%)	Controls (%)	Risk ratio
Roe and Burks 1945	New York	'Inebriates'		Male	0.0 (n = 21)	0.0 (n = 11)	(1.0)
				Female	0.0 (n = 15)	0.0 (n = 14)	(1.0)
Goodwin et al. 1973	Danish	Alcoholic (criteria employed)	1924–1947	Male	18 (n = 55)	5 (n = 78)	3.6
Goodwin et al. 1977b			1924–1947	Female	2 (n = 49)	4 (n = 47)	0.5
Cloninger et al. 1981	Sweden	Temperance Board Registrations; Medical and government records	1930–1949	Male	23.0 (n = 291)	14.7 (n = 571)	1.6
Bohman et al. 1981			1930–1949	Female	4.5 (n = 336)	2.8 (n = 577)	1.6
Sigvardsson et al. 1996			1930–1949	Male	24.1 (n = 108)	12.8 (n = 469)	1.9
			1930–1949	Female	0.9 (n = 114)	1.1 (n = 546)	0.8
Cadoret et al. 1985	Iowa	DSM-III	1938–1962	Male	61.1 (n = 18)	23.9 (n = 109)	2.6
			1938–1962	Female	33.3 (n = 12)	5.3 (n = 75)	6.3
Cadoret et al. 1987			1938–1964	Male	62.5 (n = 8)	20.4 (n = 152)	3.1

Sample identifies the origin of the subjects. Diagnosis describes the diagnostic criteria employed in the report. Birth cohort relates to the range of birth years of the subjects. Gender identifies the sex studied. The frequency of the diagnosis is reported for those with and without, a parental history of alcohol problems in the columns 'family history of parental alcoholism' and 'controls' respectively. Risk ratio is the ratio of these two frequencies

Linkage studies

Linkage studies have been used to provide a genome-wide screen for genes implicated in alcohol dependence. In spite of the likely genetic complexity, two recent linkage studies have reported positive findings and interestingly several positive linkage regions contain potential candidate genes (Long *et al.* 1998; Reich *et al.* 1998). One minimized confounding factors by examining alcohol dependence in south-western American Indians thereby reducing genetic and environmental heterogeneity (Long *et al.* 1998). Linkage was reported in two chromosomal regions, 4p and 11p. The best evidence for linkage was with an 11p marker D11S1984 (multi-point lod score ~3.1) and there was good evidence for a 4p marker, D4S3242 (multi-point lod score ~2.8). The linkage region on chromosome 4 is close to the β1 GABA (γ-aminobutyric acid) receptor gene (GABRB1) whilst that for chromosome 11 is close to the genes encoding the dopamine D4 receptor (DRD4) and tyrosine hydroxylase. The Collaborative study on the Genetics of Alcoholism (COGA) analysed genetic linkage in 105 multigenerational families identified by centres in the United States (Reich *et al.* 1998). The COGA sample is largely composed of Caucasians (approximately 74 per cent) with smaller numbers of African-Americans (approximately 17 per cent) and Hispanics (6 per cent). There was suggestive evidence of linkage on chromosomes 1 and 7, with a protective locus on chromosome 4 near the Alcohol Dehydrogenase (ADH) gene cluster. The susceptibility locus on chromosome 1, near the marker D1S1588, yielded a lod score of 2.93 using multipoint analysis; whilst on chromosome 7 the lod score was 3.49 near marker D7S1793. The protective locus on chromosome 4 is of particular interest, as associations in this region have previously been reported in oriental populations (see below) rather than Caucasians. That both studies implicated genes in an overlapping region of chromosome 4 may reflect the operation of a single gene locus or different loci acting in different populations. Following these initial reports there have been rigorous re-analyses of the COGA data comparing definitions of phenotype and methods of analysis (Birznieks *et al.* 1999; Comuzzie and Williams 1999; Curtis *et al.* 1999; Lin *et al.* 1999; Turecki *et al.* 1999). In one study the maximum number of drinks consumed in a 24-hour period was examined. This is closely related to a diagnosis of alcoholism in the sample, and demonstrated linkage to chromosome 4 in the region of the ADH cluster (Saccone *et al.* 2000). In another analysis age of onset was studied and there was increased evidence for linkage at the ADH3 functional gene polymorphism in the late onset subgroup, with no evidence for linkage in the early onset group (Kovac *et al.* 1999). Furthermore there has been extensive examination of neurophysiological measures (Begleiter *et al.* 1998). Further analyses have also implicated loci on chromosome 6 (Cantor and Lanning 1999), 8, 16 and 17 (Korczak *et al.* 1999), 10 (Daw *et al.* 1999), 15 (Windemuth *et al.* 1999), and 19 (Valdes *et al.* 1999). Furthermore methods have been applied to refine loci and isolate contributing genes, including the identification of families and individuals that contribute disproportionately to the linkage (Mitchell *et al.* 1999) and DNA pooling strategies across linkage regions (Koch *et al.* 2000).

Candidate genes

Alcohol metabolizing enzymes

ADH and ALDH2

The major path for alcohol metabolism occurs in the liver and involves the conversion of alcohol to acetaldehyde, and then to acetate. These reactions are catalysed by a group of enzymes, the alcohol dehydrogenases (ADH) and acetaldehyde dehydrogenase (ALDH2) (see Fig. 11.1). On the basis of primary sequence and kinetics mammalian ADH has been subdivided into five major classes. When their kinetic constants are considered only classes I, II and IV effectively contribute to ethanol metabolism. Class I enzymes are encoded by ADH_1, ADH_2 and ADH_3 genes. These encode monomers that can cross-hybridize with other monomers of the same class to form dimers (Pares *et al.* 2001). The allele frequencies of these class I enzymes demonstrate marked variation across different ethnic backgrounds and are given in Table 11.3.

It has been documented in oriental populations that the frequencies of $ADH_2{}^2$ and $ADH_3{}^1$ are significantly decreased in alcoholics when compared with controls (Chen *et al.* 1996; Maezawa *et al.* 1995; Higuchi *et al.* 1996; Nakamura *et al.* 1996; Shen *et al.* 1997; Tanaka *et al.* 1997; Thomasson *et al.* 1994, 1991). Associations between the ADH polymorphisms and alcohol dependence, however, have been less consistent in other populations (Gilder *et al.* 1993; Whitfield *et al.* 1998; Chen *et al.* 1999a; Borras *et al.* 2000; Ehlers *et al.* 2001). Nevertheless, the fact that the ADH genes are clustered in a region implicated in alcohol dependence by the COGA linkage study suggests a role for ADH in non-oriental populations (Reich *et al.* 1998). Variation at the ADH_2 locus appears to have the more important metabolic effect and the apparent involvement of the ADH_3 locus may be due to linkage disequilibrium between these two genes (Osier *et al.* 1999). This suggestion has been supported by the lack of association between ADH_3 and alcohol dependence in Caucasian populations and a multiple logistic regression study of ALDH2, ADH2 and ADH3 in a Chinese population (Couzigou *et al.* 1990; Gilder *et al.* 1993; Kolls *et al.* 1998; Chen *et al.* 1999a; Borras *et al.* 2000).

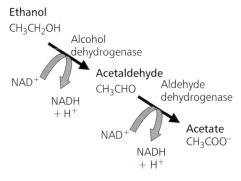

Fig. 11.1 Primary metabolic pathway of alcohol. Ethanol is metabolized by alcohol dehydrogenases to acetaldehyde. This is then metabolized by aldehyde dehydrogenase 2 to acetate.

Table 11.3 Allele frequencies and function of ADH enzymes

Allele	Peptide	K$_m$EtOH (mM) (homodimer)	Vmaxforwd (U/mg) (homodimer)	Caucasians	Orientals	Black Americans	North American Indians
ADH$_1$	α	4.2	0.6	1.00	1.00	1.00	1.00
ADH$_2$^{1}	β_1	0.049	0.23	0.90–0.96	0.32–0.38	0.85	1.00
ADH$_2$^{2}	β_2	0.94	8.6	0.04–0.10	0.62–0.68	<0.05	0.00
ADH$_2$^{3}	β_3	36	7.9	<0.05	<0.05	0.15	0.00
ADH$_3$^{1}	γ_1	1.0	2.2	0.53–0.60	0.91–0.95	0.85	0.07–0.57
ADH$_3$^{2}	γ_2	0.63	0.87	0.40–0.47	0.05–0.09	0.15	0.43–0.93
ADH$_4$	π	34	0.5	1.00	1.00	1.00	1.00

Steady state kinetic constants of homodimeric isozymes of human liver ADH. Allele describes the genetic variant that encodes the respective peptide. Kinetics of the reaction are reported for the homodimeric isoenymes and include the Michaelis constant, Km, and the limiting velocity, Vmax. The ethnic distribution is described for Caucasians, Orientals, Black Americans and North American Indians. Adapted from Ehrig and Li (1995) and (Yin and Agarwal 2001)

At intoxicating levels of alcohol ADH4, a class II enzyme, may account for up to 40 per cent of ethanol oxidation. This gene has been localized to the 4q22 ADH cluster and recently a functional polymorphism in the promoter region has been described which doubles promoter activity. Whilst this polymorphism has not yet been systematically studied in ethnically homogeneous populations, it is common in populations from the United States with an approximate frequency of 30 per cent (Edenberg *et al.* 1999).

ALDH2, the isoenzyme responsible for the majority of acetaldehyde oxidation, maps to chromosome 12q24.2. It exists in two forms which differ in activity as a result of a single nucleotide difference in exon 12 (Yoshida *et al.* 1985). The enzymatic activity is reduced virtually to zero by this change, which results in high levels of acetaldehyde and a response to alcohol similar to that produced by the drug antabuse, sometimes used in the treatment of alcoholism. This is the probable explanation of why the low activity variant protects against alcohol dependence (Thomasson *et al.* 1991; Yoshida 1992; Maezawa *et al.* 1994; Higuchi *et al.* 1995; Chen *et al.*1996; Shen *et al.* 1997). However the low activity isoenzyme is rare in many ethnic groups, including Caucasians, and therefore has no significant protective role in such groups (Goedde *et al.* 1992).

The parsimonious interpretation of these associations between alcohol dependence and polymorphisms at the loci controlling alcohol metabolizing enzymes is that they exert their influence through an effect on levels of acetaldehyde. Thus low activity variants of ADH, namely ADH$_2$^{1} and possibly ADH$_3$^{2}, and the high activity 'wild type' allele of ALDH2 both result in lower levels of this aversive compound and this predisposes individuals to alcohol dependence syndrome. Conversely, protection is afforded by high activity variants of ADH and low activity variants of ALDH2.

Table 11.4 Frequency of ALDH2 alleles in different ethnic groups

	ALDH2[1]	ALDH2[2]
Germans, Swedes, Finns, Turks	1.00	0.00
Hungarians	0.99	0.01
Indians	0.98	0.02
Koreans	0.85	0.15
Chinese	0.84	0.16
Japanese	0.76	0.24

The frequency of ALDH2[1] and ALDH2[2] are reported for different ethnic groups
Adapted from Goedde *et al.* (1992)

Cytochrome P450IIE1

Cytochrome P450IIE1 is induced by ethanol consumption and is responsible for up to 10 per cent of ethanol metabolism in chronic alcohol abusers (Lieber 1988). It has been suggested that cytochrome P450IIE1 may induce tissue damage through metabolic activation of a toxin (e.g. N-nitrosodimethylamine) or via the generation of free radicals (Sherman *et al.* 1994). Furthermore polymorphisms at the 5′ region of the gene influence the transcription rate of a reporter gene when transfected into human hepatoma cell lines, the rarer c2 allele being associated with a tenfold higher expression than the c1 allele (Hayashi 1991; Tsutumi *et al.* 1994). Further research has suggested that it may be associated with alcohol related liver damage in a Japanese sample, although this has not been consistently replicated (Tsutsumi *et al.* 1993; Maezawa *et al.* 1994; Ball *et al.* 1995; Rodrigo *et al.* 1999). As yet there have been no reported associations with alcoholism (Ball *et al.* 1995; Iwahashi *et al.* 1995; Maezawa *et al.* 1995; Higuchi *et al.* 1996; Nakamura *et al.* 1996; Tanaka *et al.* 1997; Carr. *et al.* 1996).

Dopamine receptors

Blum and colleagues (1991) reported an association between the A1 allele of a *Taq*I polymorphism close to the dopamine receptor 2 (DRD2) gene and alcohol dependence. An attraction of this finding was the link with dopaminergic theories of reward established by animal studies (Di Chiara and Imperato 1988). Thus it was proposed that alcohol and other drugs of abuse bring about direct release of dopamine in the nucleus accumbens thereby reinforcing drug-taking behaviour. Predisposition to alcohol dependence was envisaged as a genetically determined under-functioning of the system that resulted in a greater vulnerability to drugs that increased function within these reward pathways. Subsequently the DRD2 association was replicated by some (Uhl *et al.* 1993) but refuted by others (Gelernter *et al.* 1993), and it remains controversial (Noble 1998; Gelernter and Kranzler 1999; Gorwood *et al.* 2000). Large inter-population differences in the allele frequencies of this DRD2 polymorphism have been demonstrated, which suggest that associations could be generated or obscured if the case and control groups were not carefully matched for ethnic composition (Gelernter *et al.* 1993). More recent association approaches which control for different allele frequencies in different populations have failed to identify an association (Edenberg *et al.* 1998). Therefore whilst this matter is not completely resolved, the balance of the research evidence is no longer in support of an association between alcoholism and a DRD2 variant.

Recently an intriguing association between the personality trait of sensation or novelty seeking, alcohol and drug use and the dopamine 4 receptor (DRD4) has been suggested. Long alleles of the 48 base pair repeat in exon III were associated with higher novelty seeking scores. Again this has been followed by both positive and negative replication studies (Lusher *et al.* 2001). Indirect support for a role for DRD4 in alcohol dependence is the location of this gene in one of the positive linkage regions (Long *et al.* 1998). Following several negative association studies with alcohol dependence it is possible that the discrepancy may be related to severity of dependence and self-report of symptoms (Lusher *et al.* 2000).

GABA

GABA (γ-aminobutyric acid) is the major inhibitory neurotransmitter in the human brain. It acts at two receptor types, classified as type A and type B, which are defined by their pharmacological properties. $GABA_A$ receptor genes represent particularly strong candidate genes for alcohol dependence because of pharmacological similarities between alcohol and the benzodiazepines, which act at this receptor. Furthermore, cross-tolerance develops between alcohol and the benzodiazepines, which represent the drug of first choice in the treatment of the alcohol withdrawal syndrome.

A possible role for the $GABA_A$ receptors in alcohol action has been derived from *in vitro* cell models, animal studies and human research. $GABA_A$ subunit mRNA expression in cell models has suggested that the long form of the γ2 subunit is essential for ethanol enhanced potentiation of $GABA_A$ receptors, by phosphorylation of a serine contained within the extra 8 amino-acids (Loh and Ball 2000). Several animal studies have demonstrated that alterations in drug and alcohol responses may be caused by amino acid differences at the $GABA_A α6$ and $GABA_A γ2$ subunits. For example an Arg^{100}/Glu^{100} change at the $GABA_A α6$ subunit, which alters the binding efficacy of the benzodiazepine inverse agonist, Ro 15–4513, and confers benzodiazepine sensitivity was identified in alcohol-non-tolerant rats (Korpi *et al.* 1993). Several loci related to alcohol withdrawal have been identified, including a region of mouse chromosome 11 which is syntenic to that containing four $GABA_A$ subunit (β2, α6, α1 and γ2) genes on human chromosome 5q33–34 (Buck *et al.* 1997). Gene knockout studies of the role of $GABA_A α6$ and $GABA_A γ2$ subunit genes in mice have demonstrated an essential role in the modulation of other $GABA_A$ subunit expression and the efficacy of benzodiazepine binding (Loh and Ball 2000). Absence of the $GABA_A γ2$ subunit gene has more severe effects, with many of the mice dying shortly after birth.

Disappointingly few studies have examined the effects of response to alcohol in these gene knockout mice.

Several association studies of the $GABA_A$ receptor genes on 5q33–34 and alcohol dependence/alcohol related physiological phenotypes have been conducted in different ethnic populations with positive findings being reported (Radel *et al.* 1999; Sander *et al.* 1999a; Schuckit *et al.* 1999; Loh *et al.* 2000). The genes exist as clusters within the human genome and a further cluster is in the linkage region on chromosome 4, as reported by Long and colleagues (1998). Two studies have implicated the $GABA_A\beta1$ receptor, located within this region at 4p13-p12, in the development of alcohol dependence (Parsian and Zhang 1999; Sander *et al.* 1999b). Thus association studies point fairly consistently to an involvement of the $GABA_A\beta2$, $\alpha6$, $\alpha1$ and $\gamma2$ subunit genes clustered on chromosome 5 in the development of alcohol dependence and association and linkage studies further implicate the cluster of loci on chromosome 4.

Other drugs of abuse

As we have noted, historically, far less attention has been paid to the genetic epidemiology of drug abuse than to alcoholism (Pickens and Svikis 1991). There are established risk factors for drug abuse, including urban poverty, mental illness, and parental alcohol and drug abuse (Lowinson *et al.* 1997). However the extent of genetic and environmental influence has only recently been established. One problem with twin, family and adoption studies is that levels and type of drug abuse vary over time. For example a substantial increase in cocaine/crack use took place during the 1980s while heroin use fell (Lowinson *et al.* 1997). This variability in exposure must be borne in mind when interpreting twin studies.

Early studies established an elevated family history of drug abuse in the siblings of drug abusing probands (Vaillant 1966) and a series of family studies reported evidence of familial transmission (Scherer 1973; Annis1974; Fawzy *et al.* 1987; Malhotra 1983; Meller *et al.* 1988a). More recent family studies have shown that family history of drug abuse or dependence is a potent risk factor for drug abuse, and that both general and drug-specific risk factors exist for substance abuse (Mirin *et al.* 1991; Kendler *et al.* 1997; Bierut *et al.* 1998; Merikangas *et al.* 1998). Merikangas *et al.* (1998) found an eight-fold increased risk of drug disorders amongst relatives of probands with drug disorders compared to controls, which applied to a wide range of substances including opioids, cannabis, sedatives and cocaine. This familial aggregation was largely independent of alcoholism, indicating there is little co-transmission, and independent of exposure to parental drug abuse. Interestingly, there was some specificity of familial aggregation for cannabis, opiates and to a lesser extent cocaine. Bierut and colleagues (1998) examined familial transmission of alcohol, cannabis and cocaine dependence in families participating in the COGA study. They also found strong familiality for both cannabis and cocaine dependence which was independent of alcohol dependence.

Adoption studies

An adoption study (von Knorring *et al.* 1983) showed probable genetic transmission of substance abuse which overlapped with depression, but it was not until the adoption studies of Cadoret and others (1986a,b, 1995, 1996a,b; Yates *et al.* 1996) that more convincing evidence for a genetic influence on drug abuse was obtained. These studies demonstrated that substance abuse was significantly greater in adoptees whose biological parents were alcoholics, drug dependent, or had personality disorder, but the environment of rearing also seemed to play a role.

Twin studies

A series of small twin studies suggested the role of genes in drug abuse disorders (Grove *et al.* 1990; Pickens and Svikis 1991; Gynther *et al.* 1995; Jang *et al.* 1995). Pedersen (1981) found heritability of 0.28 for the abuse of tranquillisers and Pickens and colleagues (1991) found heritability of DSM-III drug abuse to be 0.31 for males and 0.22 for females. This work has been followed by large scale, population-based studies of twins, which have provided detailed information on the probable role of genes and environment in vulnerability to drug abuse disorders in general and to the abuse of specific classes of drug (Kendler 2001).

Two major population-based twin studies have examined the role of genes in the familial transmission of substance abuse, the first using the Vietnam Era Twin (VET) registry (Table 11.5) (Lin *et al.* 1996; Tsuang *et al.* 1996; Tsuang 1999, 2001) and the second using the Virginia Twin Registry (Table 11.6) (Kendler and Prescott 1998a,b;

Table 11.5 The Vietnam era veteran study of drug abuse and dependence

Source	Diagnosis		Concordance		Components of variance			
		Sex	MZ	DZ	h2	c2	n2	m2
VET	Any abuse/dependence	Male	0.63	0.44	0.34	.028	0.38	—
VET	Cannabis	Male	0.62	0.46	0.33	0.29	0.38	—
VET	Heroin/opiates	Male	0.67	0.29	0.43	‡	0.31	0.27
VET	Stimulants	Male	0.53	0.24	0.44	‡	0.49	0.07
VET	Sedatives	Male	0.44	0.25	0.38	0.06	0.56	—
VET	PCP/psychedelics	Male	0.44	0.32	0.25	0.19	0.56	—

DSM-IIIR criteria were used for diagnosis of abuse and/or dependence. VET = Vietnam Era Twin registry (Tsuang *et al.* 1996, 2001). Sample size 3,372 male twin pairs. Correlations for MZ/DZ concordance are tetrachoric. All components of variance data quoted refers to the best fitting model

‡ Best fitting model has non-additive genetic component
# concordant twins found for use only
m2 non-additive genetic component (VET only)

Table 11.6 The Virginia Twin studies of abuse and dependence. Data for drug use are not shown

Source	Diagnosis		Concordance		Components of variance			
		Sex	MZ	DZ	h2	c2	n2	m2
VTR	Cannabis abuse	Female	0.74	0.24	0.72	—	0.28	
	Dependence		0.58	0.41	0.62	—	0.38	
VTR	Cocaine abuse	Female	0.80	0.18	0.79	—	0.21	
	Dependence		0.68	0.08	0.68	—	0.35	
VTR	Hallucinogen abuse	Female	22.2	0	—	—	—	
VTR	Opiate abuse	Female	33.3	0	—	—	—	
	Dependence		50.0	0	—	—	—	
VTR	Sedative abuse	Female	22.2	0	—	—	—	
	Dependence		0	0	—	—	—	
VTR	Stimulant abuse	Female	34.5	15.4	0.72	—	0.28	
	Dependence		40.0	0	0.60	—	0.40	
VTR	Cannabis abuse	Male	0.75	0.39	0.76	—	0.24	
	Dependence		0.59	0.20	0.58	—	0.42	
VTR	Sedative abuse	Male	0.61	0.03	0.59	—	0.41	
	Dependence		0.83	0.0	0.87	—	0.13	
VTR	Stimulant abuse	Male	0.68	0.22	0.66	—	0.34	
	Dependence		0.43	0.34	–	0.39	0.61	
VTR	Cocaine abuse	Male	0.61	0.49	–	0.57	0.43	
	Dependence		0.77	0.37	0.79	—	0.21	
VTR	Opiate use	Male	0.55	0.03	0.52	—	0.48	
VTR	Hallucinogen abuse	Male	0.70	0.0	0.65	—	0.35	
	Dependence		0.80	0.0	0.79	—	0.21	
VTR	Any abuse	Male	0.76	0.40	0.77	—	0.23	
	Any dependence		0.69	0.39	0.71	—	0.29	

The VTR studies are population-based. Correlations for MZ/DZ concordance are tetrachoric. All components of variance data quoted refers to the best fitting model. ‡ Best fitting model has non-additive genetic component. # concordant twins found for use only. VTR = Virginia Twin Registry (Karkowski *et al.* 2000; Kendler *et al.* 1998a,b, 1999, 2000). Sample size 1198 male twin pairs and 1934 female twin pairs, born between 1940 and 1974

Kendler *et al.* 1999a,b). Both of these large scale studies show that:

1 Genetic factors substantially influence vulnerability to substance abuse,

2 Family environment is also important but predominantly influences initiation, and

3 Heavy use, abuse and dependence have a stronger genetic influence than occasional use (Kendler 2001).

Table 11.7 Analysis of drug abuse and dependence combined from drug treatment centres. Data for use and abuse alone are not shown

Source	Diagnosis		Concordance		Components of variance			
		Sex	MZ	DZ	h2	c2>	n2	m2
DTC	Sedative	Male	0.61	0.52	0.58	—	0.52	
	abuse + dependence	Female	0.11	0.38	0	0.29	0.71	
DTC	Opiate	Male	0.87	0.65	0.57	0.30	0.13	
	abuse + dependence	Female	0.34	0.95	0	0.60	0.40	
DTC	Cocaine	Male	0.90	0.48	0.74	0.18	0.08	
	abuse + dependence	Female	0.56	0.02	0.42	0.12	0.46	
DTC	Stimulant	Male	0.79	0.42	0.78	0	0.22	
	abuse + dependence	Female	0.88	0.19	0.73	0.12	0.16	
DTC	Cannabis	Male	0.90	0.61	0.68	0.24	0.08	
	abuse + dependence	Female	0.5	0	0.53	0	0.57	

The DTC sample is a sample of twins undergoing drug treatment. Correlations for MZ/DZ concordance are tetrachoric. All components of variance data quoted refers to the best fitting model. DTC = Drug Treatment Centre (van den Bree *et al.* 1998) Twins were clients in 16 Minnesota public or private drug treatment centres who were twins. Sample size 188 twin pairs. Abuse and/or dependence = as defined by DSM-IIIR

These studies are in general agreement with the results of a study of twins treated for substance abuse (the DTC sample) (Table 11.7) (van den Bree *et al.* 1998), and place heritability estimates for drug abuse at between 50 and 80 per cent, with the lower estimates of heritability coming from the VET study.

The Vietnam Era Twin (VET) registry

The largest twin studies of drug and alcohol abuse have involved male ex-servicemen from the USA. The VET registry contains data on over 8000 male twin pairs who served in Vietnam from 1956–1975. These twins were identified through surname and date of birth matching, and resulted in almost 16,000 potential twins from 5.5 million servicemen. This was followed by examination of service records and produced over 8000 confirmed twin pairs (Henderson *et al.* 1990). For the study of drug and alcohol use and comorbid psychiatric disorders, 5150 pairs were eligible and of these, 1874 MZ and 1498 DZ twins where both members responded to the questionnaires was the final sample (Tsuang *et al.* 2001). Twins were interviewed to produce DSM-IIIR diagnoses and information on use of alcohol, nicotine, marijuana, sedatives, stimulants, heroin/opiates and psychedelics, and the lifetime risk of dependence was about 10 per cent, similar to national levels in the USA (Warner *et al.* 1995).

Using this data, pairwise concordance rates were calculated for each of the five drug classes to estimate the extent of inherited and environmental influence on drug abuse and dependence (Table 11.5). This revealed a higher MZ compared than DZ similarity for abuse/dependence of any drug for expressed as concordance (MZ 26.2 per cent versus

DZ 16.5 per cent) and as tetrachoric correlations (r MZ 0.63 versus r DZ 0.44). Drug abuse/dependence appeared to be influenced by additive genetic, shared environment and non-shared environment, with heroin and opiates also showing an effect for non-additive genetics. Estimates of the level of additive genetic factors ranged from 25–44 per cent and unique environmental effect ranging from 31–59 per cent.

The design of the VET drug abuse and dependence study meant that the extent of shared vulnerability to different classes of drug could be assessed. Model fitting indicated a good fit to the data of a parsimonious model in which it is assumed that there is a single common vulnerability with a heritability of 31 per cent, shared environment of 25 per cent and unique environment of 44 per cent.

The Virginia Twin Registry

The Virginia Twin Registry (VTR) is a population-based register formed from a systematic review of all birth certificates in the Commonwealth of Virginia since 1918 (Kendler *et al.* 1992). The comprehensive study of psychoactive substance abuse in this population involved two phases, firstly examining 800 female–female twin pairs and secondly 1193 male–male twin pairs (Kendler and Prescott 1998a,b; Kendler 1999a,b; Karkowski *et al.* 2000; Kendler *et al.* 2000). After initial contact by post, twins were interviewed face-to-face, in almost all cases on two occasions separated by about one year.

The study examined cannabis, sedatives, stimulants, cocaine, opiates and hallucinogens, and differed from the VET study in that it considered cocaine as a separate category to stimulants (Table 11.6). Rates of abuse in both female and male samples were similar to those seen in the general population, suggesting that the twins were a representative sample.

The results overall were qualitatively similar to the two other major twin studies of drug abuse, the VET and DTC studies, but the levels of the heritability were somewhat higher in the VTR studies than in the VET studies. This may be a result of differences between the two populations; for example the twins in the VET study (all drafted into the US military) may have a different experience of drug exposure to a sample predominantly of civilians in Virginia, perhaps through different levels of exposure to different categories of drugs while in service.

One question the VTR can help to answer is whether there are differences between men and women in the genetic epidemiology of drug abuse. Estimates for women were close to the estimates for men for the most part, but family environment (C) was strongly implicated for cocaine and stimulant use in women but not men. In the DTC study, estimates for heritability were higher for men than for women (Table 11.7).

The biology of dependence on some specific groups of drug

Opiates and opioids

The use of heroin and other opiates has reached previously unparalleled proportions in many Asian and Western countries, and although less common than abuse of cannabis or stimulants, the cost of heroin addiction is high in terms of lost productivity, criminal activity and medical care (Mark *et al.* 2001). Opiates are naturally occurring alkaloid drugs originally derived from opium poppies. They are analgesic drugs which also produce feelings of euphoria and well-being, can be smoked or injected, and are highly addictive, inducing a physical dependence syndrome. Crude opium from poppies contain about 20 alkaloids, the most significant of which are morphine (10 per cent of dry weight) and codeine (0.5 per cent of dry weight).

The term opioid refers to synthetic and semi-synthetic drugs. Heroin (diacetylmorphine) is a semi-synthetic drug produced by the acetylation of morphine. Other opioids include pethidine, dipipanone, hydromorphone (Dilaudid), oxycodone (Percodan), oxymorphone (Numorphan), hydrocodone (Vicodin), meperidine (Demerol), Fentanyl, and methadone, which is used in the treatment of heroin addiction. Both heroin and codeine are prodrugs (codeine has less than 10 per cent of the analgesic activity of morphine) which are activated to morphine in the body. Both are more quickly absorbed and cross the blood–brain barrier faster than morphine itself. Codeine is converted into morphine through O-demethylation by the enzyme CYP2D6, whereas heroin is deactylated to 6-monoacetylmorphine and subsequently to morphine. Morphine is further activated to morphine-6-glucuronide by the enzyme UGT2B7 (Ritter 2000). Both of this and morphine-3-glucuronide are highly water soluble and M6G has higher affinity for the opioid receptors that morphine itself. Although M6G does not cross the blood–brain barrier efficiently, glucuronidation occurs in the brain.

There are three known opiate receptors encoded by distinct genes, mu, kappa and delta, and pharmacological evidence suggests the existence of a fourth, epsilon, which has not been isolated. Evidence suggests that the reward and reinforcement effects of opiates are mediated through the mu opiate receptor, which then increases activity of the mesolimbic dopamine system in the midbrain (Matthes *et al.* 1996; Uhl *et al.* 1999). These neurons project to the prefrontal cortex and the striatum, where specific release of dopamine occurs in the nucleus accumbens (Robbins and Everitt 1999). Dopamine release in the nucleus accumbens appears to be a key event in reinforcement and reward, in addition to the prefrontal cortex and central nucleus of the amygdala (Koob and Le Moal 2001). Vulnerabilities to abuse and addiction may result from heritable personality traits and variation in neurochemistry which may influence initiation and maintenance of drug use in relation to reward and addiction. In addition drug-specific pharmacogenetic factors related to metabolism, clearance or site of action may be important. For example CYP2D6 deficiency, as discussed in more detail later, appears to be protective against oral codeine addiction (Tyndale *et al.* 1997).

Genetic epidemiology of opiates The VET twin study gave an additive genetic component of 43 per cent a unique environmental effect of 31 per cent and a non-additive genetic effect of 26 per cent, indicating high heritability for opiate abuse and dependence

(Tsuang *et al.* 2001). Heroin/opiate abuse and or dependence showed the highest level of unique genetic vulnerability compared to other drugs of abuse, at 50 per cent, and the lowest level of shared genetic vulnerability at 30 per cent, indicating that there may be unique factors in genetic vulnerability to abuse. The heritability for use was estimated at 0.52 in both men and women, with unique environment making up the remainder of risk. The DTC study estimated heritability of 0.67 in men and 0.18 in women for use, 0.57 for abuse in men and 0.48 for dependence in men. Neither abuse nor dependence appeared to be significantly heritable in women, but power to detect genetic variance was low (Table 11.5).

Molecular genetic studies of opiates The molecular genetics of opiate addiction and abuse has been more extensively examined that that other drugs of abuse. Linkage analysis is a difficult prospect in humans, as multiply affected families are rare. However both breeding and the environment can be controlled in mice where linkage analysis has been performed on two drug-related phenotypes, morphine preference and morphine analgesia. Two loci influencing morphine preference (Berrettini *et al.* 1994) and two influencing morphine analgesia (Crabbe *et al.* 1999) have been identified. The locus on chromosome 10 for morphine preference contains the mu opioid receptor MOR. In addition extensive analysis has been performed on transgenic mouse models in which paradigms of addictive behaviour, such as conditioned place preference, tolerance, reward or withdrawal are examined (Table 11.8) and this has lead to the identification of a series of candidate neurochemical systems which could mediate addiction.

The mu opioid receptor (MOR) is the primary mediator of the reinforcement and reward effect of opiates, make the MOR gene an outstanding candidate for genetic

Table 11.8 Transgenic mice and opiate response

Gene	Comment	Ref.
Mu opioid	Loss of CPP/physical dependence	Matthes *et al.* 1996
CB1	Loss of CPP	Martin *et al.* 2000
G(z alpha)	Rapid development of tolerance	Hendry *et al.* 2000
SubP	Loss of reward/ reduction in withdrawal	Murtra *et al.* 2000
Nociceptin receptor	Loss of tolerance, increased withdrawal	Ueda *et al.* 2000; Kest *et al.* 2001
DAT	Increased reward response	Spielewoy *et al.* 2000
DRD2	No CPP in withdrawing state	Dockstader *et al.* 2001
GluR-A	Reduced tolerance and withdrawal	Vekovischeva *et al.* 2001

The table shows genes identified as modulating response to opiates when knocked out in mice. L = locomoter activity; CPP = conditioned place preference vulnerability. Sequencing of the MOR gene identified five single nucleotide polymorphisms in the gene, with one of these (Asn40Asp) showing differing affinity for beta-endorphin (Bond *et al.* 1998). A study in Hong Kong Chinese found a significant association (Szeto *et al.* 2001). However this polymorphism was not associated with heroin abuse in Chinese (Li *et al.* 2000a) or German heroin addicts (Franke *et al.* 2001). The most extensive examination of the gene identified 43 variants which were clustered into haplotypes. In African-Americans with substance dependence, one haplotype showed associated (Hoehe *et al.* 2000a). The kappa opiod receptor (KOR) has also been examined. A positive association was seen in one study (Mayer *et al.* 1997), but was not replicated in a second (Franke *et al.* 1999).

An interesting finding is the association between oral codeine abuse and CYP2D6 (Tyndale *et al.* 1997). As noted above, many opiates (e.g. codeine, oxycodone, and hydrocodone) are metabolized by cytochrome CYP2D6 to metabolites of increased activity, principally morphine. Between 4 and 10 per cent of Caucasians lack CYP2D6 activity due to inheritance of two non-functional alleles. Tyndale *et al.* found no poor metabolizers in people dependent on oral opiates, suggesting that the CYP2D6 defective genotype is a pharmacogenetic protection factor for oral opiate dependence, presumably because poor metabolizers are unable to get 'high' from oral codeine abuse.

The dopamine receptor DRD4 has also shown evidence for association with opioid dependence (Kotler *et al.* 1997). A series of studies have examined a 48-bp VNTR within the coding region of the gene, which has a series of common alleles ranging from two to seven repeats. Association is seen with 'long' repeats (five or more) in two studies (Li *et al.* 1997; Vandenbergh *et al.* 2000) although not in another (Franke *et al.* 2000).

The cannabis receptor has also been examined, since this protein is able to modulate responses to opiates. Comings *et al.* (1997) found an association between an AAT repeat and intravenous drug use, but this was not replicated in a Chinese (Li *et al.* 2000b) or American heroin abusing patients (Covault 2001).

Cannabis

Not only is cannabis (hashish, marijuana, 'dope') the most commonly abused illicit drug in developed countries, its medical use is becoming increasingly accepted (Watson *et al.* 2000). In the European Union and USA cannabis has the highest lifetime prevalence of use and dependence of all illicit drugs, having been used by close to 7 per cent of the population. Cannabis has also been controversially proposed as a gateway drug that encourages other forms of illicit drug use (Watson *et al.* 2000). It acts in the brain by binding to the central cannabinoid receptor (CB1) (Ledent *et al.* 1999) in the hippocampus, striatum and globus pallidus. Its endogenous counterpart in the human brain is the cannabinoid anandamide (Robbins and Everitt 1999). Although the existence of a physical dependence syndrome remains controversial, there is growing evidence that cannabis is an addictive drug, with the development of tolerance, dependence and a withdrawal syndrome (Farrell 1999). Cannabis dependence has some similarities to opiate dependence (Kaymakcalan 1981). For example, one of the active ingredients in cannabis, D9-THC, exerts similar effects to heroin on mesolimbic dopamine transmission through a common mu opioid receptor mechanism located in the ventral mesencephalic tegmentum in the shell of the nucleus accumbens (Tanda 1997). Opiates have reduced addictiveness in mice lacking the CB1 receptor (Ledent *et al.* 1999).

Genetic epidemiology of cannabis The VET twin data indicated that variance in liability to cannabis abuse and dependence is about equally shared between additive genetic (0.34), common (0.28) and shared (0.38) environment (Tsuang *et al.* 2001). Similar results were found by Miles *et al.* (2001). The subjective effects of cannabis use, such as suspiciousness and agitation also appear to be under genetic influence (Lyons *et al.* 1997). Kendler and Prescott (1998a) examined cannabis use, abuse and dependence in almost 2000 female twins and concluded that while use was influenced by genetic and familial–environmental factors, heavy use, abuse and dependence had a strong genetic component of between 62–79 per cent. Findings were similar in a subsequent study of male–male twin pairs (Kendler *et al.* 2000). Model fitting also suggests that a most of the genes influencing cannibis use are identical to those that predispose to use of illicit substances in general (Karkowski *et al.* 2000; Tsuang *et al.* 2001). From the VET data, cannabis also appeared to have the highest conditional probability of all drugs for the transition from exposure to frequent use (Tsuang *et al.* 1999), and this transition was significantly influenced by genes.

A subject of debate is the gateway effect, i.e. does cannabis act as a gateway for more serious illicit drug use and abuse? Tsuang *et al.* (1998) tested this hypothesis in the VET sample and found it to be a poor fit to their model, suggesting that Cannabis does not particularly facilitate involvement with 'harder' drugs.

Molecular genetic studies Few molecular genetic studies of cannabis dependence have been performed. Hoehe *et al.* sequenced the human CB1 cannabinoid receptor coding region in individuals showing acute psychotic symptoms after cannabis intake and others who did not develop any psychopathology after long-term heavy cannabis abuse. No evidence for structural mutations was obtained (Hoehe *et al.* 2000).

Stimulants and cocaine

The commonly abused stimulant drugs are principally amphetamine and its derivatives such as substituted amphetamines, but also include amphetamine-related drugs such as methylphenidate. Cocaine is also a stimulant but is often regarded as a separate category of drug in genetic studies. The rate of stimulant abuse and dependence is second only in lifetime prevalence to cannabis in the USA, at between 4 and 5 per cent. The common feature of these drugs is that their action is principally on neurotransmitter systems, through blocking reuptake at monoamine transporters and increasing neurotransmitter levels in the synapse (Robbins and Everitt 1999). There is growing evidence that the dopamine transporter (DAT), the serotonin transporter (SERT) and the noradrenaline (norepinephrine) transporter (NET) are all important for the action of psychostimulants (Xu *et al.* 2000a).

Amphetamine both inhibits reuptake and releases dopamine from presynaptic terminals by acting on both the plasma membrane DAT and vesicular monoamine transporters (VATs) inside the neuron (Jones *et al.* 1998). Cocaine is a potent reuptake inhibitor at DAT, SERT and NET, although it is not clear which of these is principally responsible for its reward and reinforcement actions. Unlike amphetamine, cocaine does not appear to act on VATs. Studies in transgenic mice indicate that both DAT and SERT can mediate cocaine's effects, but that DAT may play the more important role (Sora *et al.* 2001). Substituted amphetamines such as 3,4 methylenedioxymethamphetamine (MDMA) (ecstasy) are thought to act principally at serotonin transporters. The role of NET in the reward and reinforcing actions action of stimulants is less clear; cocaine has high affinity for NET (Giros *et al.* 1994) and transgenic mice lacking the NET gene are super-responsive to cocaine and amphetamine (Xu *et al.* 2000).

Genetic epidemiology of stimulant and cocaine use In the VET study, liability to stimulant abuse/dependence (amphetamines, cocaine) was found to be about 44 per cent heritable and the remainder of variance explained by non-shared environment (Tsuang *et al.* 2001). In both male and female twins from the Virginia Twin registry, stimulant use (not including cocaine) was about equally due to genetic and familial–environmental factors, whereas for abuse and dependence genetic factors dominated with heritability estimates of over 60 per cent (Karkowski 2000). In the VTR study cocaine abuse and dependence were highly heritable in women (0.79 and 0.68 respectively) whereas in men only dependence was significantly heritable (0.79). Analysis of the VET study data revealed that cocaine has the fastest transition from regular use to abuse/dependence, and this may have a genetic influence (Tsuang *et al.* 1999).

Molecular genetic studies Molecular genetic studies of cocaine abuse are few, and have focused on candidate gene analysis of those genes thought to mediate its reward and reinforcement actions. These candidate genes have been identified from studies of the pharmacology of stimulants, gene expression analysis and the examination of responses to amphetamine and cocaine in transgenic mice (Tables 11.8 and 11.9).

In addition, quantitative trait linkage analysis in the mouse has identified three loci causing susceptibility to cocaine-induced seizures on mouse chromosome 9, 14 and 15 (Crabbe *et al.* 1999).

Candidate gene studies in humans have failed thus far to identify genetic risk factors for cocaine abuse. The dopamine D2 receptor was positively associated with cocaine abuse in one study (Noble *et al.* 1993) but subsequent studies were negative (Gelernter *et al.* 1999; Blomqvist *et al.* 2000). A VNTR in the 3' untranslated region of the dopamine transporter gene DAT1 has also been examined. It was not associated with cocaine abuse, but was associated with cocaine induced paranoia in white subjects (Gelernter *et al.* 1994). The dopamine beta hydroxylase gene has also been associated with cocaine induced paranoia (Cubells *et al.* 2000). The 5-HT1B gene (Cigler *et al.* 2001) was not implicated in cocaine abuse but haplotypes of the mu opioid receptor

Table 11.9 Transgenic mice and cocaine response

Gene	Comment	Ref.
CD81	Altered sensitivity	(Michna, L., Brenz Verca, M. S., Widmer, D. A. et al. 2001)
DAT and SERT	No CPP	Sora et al. 2001
DAT	No difference/loss of effect	Rocha et al. 1998; Sora et al. 2001; Giros et al. 1996
SERT	No difference	Sora et al. 2001
5-HT1B	↑SA ↑LA	Rocha et al. 1997; Castanon et al. 2000; Belzung et al. 2000; Rocha et al. 1998
G(z)	Exaggerated response	Yang et al. 2000
DRD3	Increased response	Carta et al. 2000
TH	↑LA	Kim et al. 2000
GDNF	Enhanced response	Messer et al. 2000
A(2A)	Attenuated response	Chen et al. 2000
NET	↑↑A	Xu et al. 2000a
DRD1	↓LA	Xu et al. 2000b; Miner et al. 1995
GABAAb3	↑LA	Resnick et al. 1999
NNOS	↓LA ↓↓CPP	Itzhak et al. 1998
Retinoic acid receptors	Blunted response	Krezel et al. 1998
VMAT2	Supersensitivity	Wang et al. 1997
DRD4	↑LA	Rubinstein et al. 1997
NAChR beta2	↓CPP	Zachariou et al. 2001
TPA	↑LA	Ripley et al. 1999

The table shows genes identified as modulating cocaine response when knocked out in mice. L = locomoter activity; CPP = conditioned place preference

showed evidence of association with heroin/cocaine dependence in African-Americans (Hoehe et al. 2000).

Other classes of drug

Other classes of drug examined by twin, family and adoption studies include sedatives such as valium and psychedelics/hallucinogens such as LSD. Both of these classes show

Table 11.10 Transgenic mice and amphetamine response

Galpha(olf)	↓LA	Herve *et al*. 2001
Alpha2C-AR	↑LA, aggression, startle reactivity	Scheinin *et al*. 2001; Sallinen *et al*. 1998
Alpah2A-AR	↓LA	Chen *et al*. 2000a; Sallinen *et al*. 1998; Chen *et al*. 2000b
NET	↑LA	Xu *et al*. 2000
Alpha-Syn	↓LA	Abeliovich *et al*. 2000
DRD1	Decreased response	Xu *et al*. 2000b; Crawford *et al*. 1997
ATM	Hypersensitivity	Eilam *et al*. 1998
PKA RIIbeta	Abnormal LA	Brandon *et al*. 1998
5-HTT	↑LA with (+)-3, 4-methylenedioxymethamphetamine	Bengel *et al*. 1998
DRD3	Enhanced sensitivity	Xu *et al*. 1997
DAT	No response	Giros *et al*. 1996
VMAT2	Supersensitivity	Takahashi *et al*. 1997; Uhl *et al*. 2000; Fon *et al*. 1997

The table shows genes identified as modulating amphetamine response when knocked out in mice. L = locomoter activity; CPP = conditioned place preference

moderate heritability for abuse and dependence (Table 11.5). For example in the VTR, hallucinogen use, abuse and dependence showed substantial heritability (0.62, 0.65 and 0.79 respectively) in males. Heritability was lower in the VET study at 0.25 for abuse/dependence, and this is entirely made up of 'common' genetic liability to drug abuse/dependence generally. Sedatives were one of the first drugs to be shown to have a heritable component (Pedersen 1981) and both the VET (0.38) and VTR (0.59; 0.87 in males) provide evidence for abuse and/or dependence in males. In females there were only sufficient data to indicate that use of sedatives has a heritable component. There are no molecular studies specifically addressing sedatives and hallucinogens at the time of writing.

Conclusions

Family, twin and adoption studies have provided strong evidence for a substantial, genetic contribution to the development of alcohol dependence and most other forms of drug use, abuse and dependence. Environmental factors, especially individual specific experiences, are also of major importance. The transmission of genetic liability to alcohol abuse and dependence appears in the most part distinct from that of drug abuse, but there is overlap in non-genetic factors. Genetic vulnerability to drug abuse appears to be composed of two components, one specific to the drug (for example, drug specific pharmacogenetic factors) and one producing general vulnerability to drug abuse. In general, more severe phenotypes (abuse and dependence) have higher heritability than use, which tend to be influenced by shared as well as non-shared environment.

The application of molecular genetic approaches, including linkage and association, shows promise. Linkage approaches have identified chromosomal regions and association studies candidate genes implicated in protection and predisposition towards dependence. Combining information from both approaches will accelerate the identification of individual polymorphisms in genes which influence alcohol and drug dependence. These findings will explain the biological underpinning of this condition and will also help elucidate important environmental factors. Furthermore, it will provide a logical framework for the targeting and development of existing and new management approaches.

References

Abeliovich A., Schmitz Y., Farinas I. *et al.* (2000) Mice lacking alpha-synuclein display functional deficits in the nigrostriatal dopamine system. *Neuron* **25**:239–52.

Allgulander C., Nowak J., Rice J. P. (1991) Psychopathology and treatment of 30,344 twins in Sweden. II. Heritability estimates of psychiatric diagnosis and treatment in 12,884 twin pairs. *Acta Psychiatr Scand* **83**:12–15.

Annis H. M. (1974) Patterns of intra-familial drug use. *Br J Addict Alcohol Other Drugs* **69**:361–9.

Aston C. E. and Hill S. Y. (1990) Segregation analysis of alcoholism in families ascertained through a pair of male alcoholics. *Am J Hum Genet* **46**:879–87.

Ball D. M. and Murray R. M. (1994) Genetics of alcohol misuse. *Br Med Bull* **50**:18–35.

Ball D. M., Sherman D., Gibb R. *et al.* (1995) No association between the c2 allele at the cytochrome P450IIE1 gene and alcohol induced liver disease, alcohol Korsakoff's syndrome or alcohol dependence syndrome. *Drug Alcohol Depend* **39**:181–4.

Ball D. M. and Strang J. (1999) Causes of alcoholism and alcohol problems. *Medicine* **27**:1–2.

Begleiter H., Porjesz B., Reich T. *et al.* (1998) Quantitative trait loci analysis of human event-related brain potentials: P3 voltage. *Electroencephalogr Clin Neurophysiol* **108**:244–50.

Belzung C., Scearce-Levie K., Barreau S. *et al.* (2000) Absence of cocaine-induced place conditioning in serotonin 1B receptor knock-out mice. *Pharmacol Biochem Behav* **66**:221–5.

Bengel D., Murphy D. L., Andrews A. M. *et al.* (1998) Altered brain serotonin homeostasis and locomotor insensitivity to 3:4-methylenedioxymethamphetamine ('Ecstasy') in serotonin transporter-deficient mice. *Mol Pharmacol* **53**:649–55.

Berrettini W. H., Ferraro T. N., Alexander R. C. *et al.* (1994) Quantitative trait loci mapping of three loci controlling morphine preference using inbred mouse strains. *Nat Genet* **7**:54–8.

Bierut L. J., Dinwiddie S. H., Begleiter H. *et al.* (1998) Familial transmission of substance dependence: alcohol, marijuana, cocaine, and habitual smoking: a report from the Collaborative Study on the Genetics of Alcoholism. *Arch Gen Psychiatry* **55**:982–8.

Birznieks G., Ghosh S., Watanabe R. M. *et al.* (1999) The effect of phenotype variation on detection of linkage in the COGA data. *Genet Epidemiol* 17(Suppl 1):S61–6.

Blomqvist O., Gelernter J., Kranzler H. R. (2000) Family-based study of DRD2 alleles in alcohol and drug dependence. *Am J Med Genet* **96**:659–64.

Blum K., Noble E. P., Sheridan P. J. *et al.* (1991) Association of the A1 allele of the D2 dopamine receptor gene with severe alcoholism. *Alcohol* **8**:409–16.

Bohman M. (1978) Some genetic aspects of alcoholism and criminality. A population of adoptees. *Arch Gen Psychiatry* **35**:269–76.

Bohman M. (1981) The interaction of heredity and childhood environment: some adoption studies. *J Child Psychol Psychiatry* **22**:195–200.

Bohman M., Sigvardsson S., Cloninger C. R. (1981) Maternal inheritance of alcohol abuse. Cross-fostering analysis of adopted women. *Arch Gen Psychiatry* **38**:965–9.

Bond C., LaForge K. S., Tian M. *et al.* (1998) Single-nucleotide polymorphism in the human mu opioid receptor gene alters beta-endorphin binding and activity: possible implications for opiate addiction. *Proc Natl Acad Sci USA* **95**:9608–13.

Borras E., Coutelle C., Rosell A. *et al.* (2000) Genetic polymorphism of alcohol dehydrogenase in Europeans: the ADH2*2 allele decreases the risk for alcoholism and is associated with ADH3*1. *Hepatology* **31**:984–9.

Brandon E. P., Logue S. F., Adams M. R. *et al.* (1998) Defective motor behavior and neural gene expression in RIIbeta-protein kinase A mutant mice. *J Neurosci* **18**:3639–49.

Buck K. J., Metten P., Belknap J. K. *et al.* (1997) Quantitative trait loci involved in genetic predisposition to acute alcohol withdrawal in mice. *J Neurosci* **17**:3946–55.

Cadoret R. J. and Gath A. (1978) Inheritance of alcoholism in adoptees. *Br J Psychiatry* **132**:252–8.

Cadoret R. J., O'Gorman T. W., Troughton E. *et al.* (1985) Alcoholism and antisocial personality. Interrelationships, genetic and environmental factors. *Arch Gen Psychiatry* **42**:161–7.

Cadoret R. J., Troughton E., O'Gorman T. W. (1987) Genetic and environmental factors in alcohol abuse and antisocial personality. *J Stud Alcohol* **48**:1–8.

Cadoret R. J. Troughton E., O'Gorman T. W. *et al.* (1986a) An adoption study of genetic and environmental factors in drug abuse. *Arch Gen Psychiatry* **43**:1131–6.

Cadoret R. J. Adopion studies: historical and methodological critique. *Psychiatr Dev* **4**(1):45–64.

Cadoret R. J., Yates W. R., Troughton E. *et al.* (1995) Adoption study demonstrating two genetic pathways to drug abuse. *Arch Gen Psychiatry* **52**:42–52.

Cadoret R. J., Yates W. R., Troughton E. *et al.* (1996) An adoption study of drug abuse/dependency in females. *Compr Psychiatry* **37**:88–94.

Caldwell C. B. and Gottesman I. I. (1991) Sex differences in the risk for alcoholism: a twin study. *Behav Genet* **21**:563 (Abstract).

Cantor R. M. and Lanning C. D. (1999) Comparison of evidence supporting a chromosome 6 alcoholism gene. *Genet Epidemiol* **17**(Suppl 1):S91–6.

Carr L. G., Yi I. S., Li T. K. *et al.* (1996) Cytochrome P4502E1 genotypes, alcoholism, and alcoholic cirrhosis in Han Chinese and Atayal Natives of Taiwan. *Alcohol Clin Exp Res* **20**:43–6.

Carta A. R., Gerfen C. R., Steiner H. (2000) Cocaine effects on gene regulation in the striatum and behavior: increased sensitivity in D3 dopamine receptor-deficient mice. *Neuroreport* **11**:2395–9.

Castanon N., Scearce-Levie K., Lucas J. J. *et al.* (2000) Modulation of the effects of cocaine by 5-HT1B receptors: a comparison of knockouts and antagonists. *Pharmacol Biochem Behav* **67**:559–66.

Chen C. C., Lu R. B., Chen Y. C. *et al.* (1999a) Interaction between the functional polymorphisms of the alcohol-metabolism genes in protection against alcoholism. *Am J Hum Genet* **65**:795–807.

Chen C. H., Finch S. J., Mendell N. R. *et al.* (1999b) Comparison of empirical strategies to maximize GENEHUNTER lod scores. *Genet Epidemiol* **17**(Suppl 1):S103–8.

Chen J. F., Beilstein M., Xu Y. H. *et al.* (2000) Selective attenuation of psychostimulant-induced behavioral responses in mice lacking A(2A) adenosine receptors. *Neuroscience* **97**:195–204.

Chen W. J., Loh E. W., Hsu Y. P. *et al.* (1996) Alcohol-metabolising genes and alcoholism among Taiwanese Han men: independent effect of ADH2, ADH3 and ALDH2. *Br J Psychiatry* **168**:762–7.

Cigler T., LaForge K. S., McHugh P. F. *et al.* (2001) Novel and previously reported single-nucleotide polymorphisms in the human 5-HT(1B) receptor gene: no association with cocaine or alcohol abuse or dependence. *Am J Med Genet* **105**:489–97.

Cloninger C. R., Bohman M., Sigvardsson S. (1981) Inheritance of alcohol abuse. Cross-fostering analysis of adopted men. *Arch Gen Psychiatry* **38**:861–8.

Comings D. E., Muhleman D., Gade R. *et al.* (1997) Cannabinoid receptor gene (CNR1): association with i.v. drug use. *Mol Psychiatry* **2**:161–8.

Comuzzie A. G. and Williams J. T. (1999) Correcting for ascertainment bias in the COGA data set. *Genet Epidemiol* 17(Suppl 1):S109–14.

Cotton N. S. (1979) The familial incidence of alcoholism: a review. *J Stud Alcohol* **40**:89–116.

Couzigou P., Fleury B., Groppi A. *et al.* (1990) Genotyping study of alcohol dehydrogenase class I polymorphism in French patients with alcoholic cirrhosis. The French Group for Research on Alcohol and Liver. *Alcohol* **25**:623–6.

Covault J., Gelernter J., Kranzler H. (2001) Association study of cannabinoid receptor gene (CNR1) alleles and drug dependence. *Mol Psychiatry* **6**:501–2.

Crabbe J. C., Phillips T. J., Buck K. J. *et al.* (1999) Identifying genes for alcohol and drug sensitivity: recent progress and future directions. *Trends Neurosci* **22**:173–9.

Crawford C. A., Drago J., Watson J. B. *et al.* (1997) Effects of repeated amphetamine treatment on the locomotor activity of the dopamine D1A-deficient mouse. *Neuroreport* **8**:2523–7.

Cubells J. F., Kranzler H. R., McCance-Katz E. *et al.* (2000) A haplotype at the DBH locus, associated with low plasma dopamine beta-hydroxylase activity, also associates with cocaine-induced paranoia. *Mol Psychiatry* **5**:56–63.

Curtis D., Zhao J. H., Sham, P. C. (1999) Comparison of GENEHUNTER and MFLINK for analysis of COGA linkage data. *Genet Epidemiol* 17(Suppl 1):S115–20.

Daw E. W., Kumm J., Snow G. L. *et al.* (1999) Monte Carlo Markov chain methods for genome screening. *Genet Epidemiol* 17(Suppl 1):S133–8.

Dawson D. A., Harford T. C., Grant B. F. (1992) Family history as a predictor of alcohol dependence. *Alcohol Clin Exp Res* **16**:572–5.

Di Chiara G. and Imperato A. (1988) Drugs abused by humans preferentially increase synaptic dopamine concentrations in the mesolimbic system of freely moving rats. *Proc Natl Acad Sci USA* **85**:5274–8.

Dockstader C. L., Rubinstein M., Grandy D. K. *et al.* (2001) The D2 receptor is critical in mediating opiate motivation only in opiate-dependent and withdrawn mice. *Eur J Neurosci* **13**:995–1001.

Edenberg H. J., Foroud T., Koller D. L. *et al.* (1998) A family-based analysis of the association of the dopamine D2 receptor (DRD2) with alcoholism. *Alcohol Clin Exp Res* **22**:505–12.

Edenberg H. J., Jerome R. E., Li M. (1999) Polymorphism of the human alcohol dehydrogenase 4 (ADH4) promoter affects gene expression. *Pharmacogenetics* **9**:25–30.

Edwards (1982) The Treatment of Drinking Problems: A Guide for the Helping Professions. London, Grant McIntyre Medical & Scientific.

Edwards G., Marshall E. J., Cook C. C. H. (1997) The Treatment of Drinking Problems: A Guide for the Helping Professions. Cambridge, Cambridge University Press.

Edwards, G. (1996) Sensible drinking. *BMJ* **312**:1.

Ehlers C. L., Gilder D. A., Harris L. *et al.* (2001 In Press) Investigating a 'protective gene' against alcoholism. *Alcohol Clin Exp Res.*

Ehrig T., Li T. K. (1995) Metabolism of alcohol and metabolic consequences, in B. Tabakoff and P. L. Hoffmann (eds) *Biological Aspects of Alcoholism: Implications for Prevention, Treatment and Policy*. Seattle.

Eilam R., Peter Y., Elson A. *et al.* (1998) Selective loss of dopaminergic nigro-striatal neurons in brains of Atm-deficient mice. *Proc Natl Acad Sci USA* **95**:12653–6.

Farrell M. (1999) Cannabis dependence and withdrawal. *Addiction* **94**(9):1277–8.

Fawzy F. I., Coombs R. H., Simon J. M. *et al.* (1987) Family composition, socioeconomic status, and adolescent substance use. *Addict Behav* **12**:79–83.

Fon E. A., Pothos E. N., Sun B. C. *et al.* (1997) Vesicular transport regulates monoamine storage and release but is not essential for amphetamine action. *Neuron* **19**:1271–83.

Franke P., Nothen M. M., Wang T. *et al.* (2000) DRD4 exon III VNTR polymorphism-susceptibility factor for heroin dependence? Results of a case-control and a family-based association approach. *Mol Psychiatry* **5**:101–4.

Franke P., Nothen M. M., Wang T. *et al.* (1999) Human delta-opioid receptor gene and susceptibility to heroin and alcohol dependence. *Am J Med Genet* **88**:462–4.

Franke P., Wang T., Nothen M. M. *et al.* (2001) Nonreplication of association between mu-opioid-receptor gene (OPRM1) A118G polymorphism and substance dependence. *Am J Med Genet* **105**:114–19.

Gelernter J., Goldman D., Risch, N. (1993) The A1 allele at the D2 dopamine receptor gene and alcoholism. A reappraisal [see comments]. *JAMA* **269**:1673–7.

Gelernter J., and Kranzler H. (1999) D2 dopamine receptor gene (DRD2) allele and haplotype frequencies in alcohol dependent and control subjects: no association with phenotype or severity of phenotype. *Neuropsychopharmacology* **20**:640–9.

Gelernter J., Kranzler H., and Satel S. L. (1999) No association between D2 dopamine receptor (DRD2) alleles or haplotypes and cocaine dependence or severity of cocaine dependence. *Euro Biol Psychiatry* **45**:340–5.

Gelernter J., Kranzler H., and Satel S. L. *et al.* (1994) Genetic association between dopamine transporter protein alleles and cocaine-induced paranoia. *Neuropsychopharmacology* **11**:195–200.

Gilder F. J., Hodgkinson S., Murray R. M. (1993) ADH and ALDH genotype profiles in Caucasians with alcohol-related problems and controls. *Addiction* **88**:383–8.

Giros B., Jaber M., Jones S. R. *et al.* (1996) Hyperlocomotion and indifference to cocaine and amphetamine in mice lacking the dopamine transporter. *Nature* **379**:606–12.

Giros B., Wang Y. M., Suter S. *et al.* (1994) Delineation of discrete domains for substrate, cocaine, and tricyclic antidepressant interactions using chimeric dopamine-norepinephrine transporters. *J Biol Chem* **269**:15985–8.

Glass (1991) *The International Handbook of Addiction Behaviour*. London, Tavistock/Routledge.

Godfrey C. and Maynard A. (1992) *A Health Strategy for Alcohol: Setting Targets and Choosing Policies*. York, Centre for Health Economics.

Goedde H. W., Agarwal D. P., Fritze G. *et al.* (1992) Distribution of ADH2 and ALDH2 genotypes in different populations. *Hum Genet* **88**:344–6.

Goodwin D. W., Schulsinger F., Hermansen L. *et al.* (1973) Alcohol problems in adoptees raised apart from alcoholic parents. *Arch Gen Psychiatry* **28**:238–43.

Goodwin D. W., Schulsinger F., Knop J. *et al.* (1977a) Psychopathology in adopted and nonadopted daughters of alcoholics. *Arch Gen Psychiatry* **34**:1005–9.

Goodwin D. W., Schulsinger F., Knop J. *et al.* (1977b) Alcoholism and depression in adopted-out daughters of alcoholics. *Arch Gen Psychiatry* **34**:751–5.

Goodwin D. W., Schulsinger F., Krop J. *et al.* (1974) Drinking problems in adopted and nonadopted sons of alcoholics. *Arch Gen Psychiatry* **31**:164–9.

Gorwood P., Batel P., Gouya L. *et al.* (2000) Reappraisal of the association between the DRD2 gene, alcoholism and addiction. *Eur Psychiatry* **15**:90–6.

Gottesman I. I. and Carey G. (1983) Extracting meaning and direction from twin data. *Psychiatr Dev* **1**:35–50.

Grove W. M., Eckert E. D., Heston L. *et al.* (1990) Heritability of substance abuse and antisocial behavior: a study of monozygotic twins reared apart. *Biol Psychiatry* **27**:1293–304.

Gurling H. M., Murray R. M., Clifford C. A. (1981) Investigations into the genetics of alcohol dependence and into its effects on brain function. *Prog Clin Biol Res* **69**(Pt C):77–87.

Gurling H. M., Oppenheim B. E., Murray R. M. (1984) Depression, criminality and psychopathology associated with alcoholism: evidence from a twin study. *Acta Genet Med Gemellol (Roma.)* **33**:333–9.

Guze S. B., Cloninger C. R., Martin R. *et al.* (1986) Alcoholism as a medical disorder. *Compr Psychiatry* **27**:501–10.

Gynther L. M., Carey G., Gottesman I. I. *et al.* (1995) A twin study of non-alcohol substance abuse. *Psychiatry Res* **56**:213–20.

Hayashi S., Watanabe J., Kawajiri K. (1991) Genetic polymorphisms in the 5′-flanking region change transcriptional regulation of the human cytochrome P450IIE1 gene. *J Biochem (Tokyo)* **110**:559–65.

Heath A. C., Bucholz K. K., Madden P. A. *et al.* (1997) Genetic and environmental contributions to alcohol dependence risk in a national twin sample: consistency of findings in women and men. *Psychol Med* **27**:1381–96.

Henderson W. G., Eisen S., Goldberg J. *et al.* (1990) The Vietnam Era Twin Registry: a resource for medical research. *Public Health Rep* **105**:368–73.

Hendry I. A., Kelleher K. L., Bartlett S. E. *et al.* (2000) Hypertolerance to morphine in G(z alpha)-deficient mice. *Brain Res* **870**:10–19.

Herve D., Le Moine C., Corvol J. C. *et al.* (2001) Galpha(olf) levels are regulated by receptor usage and control dopamine and adenosine action in the striatum. *J Neurosci* **21**:4390–9.

Higuchi S., Matsushita S., Murayama M. *et al.* (1995) Alcohol and aldehyde dehydrogenase polymorphisms and the risk for alcoholism. *Am J Psychiatry* **152**:1219–21.

Higuchi S., Muramatsu T., Matsushita S. *et al.* (1996) Polymorphisms of ethanol-oxidizing enzymes in alcoholics with inactive ALDH2. *Hum Genet* **97**:431–4.

Hoehe M. R., Kopke K., Wendel B. *et al.* (2000a) Sequence variability and candidate gene analysis in complex disease: association of mu opioid receptor gene variation with substance dependence. *Hum Mol Genet* **9**:2895–908.

Hoehe M. R., Rinn T., Flachmeier C. *et al.* (2000b) Comparative sequencing of the human CB1 cannabinoid receptor gene coding exon: no structural mutations in individuals exhibiting extreme responses to cannabis. *Psychiatr Genet* **10**:173–7.

Hrubec Z. and Omenn G. S. (1981) Evidence of genetic predisposition to alcoholic cirrhosis and psychosis: twin concordances for alcoholism and its biological end points by zygosity among male veterans. *Alcohol Clin Exp Res* **5**:207–15.

Itzhak Y., Ali S. F., Martin J. L. *et al.* (1998a) Resistance of neuronal nitric oxide synthase-deficient mice to cocaine-induced locomotor sensitization. *Psychopharmacology (Berl)* **140**:378–86.

Itzhak Y., Martin J. L., Black M. D. *et al.* (1998b) The role of neuronal nitric oxide synthase in cocaine-induced conditioned place preference. *Neuroreport* **9**:2485–8.

Iwahashi K., Matsuo Y., Suwaki H. *et al.* (1995) CYP2E1 and ALDH2 genotypes and alcohol dependence in Japanese. *Alcohol Clin Exp Res* **19**:564–6.

Jang K. L., Livesley W. J., Vernon P. A. (1995) Alcohol and drug problems: a multivariate behavioural genetic analysis of co-morbidity. *Addiction* **90**:1213–21.

Jones S. R., Gainetdinov R. R., Wightman R. M. *et al.* (1998) Mechanisms of amphetamine action revealed in mice lacking the dopamine transporter. *J Neurosci* **18**:1979–86.

Kaij L. (1960) *Alcoholism in Twins.* Stockholm: Almquist and Wiksell.

Karkowski L. M., Prescott C. A., Kendler K. S. (2000) Multivariate assessment of factors influencing illicit substance use in twins from female–female pairs. *Am J Med Genet* **96**:665–70.

Kaymakcalan S. (1981) The addictive potential of Cannabis. *Ball Narc* **33**(2):21–31.

Kendler K. (2001) Twin studies in psychiatric disorders. *Archives of General Psychiatry* **58**:1005–14.

Kendler K. S., Davis C. G., Kessler R. C. (1997) The familial aggregation of common psychiatric and substance use disorders in the National Comorbidity Survey: a family history study. *Br J Psychiatry* **170**:541–8.

Kendler K. S., Heath A. C., Neale M. C. *et al.* (1992) A population-based twin study of alcoholism in women. *JAMA* **268**:1877–82.

Kendler K. S., Karkowski L., Prescott C. A. (1999a) Hallucinogen, opiate, sedative and stimulant use and abuse in a population-based sample of female twins. *Acta Psychiatr Scand* **99**:368–76.

Kendler K. S., Corey L. A. *et al.* (1999b) Genetic and environmental risk factors in the aetiology of illicit drug initiation and subsequent misuse in women. *Br J Psychiatry* **175**:351–6.

Kendler K. S., Neale M. C. *et al.* (2000) Illicit psychoactive substance use, heavy use, abuse, and dependence in a US population-based sample of male twins. *Arch Gen Psychiatry* **57**:261–9.

Kendler K. S., Neale M. C., Kessler R. C. *et al.* (1992) Familial influences on the clinical characteristics of major depression: a twin study. *Acta Psychiatr Scand* **86**:371–8.

Kendler K. S. and Prescott C. A. (1998a) Cannabis use, abuse, and dependence in a population-based sample of female twins. *Am J Psychiatry* **155**:1016–22.

Kendler K. S. and Prescott C. A. (1998b) Cocaine use, abuse and dependence in a population-based sample of female twins. *Br J Psychiatry* **173**:345–50.

Kendler K. S., Neale M. C. *et al.* (1997) Temperance board registration for alcohol abuse in a national sample of Swedish male twins, born 1902 to 1949. *Arch Gen Psychiatry* **54**:178–84.

Kest B., Hopkins E., Palmese C. A. *et al.* (2001) Morphine tolerance and dependence in nociceptin/orphanin FQ transgenic knock-out mice. *Neuroscience* **104**:217–22.

Kim D. S., Szczypka M. S., Palmiter R. D. (2000) Dopamine-deficient mice are hypersensitive to dopamine receptor agonists. *J Neurosci* **20**:4405–13.

Kobler J. (1974) *Ardent Spirits: The Rise And Fall Of Prohibition.* London, Michael Joseph.

Koch H. G., McClay J., Loh E. W. *et al.* (2000) Allele association studies with SSR and SNP markers at known physical distances within a 1Mb region embracing the ALDH2 locus in the Japanese, demonstrates linkage disequilibrium extending to 400 Kb. *Human Molecular Genetics.*

Kolls J. K., Lei D., Stoltz D. *et al.* (1998) Adenoviral-mediated interferon-gamma gene therapy augments pulmonary host defense of ethanol-treated rats. *Alcohol Clin Exp Res* **22**:157–62.

Koob G. F. and Le Moal M. (2001) Drug *Addiction* dysregulation of reward, and allostasis. *Neuropsychopharmacology* **24**:97–129.

Korczak J. F., Bergen A. W., Goldstein A. M. *et al.* (1999) Sib-pair linkage analyses of alcoholism: dichotomous and quantitative measures. *Genet Epidemiol* **17**(Suppl 1): S205–10.

Korpi E. R., Kleingoor C., Kettenmann H. *et al.* (1993) Benzodiazepine-induced motor impairment linked to point mutation in cerebellar GABAA receptor [see comments]. *Nature* **361**:356–9.

Koskenvuo M., Langinvainio H., Kaprio J. *et al.* (1984) Psychiatric hospitalization in twins. *Acta Genet Med Gemellol (Roma.)* **33**:321–32.

Kotler M., Cohen H., Segman R. *et al.* (1997) Excess dopamine D4 receptor (D4DR) exon III seven repeat allele in opioid-dependent subjects. *Mol Psychiatry* **2**:251–4.

Kovac I., Rouillard E., Merette C. *et al.* (1999) Exploring the impact of extended phenotype in stratified samples. *Genet Epidemiol* 17(Suppl 1):S211–16.

Krezel W., Ghyselinck N., Samad T. A. *et al.* (1998) Impaired locomotion and dopamine signaling in retinoid receptor mutant mice. *Science* **279**:863–7.

Ledent C., Valverde O., Cossu G. *et al.* (1999) Unresponsiveness to cannabinoids and reduced addictive effects of opiates in CB1 receptor knockout mice. *Science* **283**:401–4.

Li T, Liu X., Hong Z. Z. *et al.* (2000a) Association analysis of polymorphisms in the mu opioid receptor gene and heroin abuse in Chinese subjects. *Addiction Biology* **5**:179–84.

Li T, Liu X., Zhu Z. H. *et al.* (2000b) No association between (AAT)n repeats in the cannabinoid receptor gene (CNR1) and heroin abuse in a Chinese population. *Mol Psychiatry* **5**:128–30.

Li T, Xu K., Deng H. *et al.* (1997) Association analysis of the dopamine D4 gene exon III VNTR and heroin abuse in Chinese subjects. *Mol Psychiatry* **2**:413–16.

Lieber C. S. (1988) Biochemical and molecular basis of alcohol-induced injury to liver and other tissues [see comments]. *N Engl J Med* **319**:1639–50.

Lin N., Eisen S. A., Scherrer J. F. *et al.* (1996) The influence of familial and non-familial factors on the association between major depression and substance abuse/dependence in 1874 monozygotic male twin pairs. *Drug Alcohol Depend* **43**:49–55.

Lin N., Irwin M. E., Wright F. A. (1999) A multiple locus analysis of the collaborative study on the genetics of alcoholism data set. *Genet Epidemiol* 17(Suppl 1): S229–34.

Loh E. W. and Ball D. (2000) Role of the GABA$_A$2, GABA$_A$6, GABA$_A$1 and GABA$_A$2 receptor subunit genes cluster in drug responses and the development of alcohol dependence. *Neurochem Int* **37**:413–23.

Loh E. W., Higuchi S., Matsushita S. *et al.* (2000) Association analysis of the GABA$_A$ receptor subunit genes cluster on 5q33–34 and alcohol dependence in a Japanese population. *Mol Psychiatry* **5**:301–7.

Long J. C., Knowler W. C., Hanson R. L. *et al.* (1998) Evidence for genetic linkage to alcohol dependence on chromosomes 4 and 11 from an autosome-wide scan in an American Indian population. *Am J Med Genet* **81**:216–21.

Lopez A. D. and Murray C. C. (1998) The global burden of disease, 1990–2020 [news]. *Nat Med* **4**:1241–3.

Lowinson J. H., Ruiz P., Millman R. B. *et al.* (1997) *Substance Abuse: A Comprehensive Textbook, 3rd edition.* Lippincott, Williams and Wilkins.

Lusher J., Ebersole L., Ball D. (2000) Dopamine D4 receptor gene and severity of dependence. *Addiction Biology* **5**:471–4.

Lusher J. M., Chandler C., Ball D. (2001) Dopamine D4 receptor gene (DRD4) is associated with Novelty Seeking (NS) and substance abuse: the saga continues. *Mol Psychiatry* **6**:497–9.

Lyons M. J., Toomey R., Meyer J. M. *et al.* (1997) How do genes influence marijuana use? The role of subjective effects. *Addiction* **92**:409–17.

Maezawa Y., Yamauchi M., Toda G. (1994) Association between restriction fragment length polymorphism of the human cytochrome P450IIE1 gene and susceptibility to alcoholic liver cirrhosis. *Am J Gastroenterol* **89**:561–5.

Maezawa Y., Yamauchi M., Toda G. (1995) Alcohol-metabolizing enzyme polymorphisms and alcoholism in Japan. *Alcohol Clin Exp Res* **19**:951–4.

Malhotra M. K. (1983) Familial and personal correlates (risk factors) of drug consumption among German youth. *Acta Paedopsychiatr* **49**:199–209.

Mark T. L., Woody G. E., Juday T. *et al.* (2001) The economic costs of heroin addiction in the United States. *Drug Alcohol Dependl* **61**:195–206.

Martin M., Ledent C., Parmentier M. *et al.* (2000) Cocaine, but not morphine, induces conditioned place preference and sensitization to locomotor responses in CB1 knockout mice. *Eur J Neurosci* **12**:4038–46.

Matthes H. W., Maldonado R., Simonin F. *et al.* (1996) Loss of morphine-induced analgesia, reward effect and withdrawal symptoms in mice lacking the mu-opioid-receptor gene. *Nature* **383**:819–23.

Mayer P., Rochlitz H., Rauch E. *et al.* (1997) Association between a delta opioid receptor gene polymorphism and heroin dependence in man. *Neuroreport* **8**:2547–50.

McGue M. (1994) Genes, environment and the etiology of alcoholism, in R. Zucker, G. Boyd, J. Howard (eds) *The Development of Alcohol Problems: Exploring the Biopsychosocial Matrix of Risk*. Rockville.

McGue M., Pickens R. W., Svikis D. S. (1992) Sex and age effects on the inheritance of alcohol problems: a twin study. *J Abnorm Psychol* **101**:3–17.

Meller W. H., Rinehart R., Cadoret R. J. *et al.* (1998) Specific familial transmission in substance abuse. *Int J Addict* **23**:1029–39.

Merikangas K. R., Stolar M., Stevens D. E. *et al.* (1998) Familial transmission of substance use disorders. *Arch Gen Psychiatry* **55**:973–9.

Messer C. J., Eisch A. J., Carlezon W. A. Jr. *et al.* (2000) Role for GDNF in biochemical and behavioral adaptations to drugs of abuse. *Neuron* **26**:247–57.

Michna L., Brenz Verca M. S., Widmer D. A. *et al.* (2001) Altered sensitivity of CD81-deficient mice to neurobehavioral effects of cocaine. *Brain Res Mol Brain Res* **90**:68–74.

Miles D. R., van den Bree M. B., Gupman A. E. *et al.* (2001) A twin study on sensation seeking, risk taking behavior and marijuana use. *Drug Alcohol Dependl* **62**:57–68.

Miner L. L., Drago J., Chamberlain P. M. *et al.* (1995) Retained cocaine conditioned place preference in D1 receptor deficient mice. *Neuroreport* **6**:2314–16.

Mirin S. M., Weiss R. D., Griffin M. L. *et al.* (1991) Psychopathology in drug abusers and their families. *Compr Psychiatry* **32**:36–51.

Mitchell B. D., Ghosh S., Watanabe R. M. *et al.* (1999) Identifying influential individuals in linkage analysis: application to a quantitative trait locus detected in the COGA data. *Genet Epidemiol* **17**(Suppl 1):S259–64.

Murtra P., Sheasby A. M., Hunt S. P. *et al.* (2000) Rewarding effects of opiates are absent in mice lacking the receptor for substance P. *Nature* **405**:180–3.

Nakamura K., Iwahashi K., Matsuo Y. *et al.* (1996) Characteristics of Japanese alcoholics with the atypical aldehyde dehydrogenase 2*2. I. A comparison of the genotypes of ALDH2, ADH2, ADH3, and cytochrome P-4502E1 between alcoholics and nonalcoholics. *Alcohol Clin Exp Res* **20**:52–5.

Noble E. P. (1998) The D2 dopamine receptor gene: a review of association studies in alcoholism and phenotypes. *Alcohol* **16**:33–45.

Noble E. P., Blum K., Khalsa M. E. *et al.* (1993) Allelic association of the D2 dopamine receptor gene with cocaine dependence. *Drug Alcohol Depend* **33**:271–85.

Nuffield Council on Bioethics (1998) *Mental Disorders and Genetics: The Ethical Context*. London, Nuffield Council on Bioethics.

Osier M., Pakstis A. J., Kidd J. R. *et al.* (1999) Linkage Disequilibrium at the ADH2 and ADH3 Loci and Risk of Alcoholism. *Am J Hum Genet* **64**:1147–57.

Pares X., Martinez S. E., Allali-Hassani A. *et al.* (2001) Distribution of alcohol dehydrogenase in human organs: relevance for alcohol metabolism and pathology, in D. P. Agarwal and H. K. Seitz (eds) *Alcohol in Health and Disease*. New York.

Parsian A. and Zhang Z. H. (1999) Human chromosomes 11p15 and 4p12 and alcohol dependence: possible association with the GABRB1 gene. *Am J Med Genet* **88**:533–8.

Pedersen N. (1981) Twin similarity for usage of common drugs. *Prog Clin Biol Res* 69(Pt C):53–9.

Pickens R. W. and Svikis D. S. (1991) Genetic influences in human substance abuse. *J Addict Dis* **10**:205–13.

Pickens R. W., Svikis D. S., McGue M. *et al.* (1991) Heterogeneity in the inheritance of alcoholism. A study of male and female twins. *Arch Gen Psychiatry* **48**:19–28.

Plutarch (110) *The Training of Children.* http://www.fordham.edu/halsall/ancient/plutarch-education.html

Prescott C. A. (2001) The genetic epidemiology of alcoholism, in D. P. Agarwal and H. K. Seitz (eds) *Alcohol in Health and Disease.* New York.

Prescott C. A., Aggen S. H., Kendler K. S. (1999) Sex differences in the sources of genetic liability to alcohol abuse and dependence in a population-based sample of US twins. *Alcohol Clin Exp Res* **23**:1136–44.

Radel M., Vallejo R. L., Long J. C. *et al.* (1999) Sib-pair linkage analysis of GABRG2 to alcohol dependence. *Alcohol Clin Exp Res* 23:59A.

Reed T., Page W. F., Viken R. J. *et al.* (1996) Genetic predisposition to organ-specific endpoints of alcoholism. *Alcohol Clin Exp Res* **20**:1528–33.

Reich T., Edenberg H. J., Goate A. *et al.* (1998) Genome-wide search for genes affecting the risk for alcohol dependence. *Am J Med Genet* **81**:207–15.

Resnick A., Homanics G. E., Jung B. J. *et al.* (1999) Increased acute cocaine sensitivity and decreased cocaine sensitization in GABA(A) receptor beta3 subunit knockout mice. *J Neurochem* **73**:1539–48.

Ripley T. L., Rocha B. A., Oglesby M. W. *et al.* (1999) Increased sensitivity to cocaine, and over-responding during cocaine self-administration in tPA knockout mice. *Brain Res* **826**:117–27.

Ritter J. K. (2000) Roles of glucuronidation and UDP-glucuronosyltransferases in xenobiotic bioactivation reactions. *Chem Biol Interact* **129**:171–93.

Robbins T. W. and Everitt B. J. (1999) Drug addiction: bad habits add up. *Nature* **398**:567–70.

Rocha B. A., Ator R., Emmett-Oglesby M. W. *et al.* (1997) Intravenous cocaine self-administration in mice lacking 5-HT1B receptors. *Pharmacol Biochem Behav* **57**:407–12.

Rocha B. A., Fumagalli F., Gainetdinov R. R. *et al.* (1998) Cocaine self-administration in dopamine-transporter knockout mice. *Nat Neurosci* **1**:132–7.

Rocha B. A., Scearce-Levie K., Lucas J. J. *et al.* (1998) Increased vulnerability to cocaine in mice lacking the serotonin-1B receptor. *Nature* **393**:175–8.

Rodrigo L., Alvarez V., Rodriguez M. *et al.* (1999) N-acetyltransferase-2, glutathione S-transferase M1, alcohol dehydrogenase, and cytochrome P450IIE1 genotypes in alcoholic liver cirrhosis: a case-control study. *Scand J Gastroenterol* **34**:303–7.

Roe A. and Burks B. (1945) Adult adjustment of foster-children of alcoholic and psychotic parentage and the influence of the foster home. *Quarterly Journal of the Study of Alcohol.*

Romanov K., Kaprio J., Rose R. J. *et al.* (1991) Genetics of alcoholism: effects of migration on concordance rates among male twins. *Alcohol Alcohol Suppl*, 1:137–40.

Rubinstein M., Phillips T. J., Bunzow J. R. *et al.* (1997) Mice lacking dopamine D4 receptors are supersensitive to ethanol, cocaine, and methamphetamine. *Cell* **90**:991–1001.

Saccone N. L., Kwon J. M., Corbett J. *et al.* (2000) A genome screen of maximum number of drinks as an alcoholism phenotype [In Process Citation]. *Am J Med Genet* **96**:632–7.

Sallinen J., Haapalinna A., Viitamaa T. *et al.* (1998) D-amphetamine and L-5-hydroxytryptophan-induced behaviours in mice with genetically-altered expression of the alpha2C-adrenergic receptor subtype. *NeuroScience* **86**:959–65.

Sander T., Ball D., Murray R. *et al.* (1999) Association analysis of sequence variants of GABA(A) alpha6, beta2, and gamma2 gene cluster and alcohol dependence. *Alcohol Clin Exp Res* **23**:427–31.

Sander T., Samochowiec J., Ladehoff M. *et al.* (1999) Association analysis of exonic variants of the gene encoding the GABAB receptor and alcohol dependence. *Psychiatr Genet* **9**:69–73.

Scheinin M., Sallinen J., Haapalinna A. (2001) Evaluation of the alpha2C-adrenoceptor as a neuropsychiatric drug target studies in transgenic mouse models. *Life Sci* **68**:2277–85.

Scherer S. E. (1973) Self-reported parent and child drug use. *Br J Addict Alcohol Other Drugs* **68**:363–4.

Schuckit M. A., Mazzanti C., Smith T. L. *et al.* (1999) Selective genotyping for the role of 5-HT2A, 5-HT2C, and GABA alpha 6 receptors and the serotonin transporter in the level of response to alcohol: a pilot study. *Biol Psychiatry* **45**:647–51.

Shen Y. C., Fan J. H., Edenberg H. J. *et al.* (1997) Polymorphism of ADH and ALDH genes among four ethnic groups in China and effects upon the risk for alcoholism. *Alcohol Clin Exp Res* **21**:1272–7.

Sherman D. I. and Williams R. (1994) Liver damage: mechanisms and management. *Br Med Bull* **50**:124–38.

Sigvardsson S., Bohman M., Cloninger C. R. (1996) Replication of the Stockholm Adoption Study of alcoholism. Confirmatory cross-fostering analysis. *Arch Gen Psychiatry* **53**:681–7.

Sora I., Hall F. S., Andrews A. M. *et al.* (2001) Molecular mechanisms of cocaine reward: combined dopamine and serotonin transporter knockouts eliminate cocaine place preference. *Proc Natl Acad Sci USA* **98**:5300–5.

Spielewoy C., Gonon F., Roubert C. *et al.* (2000) Increased rewarding properties of morphine in dopamine-transporter knockout mice. *Eur J Neurosci* **12**:1827–37.

Strugstad E. and Sager G. (1998) [Mechanisms behind drug dependence]. Tidsskr. Nor Laegeforen **118**:1866–9.

Szeto C. Y., Tang N. L., Lee D. T. *et al.* (2001) Association between mu opioid receptor gene polymorphisms and Chinese heroin addicts. *Neuroreport* **12**:1103–6.

Takahashi N., Miner L. L., Sora I. *et al.* (1997) VMAT2 knockout mice: heterozygotes display reduced amphetamine-conditioned reward, enhanced amphetamine locomotion, and enhanced MPTP toxicity. *Proc Natl Acad Sci USA* **94**:9938–43.

Tanaka F., Shiratori Y., Yokosuka O. *et al.* (1997) Polymorphism of alcohol-metabolizing genes affects drinking behavior and alcoholic liver disease in Japanese men. *Alcohol Clin Exp Res* **21**:596–601.

Tanda G., Pontieri F. E., Di Chiara G. (1997) Cannabinoid and heroin activation of mesolimbic dopamine transmission by a common mu1 opioid receptor mechanism. *Science* **276**:2048–50.

Thomasson H. R., Crabb D. W., Edenberg H. J. *et al.* (1994) Low frequency of the ADH2*2 allele among Atayal natives of Taiwan with alcohol use disorders. *Alcohol Clin Exp Res* **18**:640–3.

Thomasson H. R., Edenberg H. J., Crabb D. W. *et al.* (1991) Alcohol and aldehyde dehydrogenase genotypes and alcoholism in Chinese men. *Am J Hum Genet* **48**:677–81.

True W. R., Heath A. C., Bucholz K. *et al.* (1996) Models of treatment seeking for alcoholism: the role of genes and environment. *Alcohol Clin Exp Res* **20**:1577–81.

Tsuang M. T., Bar J. L., Harley R. M. *et al.* (2001) The harvard twin study of substance abuse: what we have learned. *Harv Rev Psychiatry* **9**:267–79.

Tsuang M. T., Lyons M. J., Eisen S. A. *et al.* (1996) Genetic influences on DSM-III-R drug abuse and dependence: a study of 3,372 twin pairs. *Am J Med Genet* **67**:473–7.

Tsuang M. T., Lyons M. J., Harley R. M. *et al.* (1999) Genetic and environmental influences on transitions in drug use. *Behav Genet* **29**:473–9.

Tsuang M. T., Lyons M. J., Meyer J. M. *et al.* (1998) Co-occurrence of abuse of different drugs in men: the role of drug-specific and shared vulnerabilities. *Arch Gen Psychiatry* **55**:967–72.

Tsutsumi M., Takase S., Takada A. (1993) Genetic factors related to the development of carcinoma in digestive organs in alcoholics. *Alcohol Alcohol* Suppl 1(B):21–6.

Tsutsumi M., Wang J. S., Takase S. *et al.* (1994) Hepatic messenger RNA contents of cytochrome P4502E1 in patients with different P4502E1 genotypes. *Alcohol Alcohol* 29 Suppl(1):29–32.

Turecki G., Rouleau G. A., Alda M. (1999) Family density of alcoholism and linkage information in the analysis of the COGA data. *Genet Epidemiol* 17(Suppl 1):S361–6.

Tyndale R. F., Droll K. P., Sellers E. M. (1997) Genetically deficient CYP2D6 metabolism provides protection against oral opiate dependence. *Pharmacogenetics* **7**:375–9.

Ueda H., Inoue M., Takeshima H. *et al.* (2000) Enhanced spinal nociceptin receptor expression develops morphine tolerance and dependence. *J Neurosci* **20**:7640–7.

Uhl G., Blum K., Noble E. *et al.* (1993) Substance abuse vulnerability and D2 receptor genes [see comments]. *Trends Neurosci* **16**:83–8.

Uhl G. R., Li S., Takahashi N. *et al.* (2000) The VMAT2 gene in mice and humans: amphetamine responses, locomotion, cardiac arrhythmias, aging, and vulnerability to dopaminergic toxins. *FASEB J* **14**:2459–65.

Uhl G. Sora I., Wang Z. (1999) The mu opiate receptor as a candidate gene for pain: polymorphisms, variations in expression, nociception, and opiate responses. *Proc Natl Acad Sci USA* **96**:7752–5.

Vaillant G. E. (1966) Parent-child cultural disparity and drug addiction. *J Nerv Ment Dis* **142**:534–9.

Valdes A. M., McWeeney S. K., Thomson G. (1999) Evidence for linkage and association to alcohol dependence on chromosome 19. *Genet Epidemiol* 17(Suppl 1):S367–72.

van den Bree M. B., Svikis D. S., Pickens R. W. (1998) Genetic influences in antisocial personality and drug use disorders. *Drug Alcohol Depend* **49**:177–87.

Vandenbergh D. J., Rodriguez L. A., Hivert E. *et al.* (2000) Long forms of the dopamine receptor (DRD4) gene VNTR are more prevalent in substance abusers: no interaction with functional alleles of the catechol-o-methyltransferase (COMT) gene. *Am J Med Genet* **96**:678–83.

Vekovischeva O. Y., Zamanillo D., Echenko O. *et al.* (2001) Morphine-induced dependence and sensitization are altered in mice deficient in AMPA-type glutamate receptor-A subunits. *J Neurosci* **21**:4451–9.

von Knorring A. L., Cloninger C. R., Bohman M. *et al.* (1983) An adoption study of depressive disorders and substance abuse. *Arch Gen Psychiatry* **40**:943–50.

Walker A., Maher J., Coulthard M. *et al.* (2001) Drinking, in *Living in Britain: Results from the 2000/01 General Household Survey*. London.

Wang Y. M., Gainetdinov R. R., Fumagalli F. *et al.* (1997) Knockout of the vesicular monoamine transporter 2 gene results in neonatal death and supersensitivity to cocaine and amphetamine. *Neuron* **19**:1285–96.

Warner L. A., Kessler R. C., Hughes M. *et al.* (1995) Prevalence and correlates of drug use and dependence in the United States. Results from the National Comorbidity Survey. *Arch Gen Psychiatry* **52**:219–29.

Watson S. J., Benson J. A. Jr, Joy J. E. (2000) Marijuana and medicine: assessing the science base: a summary of the 1999 Institute of Medicine report. *Arch Gen Psychiatry* **57**:547–52.

Whitfield J. B., Nightingale B. N., Bucholz K. K. *et al.* (1998) ADH genotypes and alcohol use and dependence in Europeans. *Alcohol Clin Exp Res* **22**:1463–9.

Windemuth C., Hahn A., Strauch K. *et al.* (1999) Linkage analysis in alcohol dependence. *Genet Epidemiol* 17(Suppl 1):S403–7.

Xu F., Gainetdinov R. R., Wetsel W. C. *et al.* (2000a) Mice lacking the norepinephrine transporter are supersensitive to psychostimulants. *Nat Neurosci* 3:465–71.

Xu M., Guo Y., Vorhees C. V. *et al.* (2000b) Behavioral responses to cocaine and amphetamine administration in mice lacking the dopamine D1 receptor. *Brain Res* 852:198–207.

Xu M., Koeltzow T. E., Santiago G. T. *et al.* (1997) Dopamine D3 receptor mutant mice exhibit increased behavioral sensitivity to concurrent stimulation of D1 and D2 receptors. *Neuron* 19:837–48.

Yamauchi M., Maezawa Y., Mizuhara Y. *et al.* (1995) Polymorphisms in alcohol metabolizing enzyme genes and alcoholic cirrhosis in Japanese patients: a multivariate analysis [see comments]. *Hepatology* 22:1136–42.

Yang J., Wu J., Kowalska M. A. *et al.* (2000) Loss of signaling through the G protein, Gz, results in abnormal platelet activation and altered responses to psychoactive drugs. *Proc Natl Acad Sci USA* 97:9984–9.

Yates W. R., Cadoret R. J., Troughton E. *et al.* (1996) An adoption study of DSM-IIIR alcohol and drug dependence severity. *Drug Alcohol Depend* 41:9–15.

Yin S. J., Agarwal D. P. (2001) Functional polymorphism of ADH and ALDH, in D. P. Agarwal and H. K. Seitz (eds) *Alcohol in Health and Disease*. New York.

Yoshida A. (1992) Molecular genetics of human aldehyde dehydrogenase. *Pharmacogenetics* 2:139–47.

Yoshida A., Ikawa M., Hsu L. C. *et al.* (1985) Molecular abnormality and cDNA cloning of human aldehyde dehydrogenases. *Alcohol* 2:103–6.

Yuan H., Marazita M. L., Hill S. Y. (1996) Segregation analysis of alcoholism in high density families: a replication. *Am J Med Genet* 67:71–6.

Zachariou V., Caldarone B. J., Weathers-Lowin A. *et al.* (2001) Nicotine receptor inactivation decreases sensitivity to cocaine. *Neuropsychopharmacology* 24:576–89.

Chapter 12

Anxiety and eating disorders

Thalia C. Eley, David Collier, and Peter McGuffin

Symbols

GAD: Generalised Anxiety Disorder

PD: Panic Disorder

OCD: Obsessive-Compulsive Disorder

PTSD: Post-Traumatic Stress Disorder

AN: Anorexia Nervosa

BN: Bulimia Nervosa

MD: Major Depression

5-HT: Serotonin

D: Dopamine

MAO-A: Monoamine Oxidase-A

COMT: Catechol-O-Methyl Transferase

CRH: Corticotrophin Releasing Hormone

HPA axis: Hypothalamic-Pituitary-Adrenal axis

VNTR: Variable Number Tandem Repeat marker

Introduction

It has long been recognized that there is a familial component to both anxiety and eating disorders. In the past two decades research has moved on from family studies, looking at familial risk and resemblance to molecular genetic research attempting to identify specific susceptibility loci. Here we will review data from both quantitative genetic research (family and twin studies—we are not aware of any adoption studies of anxiety or eating disorders) and molecular genetics (linkage, association and animal models).

Prevalence

Anxiety disorders are the most prevalent psychiatric disorders in the western world today, and their existence has been recognized for centuries. As a group they have a lifetime prevalence of between 10 and 25 per cent (Kessler *et al.* 1994; Robins *et al.* 1988) and a six-month prevalence of between 6 and 15 per cent (Myers *et al.* 1984). The most common of the anxiety disorders is social phobia, which is experienced by 13 per cent of the population at some point during their lifetime, and all of the anxiety disorders are about twice as common in women as in men (Kessler *et al.* 1994). Eating disorders are rare in comparison. Anorexia Nervosa (AN) has been reported in the literature for over 300 years (Morton 1689) and was clearly defined as early as 1874 (Vandereycken and Van Deth 1994; Walsh and Devlin 1998), whereas Bulimia Nervosa (BN) was defined as an illness much more recently (Russell 1979). AN has a lifetime prevalence of less than 1 per cent, while the rate is around 2 per cent for BN (Wade *et al.* 1996).

Both disorders show a remarkable sex difference in prevalence, with women account-ing for between 90 and 98 per cent of all cases. Both the anxiety and eating disorders show a significant rise in prevalence during the adolescent years, with teenage girls showing particularly high levels of all these disorders.

Classification issues

Within the current classification systems there are several anxiety and eating disorders, which overlap in terms of their symptoms. There are two groups of features that are found in all the anxiety disorders, regardless of the classification system: physiological symptoms such as difficulty breathing, accelerated heart rate, sweating, dizziness and nausea; and psychological symptoms such as excessive worry and avoidance of feared situations or stimuli. The main anxiety disorders are Generalized Anxiety Disorder (GAD), Panic Disorder (PD), Phobias, Obsessive-Compulsive Disorder (OCD) and Post-Traumatic Stress Disorder (PTSD). GAD is characterized by extensive persistent worry that can lead to other problems such as difficulty concentrating or getting to sleep. PD consists of recurrent panic attacks not associated with any particular object or situation. Phobias are described as a marked fear and avoidance of an object or situation—in agoraphobia the feared situation is public places or crowds. OCD is characterized by two main sets of symptoms: obsessions (persistent, intrusive, senseless thoughts) and compulsions (repetitive, intentional but unnecessary behaviours or rituals). PTSD is diagnosed following a severe traumatic event in which the individual was exposed to an event involving actual or threatened death or serious injury. The disorder is marked by the presence of three sets of persisting symptoms following the trauma: repeated re-experiencing of the event, continuing avoidance of event-related stimuli and hyper-vigilance or arousal (e.g. inability to sleep or to concentrate).

The eating disorders share some features with the anxiety disorders, in that they also include excessive fear or worry, in this case relating to food and weight gain. The key fea-tures of AN are maintenance of body weight below normal level, intense fear of gaining weight or becoming fat and amenorrhea in females. The predominant symptoms in BN are uncontrolled eating binges, compensatory behaviours (use of laxatives, self-induced vomiting and excessive exercise) and low self-esteem, which is linked to body image. Both groups can be divided into those who show compensatory behaviours (purging subtypes) and those who do not. In addition current classifications include a category of eating disorder not otherwise specified (EDNOS) which includes subjects presenting with a milder phenotype that is unstable over time.

As can be seen from these descriptions, the classification of these disorders, in com-mon with the majority of psychiatric disorders, is made on the basis of clinical presenta-tion with little or no reference to specific aetiological influences. The lack of diagnosis-specific biological features, in combination with overlapping symptomatology and high comorbidity, has resulted in a long debate over the extent to which these dis-orders are alternate manifestations of one underlying diathesis. Behavioural genetic studies have begun to explore to what extent genetic or environmental influences on more than one disorder account for their co-occurrence.

Current diagnostic definitions such as those from the American Psychiatric Association (1994) Diagnostic and Statistical Manual of Mental Disorders (DSM-IV) represent disorders categorically, while an alternative is a dimensional approach. Such an approach focuses not only on dimensions of symptoms (Watson *et al*. 1995), but also explores the hypothesis that shared aetiological influences on these disorders are mediated via personality traits such as neuroticism. The trait or personality perspective has therefore become central to the debate, particularly regarding the aetiology of the anxiety (see Chapter 4 on the genetics of personality and Chapter 8 on personality disorders). Childhood psychiatric disorders are also covered elsewhere (see Chapter 7) and so the studies reviewed in this chapter focus on adults and mainly upon categorical disorders as defined by recent DSM and ICD diagnostic criteria. The exception to this is where diagnostic data are scarce, in which case studies of relevant trait measures are included. The chapter is divided into two major sections: a review of quantitative genetic studies of anxiety and eating disorders, followed by a review of molecular genetic data. For each diagnosis, studies addressing issues of comorbidity are reviewed after the results of univariate analyses have been explored.

Quantitative genetic approaches

Generalized anxiety disorder: family studies

Tables 12.1 to 12.9 give prevalence rates and/or morbidity risks for first-degree relatives from family studies of adult probands diagnosed with DSM-III, DSM-IIIR or DSM-IV anxiety and eating disorders. The studies are presented in the tables for the prevalence of each disorder in the relatives of probands, and where applicable, those of controls. The descriptions of the probands are given in the column headed 'proband diagnosis'. Where more than one group of probands has been described, and prevalence rates of more than one disorder given, the study is entered into each relevant diagnostic table. For example, the first family study of GAD using (modified) DSM-III (American Psychiatric Association 1980) criteria included four groups of probands: GAD and Major Depression (MD), PD and MD, Agoraphobia and MD, and MD only (Leckman *et al*. 1983). The prevalence of GAD in the relatives of each of these groups was 9.1 per cent, 10.5 per cent, 5.2 per cent and 6.2 per cent respectively, compared with 4 per cent in the relatives of controls. The prevalence of PD in each of these groups is given in the table for Panic Disorder (Table 12.2), and the prevalence of Agoraphobia is given in Table 12.3.

As can be seen from Table 12.1, the prevalence (or age–corrected morbid risk) of GAD in first-degree relatives of individuals with GAD (with or without other disorders such as PD and MD) ranges from 9 per cent to 20 per cent. In contrast, rates (or morbid risk) in relatives of controls range from around 2 per cent to 4 per cent, giving a relative risk of between 2 and 5.

Table 12.1 Family studies of generalized anxiety disorder (GAD)

Study	Proband diagnosis	System	N	N proband relatives	Controls	N	N control relatives	Method	Affected % in proband relatives	Affected % in control relatives	Proband MR	Control MR	Relative risk
Leckman et al. 1983	GAD + MD	modified DSM-III	45	810	Non-psychiatric	82	521	FH + DI	9.1	4.0	—	—	—
	PD + MD		22						10.5		—		—
	Agoraphobia + MD		10						5.2		—		—
	MD only		56						6.2		—		—
Noyes et al. 1987	GAD without PD	DSM-III	20	123	Non-anxious	20	113	DI	19.5	3.5	—	—	—
	PD		40	241					5.4		—		—
	Agoraphobia		40	256					3.9		—		—
Mendlewicz et al. 1993	GAD	DSM-III	25	102	Non-psychiatric	25	130	DI	—	—	8.9	1.9	—
	PD		25	122					—		4.0		—
	MD		25	137					—		4.9		—
Goldstein 1994	PD	DSM-IIIR	30	141	Non psychiatric	45	255	FH + DI	2.9	0.8	3.3	0.8	5.9
	PD + MD		77	442					3.9		4.0		6.2
	Early-onset MD		41	209					1.9		7.8		2.3
Skre et al. 1994	Anxiety disorder	DSM-IIIR	33	76	—	—	—	DI	22	—	—	—	—
	Mood disorders		20	45					9		—		—
Lilenfeld et al. 1998	AN[A]	DSM-IIIR	26	93	No ED 'normal weight'[B]	44	190	FH + DI	18	3.2	—	—	3.1
	BN[A]		47	177					6.8		—		2.3

[A] Female probands

[B] Male and female relatives

A meta-analysis of family and twin studies of anxiety disorders using operational-ized criteria, direct interviews of relatives, systematic ascertainment of probands and relatives, blind assessment of relatives, and which included a comparison group estimated an odds ratio of 6.1 (2.5–14.9) for generalised anxiety disorder (Hettema, Neale, & Kendler 2001).

The high comorbidity between GAD and MD has led to family studies incorporating both disorders, allowing the exploration of the origins of their association. Extant family studies of this kind suggest shared familial factors for MD and GAD, as the prevalence of GAD in relatives of probands with MD (5–10 per cent) is also significantly higher than that seen in controls (Leckman *et al.* 1983; Mendlewicz *et al.* 1993; Skre *et al.* 1994).

Generalized anxiety disorder: twin studies

The first twin study of GAD diagnosed according to modern diagnostic criteria used a rather small twin sample (85 same sex pairs) (Torgersen 1985). Levels of GAD were equal in co-twins of monozygotic (MZ) and dizygotic (DZ) co-twins, suggesting little genetic influence. However, a subsequent study with a slightly larger sample found concordance rates for MZ pairs for an anxiety disorder without panic attacks to be 34 per cent, as compared to 17 per cent in DZ pairs (Torgersen 1990). Another study from around this time again found no evidence for genetic influence on GAD (Andrews *et al.* 1990) whilst a further small study (N = 81 pairs) found higher concordance for MZ (60.0 per cent) as compared to DZ (14.3 per cent) pairs (Skre, Onstad, Torgersen, *et al.* 1993). The discrepancies between these studies are likely to result at least in part from differing sample sizes, but also differing definitions of GAD, which range from 1-month to 6-month prevalence. The results from the Virginia Twin Registry may shed some light on these conflicting data. Initial reports from this study of adult females indicated that GAD (1-month and 6-month prevalence) is moderately familial (Kendler *et al.* 1992a). This familiality was explained by genetic factors for the 1-month GAD with or without PD (heritability around 30 per cent). Results were less clear for the 6-month GAD, for which a genetic factor explained within pair similarity only when the disorder was accompanied by PD. Finally, a recent report of males only from the Vietnam Era Twin Registry produced a heritability estimate of 38 per cent for 1-month GAD (Scherrer *et al.* 2000). Finally, a study of male pairs, female pairs and opposite-sex pairs produced heritability estimates ranging from 15–20 per cent (Hettema, Prescott, & Kendler 2001). The same genetic factors influenced both males and females, but there was also a significant influence of shared environment (25 per cent) for the females only. These latter two studies were included in the meta-analysis described above which estimated heritability of GAD at 32 per cent, with female-only shared environment of 17 per cent (Hettema *et al.* 2001).

Comorbidity analyses from the Virginia Twin Registry exploring the association between GAD and MD found the same model to provide the best fit for all definitions of GAD (Kendler *et al.* 1992b). The model consisted of a shared genetic influence that entirely accounted for the comorbidity of the disorders with environmental influences being disorder specific. This shared genetic diathesis for MD and GAD has subsequently been replicated in further samples (Roy *et al.* 1995).

Panic disorder: family studies

There have been over a dozen papers published on family study data regarding PD. Early direct interview studies using DSM-III criteria gave rates of around 15–25 per cent in the first-degree relatives of probands as compared to 3.5–5 per cent in the relatives of controls. The estimated morbidity risks indicate a relative risk of between 4 and 8. Family history studies generally and not surprisingly find lower rates in relatives of both probands and controls. More recent studies using DSM-IIIR criteria have produced prevalence rates and morbidity risks in the range 6–15 per cent for first-degree relatives of probands, and around 3 per cent in relatives of controls, leading to an estimated relative risk of between 2 and 5. Female relatives are generally found to be at greater risk, in line with their increased susceptibility to these disorders (Crowe *et al.* 1983; Harris *et al.* 1983; Maier *et al.* 1993; Skre *et al.* 1994). However, studies that have specifically tested for an influence of the sex of the *proband* on the prevalence of disorder in the relatives have found no significant effects (e.g. Hopper *et al.* 1987).

The comorbidity of PD with other disorders, most notably MD and GAD, has also attracted attention. Interestingly, the combination of MD and PD is associated with a marked increase in the prevalence of disorders in relatives, who are consistently more likely to show disorders such as PD, MD and alcoholism as relatives of those with MD alone (Leckman *et al.* 1983; Maier *et al.* 1988; Mannuzza *et al.* 1994). Furthermore, rates of PD in relatives of subjects with MD alone are low, especially compared to rates of PD within relatives of PD probands (Leckman *et al.* 1983; Maier *et al.* 1988; Mendlewicz *et al.* 1993).These data suggest that while there appear to be some shared familial influences on PD and MD, there are strong and significant familial influences on PD over and above those shared with MD. This is also supported by evidence that both PD and MD are more severe when comorbid with one another than when they appear singly (Pini *et al.* 1994). The relationship between PD and GAD appears to be due to factors outside the family. This is illustrated in by data showing the low prevalence (similar to that seen in the controls) of GAD (3–5 per cent) in relatives of subjects with pure PD, and the low prevalence of PD (3–4 per cent) in relatives of probands with pure GAD (Noyes *et al.* 1987). As such, the comorbidity of these disorders appears to be largely due to non-familial factors. Moreover, it appears that there are distinct familial influences that are specifically associated with PD and not with other anxiety or mood disorders.

Panic disorder: twin studies

Panic disorder shows a clear pattern of results, and from the earliest studies onwards concordance rates have been far greater in MZ as compared to DZ pairs, which indicates non-additive genetic influence. For example, in one small early study the MZ co-twins showed levels of anxiety disorders five times as high in the DZ co-twins for the PD/Agoraphobia/panic attack group (Torgersen 1985). Similarly, in the later study from the same author, for anxiety disorder with panic attacks, concordance levels were 22 per cent and 0 per cent for MZ and DZ pairs respectively (Torgersen 1990). A further small study produced an MZ concordance rate of 41.7 per cent, and a DZ concordance of 16.7 per cent (Skre *et al.*, 1993). A similar pattern of results has been seen in the larger studies also. For example, non-additive genetic factors accounted for 24–29 per cent of the variance in 'feelings of panic' with additive factors accounting for a further 13–34 per cent (Martin *et al.* 1988). However, these studies all utilized clinical ascertainment procedures, and it may be that their results were influenced by factors affecting treatment-seeking behaviour. It is therefore of interest that data from the Virginia Twin Register, which uses an epidemiological approach incorporating interview assessments of PD, did not confirm the findings of non-additive genetic variance (Kendler *et al.* 1993b). Genetic factors were found to account for around 30–40 per cent of the variance, and these appeared to be largely additive. In contrast, a much smaller population-based sample produced concordance rates of 73 per cent and 0 per cent for MZ and DZ pairs indicating non-additive genetic influence (Perna, Caldirola, and Arancio 1997), and findings from another large epidemiological sample suggested the heritability of PD to be 44 per cent with 21 per cent non-additive genetic variance specific to PD (Scherrer *et al.* 2000). The remaining 23 per cent of the genetic influences was additive and shared with GAD. Overall these data suggest that PD is highly heritable and likely to be influenced to some extent by non-additive genetic factors.

The large majority of behavioural genetic studies examining comorbidity of PD with other disorders have utilized family data. However, the Virginia Twin Registry has explored the comorbidity of PD with phobias, GAD, bulimia, MD and alcoholism (Kendler *et al.* 1995). Two major genetic factors were found, one largely influencing GAD (22 per cent) and MD (41 per cent), which had a lesser influence on PD (12 per cent), the other influencing PD (32 per cent), phobias (33 per cent), and bulimia (29 per cent). Genetic influences on alcoholism were largely disorder specific. The predominant genetic influence on PD was therefore distinct from that on GAD.

Panic disorder has also been reported to be strongly associated with joint laxity, with PD and agoraphobia sixteen times more common in patients with joint laxity than in controls (Bulbena *et al.* 1993; Martin-Santos *et al.* 1998). This observation led to the eventual identification of a gene that may contribute to the molecular basis of panic disorder (Gratacos *et al.* 2001 see below).

Phobias: family studies

As can be seen from Table 12.3, only one group have reported data on the prevalence of *agoraphobia* in the relatives of probands with pure agoraphobia, and morbidity risks are given as 9–12 per cent in relatives of probands as compared to 4 per cent in relatives of controls (Harris *et al.* 1983; Noyes *et al.* 1986, 1987). This familial transmission was

Table 12.2 Family studies of panic disorder (PD)

Study	Proband diagnosis	System	N	N proband relatives	Controls	N	N control relatives	Method	Affected % in proband relatives	Affected % in control relatives	Proband MR	Control MR	Relative risk
Harris et al. 1983	PD Agoraphobia	DSM-III	20 20	75–108* 79–117*	Non-anxious	20	68–95*	DI	— —	—	20.5 7.7	4.2	—
Crowe et al. 1983	PD	DSM-III	41	278	Non-anxious	41	262	DI	24.5+	5.0+	17.3	1.8	—
Leckman 1983	PD + MD Agoraphobia + MD GAD + MD MD only	modified DSM-III	22 10 45 56	810	Non-psychiatric	82	521	FH + DI	3.8 2.1 0.4 2.1	0.0	— — — —	—	— — — —
Noyes et al. 1986	PD Agoraphobia	DSM-III	40 40	241 256	Non-anxious	20	113	DI	14.9 7.0	3.5	17.3 8.3	4.2	—
Noyes et al. 1987	PD Agoraphobia GAD	DSM-III	40 40 20	241 256 123	Non-anxious	20	113	DI	14.9 7.0 4.1	3.5	—	—	—
Hopper et al. 1987	PD	DSM-III	117	236 parents 283 sibs	—	—	—	FH	14.4 9.2	—	—	—	—
Reich, 1988	PD Social Phobia	DSM-III	88 17	471 76	No DSM-III Axis 1 diagnosis	10	46	FH	9.3 1.3	0.0	—	—	—
Maier et al. 1988	PD PD + MD MD	DSM-IIIR	15 15 15	62 58 56	—	—	—	DI	15 18 11	—	—	—	—
Gruppo Italiano Disturbi d'Ansia et al. 1989	PD PD + Agoraphobia	DSM-III	60 84	206 296	Community survey data**	—	—	DI + FH	4.4 10.1	1.46	—	—	—
Maier et al. 1993	PD	DSM-IIIR	40	119	Non-psychiatric	80	218	DI	—	—	7.9	2.3	2.8

Study	Diagnosis	Criteria			Comparison group			Method					
Mendlewicz et al. 1993	PD GAD MD	DSM-III	25 25 25	122 102 137	Non-psychiatric	25	130	DI	— — —	—	13.2 3.3 1.9	0.9	—
Weissman 1993	PD PD + MD Early-onset MD	DSM-III PD and RDC MD	30 77 41	141 442 209	Non psychiatric	45	255	FH + DI	14.2 3.9	0.8 8.0	18.7 4.1	1.1 11.4	16.5 9.8 3.9
Goldstein 1994	PD PD + MD Early-onset MD	DSM-IIIR	30 77 41	141 442 209	Non psychiatric	45	255	FH + DI	14.2 7.5 2.4	1.2	16.4 9.3 2.6	1.8	8.8 5.7 2.1
Mannuzza et al. 1994	PD PD + MD	DSM-IIIR	72 54	193 152	Never psychiatrically ill	77	231	DI	9 13	3	— —	—	2.8–4.1
Skre et al. 1994	Anxiety disorder Mood disorders	DSM-IIIR	33 20	76 45	—	—	—	DI	13 2	3	— –	—	—
Battaglia 1995	PD PD + somatisation	DSM-IIIR	78 23	302.3* 94.3*	Surgical	76	290.5*	FH	— —	—	6.0 6.4	1.4	—
Fyer et al. 1995	PD + Agoraphobia Simple phobia Social phobia	DSM-IIIR	49 15 39	131 49 105	Never psychiatrically ill	77	231	DI	10 0 2	3	— — —	—	3.0 — —
Fyer et al. 1996	PD PD + Social phobia Social phobia	DSM-IIIR	58 21 39	164 56 105	Never psychiatrically ill	77	231	DI	10 9 2	3	— — —	—	— — —
Lilenfeld et al. 1998	AN[A] BN[A]	DSM-IIIR	26 47	93 177	No ED 'normal weight'[+B]	44	190	FH + DI	5.4 4	0.5	— —	—	— —

[A] Female probands
[B] Male and female relatives
*Age-corrected sample sizes
+ GAD or PD
** community data reported but details of sample not given

Table 12.3 Family studies of agoraphobia

Study	Proband diagnosis	System	N	N proband relatives	Controls	N	N control relatives	Method	Affected % in proband relatives	Affected % in control relatives	Proband MR	Control MR	Relative risk
Leckman et al. 1983	Agoraphobia + MD	modified DSM-III	10	810	Non-psychiatric	82	521	FH + DI	1.0‡	1.2‡	—	—	—
	GAD + MD		45						4.5‡		—		
	PD + MD		22						3.8‡		—		
	MD only		56						2.1‡		—		
Harris et al. 1983	Agoraphobia	DSM-III	20	79–117*	Non-anxious	20	68–95*	DI	—	—	8.6	4.2	—
	PD		20	75–108*							1.9		
Moran 1985	Agoraphobia	DSM-III	60	232	—	—	—	FH	12.5	—	—	—	—
Noyes et al. 1986	Agoraphobia	DSM-III	40	256	Non-anxious	20	113	DI	9.4	3.5	11.6	4.2	—
	PD		40	241					1.7		1.9		
Noyes et al. 1987	Agoraphobia	DSM-III	40	256	Non-anxious	20	113	DI	9.4	3.5	—	—	—
	GAD		20	123					3.3		—		
	PD		40	241					1.7		—		
Gruppo Italiano Disturbi d'Ansia 1989	PD	DSM-III	60	206	Community survey data**	—	—	DI + FH	0.0	Not given	—	—	—
	PD + Agoraphobia		84	296					6.4		—		
Goldstein 1994	PD	DSM-IIIR	30	141	Non psychiatric	45	255	FH + DI	2.8	0.4	4.5	0.4	8.6
	PD + MD		77	442					1.1		1.2		2.5
	Early-onset MD		41	209					1.0		0.5		1.3
Fyer et al. 1995	PD + Agoraphobia	DSM-IIIR	49	131	Never psychiatrically ill	77	231	DI	2	1.0	—	—	—
	Simple phobia		15	49					1		—		—
	Social phobia		39	105					0		—		—

*Age-corrected sample sizes
‡Any phobia
** community data reported but details of sample not given

not specific for agoraphobia, in that similar numbers of relatives presented with PD (7–8 per cent), although fewer presented with GAD (4–5 per cent). However, this lack of specificity is seen only in one direction. That is, the relatives of PD probands had significantly greater levels of PD but not of Agoraphobia (17–21 per cent versus 2 per cent respectively) (Harris *et al.* 1983; Noyes *et al.* 1986, 1987). In the more recent reports from this same study it was noted that probands and relatives with agoraphobia reported more severe disorders in terms of symptom severity, earlier onset, more frequent complications, and a less favourable outcome than the probands and relatives with PD. Finally, a further study compared rates of agoraphobia in relatives of a pure PD group and a PD with agoraphobia group (Gruppo Italiano Disturbi d'Ansia 1989). Surprisingly, they found no cases of agoraphobia in relatives of the pure PD group, as compared to 6.4 per cent in relatives of the combined group. The highest prevalence of Agoraphobia in relatives of probands (12.5 per cent) was found in a small study utilising a clinically ascertained sample with no comparison group, which makes it difficult to interpret this result (Moran and Andrews 1985). Overall, these findings suggest that agoraphobia is a more serious variant of PD. Many studies have reported increased risk of phobias in females relatives of probands, but the one group that explored influence of sex of the *proband* on frequency of agoraphobia and PD in the relatives of probands with agoraphobia, found no effect (Gruppo Italiano Disturbi d'Ansia 1989).

Family interview studies of *simple phobia* have provided evidence for a strong familial component. The life-time prevalence of simple phobia is high in the general population, and this is reflected in the rates seen in relatives of cases (around 20–30 per cent) and controls (around 10 per cent), suggesting a relative risk of 2 to 3 (Fyer *et al.* 1990, 1995). One study considered a dimensional perspective and found that high levels of irrational fears in the absence of a simple phobia do not transmit increased risk for any psychiatric disorder in the relatives (Fyer *et al.* 1990). The comorbidity of simple phobia with MD has also been studied. Simple phobia is as common in relatives of probands with MD as relatives of anxious probands, although the lack of normal controls makes it a little hard to fully interpret these data (Skre *et al.* 1994). However, they suggest that there may be as high a level of shared familial factors for simple phobia with MD as with the other anxiety disorders. In contrast, another study found rates of simple phobia to be as low in the relatives of probands with social phobia or with PD and agoraphobia as in the relatives of controls, suggesting some level of specificity of familial factors to simple phobia (Fyer *et al.* 1995).

Family interview studies of *social phobia* reveal prevalence rates of 15–23 per cent in relatives of cases as compared to 4–6 per cent in relatives of controls, indicating a relative risk of between 2 and 6, with the majority of data supporting a relative risk at the lower end of this range (Fyer *et al.* 1993, 1995, 1996; Stein *et al.* 1998). Considering comorbidity with other disorders, the prevalence of social phobia in relatives with mood disorder is similar to that seen in relatives with any anxiety disorder (Skre *et al.* 1994). Rates of social phobia in relatives of probands with PD are significantly lower than rates seen in relatives of probands with social phobia (Reich and Yates 1988; Fyer *et al.* 1995, 1996). This, in conjunction with data described above on the rates of social phobia in relatives of probands with simple phobia, suggest that the phobias tend to 'breed true'. However, it should be noted that in contrast, rates of social phobia in relatives of those with simple phobia are as high as rates in relatives with social phobia (Fyer *et al.* 1995). As such, there may be a threshold relationship between social phobia and simple phobia, in that simple phobia increases risk for both disorders, whereas social phobia increases risk of this phobia only. Further behavioural genetic studies will help to clarify this issue. Finally, one study has explored the comorbidity between PD and phobias, and found that familial influences on PD were largely distinct from those on the phobias, with the exception of social phobia (Pini *et al.* 1994).

Phobias: twin studies

The Virginia Twin Register study explored the genetics of phobias including agoraphobia, social phobia, situational phobia and simple phobia in females (Kendler *et al.* 1992c). It was found that there were both common and specific additive genetic and non-shared environment factors for Agoraphobia, Social Phobia and Simple Phobia, with heritability estimates ranging from 30 per cent to 40 per cent. In contrast, familial resemblance for Situational Phobia, and Blood/injury Phobias in this sample were largely due to the shared environment (Kendler *et al.* 1992; Neale *et al.* 1994). A subsequent analysis of comorbidity with MD suggested shared genetic influences between all the phobias and MD, with little shared environmental influence (Kendler *et al.* 1993a). The exception to this was situational phobia, for which there was significant shared environmental influence that also predicted variance in MD.

Further analyses from the same female sample corrected for unreliability of measurement by using data from two time-points 8 years apart, produced slightly higher heritability estimates ranging from 50 per cent for Situational and Social Phobia to 61 per cent for Agoraphobia (Kendler *et al.* 1999). Interestingly in this analysis Animal Phobias were not heritable, familiality being due to shared environmental influence (34 per cent). More recently, using a sample of *male* twin pairs from a population sample, demonstrated heritability estimates from 20–37 per cent for Agoraphobia, Social Phobia, Animal Phobia, Situational Phobia and Blood/injury Phobia. Many more studies of fears and phobias have utilised trait measures, and these have also produced heritablity estimates in the 30–50 per cent range (e.g. Rose *et al.* 1981).

As described earlier, the Virginia Twin Registry has explored the comorbidity of Phobias with PD, GAD, Bulimia, MD, and Alcoholism (Kendler *et al.*, 1995) using a sample of female twin pairs. Genetic influences on Phobias were shared with PD and Bulimia but not GAD, MD or alcoholism. Finally, a multivariate genetic analysis using data from a male sample indicated a large degree of genetic overlap for Agoraphobia, Social Phobia, Animal Phobia, Situational Phobia, and Blood/injury Phobia, with between 28 per cent and 85 per cent of the genetic variance on each being due to genetic factors shared across all 5 phobias (Kendler *et al.* 2001).

Obsessive-compulsive disorder: family studies

Early family studies of OCD tended not to use explicit diagnostic criteria or structured interviews of relatives. The data from these early studies varied considerably, and are summarized in a paper presenting the first direct interview family study of obsessive-compulsive disorder (OCD) in which relatives of both probands and controls were interviewed (Black *et al.* 1992). As can be seen from Table 12.6, OCD was no more common in relatives of the probands in this study as compared to the controls. However, morbid risk for subsyndromal OCD was significantly greater in the relatives of probands (18.0 per cent) as compared to controls (12.9 per cent). More recent studies have found greater levels of OCD in relatives of probands as compared to controls both for full OCD and subsyndromal OCD (Pauls *et al.* 1995; Nestadt *et al.* 2000). A recent publication also found greater relative risks for families of probands with early onset (Nestadt *et al.* 2000). There were no cases of OCD in relatives of probands whose age at onset was greater than 18 years. However, a previous study found the reverse finding—with late onset leading to higher familiality (Black *et al.* 1992), so it is unclear whether age of onset influences the familial loading of the disorder.

Comorbidity of OCD and MD was explored in a study that discriminated between MD preceding OCD, OCD preceding MD, OCD and MD with indistinguishable onset, and pure OCD (Sciuto *et al.* 1995). Morbidity risks for OCD in first-degree relatives were around 5 per cent for all groups, except the MD leading to OCD group in which there were no cases in first-degree relatives. This suggests that MD and OCD share familial factors, except for individuals for whom MD develops into OCD where there is a different aetiology. In an earlier study, prevalence of OCD was 5 per cent in relatives of probands with anxiety disorders and only 2 per cent in relatives of probands with mood disorders (Skre *et al.* 1994). This suggests increased sharing of familial factors between OCD and other anxiety disorders as compared to depressive disorders. This confirms findings from an earlier study in which anxiety disorders were significantly more common in the relatives of OCD probands as compared to controls (30.0 per cent versus 17.1 per cent respectively), suggesting a shared familial diathesis between OCD and the other anxiety disorders (Black *et al.* 1992). The effect was most strongly seen with GAD, which was diagnosed in 21.7 per cent of the relatives of the probands as compared to 12.4 per cent for the control group.

Finally, there is evidence that Tourette's syndrome and chronic tics are more common in relatives of OCD probands (4.6 per cent) than controls (1.0 per cent), suggesting a shared familial influence on the two types of disorder (Pauls *et al.* 1995). Furthermore, the morbidity risk for tics was greater in relatives of female probands, suggesting greater genetic or shared environment influence on this association in females.

Obsessive-compulsive disorder: twin studies

Twin study data on OCD are surprisingly sparse. There have been several studies limited to individual pairs or small numbers of MZ pairs (e.g. Cryan *et al.* 1992; Lewis *et al.* 1991; McGuffin and Mawson, 1980) which are likely to be affected by a wide variety of biases including treatment seeking, and an over-emphasis on MZ concordant pairs. There has been just one small study of clinically diagnosed OCD in which 15 MZ and 15 DZ pairs were ascertained via the Maudsley twin register in London (Carey and Gottesman 1981). This found that five MZ pairs (33 per cent) were concordant compared with only one DZ pair (7 per cent). Another study presented data on twin pairs, who were clinically ascertained for a study of Tourette's syndrome (Walkup *et al.* 1988). Of the 14 pairs in the study, seven included at least one member with OCD, and of these five were concordant pairs. Unfortunately is very difficult to interpret these results because of the absence of zygosity. However they do appear to provide further, albeit weak, support for familiality of OCD.

Faced with a paucity of data on clinically treated OCD, an alternative is to assume that obsessional illness is the extreme end of a continuum of 'normal' obsessionality. There has been one study of 419 pairs of volunteer twins (Clifford *et al.* 1984) that examined the heritability of obsessional traits and symptoms and the heritability of neuroticism (N) as measured by the Eysenck Personality Questionaire (Eysenck and Eysenck 1975). There was a highly significant correlation between obsessional symptoms and N scores. The heritability estimate for obsessional symptom scores was 44 per cent and for N it was 47 per cent, with no shared environmental influence for either. The authors speculated that genetic factors may contribute to OCD by influencing both obsessional personality traits and a more general neurotic tendency which, in combination, may manifest as obsessional disorder. This interesting proposal has yet to be tested using more recent analytic approaches involving bivariate structural equation models (see chapter 2).

Post-traumatic stress disorder: family studies

The literature on the familiality of PTSD is also rather sparse. The evidence suggests a degreee of familiality for the disorder that is shared with other anxiety disorders, and to a lesser extent, with depression. This was seen in the first family study of PTSD in which a family history approach was taken for relatives of probands with PTSD, GAD or MD (Davidson *et al.* 1985). Unfortunately the GAD and MD groups were retrospectively derived and not matched to the PTSD group, and there were no normal controls. There was a high level of psychopathology in the relatives of the PTSD group (66 per cent), with depression, anxiety and alcoholism being the most prevalent. This compared to levels of psychopathology of 79 per cent and 93 per cent in the relatives of the MD and GAD groups. PTSD was diagnosed in 6 per cent of the families of the PTSD group, but not in either of the other groups. Rates of depression and anxiety in the relatives of the PTSD group were more similar to those for the GAD group than the MD group.

Table 12.4 Family studies of simple phobia

Study	Proband diagnosis	System	N	N proband relatives	Controls	N	N control relatives	Method	Affected % in proband relatives	Affected % in control relatives	Proband MR	Control MR	Relative risk
Fyer et al. 1990	Simple phobia	DSM-IIIR	15	49	Never psychiatrically ill	38	119	DI	31	11	—	—	3.3
Goldstein 1994	PD	DSM-IIIR	30	141	Non psychiatric	45	255	FH + DI	10.6	3.9	10.7	4.1	2.0
	PD + MD		77	442					7.2		7.2		1.6
	Early-onset MD		41	209					6.7		8.1		1.1
Skre et al. 1994	Anxiety disorder	DSM-IIIR	33	76	—	—	—	DI	21	—	—	—	—
	Mood disorders		20	45					24		—		—
Fyer et al. 1995	Simple phobia	DSM-IIIR	15	49	Never psychiatrically ill	77	231	DI	31	9	—	—	3.9
	Social phobia		39	105					13		—		(1.8–8.1)
	PD + Agoraphobia		49	131					10		—		—
Lilenfeld et al. 1998	AN[A]	DSM-IIIR	26	93	No ED 'normal[B] weight'	44	190	FH + DI	16.1	6.3	—	—	—
	BN[A]		47	177					9		—		—

[A] Female probands
[B] Male and female relatives

Table 12.5 Family studies of social phobia

Study	Proband diagnosis	System	N	N proband relatives	Controls	N	N control relatives	Method	Affected % in proband relatives	Affected % in control relatives	Proband MR	Control MR	Relative risk
Reich et al. 1988	Social Phobia PD	DSM-III	17 88	76 471	No DSM-III Axis 1 diagnosis	10	46	FH	6.6 0.4	2.2	—	—	—
Fyer et al. 1993	Social phobia	DSM-IIIR	30	83	Never psychiatrically ill	77	231	DI	16	5	—	—	3.12
Goldstein 1994	PD PD + MD Early-onset MD	DSM-IIIR	30 77 41	141 442 209	Non psychiatric	45	255	FH + DI	5.7 1.6 1.9	1.6	5.9 1.6 2.0	1.7	5.4 1.5 1.0
Skre et al. 1994	Anxiety disorder Mood disorders	DSM-IIIR	33 20	76 45	—	—	—	DI	7 7	—	— —	—	—
Fyer et al. 1995	Social phobia PD + Agoraphobia Simple phobia	DSM-IIIR	39 49 15	105 131 49	Never psychiatrically ill	77	231	DI	15 8 19	6	—	—	2.4 (1.2–5.0)
Fyer et al. 1996	Social phobia PD PD + Social phobia	DSM-IIIR	39 58 21	105 164 56	Never psychiatrically ill	77	231	DI	15 9 4	6	—	—	—
Stein et al. 1998	Social phobia	DSM-IIIR	23	106	No social phobia	24	74	DI	23	4	—	—	—
Lilenfeld et al. 1998	AN[a] BN[a]	DSM-IIIR	26 47	93 177	No ED 'normal weight'[b]	44	190	FH + DI	17.2 6.2	4.2	—	—	2.1 1.5

[a] Female probands

[b] Male and female relatives

Table 12.6 Family studies of obsessive-compulsive disorder (OCD)

Study	Proband diagnosis	System	N	N proband relatives	Controls	N control relatives	Method	Affected % in proband relatives	Affected % in control relatives	Proband MR	Control MR	Relative risk
Insel 1983	OCD	DSM-III	27	54 (parents)	—	—	DI + FH	0				
Rasmussen 1986	OCD	DSM-III	44	88	—	—	FH	5.0	—			
Black et al. 1992	OCD	DSM-III	32	120	Non-psychiatric	129	DI	2.5 17.5*+	2.3 12.5*+	3 21	3 16	
Skre et al. 1994	Anxiety disorder Mood disorders	DSM-IIIR	33 20	76 45	—	—	DI	5 2	—	— —	—	
Sciuto et al. 1995	OCD MD → OCD OCD → MD OCD + MD***	DSM-IIIR	112 11 12 37	445 57 44 188	—	—	DI + FH			5.7 0.0 4.6 5.9		
Pauls et al. 1995	OCD	DSM-IIIR	100	466	Non-psychiatric	113	DI + FH			10.3 7.9*+	1.9 2.0*+	
Lilenfeld et al. 1998	AN^A BN^A	DSM-IIIR	26 47	93 177	No ED 'normal weight'^B	190	FH + DI	10.8 4	1.6	—	—	3.6 1.2
Nestadt et al. 2000	OCD	DSM-IV	80	343	Community sample	300	DI + FH	11.7 16.3*+	2.7 5.7*+			

A Female probands

B Male and female relatives

*** Concurrent onset of both disorders, whereas in the previous two groups one disorder clearly preceded the other

*+ subsyndromal OCD

More recently, a family study was conducted using a combined family history and direct interview approach, which provided mixed evidence for the familiality of PTSD (Davidson *et al.* 1998). This study not only used direct interviews where possible, but also ascertained rape trauma victims (females), in contrast to previous studies, the majority of which had used combat veterans (males). There were five groups: rape survivor probands with lifetime chronic PTSD, rape survivor probands without lifetime PTSD, MDD without any anxiety disorder, anxiety disorder without any depressive disorder and non-psychiatric controls. Within the subset for whom direct family interviews were conducted there was no significant difference in the rates of anxiety between any of the groups. However, considering the entire sample (including those for whom there was only family history data), rates of anxiety disorders were higher in the relatives of all four psychiatric groups than in the healthy controls (see Table 12.7). Depression was significantly more common in the relatives of the PTSD group than the healthy controls (relatives risk: 1.66), and this group also showed increased depression as compared to the relatives of the non-PTSD rape survivor group. The lack of increase in anxiety disorders in the relatives of the PTSD group as compared to the controls may be a result of the probands all being female. It is possible that there are sex differences in the familiality of these disorders, and that while relatives of male probands are at increased risk (as seen in the veteran studies), relatives of female probands are not.

One further study explored the familiality of PTSD across the generations using a sample of 209 Khmer adolescents and their parents (Sack *et al.* 1995). Data from this study are not included in the table, which presents studies with adult probands only. However, PTSD was significantly more common in adolescents who had a parent with PTSD than those who did not (odds ratio 4.9, as compared to an odds ratio of 2.8 for depression in these adolescents). Furthermore, point prevalences of PTSD rise from around 0.15 in youths with no parental PTSD, to 0.27 in youths with one parent with PTSD, to 0.41 when both parents have a lifetime diagnosis of PTSD. This suggests a familial aspect to the aetiology of PTSD.

Post-traumatic stress disorder: twin studies

There are also very few twin studies of PTSD. One of the first of these utilized a sample of 4042 male veteran twin pairs from the Vietnam era (True *et al.* 1993). Significant genetic influences were identified for three clusters of PTSD symptoms: re-experiencing the trauma (heritability from 13–30 per cent), avoidance (30–34 per cent), and arousal (28–32 per cent). Another study found evidence for a shared genetic liability to both GAD and PTSD, in that PTSD was more common in co-twins of GAD than in co-twins of normal controls, and the prevalence of PTSD was greater in the MZ than DZ co-twins (Skre *et al.* 1993). Interestingly the Vietnam Era Twin Registry was used to study exposure to trauma, including volunteering for service in Vietnam, actual service in south east Asia, and a composite index of 18 combat experiences (Lyons *et al.* 1993). All these variables were significantly influenced by genetic factors with heritability estimates ranging

Table 12.7 Family studies of post-traumatic stress disorder (PTSD)

Study	Proband diagnosis	System	N	N proband relatives	Controls	N	N control relatives	Method	Affected % in proband relatives	Affected % in control relatives	Proband MR	Control MR	Relative risk
Davidson et al. 1985	PTSD	DSM-III	36	72 parents 141 sibs	—	—	—	FH	8.3[†] 1.4[†]	—	—	—	
	GAD		19	36 parents 57 sibs					3.8[†] 3.2[†]				
	MD		13	26 parents 31 sibs					0.0[†] 3.5[†]				
Skre et al. 1994	Anxiety disorder	DSM-IIIR	33	76	—	—	—	DI	3	—	—	—	
	Mood disorders		20	45					2		—		
Davidson et al. 1998	Rape + PTSD	DSM-IIIR	56	182	Non-psychiatric	39	159	DI + FH	7.2[++]	2.5			
	Rape, no PTSD		25	98					6.1				
	MD		31	107					6.6				
	Anxiety disorder		20	93					9.7				
Lilenfeld et al. 1998	AN[A]	DSM-IIIR	26	93	No ED 'normal weight'[B]	44	190	FH + DI	4.3	1.1	—	—	—
	BN[A]		47	177					3.9				

[A] Female probands
[B] Male and female relatives
[†] GAD, panic or phobia
[++] Diagnoses of any anxiety disorder

from 35 to 47 per cent. This suggests that heritability of PTSD may to some extent be mediated by heritability of exposure to traumatic events, an association that is likely to be mediated itself by other factors such as personality.

Anorexia nervosa and bulimia nervosa: family studies

Family studies have demonstrated familial aggregation of eating disorders, and provide some insight into the relationship between anorexia and bulimia. An aetiological overlap between anorexia and bulimia has long been suspected, because of clinical accounts of binge eating in anorexic patients, continuities between the two seen in longitudinal studies, and commonality for personality traits, mood disorders and anxiety (reviewed in Strober *et al.* 2000). Strober *et al.* (2000) reported data from a controlled family study of eating disorders as well as reviewing previous family studies. In nine out of the 10 earlier family studies of anorexia, cases were found only among relatives of AN probands but not in control relatives. In five out of six studies that included BN, the frequency of BN was elevated in the relatives of AN probands, indicating that some aetiological factors may be shared. Likewise, BN was elevated in the relatives of BN probands in three out of four family studies, and cases of anorexia were found in the families of bulimic probands in two out of four studies.

In their own new family data, Strober *et al.* (2000) found anorexia to be rare in the relatives of control probands, and estimated relative risks for AN of 11.1 in female relatives of AN probands and 11.9 in female relatives of BN probands. More BN was present in the relatives of control probands, but was still significantly familial, with increased risk for BN of 4.2 for relatives of AN probands and 4.4 for BN probands. For both AN and BN probands, familial risk fell by half for partial syndromes such as eating disorders not otherwise specified (EDNOS). Lifetime eating disorders were assessed in another recent study of probands with BN, in which a strong familial aggregation was found, with 43 per cent of sisters and 26 per cent of mothers receiving a diagnosis, mostly EDNOS (Stein *et al.* 1999). Although rare in males, a family study of the first degree relatives of male patients with anorexia has also been performed (Strober *et al.* 2001) and found elevated rates of anorexia (6.1 per cent) but not bulimia (0 per cent) in female relatives.

Patients with AN or BN are also often comorbid for OCD. For example it has been estimated that 18 per cent of AN patients (Lennkh *et al.* 1998) and 33 per cent of BN patients (Fahy *et al.* 1993; Matsunaga *et al.* 1999) have a lifetime diagnosis of OCD. OCD patients also have high rates (11–13 per cent) of eating disorders (Rubin *et al.* 1992). A case-control study of OCD patients further supports the notion that eating disorders are part of a wider 'OCD spectrum', with a threefold elevation of AN and fourfold elevation of BN in OCD probands compared to controls (Bienvenu *et al.* 2000). Curiously, two studies both found elevated rates of anorexia in males as well as females, indicating that comorbidity for eating disorders in OCD patients is not sex-specific (Bienvenu *et al.* 2000; Rubin *et al.* 1992). A recent family study explored the origins of the increased rates of OCD seen in relatives of those with eating disorders (Cavallini *et al.* 2000). Relatives of 141 families with a proband affected with an eating disorder were interviewed and diagnoses of OCD made where appropriate. The authors interpret their data as supporting a Mendelian dominant model of transmission, which they suggest provides evidence for a common genetic susceptibility for both eating disorders and OCD. However, as has been noted elsewhere in this volume (see especially Chapter 2), one should interpret the results of segregation analyses of common complex traits with caution and, a priori, a mode of transmission involving multiple genes is more likely than single gene inheritance (Plomin *et al.* 1994).

Anorexia nervosa: twin studies

As for OCD and PTSD, there are few twin studies of anorexia nervosa. Twin studies have been hampered by low power because of the comparative rarity of anorexia nervosa and the small sample size of most studies. There have been various studies of one or a handful of twin pairs (see Bulik *et al.* 2000) but only one systematic study of clinically ascertained twins. This demonstrated higher MZ than DZ concordance (Treasure and Holland 1989). Bulik *et al.* (2000) fitted an ACE model to these data, assuming prevalence of 0.75 per cent, and found evidence for an additive genetic effect of 88 per cent, with non-shared environment making up the remaining 12 per cent.

As with other low frequency disorders, an alternative approach to focusing on a clinically ascertained sample is to study population-based samples and either take a dimensional view of anorexic type symptoms or attempt to find a subset of twins who fulfil diagnostic criteria. One such study presented data from 147 female MZ twin pairs and 99 same sex female DZ pairs from a volunteer twin register (Rutherford *et al.* 1993b). The subjects completed two questionnaires, the Eating Attitudes Test (EAT) and the Eating Disorder Inventory (EDI). Total EAT scores, which are said to reflect overall susceptibility to eating disorders, had a heritability of 41 per cent. A dieting scale derived by factor analysis had a heritability of 42 per cent while 'body dissatisfaction' and 'drive for thinness' scales of the EDI had heriabilities of 52 and 44 per cent respectively. None of these scales showed significant shared environmental effects.

Population studies that have taken a categorical, disease focused approach have provided a more contradictory picture, with concordance for broadly defined AN being 10 per cent in MZ twins and 22 per cent in DZ twins from the Virginia twin register (Walters and Kendler 1995). However the authors concluded that the sample size was too small to draw clear conclusions, since only a few cases of AN were found in the sample. Subsequently two further twin studies of AN have been published. A more powerful twin study using 34142 Danish twins assessed with a self-report measure (Kortegaard *et al.* 2001) found heritability of 48 per cent for narrowly defined AN and 52 per cent for broadly defined AN. A study of 672 twins in Michigan, USA using both self-report measures and a structured instrument, found heritability to be 74 per cent and the effect of non-shared environment to be 26 per cent (Klump *et al.* 2001). A bivariate twin analysis

Table 12.8 Family studies of anorexia nervosa

Study	Proband diagnosis	System	N	N proband relatives	Controls	N control relatives	Method	Affected % in proband relatives	Affected % in control relatives	Proband MR	Control MR	Relative risk
Hudson et al. 1987												
Logue et al. 1989			24									
Kassett et al. 1989	BN		40									
Strober et al. 1990	AN	DSM-IIIR	97	387	Affective and Non-affective psychiatric disorders	469	DI + FH	4.1[C]	0[A,C] 0[B,C]	—	—	—
Stein et al. 1998	BN	DSM-IIIR	47	177	No history ED	190	DI + FH	2.2[C] 21[C,D] 33.7[C,E]	0[C] 0[C,D] 3.5[C,E]	—	—	
Lilenfeld et al. 1998	BN[A] AN[A]	DSM-IV	47	93	No history ED, 'normal weight'[B]	177	DI + FH	0.6 1.1	0	—	—	—
Strober et al. 2000	BN	DSM-IV	152	290	Never psychiatrically ill	318	DI + FH	3.4 3.4[F]	0.3 0.6[F]			11.3 5.2[F]
Bienvenu et al. 2000	OCD	DSM-IV	80	343	Community sample	300	DI + FH	9	3			
Strober et al. 2001	AN	DSM-IV	29[G]	49[C]	Never psychiatrically ill	318	DI + FH	6.1	0.3	—		20.3

[A] Female probands
[B] Male and female relatives
[C] Female first degree relatives only
[D] EDNOS
[E] Any eating disorder
[F] Partial AN
[G] Male probands only

Table 12.9 Family studies of Bulimia nervosa

Study	Proband diagnosis	System	N	N proband relatives	Controls	N	N control relatives	Method	Affected % in proband relatives	Affected % in control relatives	Proband MR	Control MR	Relative risk
Hudson et al. 1987	BN	DSM-III	69	283	Relatives of rheumatoid arthritis patients with no history ED	28	149	DI + FH	2.2	0	—	—	—
Kassett et al. 1989	BN	DSM-IIIR	40	185	No Psychiatric disorder	24	118	DI + FH	9.6	3.5	—	—	—
Halmi et al. 1991	AN	DSM-IIIR	54	169	No history ED, 'normal weight'	62	178	DI + FH	1.2	0	—	—	—
Strober et al. 1990	AN	DSM-IIIR	97	387	Affective[H] and Non-affective[I] psychiatric disorders	117	469	DI + FH	2.6[C]	0.7[A C] 1.3[B C]	—	—	—
Stein et al. 1998	BN	DSM-IIIR	47	177	No history ED	44	190	DI + FH	4.5[C] 21[D] 33.7[E]	0[C] 0[D] 3.5[E]	—	—	—
Lilenfeld et al. 1998	BN[A] AN[A]	DSM-IV	47	93	No ED, 'normal weight'[B]	190	177	DI + FH	0.6 1.1	0	—	—	—
Strober et al. 2000	BN	DSM-IV	171	297	Never psychiatrically ill	181	318	DI + FH	3.7 3.4[G]	0.3 0.6[G]	—	—	3.5 2.4[G]
Bienvenu et al. 2000	AN OCD	DSM-IV	80	343	Community sample	73	300	DI + FH	4	1	—	—	—

[A] Female probands; [B] Male and female relatives; [C] Female first degree relatives only; [D] EDNOS; [E] Any eating disorder; [G] partial BN; [H] Primary major affective disorder controls; [I] Mixed psychiatric disorder controls of AN and major depression has also been performed in which heritability of AN was estimated to be 58 per cent and liability was correlated with depression (Wade *et al.* 2000). Overall these studies indicate that AN is familial and has at least moderate and perhaps fairly high heritability, but there is no apparent role for shared environment.

Bulimia: twin studies

As BN is more common than AN, with a population prevalence estimated to be 2.5 per cent, twins studies have more power to study BN when it is defined as a present/absent disorder. Bulik *et al.* (2000) have re-analysed twin studies of BN, dividing them into clinical case series, univariate population-based studies and bivariate population-based studies. The clinical case studies (Fichter and Noegel 1990; Treasure *et al.* 1989) had the least power, with additive genetic effects accounting for 47 per cent of variance, 30 per cent shared and 23 per cent non-shared environment. Estimates were similar for univariate twin studies, with estimates ranging from 31–54 per cent for additive genetic effects but no role for shared environment. In all of these studies confidence intervals for additive genetic effects were large, and models with heritabilities set to zero could not be excluded. Furthermore, one study, taking a dimensional measure to bulimic behaviour, found little difference between the MZ and DZ intraclass correlations and their model fitting suggested that the familial component could be satisfactorily explained by shared environment and no genetic effects (Rutherford *et al.* 1993a). However Bulik *et al.* (2000) have noted that the use of measurements on more than one occasion and multivariate analysis provides reflecting greater statistical power and narrower confidence intervals. Such studies are few but result in estimates of additive genetic effects ranging from 28 per cent to 83 per cent.

The core features of bulimia, recurrent episodes of binge eating, compensatory behaviour such as self-induced vomiting and undue influence of self-evaluation by body shape and weight, have also been analysed in twin studies. Both binge eating (54 per cent) and vomiting (70 per cent) appear to have a substantial additive genetic contribution, although again a role for shared environment cannot be eliminated (Sullivan *et al.* 1998).

Finally, just one study has explored sources of comorbidity of BN with other disorders and found shared genetic influence on BN, PD and Phobias which were distinct from genetic influences on GAD, MD and alcoholism (Kendler *et al.* 1995).

Molecular genetic approaches

Several groups of candidate genes have been explored in relation to anxiety of eating disorders. These include genes involved serotonergic and dopaminergic tranmission and genes involved in the hypothalamic–pituitary–adrenal (HPA) axis. The 5-HT system contributes to variation in many physiological functions such as food intake, sleep, motor activity and reproductive activity in addition to emotional states including mood and anxiety, and is the target of uptake-inhibiting antidepressant and anti-anxiety drugs. The 5-HT transporter (5-HTT) plays a central role in neurotransmission of serotonin in the brain, and is encoded by a single gene (SLC6A4) on chromosome 17q12, making this gene a clear target for candidate gene studies (Lesch *et al.* 1996). Dopamine metabolism is implicated in the aetiology of childhood psychiatric disorders, notably attention deficit disorder, which has been associated with variation in the dopamine D4 gene (Collier *et al.* 2000). Furthermore, certain animal models of compulsive behaviour can be induced by dopaminergic drugs, suggesting a role for the dopaminergic system in OCD (Goodman *et al.* 1990). Both monoamine oxidase-A (MAO-A) and catchol-O-methyl transferase (COMT) are involved in the inactivation of serotonin, dopamine and other monoamine neurotransmitters. The COMT gene has been implicated in neuropsychological measures of executive function such as the Wisconsin Card Sorting Test. Corticotrophin releasing hormone (CRH) is central to the HPA system, and the co-regulation of CRH and adrenal and pituitary hormones has been linked to our ability to anticipate, adapt to or cope with impending future events, and resultant states of fear and anxiety (Schulkin *et al.* 1998). Furthermore, CRH influences the regulation of the immune system, sleep and appetite (Owens and Nemeroff 1991, in Schulkin *et al.* 1998).

Animal studies

The high similarity (80 per cent approx) between the mouse and human genomes means that quantitative trait loci (QTLs) found in mouse models of emotionality can be explored as candidate QTLs for human emotionality by tracing syntenic homologous regions (Crabbe *et al.* 1994). For example, three QTL regions were reported to be related to various measures of fearfulness, including open-field activity, in a study of crosses of two lines of mice selected for high and low fearfulness in the open field (Flint *et al.* 1995). Another QTL strategy uses recombinant inbred (RI) strains (Plomin and McClearn 1993) which has also indicated candidate QTLs for open-field fearfulness in mice (Phillips *et al.* 1995). Essentially this approach uses crosses between inbred strains that are effectively homozygous at all loci to produce new lines of animals. The first generation, or F1 crosses, are thus all heterozygous at all loci, but when these are mated the resultant F2 crosses show recombination of the original parental chromosomes. New inbred lines are then established from these. The attraction of maintaining RI strains is that each RI strain needs to be genotyped only once and subsequent researchers can simply look up the marker genotypes on a database. Finally animals in which a targeted disruption of a gene has been engineered, knockout mice, can be used to explore the role of particular genes.

Anxiety disorders: linkage studies

There have been very few genome screens for anxiety disorders using linkage, and those there are have concentrated on PD. One early study of PD found LOD scores described as suggestive of linkage at two locations (LOD = 1.75 for α-haptoglobin at 16q22, and LOD = 1.30 at 6p21.3) (Crowe *et al.* 1988). After accounting for sex effects the LOD score for α-haptoglobin rose to 2.27. However, a further study by the same group failed to replicate this finding (Logue *et al.* 1989). Other studies have found suggestive loci, but these are all as yet unconfirmed (e.g. Crowe *et al*. 2001; Knowles *et al*. 1998; Smoller *et al*. 2001). The association between panic/agoraphobia and joint laxity has recently led exploration of an expanded phenotype of panic/agorophobia/social phobia/joint laxity (Gratacos *et al*. 2001). Linkage with an interstitial duplication of chromosome 15q24–q26 was reported with a LOD score of 5 for the full expanded phenotype, 3.82 for joint laxity alone, and > 3 for panic, agoraphobia and social phobia, in an affecteds-only analysis.

A possible molecular basis of panic disorder and agarophobia

The interstitial duplication of 15q24–26 (DUP25), was also found to be associated with PD, agoraphobia and social phobia and to occur in more than 90 per cent of patients compared to 7 per cent of control subjects (Gratacos *et al*. 2001). The duplication appears to occur in three different forms, a telomeric duplication in direct or inverted form, and a centromeric duplication in direct form, and in a few patients duplications of both telomeric and centromeric forms. There is also mosaicism in that the duplication is found in about 60 per cent of lymphocytes and 30–60 per cent of spermatazoa from DUP25 subjects. It appears to be highly unstable and inherited in a non-Mendelian manner—bizzarely, *de novo* duplications, reversion from duplication to non-duplication and conversion between forms of the duplication are all seen—and no linkage to markers on 15q24–26 is found in families in which the deletion is segregating with panic and phobic disorders. This is a surprising and highly unusual finding, which should be treated with caution until independently replicated. If true it could be that a locus or loci elsewhere influences occurrence of the duplication and extra copies of genes in the duplicated region, such as nicotinic acid receptors or the neutrophin-3 receptor, could result in an over sensitive fear and arousal system.

Anxiety disorders: association studies

Serotonin markers

The first major result in this area utilized a trait approach to anxiety, and considered the association between a functional polymorphism in the promoter region of the serotonin transporter (5-HTTLPR) and anxiety-related personality traits, including neuroticism and harm avoidance (Lesch *et al*. 1996). The 5-HTTLPR has a common 44 bp insertion/deletion polymorphism. The short-form allele, with a population frequency of 43 per cent, reduces transcriptional efficiency of the promoter and results in decreased serotonin transporter expression. In two samples, individuals with one or two copies of the short form allele had higher neuroticism, anxiety, and depression scores than individuals homozygous for the long form allele. In addition, a within-family analysis of sibling pairs discordant for the 5-HTTLPR genotype confirmed the association in that the sibling with the short form allele had higher neuroticism scores than their siblings homozygous for the long form allele. 5-HTTLPR explained about 3–4 per cent of the total variance in these traits. This finding has been the subject of considerable attempts at replication, some of which have been positive (e.g. Katsuragi *et al.* 1999; Osher *et al.* 2000), others have not (Ball *et al.* 1997; Jorm *et al.* 1998), although many of the samples used for replication purposes have had rather small sample sizes. However, the marker has continued to be found to be associated with a number of internalizing symptoms and disorders, including both unipolar and bipolar depression (Collier *et al.* 1996), obsessive-compulsive disorder (Bengel *et al.* 1999; McDougle *et al.* 1998) and seasonal affective disorder (Rosenthal *et al.* 1998). However, it should be noted that in the two studies of OCD, it was those with a homozygous long form genotype that were at risk for OCD. Furthermore, it is interesting to note the lack of association between 5-HTTLPR and panic disorder in a reasonable sample size (Deckert *et al.* 1997). As described above, PD may be genetically distinct from other anxiety or mood disorders, suggesting that while the role of this gene may be general to anxiety disorders it does not influence PD.

There is also a VNTR in the second intron of the 5-HTT gene, the 9 repeat allele of which was associated with MD (Ogilvie *et al.* 1996). Attempts to replicate this result were largely unsuccessful (Battersby *et al.* 1996; Collier *et al.* 1996; Kunugi *et al.* 1996), though a Japanese replication found a significant association between the 5-HTT VNTR and anxiety disorders (GAD, PD, OCD, and phobias), with the result being particularly strong for GAD and OCD (Ohara *et al.* 1999). However, it should be noted that the association was with a different allele (12 repeats) from that in the initial study.

There is preliminary evidence suggesting association and linkage between a polymorphism in the promoter region of the SLC6A4 gene, which encodes for serotonin transporter protein and OCD (McDougle *et al.* 1998). Positive linkage disequilibrium has also been reported between a silent G to C substitution polymorphism in the coding region of the 5-HT$_{1\beta}$ receptor gene in a TDT analysis of parents and siblings of probands with OCD (Mundo *et al.* 2000). The G allele was preferentially transmitted to the affected subjects.

Dopamine markers

Following the initial report of an association between the 48 bp repeat allele in the dopamine receptor D4 (DRD4) and novelty-seeking, in which increased levels of the 7-repeat allele were associated with high scores for novelty-seeking, several replications and extensions were attempted (Benjamin *et al.* 1996; Ebstein *et al.* 1996). A study of OCD probands found the 7-repeat allele to be *more* common in probands with OCD than controls, and was significantly more common in OCD probands with tics than those without (Cruz *et al.* 1997). This is counter-intuitive, given that subjects with OCD might be predicted to have lower than average levels of novelty-seeking. A subsequent association study of several dopamine system genes and OCD using over 100 probands and matched controls reported a positive association with DRD4 (Billett *et al.* 1998). However, in this study the 7 repeat allele did not show group differences—subjects with OCD were significantly more likely to possess the 4 repeat allele than the controls (67 per cent versus 56 per cent), and significantly less likely to possess the 2 repeat allele. Furthermore, the prevalence of the 2, 4 genotype was almost three times lower in the OCD probands than the healthy controls.

There are very few molecular genetic studies of PTSD. One study explored the role of the Dopamine D2 receptor (DRD2) as this had previously been shown to play a role in ADHD, Tourette's syndrome, conduct disorder and substance abuse, each of which shares symptoms with PTSD—notably hyper-arousal and reactivity to stress (Comings *et al.* 1996). The study utilized veterans exposed to severe combat, ascertained through an addiction clinic. Unfortunately this meant that all the subjects also had comorbid alcohol or drug dependence. However, the study, which included a replication sample showed significantly higher rates of the A1 allele in those subjects diagnosed with PTSD as compared to those found not to show the disorder (58 per cent versus 13 per cent in the original sample and 62 per cent versus 0 per cent in the replication group). The lack of the A1 allele in the non-PTSD group is almost as informative as the high proportion of those with PTSD who do have the A1 allele, given the high level of stress to which all subjects had been exposed. The authors interpreted this result as demonstrating the absence of the DRD2 A1 allele to be a protective factor against response to extreme stress.

COMT

There is a common functional allele that results in a three- to fourfold increase in enzyme activity in the synaptic cleft enzyme catechol-ortho-methyl-transferase (COMT) (Lotta *et al.* 1995) which has been a particular focus in studies of OCD. In one study, the low activity allele was significantly more common in probands than controls, although the effect is only seen for homozygous genotypes, i.e. the effect appears to be recessive (Karayiorgou *et al.* 1997). This finding was particularly strong in males, and the disease by sex interaction was highly significant. A subsequent within-family study from the same group replicated the sexually dimorphic association, in that the low activity allele was preferentially transmitted to the OCD probands (Karayiorgou *et al.* 1999). The study extended this finding to incorporate a G to T substitution polymorphism in exon 8 of the MAOA gene. The G allele has been previously linked with high MAO-A activity, and was also preferentially transmitted. This association was particularly marked amongst males with comorbid MD, which made up over 50 per cent of the sample. An alternate single base pair substitution polymorphism in MAO-A that also influences levels of enzyme activity was examined in an independent case-control study of OCD (Camarena *et al.* 1998). In this study the low activity allele was significantly more common amongst the *female* probands. However, it should be noted that the sample size in this study was rather small (N = 41 probands). In contrast, no association was found with any anxiety disorders (GAD, PD, OCD, Phobias) in a Japanese population of 108 probands and 135 controls (Ohara *et al.* 1998), suggesting further studies are required to clarify this result.

Other markers

Another interesting recent finding is that a mutation in the promoter region of the cholecystokinin (CCK) gene was significantly more prevalent in individuals with panic disorder than in controls (Wang *et al.* 1998). CCK is an important candidate because CCK4 induces panic attacks when injected into healthy individuals (Harro *et al.* 1993).

Anxiety disorders: animal studies

Two studies recently showed elevated levels of anxiety-like behaviours in mice lacking a functional 5-HT$_{1A}$ receptor (Heisler *et al.* 1998; Parks *et al.* 1998). Homozygous mutants displayed high levels of anxiety in a number of different paradigms including open field, elevated-zero maze, and forced swim stress tests.

Decreased anxiety and impaired response to stress have been demonstrated in mice with a modified glucocorticoid receptor gene, resulting in specific lack of glucocorticoid receptor function in the nervous system impairing function in HPA-axis regulation (Tronche *et al.* 1999). Similarly, CRF receptor 1-deficient mice display decreased anxiety using the open field, and impaired stress response following disruption of the HPA axis (Contarino *et al.* 1999; Smith *et al.* 1998). Furthermore, CRH-binding protein deficient mice display increased anxiety (Karolyi *et al.* 1999). CRH-binding protein acts as a negative regulator of CRH, playing a role in CRH clearance or degradation. Finally, drug-induced anxiety in rats has been demonstrated to induce CCK gene expression in brain structures implicated in anxiety (Pratt and Brett 1995).

Anxiety has also been unexpectedly observed in mice after knockout of several genes. This suggests that anxiety may be even more a core feature of mammalian behaviour than previously imagined. Mice with deregulation of the opioid system (Konig *et al.* 1996), lack of angiotensin II receptors (Okuyama *et al.* 1999), and puromycin-sensitive aminopeptidase gene-deficiency (Osada *et al.* 1999); all demonstrate anxiety as one aspect of their behavioural phenotype. These systems will all require further characterization before it becomes clear whether they play an important role in clinically diagnosed anxiety disorders in humans. For a more detailed review of animal models of anxiety see Flint (2002).

Eating disorders: association studies

A variety of candidate genes have been examined in eating disorders, principally focusing on the serotonin system. Collier *et al.* (1997) reported association between a polymorphism in the promoter of the 5-HT2A gene and AN (-1348G/A), which was replicated in AN and control populations from the USA and Italy (Enoch *et al.* 1998; Sorbi *et al.* 1998) but not in UK and German subjects (Campbell *et al.* 1998; Hinney *et al.* 1997). A subsequent meta-analysis (Ziegler *et al.* 1999) showed no overall statistically significant association with AN and the 5-HT2A gene. Studies in both BN and AN found no evidence for involvement of polymorphisms in the 5-HT2C gene (Burnet *et al.* 1999; Lentes *et al.* 1997), including a Cys23Ser variant (Lappalainen *et al.* 1995) which may have functional consequences for the receptor (Lappalainen *et al.* 1999; Quested *et al.* 1999).

Eating disorders: animal studies

Although there are no ideal animal models of eating disorders, mouse genetics and animal breeding experiments have provided two useful models which might provide insight into the aetiology of eating disoders, the anorexia (*anx*) mouse and thin sow syndrome. There are also genetic models of hypophagia and leanness, such as the transgenic mice lacking melanin concentrating hormone (MCH).

The *anx* mouse is an autosomal recessive lethal mutation, characterized by an emaciated appearance and neurological symptoms (Johansen and Schalling 2001). Mice die because they fail to eat enough to survive. Neurobiological analysis of these mice reveals drastic and selective serotonergic hyperinnervation of the hypothalamus (Son *et al.* 1994) and altered processing of neuropeptide Y. The *anx* mutation and 24-hour food-deprivation of wild type mice result in differential effects on the expression of TH, NET, and 5HTT genes. Decreased 5HTT expression in the anx mouse is consistent with upregulation of serotonergic neurotransmission that may accompany 5HT hyperinnervation. The *anx* mutation has not yet been cloned.

A second animal model of anorexia is thin sow syndrome. Pigs bred for leanness can develop irreversible self-starvation and emaciation (Kyriakis 2001) and a similar syndrome can be seen in sheep and goats. The prevalence is from 6–30 per cent. Like human anorexia nervosa, thin sow syndrome appears to have a psychosocial aspect, with early maternal separation and mixing with unfamiliar animals acting as a trigger. There is also a genetic component, since some strains are more vulnerable. The illness may involve the serotonin system since 5HT2 antagonist drugs such as amperozide are reported to prevent and treat the symptoms (Kyriakis 2001).

Conclusions

Controlled family studies have shown that anxiety disorders and eating disorders are familial, and that this familiality is largely a result of additive genetic effects has been demonstrated by twin studies. However, environment, particularly individual specific environment, is also of major aetiological importance and heritability estimates are generally in the 30–50 per cent range. It is also apparent that anxiety and eating disorders are complex disorders, i.e. their aetiology is composed of the effects of multiple genes of small effect that co-act or interact with a variety environmental factors.

Family studies indicate genetic overlap between many of these disorders and other psychiatric illnesses, particularly depression, indicating that the same genes may influence several phenotypes (pleiotropy). Which disorder someone has will depend on the mix of susceptibility genes, and the interplay with the environment. Much research is now aimed at identifying the unique environmental factors that influence risk. However environmental risk factors implicated in anxiety and eating disorders are common in the population; the puzzle is why so few people develop these disorders. The challenge will be the integration of genetic and environmental information in order to determine how they co-act and interact in determining disease.

At present the molecular genetic data in this area is somewhat limited, and further exploration is required. However, it appears likely that as samples increase in size and methods progress, linkage and association between specific markers and anxiety and eating disorders will be identified.

References

American Psychiatric Association (1980) *Diagnostic and Statistical Manual of Mental Disorders*, 3rd edn. Washington, DC: American Psychiatric Association.

American Psychiatric Association (1994) *Diagnostic and Statistical Manual of Mental Disorders*, 4th edn. Washington, DC: American Psychiatric Association.

Andrews G., Stewart G., Allen R. *et al.* (1990) The genetics of six neurotic disorders: a twin study. *Journal of Affective Disorders* **19**:23–9.

Ball D. M., Hill L., Freeman B. *et al.* (1997) The serotonin transporter gene and peer-rated neuroticism. *Neuro Report* **8**:1301–4.

Battaglia M., Bernardeschi L., Politi E. *et al.* (1995) Comorbidity of Panic and Somatisation Disorder: A genetic-epidemiological approach. *Comprehensive Psychiatry* **36**:411–20.

Battersby S., Ogilvie A. D., Smith C. A. *et al.* (1996) Structure of a variable number tandem repeat of the serotonin transporter gene and association with affective disorder. *Psychiatric Genetics* **6**:177–81.

Bengel D., Greenberg B. D., Cora-Locatelli G. *et al.* (1999) Association of the serotonin transporter promoter regulatory region polymorphism and obsessive-compulsive disorder. *Molecular Psychiatry* **4**:463–6.

Benjamin J., Li L., Patterson C. *et al.* (1996) Population and familial association between the D4 dopamine receptor gene and measures of novelty seeking. *Nature Genetics* **12**:81–4.

Bienvenu O. J., Samuels J. F., Riddle M. A. *et al.* (2000) The relationship of obsessive-compulsive disorder to possible spectrum disorders: Results from a family study. *Biological Psychiatry* **48**:287–93.

Billett E. A., Richter M. A., Sam F. *et al.* (1998) Investigation of dopamine system genes in obsessive-compulsive disorder. *Psychiatric Genetics* **8**:163–9.

Black D. W., Noyes R. J., Goldstein R. B. *et al.* (1992) A family study of obsessive-compulsive disorder. *Archives of General Psychiatry* **49**:362–8.

Bulbena A., Duro J. C., Porta M. *et al.* (1993) Anxiety disorders in the joint hypermobility syndrome. *Psychiatry Research* **1**:59–68.

Bulik C. M., Sullivan P. F., Wade T. D. *et al.* (2000) Twin studies of eating disorders: A review. [Review] [86 refs]. *International Journal of Eating Disorders* **27**:1–20.

Burnet P. W., Smith K. A., Cowen P. J. *et al.* (1999) Allelic variation of the 5-HT2C receptor (HTR2C) in bulimia nervosa and binge eating disorder. *Psychiatric Genetics* **9**:101–4.

Camarena B., Cruz C. de, I. F. J., and Nicolini H. (1998) A higher frequency of a low activity-related allele of the MAO-A gene in females with obsessive-compulsive disorder. *Psychiatric Genetics* **8**:255–7.

Campbell D. A., Sundaramurthy D., Markham A. F. *et al.* (1998) Lack of association between 5-HT2A gene promoter polymorphism and susceptibility to anorexia nervosa. *Lancet* **351**:499.

Carey G. and Gottesman I. I. (1981) Twin and family studies of anxiety, phobic and obsessive disorders, in *Anxiety: New Research and Changing Concepts*. New York: Raven Press.

Cavallini M. C., Bertelli S., Chiapparino D. *et al.* (2000) Complex segregation analysis of obsessive-compulsive disorder in 141 families of eating disorder probands, with and without obsessive-compulsive disorder. *Am J Med Genet* **96**:384–91.

Clifford C. A., Murray R. M., and Fulker D. W. (1984) Genetic and environmental influences on obsessional traits and symptoms. *Psychological Medicine* **14**:791–800.

Collier D., Curran S., and Asherson P. (2000) Mission: Not impossible? Candidate gene studies in child psychiatric disorders. *Molecular Psychiatry* **5**:457–60.

Collier D. A. and Sham P. C. (1997) Catch me if you can: are catechol- and indoleamiine genes pleitropic QTLs for common mental disorders? *Molecular Psychiatry* **2**:181–3.

Collier D. A., Stöber G., Heils A. *et al.* (1996) A novel functional polymorphism within the promoter of the serotonin transporter gene: Possible role in susceptibility to affective disorders. *Molecular Psychiatry* **1**:453–60.

Comings D. E., Muhleman D., and Gysin R. (1996) Dopamine D2 receptor (DRD2) gene and susceptibility to post-traumatic stress disorder: A study and replication. *Biological Psychiatry* **40**:368–72.

Contarino A., Dellu F., Koob G. F. *et al.* (1999) Reduced anxiety-like and cognitive performance in mice lacking the corticotropin-releasing factor receptor 1. *Brain Research* **835**:1–9.

Crabbe J. C., Belknap J. K., and Buck K. J. (1994) Genetic animal models of alcohol and drug abuse. *Science* **264**:1715–23.

Crowe R. R., Goedken R., Samuelson S. *et al.* (2001) Genomewide survey of panic disorder. *Am J Med Genet* **105**:105–9.

Crowe R. R., Noyes R., Pauls D. L. *et al.* (1983) A family study of panic disorder. *Archives of General Psychiatry* **40**:1065–9.

Crowe R. R., Noyes R., Persico T. *et al.* (1988) Genetic studies of panic disorder and related conditions, in D. L. Dunnel, E. S. Gershon, J. E. Barrett (eds) *Relatives at Risk for Mental Disorders.* New York: Raven Press Ltd.

Cruz C., Camarena B., King N. *et al.* (1997) Increased prevalence of the seven-repeat variant of the dopamine D4 receptor gene in patients with obsessive-compulsive disorder with tics. *Neuroscience Letters* **231**:1–4.

Cryan E. M., Butcher G. J., and Webb M. G. (1992) Obsessive-compulsive disorder and paraphilia in a monozygotic twin pair. *British Journal of Psychiatry* **161**:694–8.

Davidson J., Swartz M., Storck M. *et al.* (1985) A diagnostic and family study of posttraumatic stress disorder. *American Journal of Psychiatry* **142**:90–3.

Davidson J. R., Tupler L. A., Wilson W. H. *et al.* (1998) A family study of chronic post-traumatic stress disorder following rape trauma. *Journal of Psychiatric Research* **32**:301–9.

Deckert J., Catalano M., Heils A. *et al.* (1997) Functional promoter polymorphism of the human serotonin transporter: lack of association with panic disorder. *Psychiatric Genetics* **7**:45–7.

Ebstein R., Novick O., Umansky R. *et al.* (1996) Dopamine D4 receptor (D4DR) exon III polymorphism associated with the human personality trait of novelty seeking. *Nature Genetics* **12**:78–80.

Enoch M., Kaye W., Rotondo A. *et al.* (1998) 5-HT2A promoter polymorphism ≠1438G/A, anorexia nervosa, and obsessive-compulsive disorder. *Lancet* **351**:1785–6.

Eysenck H. J. and Eysenck S. B. G. (1975) *Manual of the Eysenck Personality Questionnaire.* London, UK: Hodder and Stoughton.

Fahy T. A., Osacar A., and Marks I. M. (1993) History of eating disorders in female patients with obsessive-compulsive disorder. *International Journal of Eating Disorders* **14**:439–43.

Fichter M. M. and Noegel R. (1990) Concordance for bulimia nervosa in twins. *International Journal of Eating Disorders* **9**:255–63.

Flint J. (2002) Animal models of anxiety. In R. Plomin, J. C. DeFries, I. W. Craig *et al.* (eds), *Behavioral genetics in the postgenomic era* (pp. 425–41). Washington, D.C.: APA Books.

Flint J., Corley R., DeFries J. C. *et al.* (1995) A simple genetic basis for a complex psychological trait in laboratory mice. *Science* **269**:1432–5.

Fyer A. J., Mannuzza S., Chapman T. F. *et al.* (1993) A direct interview family study of social phobia. *Archives of General Psychiatry* **50**:286–93.

Fyer A. J., Mannuzza S., Chapman T. F. *et al.* (1996) Panic disorder and social phobia: Effects of comorbidity on familial transmission. *Anxiety* **2**:173–8.

Fyer A. J., Mannuzza S., Chapman T. F. *et al.* (1995) Specificity in familial aggregation of phobic disorders. *Archives of General Psychiatry* **52**:564–73.

Fyer A. J., Mannuzza S., Gallops M. S. *et al.* (1990) Familial transmission of simple phobias and fears. A preliminary report. *Archives of General Psychiatry* **47**:252–6.

Goldstein R. B., Weissman M. M., Adams P. B. *et al.* (1994) Psychiatric disorders in relatives of probands with panic disorder and/or major depression. *Archives of General Psychiatry* **51**:383–94.

Goodman W. K., McDougle C. J., Price L. H. *et al.* (1990) Beyond the serotonin hypothesis: A role for dopamine in some forms of obsessive compulsive disorder? *Journal of Clinical Psychiatry* **51**(Suppl):36–43.

Gratacos M., Nadal M., Martin-Santos R. *et al.* (2001) A polymorphic genomic duplication on human chromosome 15 is a susceptibility factor for panic and phobic disorders. *Cell* **106**:367–9.

Gruppo Italiano Disturbi d'Ansia (1989) Familial analysis of panic disorder and agoraphobia. *Journal of Affective Disorders* **17**:1–8.

Halmi K. A., Eckert E., Marchi P. *et al.* (1991) Comorbidity of psychiatric diagnoses in anorexia nervosa. *Archives of General Psychiatry* **48**:712–18.

Harris E. L., Noyes R. J., Crowe R. R. *et al.* (1983) Family study of agoraphobia. Report of a pilot study. *Archives of General Psychiatry* **40**:1061–4.

Harro J., Vasar E., and Bradwejn J. (1993) CCK in animal and human research on anxiety. *Trends Pharmacol Sci* **14**:244–9.

Heisler L. K., Chu H. M., Brennan T. J. *et al.* (1998) Elevated anxiety and antidepressant-like responses in serotonin 5-HT1A receptor mutant mice [see comments]. *Proceedings of the National Academy of Sciences of the United States of America* **95**:15049–54.

Hettema J. M., Neale M. C., and Kendler K. S. (2001) A review and meta-analysis of the genetic epidemiology of anxiety disorders. *American Journal of Psychiatry* **158**:1568–78.

Hettema J. M., Prescott C. A., and Kendler K. S. (2001) A population-based twin study of generalized anxiety disorder in men and women. *Journal of Nervous & Mental Disease* **189**:413–20.

Hinney A., Ziegler A., Nothen M. *et al.* (1997) 5-HT2A receptor gene polymorphisms, anorexia nervosa, and obesity. *Lancet* **1**:1324–5.

Hopper J. L., Judd F. K., Derrick P. L. *et al.* (1987) A family study of panic disorder. *Genetic Epidemiology* **4**:33–41.

Hudson J. I., Pope H. G. J., Jones J. M. *et al.* (1987) A controlled family history study of bulimia. *Psychological Medicine* **17**:883–90.

Insel T. R., Hoover C., and Murphy D. L. (1983) Parents of patients with obsessive-compulsive disorder. *Psychological Medicine* **13**:807–11.

Johansen J. and Schalling M. (2001) The anorexia mouse, in J. B. Owen, J. L. Treasure, D. A. Collier (eds) *Animal Models: Disorders of Eating Behaviour and Body Composition*. Dordrecht: Kluwer.

Jorm A. F., Henderson A. S., Jacomb P. A. *et al.* (1998) An association study of a functional polymorphism of the serotonin transporter gene with personality and psychiatric symptons. *Molecular Psychiatry* **3**:449–51.

Karayiorgou M., Altemus M., Galke B. L. *et al.* (1997) Genotype determining low catechol-O-methyltransferase activity as a risk factor for obsessive-compulsive disorder. *Proceedings of the National Academy of Sciences of the United States of America* **94**:4572–5.

Karayiorgou M., Sobin C., Blundell M. L. *et al.* (1999) Family-based association studies support a sexually dimorphic effect of COMT and MAOA on genetic susceptibility to obsessive-compulsive disorder. *Biological Psychiatry* **45**:1178–89.

Karolyi I. J., Burrows H. L., Ramesh T. M. *et al.* (1999) Altered anxiety and weight gain in corticotropin-releasing hormone-binding protein-deficient mice. *Proceedings of the National Academy of Sciences of the United States of America* **96**:11595–600.

Kassett J. A., Gershon E. S., Maxwell M. E. *et al.* (1989) Psychiatric disorders in the first-degree relatives of probands with bulimia nervosa. *American Journal of Psychiatry* **146**:1468–71.

Katsuragi S., Kunugi H., Sano A. *et al.* (1999) Association between serotonin transporter gene polymorphism and anxiety-related traits. *Biological Psychiatry* **45**:368–70.

Kendler K. S., Neale M. C., Kessler R. C. *et al.* (1992a) Generalized anxiety disorder in women. A population-based twin study. *Archives of General Psychiatry* **49**:267–72.

Kendler K. S., Neale M. C., Kessler R. C. *et al.* (1992b) Major depression and generalized anxiety disorder. Same genes, (partly) different environments? *Archives of General Psychiatry* **49**:716–22.

Kendler K. S., Neale M. C., Kessler R. C. *et al.* (1992c) The genetic epidemiology of phobias in women: The interrelationship of agoraphobia, social phobia, situational phobia, and simple phobia. *Archives of General Psychiatry* **49**:273–81.

Kendler K. S., Neale M. C., Kessler R. C. *et al.* (1993a) Major depression and phobias: The genetic and environmental sources of comorbidity. *Psychological Medicine* **23**:361–71.

Kendler K. S., Neale M. C., Kessler R. C. *et al.* (1993b) Panic disorder in women: A population-based twin study. *Psychological Medicine* **23**:397–406.

Kendler K. S., Walters E. E., Neale M. C. *et al.* (1995) The structure of the genetic and environmental risk factors for six major psychiatric disorders in women: Phobia, generalized anxiety disorder, panic disorder, bulimia, major depression, and alcoholism. *Archives of General Psychiatry* **52**:374–83.

Kendler K. S., Karkowski L. M., and Prescott C. A. (1999) Fears and phobias: Reliability and heritability. *Psychological Medicine* **29**:539–53.

Kendler K. S., Myers J., Prescott C. A. *et al.* (2001) The genetic epidemiology of irrational fears and phobias in men. *Archives of General Psychiatry* **58**:257–65.

Kessler R., McGonagle K. A., Zhao C. B. *et al.* (1994) Lifetime and 12-month prevalence of DSM-III-R psychiatric disorders in the United States: Results from the National Comorbidity Study. *Archives of General Psychiatry* **51**:8–19.

Klump K. L., Miller K. B., Keel P. K. *et al.* (2001) Genetic and environmental influences on anorexia nervosa syndromes in a population-based twin sample. *Psychological Medicine* **31**:737–40.

Knowles J. A., Fyer A. J., Vignal A. *et al.* (1998) Results of a genome-wide genetic screen for panic disorder. *Am J Med Genet* **81**:139–47.

Konig M., Zimmer A. M., Steiner H. *et al.* (1996) Pain responses, anxiety and aggression in mice deficient in pre-proenkephalin. *Nature* **383**:535–8.

Kortegaard L. S., Hoerder K., Joergensen J. *et al.* (2001) A preliminary population-based twin study of self-reported eating disorder. *Psychological Medicine* **31**:361–5.

Kunugi H., Tatsumi M., Sakai T. *et al.* (1996) Serotonin transporter gene polymorphism and affective disorder. *Lancet* **347**:731–3.

Kyriakis S. C. (2001) Anorexia-like wasting syndromes in pigs, in J. B. Owen, J. L. Treasure, D. A. Collier (eds) *Animal Models: Disorders of Eating Behaviour and Body Composition*. Dordrecht: Kluwer.

Lappalainen J., Long J. C., Virkkunen M. *et al.* (1999) HTR2C Cys23Ser polymorphism in relation to CSF monoamine metabolite concentrations and DSM-III-R psychiatric diagnoses. *Biological Psychiatry* **46**:821–6.

Lappalainen J., Zhang L., Dean M. *et al.* (1995) Identification, expression, and pharmacology of a Cys23-Ser23 substitution in the human 5-HT2c receptor gene (HTR2C) *Genomics* **20**(27):274–9.

Leckman J. F., Weissman M. M., Merikangas K. R. *et al.* (1983) Panic disorder and major depression. Increased risk of depression, alcoholism, panic, and phobic disorders in families of depressed probands with panic disorder. *Archives of General Psychiatry* **40**:1055–60.

Lennkh C., Strnad A., Bailer U. *et al.* (1998) Comorbidity of obsessive compulsive disorder in patients with eating disorders. *Eating and Weight Disorders* **3**:37–41.

Lentes K. U., Hinney A., Ziegler A. *et al.* (1997) Evaluation of a Cys23Ser mutation within the human 5-HT2C receptor gene: No evidence for an association of the mutant allele with obesity or underweight in children, adolescents and young adults. *Life Science* **61**:PL9–16.

Lesch K. P., Bengel D., Heils A. *et al.* (1996) Association of anxiety-related traits with a polymorphism in the serotonin transporter gene regulatory region. *Science* **274**:1527–31.

Lewis S. W., Chitkara B., and Reveley A. M. (1991) Obsessive-compulsive disorder and schizophrenia in three identical twin pairs. *Psychological Medicine* **21**:135–41.

Lilenfeld L. R., Kaye W. H., Greeno C. G. *et al.* (1998) A controlled family study of anorexia nervosa and bulimia nervosa: Psychiatric disorders in first-degree relatives and effects of proband comorbidity. *Archives of General Psychiatry* **55**:603–10.

Logue C. M., Crowe R. R., and Bean J. A. (1989) A family study of anorexia nervosa and bulimia. *Comprehensive Psychiatry* **30**:179–88.

Lotta T., Vidgren J., Tilgmann C. *et al.* (1995) Kinetics of human soluble and membrane-bound catechol O-methyltransferase: A revised mechanism and description of the thermolabile variant of the enzyme. *Biochemistry* **34**:4202–10.

Lyons M. J., Goldberg J., Eisen S. A. *et al.* (1993) Do genes influence exposure to trauma: A twin study of combat. *American Journal of Medical Genetics (Neuropsychiatric Genetics)* **48**:22–7.

Maier W., Buller R., and Hallmayer J. (1988) Comorbidity of panic disorder and major depression: Results from a family study, in I. Hand and H. U. Wittchen (eds) *Panic and Phobias*. Berlin Heidelberg: Springer-Verlag.

Maier W., Lichtermann D., Minges J. *et al.* (1993) A controlled family study in panic disorder. *Journal of Psychiatric Research* **27**:79–87.

Mannuzza S., Chapman T. F., Klein D. F. *et al.* (1994) Familial transmission of panic disorder: Effect of major depression comorbidity. *Anxiety* **1**:180–5.

Martin-Santos R., Bulbena A., Porta M. *et al.* (1998) Association between joint hypermobility syndrome and panic disorder. *American Journal of Psychiatry* **155**:1578–83.

Martin N. G., Jardine R., Andrews G. *et al.* (1988) Anxiety disorders and neuroticism: Are there genetic factors specific to panic? *Acta Psychiatrica Scandinavica* **77**:698–706.

Matsunaga H., Kiriike N., Iwasaki Y. *et al.* (1999) Clinical characteristics in patients with anorexia nervosa and obsessive-compulsive disorder. *Psychol Med* **29**:407–14.

McDougle C. J., Epperson C. N., Price L. H. *et al.* (1998) Evidence for linkage disequilibrium between serotonin transporter protein gene (SLC6A4) and obsessive compulsive disorder. *Molecular Psychiatry* **3**:270–3.

McGuffin P. and Mawson D. (1980) Obsessive-compulsive neurosis: Two identical twin pairs. *British Journal of Psychiatry* **137**:285–7.

Mendlewicz J., Papadimitriou G., and Wilmotte J. (1993) Family study of panic disorder: Comparison with generalized anxiety disorder, major depression, and normal subjects. *Psychiatric Genetics* **3**:73–8.

Moran C. and Andrews G. (1985) The familial occurrence of agoraphobia. *British Journal of Psychiatry* **146**:262–7.

Morton (1689) *Pathisologica—Or a Treatise of Consumption*. London: Smith and Walford.

Mundo E., Richter M. A., Sam F. *et al.* (2000) Is the 5-HT(1Dbeta) receptor gene implicated in the pathogenesis of obsessive-compulsive disorder? *American Journal of Psychiatry* **157**:1160–1.

Myers J. K., Weissman M. M., Tischler G. L. *et al.* (1984) Six-month prevalence of psychiatric disorders in three communities. *Arch Gen Psychiatry* **41**:959–67.

Neale M. C., Walters E. E., Eaves L. J., *et al.* (1994) Genetics of blood-injury fears and phobias: A population-based twin study. *American Journal of Medical Genetics (Neuropsychiatric Genetics)* **54**:326–34.

Nestadt G., Samuels J., Riddle M. *et al.* (2000) A family study of obsessive-compulsive disorder. *Archives of General Psychiatry* **57**:358–63.

Noyes R. J., Clarkson C., Crowe R. R. *et al.* (1987) A family study of generalized anxiety disorder. *American Journal of Psychiatry* **144**:1019–24.

Noyes R. J., Crowe R. R., Harris E. L. *et al.* (1986) Relationship between panic disorder and agoraphobia. A family study. *Archives of General Psychiatry* **43**:227–32.

Ogilvie A. D., Battersby S., Bubb V. J. *et al.* (1996) Polymorphism in serotonin transporter gene associated with susceptibility to major depression. *Lancet* **347**:731–3.

Ohara K., Nagai M., Suzuki Y. *et al.* (1998) No association between anxiety disorders and catechol-O-methyltransferase polymorphism. *Psychiatry Research* **80**:145–8.

Ohara K., Suzuki Y., Ochiai M. *et al.* (1999) A variable-number-tandem-repeat of the serotonin transporter gene and anxiety disorders. *Progress in Neuro-Psychopharmacology and Biological Psychiatry* **23**:55–65.

Okuyama S., Sakagawa T., Chaki S. *et al.* (1999) Anxiety-like behavior in mice lacking the angiotensin II type-2 receptor. *Brain Research* **821**:150–9.

Osada T., Ikegami S., Takiguchi-Hayashi K. *et al.* (1999) Increased anxiety and impaired pain response in puromycin-sensitive aminopeptidase gene-deficient mice obtained by a mouse gene-trap method. *Journal of Neuroscience* **19**:6068–78.

Osher Y., Hamer D., and Benjamin J. (2000) Association and linkage of anxiety-related traits with a functional polymorphism of the serotonin transporter gene regulatory region in Israeli sibling pairs. *Molecular Psychiatry* **5**:216–19.

Owens M. J. and Nemeroff C. B. (1991) Physiology and pharmacology of corticotropin-releasing factor. *Pharmacological Reviews* **43**:425–73.

Parks C. L., Robinson P. S., Sibille E. *et al.* (1998) Increased anxiety of mice lacking the serotonin1A receptor. *Proceedings of the National Academy of Sciences of the United States of America* **95**:10734–9.

Pauls D. L., Alsobrook J. P., Goodman W. *et al.* (1995) A family study of obsessive-compulsive disorder. *American Journal of Psychiatry* **152**:76–84.

Perna G., Caldirola D., and Arancio C. (1997) Panic attacks: A twin study. *Psychiatry Research* **66**:69–71.

Phillips T. J., Huson M., Gwiazdon C. *et al.* (1995) Effects of acute and repeated ethanol exposures on the locomotor activity of BXD recombinant inbred mice. *Alcoholism: Clinical and Experimental Research* **19**:269–78.

Pini S., Goldstein R. B., Wickramaratne P. J. *et al.* (1994) Phenomenology of panic disorder and major depression in a family study. *Journal of Affective Disorders* **30**:257–72.

Plomin R. and McClearn G. E. (1993) Quantitative trait loci (QTL) analyses and alcohol-related behaviors. *Behavior Genetics* **23**:197–211.

Plomin R., Owen M. J., and McGuffin P. (1994) The genetic basis of complex human behaviors. *Science* **264**:1733–9.

Pratt J. A. and Brett R. R. (1995) The benzodiazepine receptor inverse agonist FG 7142 induces cholecystokinin gene expression in rat brain. *Neuroscience Letters* **184**:197–200.

Quested D. J., Whale R., Sharpley A. L. *et al.* (1999) Allelic variation in the 5-HT2C receptor (HTR2C) and functional responses to the 5-HT2C receptor agonist, m-chlorophenylpiperazine. *Psychopharmacology (Berl)* **144**:306–7.

Rasmussen S. A. and Tsuang M. T. (1986) Clinical characteristics and family history in DSM-III obsessive-compulsive disorder. *American Journal of Psychiatry* **143**:317–22.

Reich J. and Yates W. (1988) Family history of psychiatric disorders in social phobia. *Comprehensive Psychiatry* **2**:72–5.

Robins L., Wing J. K., Wittchen H. U. *et al.* (1988) The Composite International Diagnostic Interview: An epidemiological instrument suitable for use in conjunction with different diagnostic systems and in different cultures. *Archives of General Psychiatry* **45**:1067–9.

Rose R. J., Miller J. Z., Pogue-Geile M. F. *et al.* (1981) Twin-family studies of common fears and phobias, in L. Gedda, P. Parisi, W. E. Nance (eds) *Twin Research 3: Intelligence, Personality and Development*. New York: Alan R. Liss Inc.

Rosenthal N., Mazzanti C., Barnett R. *et al.* (1998) Role of serotonin transporter promoter repeat length polymorphism (5-HTTLPR) in seasonality and seasonal affective disorder. *Molecular Psychiatry* **3**:175–7.

Roy M. A., Neale M. C., Pedersen N. L. *et al.* (1995) A twin study of generalized anxiety disorder and major depression. *Psychological Medicine* **25**:1037–49.

Rubin C., Rubenstein J. L., Stechler G. *et al.* (1992) Depressive affect in 'normal' adolescents: Relationship to life stress, family, and friends. *American Journal of Orthopsychiatry* **67**:430–41.

Russell G. (1979) Bulimia nervosa: an ominous variant of anorexia nervosa. *Psychological Medicine* **9**:429–48.

Rutherford J., McGuffin P., Katz R. J. *et al.* (1993) Genetic influences on eating attitudes in a normal female twin population. *Psychological Medicine* **23**:425–36.

Sack W. H., Clarke G. N., and Seeley J. (1995) Post-traumatic stress disorder across two generations of Cambodian refugees. *Journal of the American Academy of Child & Adolescent Psychiatry* **34**:1160–6.

Scherrer J. F., True W. R., Xian H. *et al.* (2000) Evidence for genetic influences common and specific to symptoms of generalized anxiety and panic. *Journal of Affective Disorders* **57**:25–35.

Schulkin J., Gold P. W., McEwen B. S. (1998) Induction of corticotropin-releasing hormone gene expression by glucocorticoids: implication for understanding the states of fear and anxiety and allostatic load. [Review] [150 refs]. *Psychoneuroendocrinology* **23**:219–43.

Sciuto G., Pasquale L., Bellodi L. (1995) Obsessive compulsive disorder and mood disorders: a family study. *Am J Med Genet* **60**:475–9.

Skre I., Onstad S., Edvardsen J. *et al.* (1994) A family study of anxiety disorders: familial transmission and relationship to mood disorder and psychoactive substance use disorder. *Acta Psychiatrica Scandinavica* **90**:366–74.

Skre I., Onstad S., Torgersen S. *et al.* (1993) A twin study of DSM-III-R anxiety disorders. *Acta Psychiatrica Scandinavica* **88**:85–92.

Smith G. W., Aubry J. M., Dellu F. *et al.* (1998) Corticotropin releasing factor receptor 1-deficient mice display decreased anxiety, impaired stress response, and aberrant neuroendocrine development. *Neuron* **20**:1093–102.

Smoller J. W., Acierno J. S. Jr, Rosenbaum J. F. *et al.* (2001) Targeted genome screen of panic disorder and anxiety disorder proneness using homology to murine QTL regions. *Am J Med Genet* **105**:195–206.

Son J. H., Baker H., Park D. H. *et al.* (1994) Drastic and selective hyperinnervation of central serotonergic neurons in a lethal neurodevelopmental mouse mutant, Anorexia (anx). *Brain Research, Molecular Brain Research* **25**:129–34.

Sorbi S., Nacmias B., Tedde A. *et al.* (1998) 5-HT$_{2A}$ promoter polymorphism in anorexia nervosa. *Lancet* **351**:1785.

Stein M. B., Chartier M. J., Hazen A. L. *et al.* (1998) A direct-interview family study of generalized social phobia. *American Journal of Psychiatry* **155**:90–7.

Strober M., Lampert C., Morrell W. *et al.* (1990) A controlled family study of anorexia nervosa: Evidence of familial aggregation and lack of shared transmission with affective disorders. *International Journal of Eating Disorders* **9**:239–53.

Strober M., Freeman R., Lampert C. *et al.* (2000) Controlled family study of anorexia nervosa and bulimia nervosa: Evidence of shared liability and transmission of partial syndromes. *American Journal of Psychiatry* **157**:393–401.

Strober M., Freeman R., Lampert C. *et al.* (2001) Males with anorexia nervosa: A controlled study of eating disorders in first-degree relatives. *International Journal of Eating Disorders* **29**:263–9.

Sullivan P. F., Bulik C. M., and Kendler K. S. (1998) Genetic epidemiology of binging and vomiting. *British Journal of Psychiatry* **173**:75–9.

Torgersen S. (1985) Hereditary differentiation of anxiety and affective neuroses. *British Journal of Psychiatry* **146**:530–4.

Torgersen S. (1990) Comorbidity of major depression and anxiety disorders in twin pairs [see comments]. *American Journal of Psychiatry* **147**:1199–202.

Treasure J. and Holland A. (1989) Genetic vulnerability to eating disorders: evidence from twin and family studies, in H. Remschmidt and M. Schmidt (eds) *Child and Youth Psychiatry: European Perspectives.* New York: Hogrefe and Huber.

Tronche F., Kellendonk C., Kretz O. *et al.* (1999) Disruption of the glucocorticoid receptor gene in the nervous system results in reduced anxiety. *Nature Genetics* **23**:99–103.

True W. R., Rice J., Eisen S. A. *et al.* (1993) A twin study of genetic and environmental contributions to liability for posttraumatic stress symptoms. *Archives of General Psychiatry* **50**:257–64.

Vandereycken W. and Van Deth R. (1994) *From Fasting Saints to Anorexic Girls.* New York: New York University Press.

Wade T., Heath A. C., Abraham S. *et al.* (1996) Assessing the prevalence of eating disorders in an Australian twin population. *Australian and New Zealand Journal of Psychiatry* **30**:845–51.

Wade T. D., Bulik C. M., Neale M. *et al.* (2000) Anorexia nervosa and major depression: Shared genetic and environmental risk factors. *American Journal of Psychiatry* **157**:469–71.

Walkup J. T., Leckman J. F., Price R. A. *et al.* (1988) The relationship between obsessive-compulsive disorder and Tourette's syndrome: A twin study. *Psychopharmacology Bulletin* **24**:375–9.

Walsh B. T. and Devlin M. J. (1998) Eating disorders: progress and problems. *Science* **280**:1387–90.

Walters E. E. and Kendler K. S. (1995) Anorexia nervosa and anorexic-like syndromes in a population-based female twin sample. *American Journal of Psychiatry* **152**:64–71.

Wang Z., Valdes J., Noyes R. *et al.* (1998) Possible association of a cholecystokinin promotor polymorphism (CCK-36CT) with panic disorder. *Am J Med Genet* **81**:228–34.

Watson D., Clark L. A., Weber K. *et al.* (1995) Testing a tripartite model: II. Exploring the symptom structure of anxiety and depression in student, adult, and patient samples. *Journal of Abnormal Psychology* **104**:15–25.

Weissman M. M., Wickramaratne P., Adams P. B. *et al.* (1993) The relationship between panic disorder and major depression. A new family study. *Archives of General Psychiatry* **50**:767–80.

Ziegler A., Hebebrand J., Görg T. *et al.* (1999) Further lack of association between the 5-HT2A gene promoter polymorphism and susceptibility to eating disorders and a meta-analysis pertaining to anorexia nervosa. *Molecular Psychiatry* **4**:410–17.

Chapter 13

The dementias

Malcolm B. Liddell, Julie Williams, and
Michael J. Owen

Introduction

The most successful applications of molecular genetic approaches to mental disorders have been to the various forms of dementia. Dementia can occur as a result of more than 55 different diseases. Consequently, the focus of this chapter will be upon the most common disorders; Alzheimer's disease and vascular dementia, and on those in which advances have been especially striking; frontotemporal dementia, Huntington's Disease and the transmissible spongiform encephalopathies.

Twelve genes and over eighty mutations known to contribute to dementia have already been identified. In some cases the possession of a single mutation results in the certain development of dementia, with an onset that can be predicted almost to the year, while other susceptibility alleles increase risk, but are neither necessary nor sufficient to cause dementia. In some genes only a single mutation predisposes to disease, whereas in others many mutations are found. There are also complex relationships between mutations and the clinical phenotypes, with both pleiotropy and heterogeneity observed. It is encouraging that the progress made in identifying genes contributing to dementia has already led to profound advances in understanding disease mechanisms and holds the promise of imminent therapeutic advance. It is also likely that, by blazing the trail in psychiatric genetics, dementia research will have important implications for work in other areas, both in respect of indicating what sort of approaches are likely to be successful in our attempts to find genes and in pointing out the possible clinical and ethical consequences of genetics research.

Alzheimer's disease

The prevalence of Alzheimer's disease and dementia

Alzheimer's disease (AD) is the commonest cause of both pre-senile and senile dementia. Autopsy studies suggest that AD accounts for some 42–76 per cent of all dementia, vascular dementia (VaD) for 24–36 per cent of cases, and other causes for the remaining 8–20 per cent of cases (Tomlinson *et al.* 1970); (Jellinger *et al.* 1990); (Brun and Gustafson 1993). In reality, particularly in the elderly, clinical dementia is often caused by a combination of AD and vascular pathology (Kosunen *et al.* 1996; Gearing *et al.* 1995;

Morris *et al.* 1988; Lim *et al.* 1999; Snowdon *et al.* 1997; Holmes *et al.* 1999; Polvikoski *et al.* 2001a).

The prevalence of AD and dementia increases with age, but may level off amongst the very elderly (Table 13.1). This may not be the result of a true fall in the incidence of new cases, but a consequence of the very old not surviving as long with dementia (Drachman 1994). Indeed, two nearly completely ascertained community-based studies found that the prevalence of dementia in centenarians was very high: 70 per cent in the Japanese and 88 per cent in the Dutch (Asada *et al.* 1996; Blansjaar *et al.* 2000).

The large Rotterdam study suggests that the risk at age 55 of developing dementia in the following 35 years is 0.26 for a woman and 0.15 for a man, 73 per cent of the dementia being attributable to AD (Ott *et al.* 1998). The large continuing Framingham study yields somewhat lower lifetime risks of AD after the age of 65: 6.3 per cent for men and 12 per cent for women with a corresponding risk for all dementia of 10.9 per cent and 19 per cent respectively (Seshadri *et al.* 1997). Many epidemiological studies suggest that women are at increased risk of AD. The reasons for this are not clear, but increased longevity compared to men, increased survival with the disease and some increase in intrinsic vulnerability probably all play a part.

Diagnosis of AD

The received wisdom is that a definitive diagnosis of AD can only be made after *post mortem* examination of the brain. Unfortunately, the acccuracy or, more particularly, the relevance of a neuropathological diagnosis of AD can be confounded by the sometimes poor correlation between the numbers of Senile Plaques (SPs) and Neurofibrillary tangles (NFTs) and the severity of the dementia in life. This discrepancy may come about because of the aggravation of the clinical impact of AD by cerebral vascular disease, as was demonstrated by Snowdon and colleagues in the 'Nun Study' (Snowdon *et al.* 1997). Thus, even apparently significant AD pathology may be clinically silent in

Table 13.1 Prevalence of dementia by age. From Ritchie and Kildea, 1995

Age	Prevalence of Dementia (%)
65–69	1.53 (1.16–2.60)
70–74	3.54 (3.01–4.07)
75–79	6.80 (6.18–7.42)
80–84	13.57 (12.54–14.58)
85–89	22.26 (21.00–23.52)
90–94	31.48 (29.04–33.92)
95–99	44.48 (36.28–52.68)

the absence of cerebral vascular disease; whereas, in contrast, severe dementia can result from low-grade AD pathology occurring with even one strategically placed small brain infarct. Indeed, it may have been more than an incidental finding when, in his 1907 paper, Alzheimer made reference to arteriosclerotic changes in the large cerebral vessels of his index case (Wilkins and Brody 1969). Nevertheless, the absence of neuropathological confirmation of the diagnosis of AD is a potential complication for genetic research where in many instances only a clinical diagnosis is available. So, in practice just how accurate is a clinical diagnosis of probable AD when compared with a standardized neuropathological diagnosis of AD?

Using the NINCDS-ADRDA criteria (McKhann *et al.* 1984), on recognized cases of dementia referred on to specialist dementia assessment centres, the correlation between a clinical diagnosis of 'Probable AD' and neuropathologically confirmed AD is greater than 80 per cent, whereas the clinico-pathological correlation for 'Possible AD' is around 70 per cent (Gearing *et al.* 1995). In a large study of 2,188 dementia sufferers Mayeux and colleagues found that the Positive Predictive Value (PPV) of a clinical diagnosis of AD was close to 90 per cent, whereas the Negative Predictive Value (NPV) was 64 per cent and, furthermore, the inclusion of ApoE genotyping produced very little additional benefit (Mayeux *et al.* 1998). In a much smaller community register sample of 80 deceased individuals with dementia, Holmes and colleagues found a PPV of 76 per cent (NPV 64 per cent) when NINCDS criteria were asked to predict 'pure' AD in the absence of any other significant pathology, but rose to 92 per cent (NPV 17 per cent) when asked to predict AD + other significant pathology (Holmes *et al.* 1999). This study found that AD pathology was associated with other pathologies (principally either vascular infarction or Lewy bodies) in 30 per cent of the cases, a finding which explains the NPV of 17 per cent with the more flexibly applied clinical criteria. Therefore, it seems reasonable to conclude that NINCDS clinical criteria are good at predicting the presence of AD pathology, but not good excluding other pathologies that might be contributing to the syndrome of dementia. However, for genetic studies of a disease of later life, when multiple pathologies are to be expected, it would seem that modern clinical diagnostic criteria for AD are sufficient to ensure that diagnostic inaccuracy is not a major confounding factor in subsequent genetic analyses.

The Neuropathology of AD

Macroscopically the AD brain at end stage is fairly uniformly atrophic with enlarged ventricles. Initially, neurodegeneration preferentially involves the medial temporal lobe structures, in particular the entorhinal cortex and hippocampus, a finding which explains why amnesia is often the earliest symptom of the disorder (Tomlinson 1982; Pearson and Powell 1989). However, the characteristic neuropathology of Senile Plaques (SPs) and Neurofibrillary tangles (NFTs), as first described by Alois Alzheimer in 1907, are revealed after microscopic examination of silver-stained sections of brain (Alzheimer's 1907 paper is reviewed by Wilkins and Brody (1969)).

SPs are spherical or multilocular aggregations of a material with the staining properties of amyloid. The plaques are surrounded by abnormal nerve cell processes (neurites) and surrounding these there is often a glial cell reaction: for a review see Selkoe (1991) and Dickson (1997). The amyloid material is chiefly composed of accreted fibres of a 40–42 amino acid peptide known variously as Aβ, β-amyloid, or β-peptide. Aβ is derived from the Amyloid Precursor Protein (APP) (Selkoe 1991). The longer, 42 amino acid peptide is more prone to aggregation and, hence, amyloid formation than the shorter forms and seems to play an important role in the pathogenesis of highly genetic forms of the disorder (Iwatsubo *et al.* 1994; Lemere *et al.* 1996; Walsh *et al.* 1997; Harper *et al.* 1997). In AD, Aβ is also found in the walls of blood vessels both in the brain and throughout the body (Joachim *et al.* 1989), where it is known as amyloid angiopathy.

The other main pathological finding is the large number of NFTs which are intraneuronal flame-shaped structures chiefly composed of abnormally hyperphosphorylated Tau protein (Spillantini and Goedert 1998a; Tolnay and Probst 1999). The principle constituents of NFTs are paired helical filaments (PHFs), which are 10 nm twisted ribbons of hyperphosphorylated Tau protein. NFTs appear to be toxic to the neuron for in advanced AD numerous 'Tombstone tangles' are seen, which are large extracellular NFTs, now devoid of cellular material. For the medial temporal lobe structures there is some support for the idea that SPs form in the regions corresponding to the terminal arborizations of projection fibres originating from neurons containing NFT (Hyman *et al.* 1990).

Both SPs and NFTs are found to a lesser degree and largely confined to the medial temporal lobe structures in normal ageing. It is not known if these changes actually represent the early stages of AD. West and colleagues concluded from a careful stereological study that the patterns of neuronal loss seen in AD and advanced healthy old age was quantitatively and qualitatively different (West *et al.* 1994). On the other hand, community-based studies suggest that the prevalence of dementia continues to rise, even amongst centenarians, and that most of the dementia is attributable to AD (Asada *et al.* 1996; Blansjaar *et al.* 2000).

Middle-aged individuals with Down syndrome who come to autopsy almost invariably show the neuropathological features of AD. The APP gene is located on chromosome 21 and it is believed that trisomy 21 results in overproduction of Aβ (Heston and Mastri 1977; Rumble *et al.* 1989). *Post mortem* studies of young people with Down syndrome suggest that the Aβ deposition in senile plaques is preceded for many years by diffuse deposits of 42 amino acid Aβ, which occurs without inciting gliosis, NFT formation, or scarring (Mann 1989; Motte and Williams 1989). Such diffuse Aβ deposits are also commoner in the brains of healthy elderly people than are classical SPs. It is, therefore, inferred that diffuse Aβ deposits represent the earliest, non-toxic, stage in the elaboration of SPs. Diffuse Aβ deposits occurring with mature SPs are seen in AD, both in areas classically involved in the neurodegeneration, and in regions such as the cerebellum, which never seem to advance beyond the diffuse Aβ stage and, perhaps as a result, never show NFT formation and neurodegeneration.

Risk Factors for AD

Epidemiological studies have shown that the chief risk factors for AD are increasing age, a family history of the disorder, female gender and Down Syndrome. Other risk factors that have been identified include: basic level education only; smoking; hypothyroidism (Launer *et al.* 1999). Possible protective factors are a favourable family history (Silverman *et al.* 1994; Payami *et al.* 1994a), living as a rural Nigerian (Hendrie *et al.* 1995; Hendrie *et al.* 2001), taking treatment doses of non-steroidal anti-inflammatory drugs (Wyss-Coray and Mucke 2000), and, for women, oestrogen in the form of HRT (Tang *et al.* 1996; Barrett-Connor 1998).

Family studies in AD

Family studies in AD are complicated by a number of factors. First, clinical diagnosis is not completely accurate and lack of *post mortem* confirmation is of particular concern for distant relatives for whom detailed clinical information may be lacking. Second, the late onset of the disorder means that many at-risk individuals die before reaching the age of risk for AD. Third, family members may be studied before reaching the age of risk and classified as healthy when, in fact, some years later they go on to develop AD. To a certain extent these confounding factors operate in different directions and may cancel one another out. Also, the question of diagnosis may not be such an issue because, in the elderly, dementia without any AD component is probably rare, at least in old age (Holmes *et al.* 1999; Polvikoski *et al.* 2001).

In the 1980s and early 1990s there were a number of family studies which attempted to identify the inheritance pattern of AD (Heston *et al.* 1981; Mohs *et al.* 1987; Breitner *et al.* 1988; Huff *et al.* 1988; Martin *et al.* 1988; Fitch *et al.* 1988; Zubenko *et al.* 1988; Farrer *et al.* 1989; Farrer *et al.* 1990; Mayeux *et al.* 1991; Farrer *et al.* 1991; Silverman *et al.* 1994; Payami *et al.* 1994b). These studies also tried to quantify the risk to first degree relatives in families with a history of AD compared to the family members of control individuals. There have also been a few studies based on more representative community samples (Hofman *et al.* 1989; Mayeux *et al.* 1993; van Duijn *et al.* 1993).

Overall the findings from studies based in memory clinics (that are likely to be centres of secondary referral) suggest that 30–48 per cent of AD probands have a history of affected first-degree relatives compared to 13–19 per cent of controls. This translates to 6–14 per cent of the relatives of AD patients having a history of AD as compared to 3.5–7 per cent of the first-degree relatives of healthy controls. Kaplan-Meier life table analysis has been used in these studies to infer that the cumulative risk of dementia by age 90 varies between 30 and 50 per cent compared to between 10 and 23 per cent in control relatives. However, as has been pointed out by Breitner, owing to competing causes of death, only about one third of this theoretical familial predisposition to AD is realized in the usual life span. This translates to an actual predicted risk of developing AD in the first-degree relatives of AD probands of 15 –19 per cent as compared to 5 per cent in controls (Breitner 1991). However, the genetic risk may be considerably increased, perhaps to as much as 54 per cent ($\pm$ 10.9 per cent) by age 80, in the off-spring of parents who both have had the misfortune to develop AD (Bird *et al.* 1993; Lautenschlager *et al.* 1996).

Large family studies using consecutively ascertained probands suggest that most of the familial component of risk to the relatives of AD probands is expended by the end of the ninth decade, after which time the risk is very similar to that in controls (Silverman *et al.* 1994; Lautenschlager *et al.* 1996; Silverman *et al.* 2000). A number of other family history studies have suggested that familial factors are more prominent when onset is earlier (McGuffin *et al.* 1994). Thus, much of AD in late old age may be considered as not very familial and merely an expression of one of the ways the ageing process is manifested in those few survivors into late old age.

There is also evidence that the risk of developing AD in the relatives of non-demented 85 year olds may be decreased. In one study (Payami *et al.* 1994a) the cumulative risk to relatives of AD probands by age 85 was 36 per cent, the risk to relatives of randomly ascertained controls was 10 per cent, and the risk to the relatives of 43 individuals selected to be examples of highly successful ageing, was only 4 per cent by 85 years! Silverman and colleagues have reported similar findings (Silverman *et al.* 1999).

Twin studies in AD

A number of register based twin studies suggest that genes are important in AD. The Swedish Study of Dementia in Twins found 49 cases of AD in twins, which yielded probandwise concordance rates of 73 per cent in 9 MZ twin pairs and 32 per cent in 34 DZ twin pairs (Gatz *et al.* 1997). A Norwegian study found probandwise concordance rates for AD of 83 per cent in 12 MZ twin pairs and 46 per cent in 26 DZ twin pairs. A Finnish study yielded probandwise concordance rates for AD of 31 per cent in 43 MZ and 9 per cent in 41 DZ twin pairs (Raiha *et al.* 1996). The Swedish and Norwegian data suggest a heritability of around 0.8, whereas the Finnish data suggest a somewhat lower heritability in the region of 0.44. The reasons for these different heritability estimates are not clear. It may be relevant that the Finnish study employed hospital discharge data, whereas the other samples were ascertained from community settings. In the USA the Ageing Veteran Twins Study is underway, but the cohort is not yet old enough to allow the concordance rate of AD in twin pairs to be estimated with any accuracy (Breitner *et al.* 1995).

Autosomal dominant familial Alzheimer's disease (FAD)

Much of the initial work on the molecular genetics of AD, and most of the advances, have come from studies of rare families in which the disease is of unusually early onset and is transmitted in an autosomal dominant fashion. Such cases are often known as Familial Alzheimer's Disease (FAD), although most cases of AD with a family history, especially where the onset is after 65 years, do not have autosomal dominant forms of the disorder. These families account for only 13 per cent of early onset (<65 years) or less than 0.01 per cent of all AD (Campion *et al.* 1999). However, the identification of mutations in three genes that cause FAD has already had a major impact on understanding of the pathogenesis of AD.

The amyloid precursor protein gene and FAD

The search for the genes responsible for FAD started on chromosome 21, when the APP gene was cloned and localized to the long arm of this chromosome (for reviews of the APP linkage and localization studies see Hardy *et al.* (1989) and Tanzi (1991)). The initial reports of linkage to chromosome 21 were not wholly consistent and appeared to exclude the APP gene. However, the confusion was partially resolved when it became apparent that FAD is genetically heterogeneous, with only a proportion of families showing linkage to chromosome 21.

The candidacy of the APP gene was greatly strengthened by the demonstration that hereditary cerebral haemorrhage with amyloidosis—Dutch type (HCHWA-Dutch), a rare cause of familial cerebral haemorrhage, was caused by mutations within exon 17 of the APP gene, which, together with exon 16, encodes the Aβ peptide (Levy *et al.* 1990; Van Broeckhoven *et al.* 1990). This finding prompted the St Mary's Hospital group in London to sequence exon 17 of the APP gene in members of a large British pedigree with FAD that showed clear linkage to chromosome 21 and the APP locus. Affected members of this family were shown to carry a C to T transition causing a valine to isoleucine substitution at amino acid 717 (Goate *et al.* 1991).

To date there have been reports of seven missense mutations definitely associated with AD, three missense mutations definitely associated with HCHWA-D, and one mutation associated with neuropathological features of both AD and HCHWA-D (for a review see (Lendon *et al.* 1997; Martin 1999) (http://www.ncbi.nlm.nih.gov/htbin-post/omim/dispmian?104760). The type and position of the mutations in and surrounding Aβ are depicted in Fig. 13.1. There have been sixteen independent isolates of the original codon 717 valine to isoleucine mutation in different FAD families from around the world. The remaining FAD mutations have only been reported in single pedigrees. Five other missense mutations have been described within the APP gene, four within the Aβ coding region, and one within exon 7, the alternatively spliced protease inhibitor domain, but it is uncertain if these mutations actually cause AD (http://www.ncbi.nlm.nih.gov/htbin-post/omim/dispmian?104760) (Liddell *et al.* 1995).

APP is a multidomain glycoprotein with one membrane-spanning domain. APP has manifold but incompletely understood functions including an action as a type of neuronal stress protein, a neuronal trophic protein, and as an inhibitor of the coagulation cascade.It is subject to variable proteolytic processing by the α, β, and γ secretases. The secretase mediated processing of APP is portrayed in Fig. 13.2 (see Buxbaum *et al.* 1998; Selkoe 1999; Lammich *et al.* 1999; Luo *et al.* 2001; Cai *et al.* 2001).

However, despite their rarity, the FAD mutations underscore the importance of Aβ metabolism to the pathogenesis of AD. They are clustered adjacent to either the β or γ—secretase cleavage sites (Fig.13.1) and have been shown in cell culture studies and

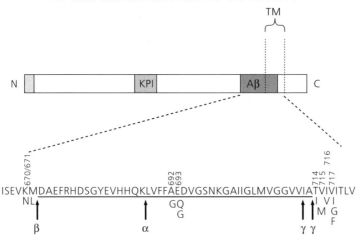

Fig. 13.1 Diagramatic representation of the APP protein with the Aβ peptide region magnified and underlined to show the siting of mutations associated with FAD or HCHWA-D and the positions of the α,β and γ—secretase cleavage sites. The mutations found at amino acid residues 670/6671, 714, 716, and 717 are all associated with FAD. The 693 residue mutations are associated with HCHWA-D, and the 692 residue mutation with a hereditary condition with features of both AD and HCHWA-D. All these mutations perturb Aβ metabolism to facilitate the deposition of Aβ as amyloid—see text for further details.

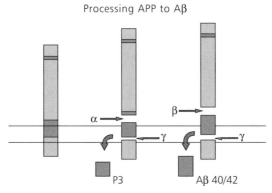

Fig. 13.2 Membrane bound APP can be cleaved by an α-secretase, which cuts within the mid-dle of the Aβ peptide to generate a secreted form of APP. The remaining membrane bound stub is cleaved by γ-secretase, which cuts within the cell membrane, close to the carboxy-terminus of Aβ, to yield a carboxy terminal Aβ fragment and the membrane bound cytoplasmic portion of APP. Second, APP can be processed by the β and γ-secretases to yield another, shorter secreted APP, the cytoplasmic APP fragment, and intact, full-length Aβ. It is thought that most APP escapes processing by the secretases, but that the APP so processed is predominately cleaved by α-secretase. Thus, Aβ normally results from a minor pathway of APP metabolism. Aβ is produced, in part, in the secretory pathway after processing by β and γ-secretases in the endoplasmic reticulum and Golgi and, in part, by the processing of re-internalized APP in late endosomes. The α-secretase has not been definitely identified, but in mammalian cells at least two members of the ADAM family of metalloproteases are implicated (Lammich *et al.* 1999; Buxbaum *et al.* 1998). The β-secretase has recently been identified as BACE1 (Luo *et al.* 2001; Cai *et al.* 2001). The γ-secretase has not been definitely identified, although the Presenilin proteins are very strong candidates (e.g. see Selkoe 1999).

transgenic mice to enhance cleavage at these sites (Lendon *et al.* 1997). The Swedish 'double' mutation located adjacent to the β-secretase site gives rise to a several-fold increased production of both the more common Aβ40 and the longer pro-amyloidgenic 42 amino acid Aβ peptide. The codons 716 and 717 mutations located close to the γ-secretase site seem to result in the preferential production of the 42 amino acid Aβ peptide (Lendon *et al.* 1997; Eckman *et al.* 1997). The Val715Met and Thr714Ile mutations differ from the others because when over-expressed in tissue culture, increased amounts of N-terminally truncated Aβ ending at position 42 are produced (Ancolio *et al.* 1999; Kumar-Singh *et al.* 2000). If applicable to the *in vivo* situation, this finding suggests that full-length Aβ may not be necessary for the formation of SPs and, furthermore, that the amino terminus of Aβ is unnecessary for pathogenesis. The Ala692Gly mutation may interfere with action of α-secretase and shift APP metabolism towards the production of Aβ.

The mutations associated with HCHWA-D appear to give rise to the accelerated deposition of the 40 amino acid Aβ peptide in the walls of cerebral vessels, but without actually causing an increase in Aβ production. Recent work by Nilsberth and colleagues (2001) helps explain this apparent anomaly (reviewed by Haass and Steiner (2001)). They studied the so-called Glu693Gly 'Arctic mutation' and found that the levels of Aβ40 and Aβ42 were reduced both in the culture media of cells transfected with this mutated APP cDNA and in the plasma of carriers of the mutation. Nilsberth and colleagues (Nilsberth *et al.* 2001) demonstrated that the Arctic mutation gave rise to the increased and accelerated production of Aβ protofibrils, which are the direct precursor of Aβ amyloid deposits and are themselves directly neurotoxic (Walsh *et al.* 1999; Hartley *et al.* 1999). It seems quite probable that this mechanism will also explain the toxicity associated with the Val715Met and Thr714Ile mutations which give rise to N-terminally truncated Aβ.

Presenilin genes and FAD

APP mutations explain only about 5 per cent of early onset FAD. The majority of the non-chromosome 21 linked FAD families were shown to be linked to a locus on the long arm of chromosome 14 at 14q24.3 (for reviews see (Cruts *et al.* 1996; Lendon *et al.* 1997; Nishimura *et al.* 1999) (http://www.ncbi.nlm.nih.gov/htbin-post/omim/dispmim?104311). Eventually four different missense mutations were identified in transcript S182. When S182 was fully sequenced it was found to be a novel gene, which was named Presenilin-1 (PS-1). Currently some 73 missense and 2 splicing defect mutations mutations within the PS-1 gene are known and these account for some 70 per cent of early-onset FAD (see Nishimura *et al.* 1999; Tandon *et al.* 2000). The mutations tend to occur in evolutionary conserved residues; in the proposed trans-membrane regions and in the loop regions connecting some of the transmembrane domains; but, in fact, mutations are found throughout the protein. PS-1 and FAD associated mutations are portrayed in Fig. 13.3.

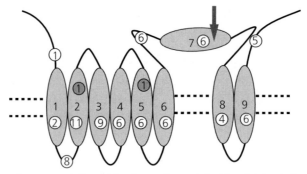

Fig. 13.3 Schematic representation of the Presenilin protein. The eight putative trans-membrane domains are shown. The arrow represents the approximate location of the activation cleavage site. Numbers in white circles indicate the number of Presenilin 1 mutations associated with a particular domain of the protein. The grey circles indicate the number and approximate mutations described in Presenilin 2.

Yet another form of FAD, found in descendants of a Volga-German family, was explained by a mutation (Asn141Ile) in the Presenilin-2 gene, which is highly homologous to PS-1, but is located on chromosome 1q42.1 (see Cruts *et al.* 1996; Lendon *et al.* 1997; Nishimura *et al.* 1999) (http://www.ncbi.nlm.nih.gov/htbin-post/omim/dispmim? 600759). This gene was isolated by a combination of genetic linkage analysis and by what was at at the time the novel strategy of searching the emerging genome data bases for mapped DNA clones bearing genes homologous to PS1. Subsequent screening of FAD families for PS-2 mutations has identified an Italian pedigree with a Met239Val mutation, so, overall, PS-2 mutations explain only a tiny fraction of AD. It is worth noting that about one third of clearly autosomal dominant inherited AD are not associated with mutations in either the APP or PS genes, which implies the existence of additional loci for FAD (Campion *et al.* 1999).

The Presenilin proteins are, considered as a whole, 60 per cent homologous at the amino acid level. This homology increases to 90 per cent if the transmembrane domains are considered in isolation. Homologous proteins exist across different phylla where they are implicated in the evolutionarily conserved Notch signalling pathway (Selkoe 2000a). Hydrophobicity plots suggest that the Presenilins have eight transmembrane (TM) domains and that the sixth and seventh TM domains are separated by a large intracytoplasmic loop. The proteins are synthesized as holoproteins of about 50 kDa that are found in a 180 kDa complex in the ER. Activation of the presenilins appears to occur by an endoproteolytic cleavage in the TM6-TM7 loop cytoplasmic domain to generate an N-terminal fragment of 35 kDa and a C-terminal fragment of 20 kDa. The introduction of mutated PS-1 genes into different cell lines indicates that two conserved aspartate residues in TM6 and TM7 are essential both for this endoproteolytic cleavage and for the activity of PS-1 (see review by Selkoe (1999)). Activated PS-1 is found as a 250 kDa multimeric complex with a variety of other proteins, principally in the Golgi apparatus. The levels of activated Presenilins appear to be tightly regulated with the protein that is excess to requirements being degraded in the proteasome.

Presenilin mutations and the function of the presenilins

There is good evidence that the presenilins are intimately involved in the γ-secretase cleavage step in APP processing, and that PS1 FAD mutations cause increased production of Aβ-42 due to enhanced γ-secretase activity (see Selkoe 1999). Moreover evidence is accumulating that that they may even be the active proteases in the intramembraneous γ-secretase complex (Wolfe *et al.* 1999; Li *et al.* 2000a,b). These findings fit in neatly with the data from APP mutations and support the view that abnormal metabolism of APP leading to deposition of Aβ is the central pathogenic process in FAD. However, the Presenilins have also been implicated in a number of other processes including the notch signalling pathway (Selkoe 2000b), the stabilization of β-Catenin (Anderton *et al.* 2000) and intracellular calcium homeostasis (Leissring *et al.* 2000; Grilli *et al.* 2000). It is not yet clear whether disturbances of these or other processes contribute to FAD in PS1 mutation carriers.

Transgenic mouse models of AD

The development of transgenic mice bearing mutant human APP and PS genes was accomplished both in the hope of developing an animal model of AD and also as a means of testing the amyloid hypothesis. Essentially, although such animals develop immunoreactive APP plaques and, in some lines, neuritic SP-like plaques, no *singly* transgenic mouse has been induced to develop the full pathology of AD (Games *et al.* 1995; Hsiao *et al.* 1996; Duff *et al.* 1996). Many of these animals show problems with spatial memory as they age. The accumulation of AD pathology is accelerated in a double transgenic line that overexpresses both the mutant PS-1 Met146Leu and the APPLys670Asn, Met671Leu 'Swedish' mutation (Borchelt *et al.* 1997; Holcomb *et al.* 1998). These mice show high levels of Aβ production and deposition of Aβ plaques by three months of age that increase 178-fold by one year.

The most complete recapitulation of AD neuropathology is achieved in the TAPP mouse, which is transgenic both for a FTD associated Tau mutation and for the 'Swedish Double' APP mutation (Lewis *et al.* 2001). These mice develop neuritic plaques *and* NFTs in the medial temporal lobe structures. The parent bearing the Tau mutation on its own develops NFTs in various parts of the brain and brain stem, but not the medial temporal lobe. Therefore, it is hypothesized that Aβ deposition, which is predominantly in the cortex and hippocampus, facilitates the emergence of NFTs, a sequence of events that is in accord with the amyloid hypothesis.

The pathogenesis of AD: Baptists, Taoists and the amyloid cascade hypothesis

There has for a number of years been a polarization of views between those championing the primacy of SPs and Aβ deposition in the pathogenesis of AD and those who believe that NFTs and abnormalities of Tau are the critical events. These factions have been mnemonically dubbed the Baptists—for Aβ or β-amyloid—and the Taoists—for Tau. The chief supporting arguments for the primary role of Aβ in the causation of AD are as follows. First, amyloid deposition is always indicative of disease, though there is circularity to this since amyloid deposition is required to make a diagnosis. Second, Down's dementia is preceded by and apparently caused by Aβ deposition. Third, frontotemporal dementia due to Tau mutations is characterized by large numbers of NFTs but no Aβ deposits, so NFT formation does not by itself lead to Aβ deposition. Fourth, NFTs are a non-specific marker of neuronal dysfunction: Aβ is only found in AD, Down's dementia, and, to a much lesser extent, as a consequence of normal ageing. Fifth, mutations in APP cause AD and both PS1 and APP mutations affect APP metabolism.

In contrast the arguments in support of NFTs as being of primary importance are as follows. First, the studies by Braak and Braak suggest that NFTs in neurones of the perforant pathway in the transentorhinal cortex are the first neuropathological indication of incipient AD and that senile plaques are a later finding (Braak and Braak 1991). Second, the severity of dementia is better correlated with the numbers of NFTs rather than SPs (Wilcock and Esiri 1982; Arriagada et al. 1992).

The finding that mutations in APP cause AD and both PS1 and APP mutations affect APP metabolism has provided strong and many would say crucial support for a primary role of Aβ deposition in AD. The resulting 'amyloid-cascade hypothesis' (Hardy and Allsop 1991) is illustrated in simplified, diagrammatic form in Fig. 13.4. The essence of the argument is that, although initially APP production may represent a protective, homeostatic response, either over a lifetime, or as result of a number of disorders that lead to increased production of Aβ, there is the accumulation of Aβ as neuritic plaques. The aggregated Aβ is neurotoxic and may further aggravate neuronal

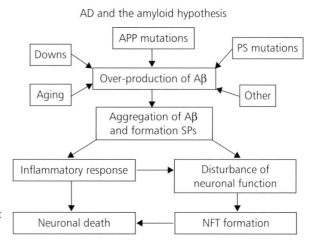

Fig. 13.4 A simplified portrayal of the Amyloid cascade hypothesis. See text for further details.

function by inciting a chronic inflammatory reaction in the neuropil surrounding the senile plaques. Eventually NFTs form in the neurons damaged by Aβ, which, after a period of time, die, leaving behind 'tombstone tangles'.

The amyloid hypothesis may not be the complete explanation of the pathogenesis of AD, but it is testable because, at least in part, it suggests potential treatments for AD could involve either reducing the production, accelerating the clearance, or inhibiting the aggregation of Aβ. Remarkably, it looks as if amyloid deposition can be prevented, and its removal facilitated, at least in transgenic mice over expressing either the Va1717Phe mutated APP gene (Schenk *et al.* 1999), or in a doubly transgenic line over-expressing both the mutant PS-1 Met146Leu and the APPLys670Asn, Met671Leu 'Swedish' mutation (Morgan *et al.* 2000), if the mice are immunized with, and raise antibodies to, Aβ. Both teams of investigators found that immunized mice failed to develop the expected deficits in spatial memory seen in their non-immune littermates. Trials are currently underway to test the feasibility of Aβ vaccination in treating AD in humans.

Apolipoprotein E and AD

Apolipoprotein E (ApoE) is a protein with roles in lipid metabolism and tissue repair. Its primary site of biosynthesis is the liver, but the second major site of synthesis is the brain (Mahley 1988; Siest *et al.* 1995). Like the APP, the synthesis of ApoE is up-regulated after the nervous system has been damaged. There are 3 commonly occurring poly-morphic forms of ApoE known as ApoE2, ApoE3, and ApoE4, which originate from the ε2, ε3 and ε4 alleles of the gene. The key observation, originally made by Roses' Duke University, North Carolina, group, is that the frequency of ε4 in AD patients, at 0.3 to 0.5, is greater than the ε4 frequency of 0.1 to 0.15 in age-matched or population controls (Strittmatter *et al.* 1993). Some studies suggest that the ε2 allele is under-represented in AD and may, by inference, be protective. Consistent with this idea is the finding that healthy centenarians have a higher ε2 frequency than general population controls, perhaps supporting the notion that possession of the ε2 allele protects against the development of dementia (Farrer *et al.* 1997). This is a rather surprising finding in view of the fact that APOE2/2 is the most common genotype associated with type III hyperlipidaemia.

The association of ε4 with AD has been replicated in many laboratories around the world. In a recent meta-analysis of over 40 studies (representing 30,000 APOE alleles), Farrer and colleagues (1997) observed elevated frequencies of the APOE 4 allele among AD patients in every group, although the association was stronger among Caucasians (odds ratios (OR) of 2.7–3.3 and 12.5–14.9 in ε3–ε4 and ε4–ε4 subjects, respectively) and the Japanese (ORs of 5.6 and 33.1, respectively), but weaker in African-Americans (OR = 1.1 and 5.7, respectively). The data from this meta-analysis are summarized in Table 13.2. The ε4 association does not appear in Nigerians, who appear to have a lower risk of developing dementia than do Caucasians or urban North American

Table 13.2 Odds Ratios for the 6 different APOE genotypes in 4 different populations. Data from a meta-analysis of nearly 30,000 APOE alleles carried out by Farrer et al., 1997—see text for further details

Odds Ratio	Caucasian: General Population	African-Americans	Hispanics	Japanese
ε3/ε3	1.0 (Reference)	1.0 (Reference)	1.0 (Reference)	1.0 (Reference)
ε2/ε2	0.9 (0.3–2.8)	2.4 (0.3–22.7)	2.6 (0.2–33.3)	1.1 (0.1–17.2)
ε2/ε3	0.6 (0.5–0.9)	0.6 (0.4–1.7)	0.6 (0.3–1.3)	0.9 (0.4–2.5)
ε2/ε4	1.2 (0.8–2.0)	1.8 (0.4–8.1)	3.2 (0.9–11.6)	2.4 (0.4–15.4)
ε3/ε4	2.7 (2.3–3.2)	1.1 (0.7–1.8)	2.2 (1.3–3.4)	5.6 (3.9–8.0)
ε4/ε4	12.5 (8.8–17.7)	5.7 (2.3–14.1)	2.2 (0.7–6.7)	33.1 (13.6–80.5)

blacks (Ogunniyi et al. 2000). Allele ε2 was observed to have a protective effect in all ethnic groups (OR = 0.6, ε2–ε3) and they also demonstrated APOE 4 to be a risk factor for AD in persons as young as 40, but that this risk diminished in those over the age of 70. There also appeared to be a greater risk of AD in women attributable to increased risk in those with the ε3–ε4 genotype, although it was noted that the increased risk of AD in women is unlikely to be due to APOE alone and was also not observed in the Japanese. However, unlike mutations causing FAD, which behave generally in an autosomal dominant manner, APOE is a risk factor, which is neither necessary nor sufficient to cause AD. For example, at least one third of individuals with AD lack an ε4 allele (Strittmatter et al. 1993; Farrer et al. 1997) and up to 50 per cent of individuals who have a double dose of ε4 survive to age 80 without developing Alzheimer's disease (Henderson et al. 1995; Myers et al. 1996). In addition, factors such as serious head injury, smoking, cholesterol level, and estrogen may also modify APOE-related risk of AD development (van Duijn et al. 1994; Payami et al. 1994b; van Duijn et al. 1995; Farrer et al. 1995; Mayeux et al. 1995; Jarvik et al. 1995; Slooter et al. 1999; Martin 1999).

There is an emerging consensus that apoE genotype determines when rather than whether AD develops. For example, the Cache County study (Meyer et al. 1998; Breitner et al. 1999) on an elderly population of almost 5000, found 22 AD cases amongst 141 ε4 homozygotes, 118 among 1,452 ε4 heterozygotes and 80 amongst 3,339 not bearing the ε4 allele. There appeared to be a plateau in each group's survival curve beyond which no new cases of AD were seen. For ε4 homozygotes no new cases were seen after the age of 84, with nine individuals surviving without AD for a combined total of 37 years. For ε4 heterozygotes the last AD onset was age 94, with four long-term survivors. The last onset of dementia in individuals without the ε4 allele was age 95, with 31 surviving free of dementia for a combined total of 100 years thereafter. This differential effect of ApoE genotype on the age of maximum risk of AD is also suggested in the ApoE meta-analysis of Farrer et al. (Farrer et al. 1997).

Finally, association with AD has been reported between polymorphisms in the APOE promoter at positions −491, and polymorphisms within a potential Ch 1/E47 transcription factor binding consensus sequence (Artiga *et al.* 1998). Attempts to replicate these associations have resulted in some positive studies but also some negative (Town *et al.* 1998; Zurutuza *et al.* 2000).

Putative effects of ApoE on APP and Tau biochemistry

Exactly how ε4 and ApoE4 influences the pathophysiology of AD is still unknown, but bearers of this risk allele appear to get AD earlier and develop a heavier amyloid burden (for a recent accessible review on ApoE and AD see Saunders (2000)). Early studies suggested that purified delipidated apoE4 bound with greater avidity to synthetic Aβ *in vitro* than did apoE3 (Strittmatter *et al.* 1993). In contrast, subsequent studies which employed different preparations of apoE either failed to show this effect, or even demonstrated the preferential binding of apoE3 to Aβ (LaDu *et al.* 1995; Chan *et al.* 1996; Haas *et al.* 1997).

Perhaps the clearest demonstration of the relevance of apoE to Aβ deposition comes from the progeny of apoE knockout mice crossed with transgenic mice over-expressing a mutant APP gene. These mice over-expressed APP and over-produced Aβ, yet, in the absence of apoE, no deposits of Aβ were seen at six months of age. In contrast, the transgenic mice with intact murine APOE genes showed abundant fibrillar Aβ deposits at the same age (Bales *et al.* 1997; Holtzman *et al.* 2000). Moreover, doubly transgenic mice bearing both an APP mutation and human APOE4 showed greater Aβ deposition than those bearing APOE3 (Holtzman *et al.* 2000).

There is also some evidence that apoE can affect the metabolism of Tau and, thus, influence the formation of NFTs. First, the LDL-receptor binding domain of apoE also allows apoE to bind to the microtubule binding repeat region of Tau. ApoE3 appears to bind more avidly and easily than apoE4 and, thus, may be able to stabilize Tau and slow down NFT formation (Strittmatter *et al.* 1994). Neither apoE isoform was able to bind to hyperphosphorylated Tau (Huang *et al.* 1994). Second, the VLDL receptor and the LRP8 receptor bind two other proteins as well as apoE: reelin and mammalian disabled-1. Binding of reelin to these receptors leads to the phosphorylation of cytosolic disabled-1, which indirectly leads to reduction of Tau phosphorylation. ApoE appears to compete with reelin for occupation of these receptors, with apoE3 and apoE4 showing much stronger affinities than apoE2. Thus, it may be that in APOE3 and APOE4 individuals the reelin pathway is underactive, and that this predisposes towards the hyperphosphorylation of Tau and the formation of NFTs (D'Arcangelo *et al.* 1999; Hiesberger *et al.* 1999).

Clinical correlates of ApoE genotype

One study suggests that AD patients with ε4 deteriorate more rapidly than AD patients without this allele, but other studies do not show this effect (Corder *et al.* 1995; Kurz *et al.* 1996; Growdon *et al.* 1996; Stern *et al.* 1997; Craft *et al.* 1998). It has also been suggested that ε4 is a marker for a poor response to treatment with Acetylcholinesterese inhibitors (Poirier *et al.* 1995; Farlow *et al.* 1998), although subsequent studies have not confirmed this original finding (e.g. Farlow *et al.* 1999; Wilcock *et al.* 2000). ApoE4 is associated with a poorer recovery from acute brain injury, whether due to head injury, intracerebral haemorrhage, or resulting from cardiopulmonary bypass (Saunders 2000). The risk of developing AD is increased tenfold in those survivors of traumatic-head injury who are bearers of ε4, whereas the risk is only doubled in association with ε4 alone, and not increased at all in head injury survivors who do not possess ε4 (Mayeux *et al.* 1995). The ε4 allele is also associated with a worse prognosis in cases of dementia pugilistica (Jordan *et al.* 1997). Finally, there is evidence that glucose metabolism is reduced in middle-aged carriers of ε4. Using PET and ^{18}F-deoxyglucose it was found using volunteers from the normal population that ε4/ε4 carriers had significantly reduced cerebral glucose metabolism as compared to ε3/ε3 or ε3/ε2 volunteers. These subjects showed no cognitive deficits and were, on average, 20 years younger than the expected men age at onset of AD for ε4/ε4 homozygotes (Reiman *et al.* 1996).

The potential use of ApoE genotyping for predictive testing and diagnosis is discussed in Chapter 15.

Other genes for late-onset AD

Recent simulation studies have estimated that at least four genes contribute to age of onset variance with an equal effect or greater than APOE (Daw *et al.* 1999). The observed contribution of APOE to the total variance in age of onset, was between 7–9 per cent, whereas the contribution of the largest Quantitative Trait Loci (QTLs) ranged from 5 per cent to over 50 per cent. Consequently, it seems that additional genes contributing to late onset AD remain to be identified. Schellenberg and colleagues (2000) recently reviewed evidence for a further 40 candidate genes tested for association with AD. Table 13.3 contains a selection of these genes, which showed an initial positive association and demonstrate some evidence of replication. However, none of these genes can be said to show confirmed evidence of association. It is possible that a number will prove to be genuinely associated but may have a limited effect.

Further evidence for the existence of new genes for late onset AD has come from large-scale linkage studies of families with late-onset AD (Pericak-Vance *et al.* 1997, 2000; Kehoe *et al.* 1999; Myers *et al.* 20000). A two-stage genome scan involving 430 affected sibling pairs (ASPs) with probable/definite AD (Kehoe *et al.* 1999; Myers *et al.* 2000) found significant linkage to chromosome 10 (see Fig. 13.5) and suggestive linkage to regions on chromosomes 1, 5, 9, and 19 (near APOE). Limiting their sample to ASPs with definite AD, Pericack-Vance and colleagues (2000) detected significant linkage to a region on chromosome 9 which corresponds with a region of suggestive linkage cited above. Olson and colleagues have re-analysed the data from Kehoe *et al.* (1999) using the covariate-based affected sib pair linkage method (Olson *et al.* 2001). This analysis suggests that there is evidence for linkage to markers on chromosome 21 close to APP

Table 13.3 From Schellenberg *et al.* 2000. Possible Candidate Gene Associations with AD

Gene/locus	Chromosome	Positive Studies	Negative Studies
Butyrylcholinesterase	3q	2	12
HLA-A2	6p	5	4
HLA-DR	6p	2	1
VLDLR	9p	4	8
Estrogen receptor	11q	2	0
2-Macroglobulin	11q	4	10
LRP	12q	3	5
Presenilin-1 LOAD	14q	6	16
Dihydrolipoamide			
Succinyltransferase	14q	3	1
−1-antichymotrypsin (ACT)	14q	223	22
Serotonin transporter	17q	2	0
Angiotensin-converting enzyme	17q	3	1
CYP2D6	22q	2	3
IDE	10q	1*	1**

*Bertram *et al.* (2000)

**Abrahams *et al.* (submitted)

for disease of especially late onset, an observation which suggests that variation in and around the APP gene may well be of significance in this group of cases.

Vascular dementia

Epidemiology

Vascular dementia (VaD) is usually found to be the second most common form of dementia after AD, although in populations with a high prevalence of hypertension, such as Japan, it may, in fact, exceed AD in frequency. The prevalence estimates are broad being anywhere between 9 and 39 per cent for VaD and 11–43 per cent for mixed AD and VaD (Kase *et al.* 1989; Kase 1991). Most studies suggest that males are more often affected than females except in late old age, when the condition may be commoner in women. The Rotterdam study (Ott *et al.* 1995) found that VaD accounted for some 16 per cent of dementia and that the prevalence increased with increasing age, as is summarized in Table 13.4. Epidemiological studies have identified a number of risk factors for VaD, most of which are also risk factors for arteriosclerosis and hypertension and, therefore, stroke. The chief culprits are, of course, a history of hypertension, diabetes, smoking, hyperlipidaemia, and hepercholesterolaemia, and atrial fibrillation. Non-vascular risk factors include psychological stress in early life, low educational

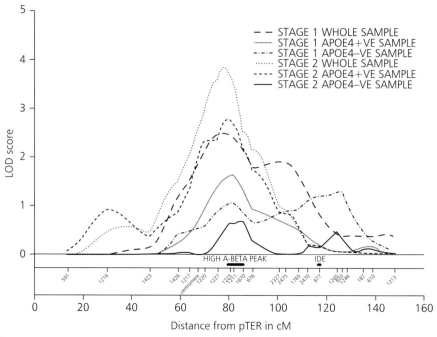

Fig. 13.5 Genetic Marker sharing on chromosome 10 amongst AD sibling pairs.

Table 13.4 The prevalence of VaD by age as determined in the Rotterdam Study (Ott *et al.*, 1995). Total sample size 7,528 subjects aged 55–106 years with 474 cases of dementia due to all causes. This table summarises the findings for VaD

	Prevalence of VaD (%)	
Age in Years	Men	Women
65–74	3 (0.3)	3 (0.2)
75–84	12 (2.1)	22 (2.0)
>85	3 (2.2)	28 (4.9)

attainment, high alcohol intake, and occupational exposure to pesticides and liquid plastic or rubber (Skoog 1998). It is interesting that many of the classical risk factors for vascular disease have also been shown to be risk factors for AD (Stewart 1998). In this context perhaps what has been elicited are risk factors for the *syndrome* of dementia, which, of course, is often the result of several pathologies.

Pathology

VaD results from a number of different pathological processes that singly or in combination disturb either the patency or integrity of the blood vessels. Such processes chiefly include strokes, either due to infarction or haemorrhage, and leukoaraiosis, also known as White Matter Lesions (WML). Large vessel atherosclerotic disease is responsible for some 30 per cent of all ischaemic strokes, 20 per cent are lacunar strokes resulting from the occlusion of small penetrating arteries, and some 30 per cent are due cerebral embolism, usually from a cardiac focus (Hademenos *et al.* 2001).

Leukoaraiosis results from partial demyelination of the deep cortical white matter and is associated with hyaline arteriosclerosis of the small penetrating arteries and arterioles. It is associated with hypertension and other risk factors for vascular disease in mid-life, whereas dementia is associated with hypotension in later life (Skoog 1998; Kivipelto *et al.* 2001). It is possible that the arteriosclerotic deep perforating arteries and arterioles lose the capacity to autoregulate cerebral blood flow, and it is this that predisposes towards cerebral ischaemia. Perhaps the most extreme form of leukoaraiosis is Binswanger's disease, which results from severe hypertension leading to lacunar infarcts affecting the deep perforating arteries and resulting white matter degeneration and the emergence of a subcortical dementia.

Population studies suggest that leukoaraiosis and strokes contribute independently to cognitive impairment (Skoog 1998).The frequency of leukoaraiosis increases with age and is also found more often in AD patients than in healthy controls. It is possible that in many cases leukoaraiosis, strokes, and AD pathology interact to bring about the syndrome of dementia. The 'Nun-Study', an intriguing and epidemiologically tightly defined study carried out on, and with the permission of, the School Sisters of Notre Dame, demonstrated that dementia was much more likely when early AD changes occurred together with one or two strategically placed infarcts, than when the AD changes were found in isolation (Snowdon *et al.* 1997). Snowdon *et al.* also found that the frequency of strokes was the same whether minimal AD changes were present or not, a finding which argues that AD does not of itself predispose towards stroke (Snowdon *et al.* 1997). Two community based studies by Ferrucci *et al.* and Gale *et al.* found that the risk of stroke was increased in those elderly with borderline cognitive impairment, some of whom may well have been in the early stages of AD (Gale *et al.* 1996; Ferrucci *et al.* 1996).

Genetics of VaD

Apart from rare autosomal dominant conditions, the genetics of VaD is essentially the genetics of arteriosclerosis, atheroma formation, ischaemic heart disease and hypertension. Put another way, VaD is simply one of the long-term complications of vascular disease.

Hademenos and colleagues (2001) have recently reviewed the genetics of cerebrovascular disease and stroke. Eleven genetic epidemiological studies of family history and stroke suggest an increased risk to offspring of parents affected by cerebrovascular disease and stroke (Hademenos *et al.* 2001) and supplementary information on *Neurology* web site). In some studies the effect is more marked in the children of affected mothers, and in others, in the offspring of affected fathers. Two studies suggested that black people are more vulnerable to stroke than white.

Family studies also support the role of hereditary factors in increasing the risk of stroke in offspring. Graffagnino *et al.* found a family history of stroke in 47 per cent of 85 patients and 24 per cent of 86 control subjects (the histories for ischaemic heart disease were, respectively, 73 per cent and 53 per cent (Graffagnino *et al.* 1994). Kubota *et al.* found that family history was the strongest independent risk factor for subarachnoid haemorrhage, was also associated with an increased risk of intracerebral haemorrhage, but was not a significant risk factor for cerebral infarction (Kubota *et al.* 1997). In fact, there is good evidence from a number of studies for the familial aggregation of intracranial aneurysms and subarachnoid haemorrhage and, to a lesser extent, for intracerebral haemorrhage (ter Berg *et al.* 1992; Schievink *et al.* 1995, 1997; Raaymakers 1999). Recently, Onda and colleagues found genetic linkage for intracranial aneurysms on chromsomome 7 in 104 affected Japanese sibling pairs (Onda *et al.* 2001). The best evidence of linkage was close to a good candidate, the elastin gene, and a haplotype between the intron-20 and intron-23 SNPs was found to be strongly associated with IA, particularly in homozygotes.

Twin studies

A number of twin studies on the genetics of vascular risk factors have been published.

The Swedish Adoption/Twin Study of Ageing suggested heritabilities ranging from 0.28 to 0.78 for the serum levels of each lipid measured, but also found that the environment of rearing accounted for between 0.15 and 0.36 of the variance of total cholesterol (Heller *et al.* 1993). The same study found that when younger twins were compared with older twins, the heritability of apolipoprotein B and triglyceride levels decreased with age (Heller *et al.* 1993). A large US twin study found a moderate genetic influence on lifetime smoking practices (Carmelli *et al.* 1992) and another study estimated that some 2/3 of the risk of cardiovascular disease associated with cigarette smoking is attributable to genetic influence (Khaw and Barrett-Connor 1986). Thus, these studies suggest that both genetic and environmental factors influence the risk of developing cardiovascular disease, but that, at least for some risk factors, the genetic loading decreases with age.

In a study of Finnish twins Räihä *et al.* found probandwise concordance rates for VaD of 31 per cent in MZ versus 12.5 per cent in DZ twins, but the twins were not interviewed, the information being collected from the hospital discharge records (Räihä *et al.* 1996). In a US mailed questionnaire study Brass *et al.* found an increased concordance for stroke in MZ (18 per cent) versus DZ (4 per cent) twins for stroke occurring in middle-age and early old age (Brass *et al.* 1992). However, a telephone interview follow-up found that the concordance rates were more similar at 13 per cent versus 8 per cent (quoted as a personal communication in Plassman and Breitner (1996)). In contrast the Norwegian register based study of elderly twins, in which the participants were interviewed and examined clinically, found identical probandwise concordance rates of 29 per cent in MZ and DZ twins in those twins judged to have VaD (Bergem *et al.* 1997).

Thus these twin studies do not give a clear picture of the heritability of either VaD or stroke. The study by Bergem *et al.* (1997) involving 7 MZ and 17 DZ twin pairs is the smallest, but is the most methodologically sound, and suggests that environmental factors dominate in VaD. The Finnish twin study (Räihä *et al.* 1996) of 11 MZ and 30 DZ twin pairs almost certainly fails to ascertain mild dementia, although, as the authors point out, it is extremely unlikely that this would be affected by the zygosity of the twins, so the MZ/DZ ratios should be unaffected. Thus the Finnish study suggests that genes are as important in conferring risk of developing VaD as they are of AD. However, it is possible that the Finnish VaD cases were 'contaminated' with AD or mixed cases, which, if AD is truly more heritable than VaD, would have reduced the MZ/DZ ratio owing to the greater number of DZ twin pairs in the sample. The community point prevalence dementia study of Polvikosky and colleagues, which is discussed in detail below, would seem particularly relevant here (Polvikoski *et al.* 2001). The US study examined the heritability of stroke and not dementia, and has not been reported in full, but may suggest that genes are important risk factors for stroke occurring at a relatively young age, but that they wane in importance at advanced ages.

Rare single gene disorders causing VaD

There are a number of autosomal dominantly inherited forms of VaD in which the causative mutations have been identified. It is not thought that they highlight any metabolic pathways that are relevant to common, less highly heritable forms of VaD.

Heritable cerebral haemorrhage with amyloidosis—Dutch type (HCHWA-D)

This is characterized by the inheritance of a predisposition to develop multiple cerebral haemorrhages starting around the ages 40 to 50 years, which progress to increasing neurological deficit, dementia, and death. The cause of the cerebral haemorrhage is the deposition of b-amyloid peptide, the same peptide that is found in the senile plaques of Alzheimer's disease, in the walls of the cerebral blood vessels. Missense mutations have been found in the central portion of the β-amyloid segment coded for by exon 17 of the APP gene (Levy *et al.* 1990; Van Broeckhoven *et al.* 1990).

Hereditary cerebral haemorrhage with amyloidosis—Icelandic type (HCHWA-I)

This disorder occurs in people of Icelandic origin and is characterized by recurrent cerebral haemorrhages starting in early adulthood and deposition of a mutated form of the protease inhibitor cystatin C in the walls of cerebral blood vessels. The gene for cystatin C is found on chromosome 21 and a leucine −68 to glutamine mutation in exon 2 appears to be causal (Abrahamson *et al.* 1987). It appears that the mutant protein is unstable and readily dimerizes and forms insoluble aggregates (Abrahamson and Grubb 1994). This process is highly temperature dependent with a 60 per cent increase (at physiological concentrations of the mutant protein) in the rate of dimerization seen as the temperature increased from 37 to 40°C. This finding suggests that the prompt treatment of febrile illnesses in carriers of HCHWA-I might help delay the build up of L68Q cystatin C aggregates.

Cerebral autosomal dominant arteriopathy with subcortical infarcts and leukoencephalopathy (CADASIL)

This condition is manifest as recurrent subcortical infarcts, leading to pseudobulbar palsy and dementia (Kalimo *et al.* 1999). It can affect individuals from the twenties upwards and is not associated with any of the traditional vascular risk factors. In one-third of patients the condition is first manifest as migraine with aura. CADASIL may be diagnosed, even before the first stroke, on the basis of characteristic hyperintensities in T2-weighted MRI scans.

The pathological findings are of multiple small infarcts involving the white matter or deep grey matter of the brain, associated with thickening and fibrosis of the walls of the small and medium-sized penetrating arteries. Histopathological findings are of the accumulation of basophilic and PAS positive granular material in the thickened tunica media of the vessels. These histopathological findings are not confined to the cerebral blood vessels (although the symptoms are entirely neurological) and dermal biopsy can be used to confirm the diagnosis.

CADASIL was mapped to chromosome 19p13.1–13.2 by Tournier-Lasserve *et al.* (1993) and subsequently mutations in the *Notch 3* gene were identified that segregated with the disease in affected families (Tournier-Lasserve *et al.* 1993; Joutel *et al.* 1996). The *Notch 3* gene contains 33 exons encoding a Notch 3 protein of 2321 amino acids with a single transmembrane domain. The extracellular portion of Notch 3 comprises 34 epidermal growth factor (EGF) domains followed by three notch/lin-12 repeats. The cytoplasmic portion of Notch 3 consists of six cdc10/ankyrin repeats. Over 90 per cent of CADASIL results from missense mutations within the EGF repeats of Notch 3. Some 26 different point mutations have been described, 70 per cent of these being located either within exons 3 or 4. The mutations result in either a loss or gain of a cysteine residue so that an EGF repeat contains 5 or 7 cysteines as opposed to the usual six (Joutel *et al.* 1997).

Other autosomal dominant forms of VaD

A syndrome of hereditary endotheliopathy with retinopathy and stroke (HERNS) has been described in three generations of a Chinese American family (Jen *et al.* 1997). The vascular basement membrane is multilaminated and dysfunction is seen in a number of organs, including the retina, brain, and kidneys. Migraine and strokes are seen from the third to fourth decades of life. Iglesias *et al.* have described another pedigree showing the possibly dominantly inherited syndrome of bilateral occipital calcifications, haemorrhagic strokes, leukoencephalopathy, dementia, and external carotid dysplasia. Dermal biopsies show that the syndrome is associated with pathological changes in the basal lamina of capillaries (Iglesias *et al.* 2000). No genetic linkage data are available as yet for these syndromes.

Apolipoprotein E and VaD

Many studies suggest that APOE ε4 increases the risk of dementia associated with stroke. Slooter and colleagues (Slooter *et al.* 1997) found in a population based study from communities in Rotterdam and New York that the APOE ε4 allele was a risk factor for dementia with stroke, whether due to VaD or mixed VaD and AD. They estimated that the risk attributable to APOE ε4 was 33 per cent for VaD and 44 per cent for mixed VaD and AD. The Gothenburg longitudinal study (Skoog *et al.* 1998) found that APOE ε4 was associated with an increased risk of AD and mixed dementia, but not pure VaD. The same study also found that leukoaraiosis was an independent risk factor for the development of dementia, and that the risk of dementia was substantially increased (OR 6.1) in those individuals who had leukoaraiosis and were also APOE ε4 carriers. A previous study in this sample group (Skoog *et al.* 1994) had demonstrated an increased frequency of leukoaraiosis in AD (64 per cent) and VaD (70 per cent) sufferers as compared to a healthy aged control group (34 per cent). Hirono and colleagues found that leukoaraiosis was not significantly associated with APOE ε4, whereas hypertension and lacunar infarcts were (Hirono *et al.* 2000). A substantial number (370) of the survivors of the MRC trial of the treatment of hypertension in older adults were contacted by Prince and colleagues (2000) and assessed for the presence of dementia, APOE ε4, cholesterol and other risk factors for vascular disease. They found an association between APOE ε4 and both dementia (OR 2.4) and AD (OR 3.4) and that the association increased after logistic regression was used to control for vascular risk factors. The authors concluded that the presence of APOE ε4 increases the risk of AD and dementia independently of its effect on dyslipidaemia and atherogenesis (Prince *et al.* 2000). The studies summarized so far are weakened because the clinical diagnosis of the different types of dementia was not confirmed by *post mortem* examination of the brain. One recent study has addressed this deficiency.

Polvikoski and colleagues established the point prevalence of neuropathologically defined AD and other dementias, together with APOE genotype in 88 per cent of the 85 year old and older residents of Vantaa, Finland (Polvikoski *et al.* 2001). The population was first examined in 1991 and followed up thereafter. On 1 April 1991 the point prevalence of dementia was 38 per cent: individuals who developed dementia after this time point were not included in the study. Neuropathological diagnoses were obtained in 118 of the 198 genotyped demented subjects and 62 of the 201 individuals who had died non-demented by December 1999. The main findings were as follows: first, the point prevalence for clinically diagnosed AD and clinically diagnosed VaD were 16 per cent and 19 per cent, respectively. Estimated neuropathologically diagnosed AD was

33 per cent in all 85+ subjects. Second, APOE ε4 frequencies were 0.47 in clinically diagnosed AD, 0.4 in clinically diagnosed VaD+ other, and 0.2 in controls. Neuropathological criteria for AD were met in 72 per cent APOE ε4 positive and 18 per cent APOE ε4 negative clinically diagnosed VaD or other dementias: in controls the figures were 42 per cent and 16 per cent, respectively. Third, the clinical diagnosis of AD was not very accurate: some 35 per cent of clinically diagnosed AD did not have AD; and, some 53 per cent of neuropathologically diagnosed AD were found in the VaD and other dementia group, or in controls.

The implications of this study would appear to be that very early or subclinical AD is poorly diagnosed, that dementia with AD neuropathology is very common, and that the apparent association of APOE ε4 with VaD probably results from an association with underlying and undiagnosed AD. In fact, previous autopsy series had either suggested no association between VaD and APOE ε4, or an association only in cases of mixed VaD and AD (Saunders and Roses 1993; Betard *et al.* 1994).

The frontotemporal dementias

Frontotemporal dementia probably accounts for 10 per cent or less of dementias. It is currently widely believed that frontotemporal dementia can be subtyped into Classical Frontotemporal dementia (Pick's Disease), Primary Progressive Aphasia, and Semantic dementia (Neary *et al.* 1998). The three subtypes can be related by the similarity of the underlying histopathology, the different clinical spectra stemming from the differential involvement of either the frontal or temporal lobes. Some 50 per cent of FTD is familial and the different subtypes can be found in different members of the same pedigree. Mutations in the Tau gene on chromosome 17 are found in some families, and some other FTD families show linkage to chromosome 3. The onset of FTD is typically 45–65, although it can occur in the elderly and has been recorded in a 21-year-old. The disease duration is typically eight years, with a range of 4 to 20 years.

Neuropathology of FTD

Atrophy of the frontal and anterior aspects of the temporal lobes is pronounced. The lateral ventricles are enlarged and the brain sometimes less than 1000 g. In classical FTD the atrophy is symmetrical, although in cases where the behaviour is disinhibited, the atrophy is particularly marked in the inferior aspects of the frontal lobes and the anterior temporal gyrus. In PPA the atrophy is asymmetrical, being most pronounced in the left temporal and frontal lobes around the Sylvian fissure. In semantic dementia the atrophy involves the temporal lobes in an asymmetric fashion. Where FTD is complicated by Parkinsonism the substantia nigra is atrophic and pallid. In cases where stereotyped compulsive behaviour is prominent, marked atrophy of the striatum is seen.

There are two main types of histopathology seen in FTD: microvacuolar degeneration, and neuronal loss with pronounced gliosis, sometimes with Pick bodies. The microvacuolar type is characterized by vacuolation of the neuropil in the affected regions accompanied by the shrinkage of nerve cell bodies and their dendrites and loss of pyramidal neurons in layers II and III of the cortex. There is minimal astrocytic reaction which is confined to the subpial regions and the border between grey and white matter. In the gliotic type the pyramidal cell loss is accompanied by a florid transcortical astrcytosis. Sometimes Pick bodies are found in pyramidal cells of the cortex and the granule cells of the dentate gyrus. Pick bodies are rounded intraneuronal aggregates that are immunoreactive to ubiquitin. In many cases of FTD, particularly in those that are due to mutations in the Tau gene, there are widespread NFTs in the cortex and many subcortical structures. However, in FTD NFT deposition is not accompanied by SPs or diffuse deposits of Aβ. Additionally, Motor Neurone Disease pathology can complicate the two main histopathological types (for reviews see Heutink (2000) Mann (2000)).

The molecular biology of Tau

In normal nerve cells 6 isoforms of Tau exist which result from the alternative splicing of a single messenger RNA (mRNA) transcript produced from the Tau gene (see Fig. 13.6). These differ first, in terms of the inclusion of certain N-terminal domains and secondly in the inclusion of either 3 or 4 imperfect repeat domains within the carboxy-terminal region of the molecule (see reviews by Spillantini and Goedert (1998) and Heutink (2000)). The repeat domains are important because they act as the microtubule-binding region and thus play a precise role in the maintenance of microtubule structure within the cell. A little more than half the Tau molecules found in normal neurons are of the 3-repeat Tau form with a remainder being of the 4-repeat form (Spillantini and Goedert 1998). If either the 3- or 4-repeat isoforms fail to function, or if the ratio of the two alters within the cell, microtubule formation and the stability of existing microtubules become compromised. In addition, any excess of unused Tau of either form, can be bundled into a tangle (indigestible residue) which will choke the cell and impair its function (Mann 2000).

The molecular genetics of FTD

In 1998 Spillantini and colleagues demonstrated linkage to chromosome 17 in families with frontotemporal dementia with Parkinsonism (Spillantini and Goedert 1998). Soon, other families were reported that also showed linkage to FTD in this region and shortly after mutations in the Tau gene, located within the region of overlapping linkage, were identified (Hutton *et al*. 1998; Poorkaj *et al*. 1998; Clark *et al*. 1998; Spillantini and Goedert 1998; Goedert *et al*. 1999). It is now clear that a substantial proportion of FTD is caused by mutations in the Tau gene. For example, in a nationwide population based study Rizzu and colleagues (1999) found pathogenic mutations in the Tau gene in 40 per cent of familial cases. Others have reported frequencies between 13.6 per cent and 29 per cent (Dumanchin *et al*. 1998; Houlden *et al*. 1999). However, Tau mutations do not appear to be a common cause of general dementia, as none have been identified in

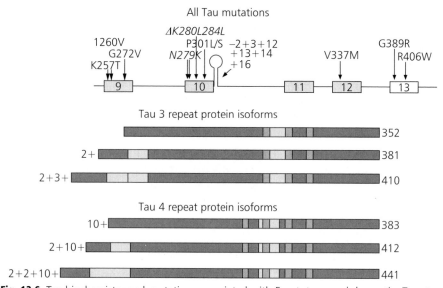

Fig. 13.6 Tau biochemistry and mutations associated with Frontotemporal dementia. Twenty mutations resulting in forms of FTD have been identified in the Tau gene, although only 14 have been published in detail. All disease-causing mutations occur within the C-terminal region of the gene and most within or adjacent to the microtubule binding domain. The mutations in exons 9, 10, 12, and 13 (i.e. ΔK280, G272V, P301L, P301S, V337M, and R406W) will result in Tau proteins with conformational changes either within or near the microtubule-binding region. Thus impairing the efficiency with which Tau can interact with tubuline and thereby destabling microtubules (Rizzu *et al.* 1999; Hasegawa *et al.* 1998). The exonic mutations of V337M and R406W result in abnormal tau made up of a mixture of 3- and 4-repeat isoforms (Hutton *et al.* 1998; Poorkaj *et al.* 1998; Spillantini and Goedert 1998; Rizzu *et al.* 1999; Hong *et al.* 1998), which results in structures identical to the paired helical filaments observed in AD (Reed *et al.* 1997). The intronic mutations identified are found within the stem loop structure of a splice accepter domain and act to increase splicing of exon 10 (Hutton *et al.* 1998; Poorkaj *et al.* 1998; Spillantini and Goedert 1998). This appears to result in an increased proportion of mRNA transcripts containing exon 10, producing an imbalance in the relative formation of tau isoforms in favour of those containing the 4-repeats. Similarly the N279K and S305N mutation may enhance splicing of exon 10 (Morris *et al.* 1999; Hong *et al.* 1998; Clark *et al.* 1998; Hasegawa *et al.* 1999), producing an excess of 4-repeat Tau. These mutations as well as P301L and P301S and the entronic mutations, produce tangles composed of predominantly) 4-repeat Tau (Hutton *et al.* 1998; Spillantini and Goedert 1998; Hong *et al.* 1998; Clark *et al.* 1998; Hasegawa *et al.* 1999), which occur as flat ribbons rather than the paired helical filaments observed in AD. Mutations outside exon 10 appear to produce only neuronal pathology, whereas the intronic mutations, produce pathology within glia, as well as in neurons. Finally, the mutations K257T and G389R produce a Pick-type histology (Pickering-Brown 2000) they affect both 3 and 4 repeat Tau which results in their accumulation as intracytoplasmic Pick-like bodies rather than paired helical filaments. They may also confer a reduced ability to bind microtubules and to promote their efficient assembly. Mutation G272V also causes the production of Pick like bodies (Rizzu *et al.* 1999; Spillantini and Goedert 1998).

early onset AD, Parkinson's disease or sporadic FTD (Rizzu *et al.* 1999; Roks *et al.* 1999; Zareparsi *et al.* 1999). Moreover, in the majority of familial FTD cases (between 60–75 per cent) the genetic defect remains unknown. Other genes which contribute to FTD remain to be detected. Indeed, linkage has already been reported to chromosome 3 in a single family (Brown *et al.* 1995).

The various FTD associated Tau mutations are portrayed in Fig. 13.6 and further details of these mutations are given in the legend to this figure. It has been postulated that all the mutations in some way result in a toxic gain of function caused by the excess deposition of hyperphosphorylated Tau protein (Heutink 2000). Three-repeat Tau binds to microtubules at different sites to those bound by 4-repeat isoforms (Goode and Feinstein 1994). Since the splicing mutations cause an excess of specific isoforms of the protein, the result could be a shortage of available binding sites for the over-expressed Tau isoform, leading to an excess of free Tau. In addition, the partial loss of function postulated to be the effect of the coding mutations, may also result in excess free Tau. The resulting free Tau could then hyperphosphorylate and/or assemble into filaments and/or aggregates (Nacharaju *et al.* 1999; Arrasate *et al.* 1999; Goedert *et al.* 1999). There is evidence that Tau aggregates become tagged for degradation by the ubiquitin degradation pathway in proteasomes, and it has been suggested that because these aggregates resist degradation they may interfere with normal ubiquitin recycling and disrupt the proteasome (Cummings *et al.* 1998). This would disrupt the normal functioning if the cell making the neuron vulnerable to other stress factors. However, whether Tau mutations cause a loss of function or a toxic gain of function or a combination of both, has yet to be fully established.

Finally, the finding that Tau mutations contribute to FTD has generated interest in their possible role in other forms of dementia which show extensive Tau pathology such as cortico–basal degeneration (CBD), progressive supranuclear palsy (PSP) and the amyotrophic lateral sclerosis/Parkinsonism/dementia complex of Guam. Genetic association studies have found no evidence of a relationship between polymorphisms in the Tau gene and these disorders with the exception of PSP. Linkage to chromosome 17 for familial forms of PSP has been excluded (Hoenicka *et al.* 1999), but several studies have reported allelic association with the intronic polymorphism after exon 9 and sporadic PSP cases (Conrad *et al.* 1997; Bennett *et al.* 1998; Oliva *et al.* 1998; Higgins *et al.* 1998; Baker *et al.* 1999; Morris *et al.* 1999; Bonifati *et al.* 1999). The association was found to an extended haplotype (H1/H1) spanning the whole Tau gene including the promoter region (Ezquerra *et al.* 1999). This has rendered the search for the biologically relevant defect difficult and so far unsuccessful. However, the finding that only the 4-repeat Tau contributes to NFT Tau pathology in PSP cases, indicates that mutations involved with the splicing ratio of exon 10 maybe of importance (Sergeant *et al.* 1999). It has also been speculated (Heutink 2000) that the relevant mutation, maybe within an as yet unknown element, may regulate this splicing and that a similar mechanism might contribute to other dementias where Tau pathology consists mainly of 3-repeat (Pick's Disease) or 4-repeat (CBD) isoforms (Feany and Dickson 1996; Feany *et al.* 1996; Delacourte 1999).

Huntington's disease and other trinucleotide repeat disorders

The trinucleotide repeat (TNR) disorders are a group of conditions in which the primary genetic lesions are expansions of a variety of different TNRs in at least 14 unrelated genes. All principally involve the central nervous system, with an onset usually in mid-life, although severe congenital forms of some can occur: for some recent reviews see (Gusella and MacDonald 2000; Cummings and Zoghbi 2000; Gutekunst *et al.* 2000). Inheritance is autosomal dominant in 9, X-Linked dominant in 4 and autosomal recessive in only one, Friedreich's ataxia. As a group the TNR disorders show interesting departures from classical Mendelian patterns of inheritance. Apart from non-penetrance and variable expressivity in some, the principal deviation is that they tend to show *anticipation*; that is they tend to become more severe and occur earlier in life in successive generations of a pedigree. TNRs tend to be unstable because they can expand during gametogenesis, with the result that the children of an affected person are at risk of inheriting longer repeats. In general, longer repeats result in a more severe phenotype and an earlier age of onset. More rarely, the repeats can contract during gametogenesis, which either results in the disorder appearing to miss a generation, or to be very much less severe.

The TNR disorders are not very common, although Fragile X-associated mental handicap (FRAXA) is, after Down syndrome, the most common cause of mental retardation (see Chapter 5). Other than FRAXA, the TNR disorder that a psychiatrist is most likely to see is Huntington's Disease (HD).

The TNR disorders are subclassified into those in which the TNR is in a non-coding part of the gene which is not, therefore, expressed in the protein; and those in which the repeat occurs in an exon, and which is manifest in the protein as a run of, depending on the disease, iterated glutamine or alanine residues. The eight disorders in which polyglutamine tracts are expanded, and in which the mutant proteins otherwise have no homology with one another, are thought to result in similar disturbances of cellular function, and so are often known as polyglutamine (polyQ) disorders.

Huntington's disease—clinical features and neuropathology

Hereditary chorea, later known eponymously as Huntington's disease (HD), was first described in a family from Eastern Long Island, New York State, by George Huntington in 1872. The condition had previously been recognized by Huntington's father and grandfather. In the 1930s this North American pedigree was traced back to two bothers who originated from the County of Suffolk, England (for a review see OMIM 143100).

It is an autosomal dominant disorder, affecting both sexes equally. Its mode of onset is between 30 and 40 years, although a severe childhood form can occur and, at the other extreme, some cases do not manifest until after the age of 60. Its prevalence is around 5 per 100 000. The usual picture is of a progressive generalized chorea complicated in the later stages by deterioration of the personality and intellect. CT or MRI scans can show characteristic atrophy of the caudate nucleus. Death occurs from 10 to

15 years after the onset of symptoms. There is no effective treatment, although the chorea can be controlled to some extent by the use of dopamine blocking or dopamine depleting drugs.

HD is associated with considerable psychiatric morbidity, both in affected individuals and their families. Early studies suggested that behavioural disturbance, affective disorder, or even psychosis often complicated the prodromal period; and that these psychiatric symptoms were a harbinger of the disease. However, more recent studies cast doubt on these observations and suggest that psychiatric morbidity is equally increased in those who do not go on to develop HD as in those who do (Shiwach and Norbury 1994), suggesting that the higher rates of psychiatric morbidity seen in HD families probably result from belonging to an affected family, and of living with the knowledge increased risk. However, once HD is manifest, complicating psychiatric illness is very common with risks of 39 per cent for depression; 72 per cent personality change, and 9 per cent schizophrenia (Shiwach 1994).

It is not certain whether subtle neuropsychological dysfunction antedates the onset of HD. One small study found no differences between confirmed carriers and non-carriers (Giordani *et al.* 1995); whereas another study found that confirmed carriers did worse, particularly on tests of attention, learning and planning functions (Rosenberg *et al.* 1995). It may be pertinent that the subjects tended to be younger in the study by Giordani and colleagues. Certainly, once HD has declared itself, neuropsychological tests are abnormal, with most decline being seen in tests of psychomotor skill (Bamford *et al.* 1995; Kirkwood *et al.* 1999; Kirkwood *et al.* 2000).

At the time of death the HD brain is atrophic with dilated ventricles. Atrophy and neuronal loss is particularly marked in the neostriatum and the frontal lobes. The first signs of neurodegeneration and cell loss are in the enkephalin positive medium spiny projection neurones found in the head of the caudate nucleus and putamen. These neurones utilize GABA and are inhibitory, which suggests that the chorea is due to an imbalance between the dopaminergic and GABAergic input to the neostriatum.

Huntingtin—the gene

HD was one of the first human genetic disorders to be targeted by the then new recombinant DNA technology, and linkage was rapidly found to markers on the short arm of chromosome 4 (Gusella *et al.* 1983). However, it took 10 years of intense international cooperation and endeavour before the HD Research collaborative Group were able to report the cloning of the HD gene. The gene is on chromosome 4p16.3 and consists of 67 exons ranging in size from 48 bp to 341 bp, which encode a protein, known as huntingtin, which is approximately 350 kD in size. Huntingtin has no known homology (outside of the polyQ tract) with any other protein and its normal function is still unknown. It is expressed ubiquitously, is normally located within the cytoplasm and may, possibly, interact with cytoskeletal components.

In HD a CAG repeat located in the first exon of the gene is expanded from a normal range of 9–37 to 37–121 repeats. The CAG codes for glutamine (Q), which is expressed in the protein as a polyQ tract. Mutant, repeat-expanded huntingtin appears to be expressed at comparable levels to the normal huntingtin. It would appear that HD is a truly dominant condition because individuals homozygous for the HD mutation do not have more severe disease than heterozygotes (Wexler *et al.* 1987; Myers *et al.* 1989). Moreover, since inactivation of one huntingtin gene does not lead to disease in either humans or experimental mice, and in HD, equivalent levels of expression are seen from both the normal and mutated huntingtin genes, it seems that HD is the result of a toxic gain of function.

CAG repeat instability in HD

In several large studies, the normal range of repeats was from 9–37 with a mean around 19. In HD the repeat length ranged from 37–86 with a mean of 46.4. A later study looked in detail at a large cohort of individuals who carried between 30 and 40 CAG repeats in the huntingtin gene (Rubinsztein *et al.* 1996). It was found that nobody with 35 repeats or less developed HD, that most people with 36–39 repeats were clinically affected, but that 10 individuals with 36–39 repeats, ranging in age from 67–95 years, were apparently normal. It may be, therefore, that HD may not be fully penetrant for repeats in the 36–39 range. At the other extreme Nance and co-workers (Nance *et al.* 1999) describe a juvenile case of HD with onset at age two-and-a-half associated with a polyglutamine tract of 250 residues!

In HD the CAG repeat length is the same in different tissues: in other words, the huntingtin gene does not show somatic instability in contrast to FRAXA in which the CGG repeat varies in length in different tissues. On the other hand, CAG repeat length typing of *single* sperm indicates that the huntingtin gene is unstable in the male germline. As the pathologic repeats increased in size both the frequency and size of germline expansions increases. For normal alleles, the frequency of repeat contraction is higher for longer repeats, at least up to 36 CAG repeats, after which the contraction frequency falls as the expansion frequency rises (Leeflang *et al.* 1995).

Alleles of between 29–35 repeats have a frequency of approximately 1 per cent in Caucasians. There is evidence that these intermediate alleles are unstable and can act as pre-mutation alleles because they are found in the parents of patients with sporadic HD, all of whom have repeat expansions well within the pathological range (Goldberg *et al.* 1993). A study by Chong and colleagues used single sperm CAG typing to look at the frequency of repeat expansion from between 29 and 35 CAG repeats in normal men as opposed to men from families with sporadic HD (Chong *et al.* 1997). They found that intermediate alleles were more likely to expand into the pathological range when they originated from the sporadic HD families; the actual difference in frequencies being 10 per cent as opposed to 6 per cent. Two sequence differences in 12 nucleotides separating the CAG repeat from a downstream, non-coding CCG repeat seemed to account, at least in part, for the difference.

HD family studies have shown that the age of onset tends to decrease and the severity of the illness increase in successive generations, and that this effect is almost only seen in transmission from affected fathers. This phenomenon of genetic anticipation seen in HD is explained by the paternal germline instability of the CAG repeat. A number of studies have shown an inverse relationship between the CAG repeat length and the age of onset, with the change in the age of onset between the fathers and their children being correlated with the expansion of the repeat (Ranen *et al.* 1995; Trottier *et al.* 1995). Anticipation due to maternally derived repeat expansions has been described by Laccone and Christian (2000). In this instance a woman with an allele of 36 repeats and no family history of HD went on to have two daughters, both of whom developed HD in early adult life. One of the sisters had 66 repeats and the other, 57 repeats. The authors suggested that the mother was probably a germline mosaic and that repeat instability might occur in the female germline, but that oocytes carrying long repeats might experience negative selection (Laccone and Christian 2000).

CAG repeat length accounts for up to 60 per cent of the variance of the age of onset in HD, though at the lower end of the disease repeat spectrum only some 20 per cent of the variance is accounted for. This suggests that other genes or environmental factors may influence the age of HD onset, at least in those patients with fewer than 50 repeats. Rubinszstein *et al.* (1997) and subsequently MacDonald *et al.* (1999) found that up to 13 per cent of the variance in the age of onset could be attributed to a polymorphic TAA repeat in the 3'-UTR of the GluR6 kainate receptor, a receptor which renders neurons vulnerable to excitotoxic cell death (Rubinsztein *et al.* 1997; MacDonald *et al.* 1999).

The mechanism by which the expanded polyQ tract causes neuronal dysfunction and death in HD and other CAG repeat disorders is the object of much current investigation. Insights have been obtained mainly from expressing engineered polyQ tract proteins in cell culture systems and from transgenic mouse studies. Overall there is evidence that mutated polyQ proteins have a different subcellular distribution, form aggregates with themselves and other proteins, that these aggregates stress the cell's waste disposal proteolytic pathways, and that the long polyQ tracts, by binding to transcription factors and other regulatory proteins, adversely affect transcription of certain genes (for reviews of the cell biology of Huntington's Disease see Sisodia (1998); Gusella and MacDonald (2000); Cha (2000); Cummings and Zoghbi (2000)). Evidence which suggests that at least some of the neurodegeneration seen in HD is actually the result of loss of normal huntingtin function is also emerging, the normal protein being inactivated as a result of forming aggregates with the mutated huntingtin (Cattaneo *et al.* 2001).

The transmissible spongiform encephalopathies

The transmissible spongiform encephalopathies (TSEs) are disorders of a variety of different animal species, including man, that are thought to be due to an unusual infectious agent, now believed by the majority of workers in the field to be an abnormally folded form of a cellular protein, the prion protein. This is hypothesized to induce native, normally folded prion protein to adopt the more stable abnormal conformation, thus intitiating a chain reaction, which is thought to lead to toxic, insoluble aggregates of Prion protein (Fig. 13.7) (Horwich and Weissman 1997; Prusiner 1982, 1993, 1997; Liberski 1995; Haywood 1997; Johnson and Gibbs Jr 1998; Hegde *et al.* 1998; Telling 2000). This hypothesis explains why TSE agents apparently replicate without the need of nucleic acid. Pruisner (1982) coined the acronym 'prion', which stands for 'Proteinaceous infectious particles which replicate without the need of

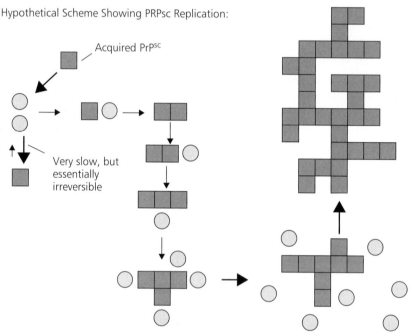

Hypothetical Scheme Showing PRPsc Replication:

Acquired PrP^{SC}

Very slow, but essentially irreversible

Fig. 13.7 Diagramatic representation of the conversion of PrP^C to PrP^{SC}. PrP^C is represented by the orange circles and PrP^{SC} by the green squares. PrP^C has 40% α-helix and 3% β-sheet whereas PrP^{SC} has 30% α-helix and 40% β-sheet. It is proposed that PrP^C can very occasionally 'flip' into a more stable β-sheet rich PrP^{SC} form which has the potential to interact with native PrP^C and greatly facilitate its conversion into PrP^{SC}, thus initiating a chain reaction, which leads to the formation of toxic aggregates of PrP^{SC}. It is thought that Prion gene mutations can lower the activation energy of the conversion of PrP^C to PrP^{SC}, thus greatly increasing the likelihood that the conversion takes place within the average lifespan. Alternatively, PrP^{SC} can be acquired, either by ingestion or innoculation, form the environment, and this foreign PrP^{SC} initiates the chain reaction. The species barrier occurs because a certain degree of similarity/identity is required between the host PrP^C and invading PrP^{SC} for the interaction to occur (see reviews by (Prusiner 1982))(Haywood 1997; Horwich and Weissman 1997; Johnson and Gibbs, Jr. 1998; Liberski 1995; Prusiner 1993; Prusiner 1997; Telling 2000). This schema is an oversimplification. For example, neurodegeneration may occur without the formation of aggregates of PrP^{SC}, rather that C-terminal fragments of PrP^{SC} become lodged in the membrane of the endoplasmic reticulum, and that it is this event which proves toxic to the neurone (Hegde *et al.* 1998).

nucleic acid' (Prusiner 1982). The normal soluble cellular form of the prion protein is abbreviated to PrP^C and the protease resistant infectious form of the prion protein as PrP^{SC}. Different TSE strains can be biochemically 'typed' by digesting infected brain tissue with proteases, running the digested material out on SDS-PAGE, Western blotting, and visualizing the PrP^{SC} by using antobodies to prion protein.

Affected animals generally appear distressed, and show ataxia and other motor signs before showing signs consistent with the progressive loss of higher neurological functions. Although the incubation periods of TSEs are characteristically very long, once symptoms appear, susequent deterioration is relentless, death typically occurring after a matter of months. Neuropathological studies show profound loss of neurons, widespread vacuolation of the neuropil (status spongiosis) and pronounced reactive gliosis. Gross atrophy is not usually marked and signs of inflammation are absent. Often there are extracellular aggregates of a protease resistant form of the prion protein with the staining properties of amyloid. TSEs can appear as inherited, congenital or transmissible diseases. Occasionally a TSE affecting one species can cross the 'species barrier' and infect a new species: such a scenario is thought to explain the transmission of Bovine Spongiform Encephalopathy from cattle into humans, where it gives rise to a condition known as new variant CJD (vCJD).

The TSEs have been the subject of a number of excellent reviews, to which the interested reader is referred for more detailed information (Prusiner 1993, 1997; Liberski 1995; Horwich and Weissman 1997; Haywood 1997; Westaway *et al.* 1998; Telling 2000).These reviews give good accounts of the testing of the various predictions of the prion hypothesis using transgenic mice.

Varieties of TSEs

The archetypal prion disorder is scrapie, a disease of sheep, which has been known for hundreds of years. Affected animals show either ataxia of the hind limbs, or signs of an intense pruritis, which leads the afflicted animal to rub or scrape against the sides of its enclosure. Icelandic shepherds thought that it was an infectious disease. However, when subsequent observations suggested that certain breeds of sheep, such as Cheviot, Swaledale and Suffolk, were more vulnerable than others, it began to be thought of as a genetic disorder. Cuillé and Chelle established the infectious potential of scapie in 1936 when they demonstrated that it could be transmitted to healthy sheep by injection of bodily material from affected sheep (Cuillé and Chelle 1936). Subsequently it was found that the scrapie agent could multiply in, and be passaged indefinitely in, sheep and goats. Early work established the small size and extreme robustness of the scrapie agent and the absence of a host immune response to the infection. Further advances followed when it was discovered that it was possible to infect and propagate the scrapie agent in experimental mice and hamsters. Much of our current understanding of the transmissibility and pathogenesis of TSEs is based on work carried out with rodent-adapted scrapie strains.

TSEs affecting mankind—Kuru

Zigas and Gadjusek brought Kuru to the attention of the western world in 1957 (Zigas and Gajdusek 1957). It used to be the major cause of death in women, young adults and children among the Fore-speaking highlanders of Papua New Guinea. Kuru, which translates as 'trembling', is thought to have resulted from either ingestion, or inoculation through skin abrasions, of infected brain tissue during the course of funerary or endocannibalistic rites involving dead relatives. It was the women, together with the infants sitting on their laps during the rites who were responsible for these rites, an observation that explains the epidemiology of the disorder. Affected individuals suffered prodromal symptoms of malaise before developing ataxia, a shivering-like tremor and cerebellar signs. Those afflicted, once the disease had become manifest, often showed an incongruous hilarity. Symptoms progressed over a matter of months to a state of recumbency, inanition and stupor, before death occurred after a disease duration of between one year and eighteen months. Children succumbed more swiftly and often developed bulbar signs. The disease is now all but extinct owing to the banning of the traditional funerary rites. No cases of maternal (vertical) transmission were ever documented and the consensus is that Kuru was infectious rather than inherited.

The neuropathology of Kuru is characterized by the occurrence of abundant amyloid plaques, together with status spongiosis and reactive gliosis. The pathological similarities between Kuru and scrapie were noted by Hadlow in 1959, who suggested that transmission into primates should be attempted (Hadlow 1959). This was first achieved in 1966 when it was demonstrated that chimpanzees developed a disorder similar to Kuru some 18–21 months after intracerebral inoculation with infected brain (Beck *et al.* 1966; Gajdusek *et al.* 1966). It has not been possible to transmit Kuru to rodents.

CJD

Creutzfeld-Jakob disease (Creutzfeldt 1920; Jakob 1921) is a human TSE occurring at about 0.5 cases/million people/year, the majority of cases being in the seventh decade. Some 10 per cent of CJD cases are familial, showing autosomal dominant inheritance. Those affected show non-specific mental and physical symptoms before developing a rapidly progressive dementia complicated by a number of possible neurological signs. Such ancillary signs commonly include myoclonus, pyramidal and extrapyramidal signs, and cerebellar signs: occuring more rarely are cortical blindness, ocular palsies, vestibular dysfunction, lower motor neurone signs, sensory deficits, autonomic abnormalities and seizures. Classically the myoclonus is accompanied by biphasic or triphasic spike and wave complexes on EEG. The demonstration of such complexes is almost diagnostic of the disorder, but they may only occur briefly during the course of the disease, or may not occur at all. Death normally occurs within one year and there is no effective treatment.

Apart from the demonstration of the characteristic EEG changes, the clinical diagnosis may be supported by MRI scanning and immunoassay of the neural 14–3–3 protein in CSF. Ordinary CT scanning is not very helpful. MRI scanning has been reported to show hyperintense signals in the basal ganglia on T2-weighted images in 79 per cent of patients (Finkenstaedt *et al.* 1996). Although not absoutely specific for CJD, the demonstration of elevated 14–3–3 protein, in the absence of a history of strokes or encephalitis, is highly supportive of a diagnosis of CJD (Hsich *et al.* 1996).

The neuropathological similarities between Kuru and CJD were first noted by Klatzo and colleagues in 1959, although only some 10 per cent of CJD brains show amyloid deposits (Klatzo *et al.* 1959). CJD was transmitted to chimpanzees in 1968 and subsequently it has been possible to transmit some isolates of CJD to rodents.

Iatrogenic CJD

There have been close to 200 cases of CJD resulting from medical procedures (Johnson and Gibbs Jr 1998). The chief sources of infection have been human growth hormone preparations derived from pooled cadaveric pituitaries, corneal transplants, dura mater grafts and from contaminated surgical instruments used in neurosurgical procedures. Scrapie can be transmitted experimentally and possibly in the field by inoculation of the conjunctiva, and high titres of the scrapie agent are found in the corneas of infected animals. At one time it was thought that the higher incidence of CJD in Libya was the result of the dietary practice of eating sheep eyeballs. However, it is now thought more probable that a clustering of familial CJD associated with a mutation within the prion gene is the most likely explanation (Hsiao *et al.* 1991)

Gerstmann-Strässler-Scheinker disease and fatal familial insomnia

These are very rare conditons, showing autosomal dominant inheritance, caused by mutations within the prion gene. It is usually possible to transmit these disorders to rodents.

Gerstmann-Strässler-Scheinker (GSS) disease usually becomes manifest as a cerebellar ataxia in the third or fourth decade of life. Dementia occurring together with abnormalities of gaze and dysarthria develops after several years. Death usually occurs after four or five years. Some forms of GSS present as a dementing illnes, usually in association with Parkinsonian signs. In this form it can be mistaken for Alzheimer's Disease (AD).

Fatal Familial Insomnia is an extremely rare autosomal dominant condition. It manifests in middle age as a progressively worsening insomnia associated with motor signs and autonomic dysfunction. Death occurs in six months to three years.

'New variant' CJD and BSE

Bovine Spongiform Encephalopathy (BSE), popularly known as 'mad cow disease', has devastated British beef farming and has crossed the species barrier to cause novel TSEs in other animal hosts, including man, where it is known as 'New variant' CJD (vCJD). Affected animals show ataxia of the hind limbs, hyperaesthesia, and apprehension. Most animals were infected as calves and, with an incubation period of 2.5 to 8 years, the modal period of occurrence is 5 years. Neuropathological changes typical of a TSE are seen in the hind brain.

Since 1986 there have been 178,378 confirmed cases of BSE in the UK. The epidemic peaked in 1992 at 36,680 cases, but is still lingering, with 2,254 cases being reported in 1999 and 572 confirmed cases by October 2001 (www.defra.gov.uk/animalh/bse). Although the data are ambiguous, it is possible that maternal transmission accounts for the persistence of BSE. A maternal transmission rate of 10 per cent was seen in an experimental cohort kept for seven years, although, owing to the slaughter of cattle very early in the incubation period, and the likelihood of transmission to calves from the dams being very low early in the incubation period, it is probable that the rate of maternal transmission on the farm is very much lower.

The origin of the BSE epidemic would appear to be meat-and-bone meal (MBM) derived from the rendering down of the carcasses of a variety of animals to produce food supplements for dairy cattle and other animals (for a review see Collee and Bradley 1997a, 1997b). Although feeding such supplements had been routine practice since the 1940s, in 1981/82 the rendering procedure was simplified by omitting the final hydrocarbon solvent extraction and subsequent steam assisted solvent recovery steps. This was done for safety reasons and also because the market for tallow (rendered down fats) declined. Because the first cases of BSE were reported some five years after the rendering process became less stringent, it is possible that there is a causal link between the change in meat-and-bone meal production and the emergence of BSE.

It is not known whether the original source of the BSE agent was bovine, ovine, or some other animal. Sheep scrapie experimentally transmitted to cattle does not produce the same symptoms or neuropathology as does the BSE agent, which suggests that BSE may not due to scrapie. A report that an outbreak of transmissible mink encephalopathy was traced to a dead steer that was subsequently fed to the captive mink, suggests the possibility that cattle are susceptible to their own form of TSE and that BSE is of bovine origin (Marsh and Bessen 1993). On the other hand, only 29 ovine strains of scrapie have been characterized, so there is every possibility that BSE may, in fact, be due to a previously unknown sheep strain (Coghlan 2001).

There is now incontrovertible evidence that the BSE agent can infect other animals, including man. BSE can be transmitted to a variety of different animal species both by perenteral inoculation or ingestion of brain material from infected cattle. In every case the neuropathological findings are similar enough to suggest that the BSE agent transmits as a new strain of TSE. Western blot studies of protease and detergent resistant prion protein from the brains of infected animals suggest that the BSE agent can cross the species barrier and keep its characteristic biochemical signature.

Although, it is not formally proven, it is very likely that a new form of TSE, known as new-variant CJD (vCJD), affecting, for the most part, young people, has resulted from eating commercially prepared meat products which, before the November 1989 ban on the use of cattle offal in human food, often contained CNS material as a binding agent. It has been estimated that some 450,000 infected cattle entered the human food chain before this ban was instituted (Collee and Bradley 1997a,b). The first case of vCJD was reported in 1994 and, as of October 2001, there have been 104 definite and probable cases of vCJD reported to the CJD surveillance unit in Edinburgh (www.doh.gov.uk/cjd/stats). The latest figures suggest that the rate of annual increase may be levelling off. In addition, the fact that an anonymous series of 3000 tonsil and appendix specimens examined for the presence of the vCJD PrP^{SC} failed to yield any positives, suggests that the disease is (hopefully) not widespread in the general population (Ironside *et al.* 2000).

vCJD cases have all been young adults or people in early middle age. The illness has presented with a prodromal phase of non-specific neuropsychiatric symptoms and dysaesthesia lasting some six months, before neurological symptoms of ataxia and tremor first appear. Many victims have first been referred to the psychiatric services. The disease then progresses towards a dementia occurring with progressive motor and sensory deficits. The course of the illness is more protracted than with sporadic CJD but, even so, death occurs within two years of the onset of symptoms. Although the EEG is abnormal, the triphasic spike and wave complexes characteristic of CJD are not seen. There are reports that MRI scanning may be helpful in establishing the diagnosis because, for unknown reasons, bilateral hyperintense signals are seen in the pulvinar regions. The disease may be diagnosed by tonsil biopsy (Hill *et al.* 1997). Neuropathological findings are of pronounced vacuolation and dense amyloid staining deposits of prion protein. So far, all reported cases have been homozygous for methionine at the common polymorphic site at codon 129 of the prion protein gene.

The prion protein gene

The human PrP is located on the short arm of chromosome 20p12-pter (Sparkes *et al.* 1986; Liao *et al.* 1986). Surprisingly it was found that the PrP mRNA levels were the same in normal as in infected brain. The gene is widely expressed and encodes a sialoglycoprotein that is bound to the plasma membrane via a glycolphosphatidylinosito (GPI) anchor. The coding region is contained within one exon and is separated from the 5′ untranslated region by an intron of at least 10 kb. In some species the 5′ untranslated region is split into two exons by a short intron. Gene transfection studies into mice have suggested that the large intron preceeding the coding exon contains an enhancer vital to the expression of the prion gene.

The PrP gene is represented in diagrammatic form in Fig. 13.8. Polymorphisms are depicted above, and disease associated mutations below the gene sequence. Usually there are five octarepeat sequences (codons 51 to 91), sometimes four, and, rarely, more than five. Ten or more octarepeats are associated with familial CJD. The M129V polymorphism is benign but can influence the clinical manifestation of the D178N mutation and homozygosity for M or V can increase susceptibility to acquired prion diseases (see below). There is a benign E219K polymorphism found in people of Japanese descent and there is a suggestion that heterozygotes for this polymorphism are under represented in sporadic CJD. Amino acid polymorphisms have also been described for the mouse and sheep PrP which influence susceptibility to infection with scrapie.

Mutations in familial CJD and GSS

In the region of 20 mutations have been described that are associated with, and probably cause, the various autosomal dominant prion disorders. Some examples are described in the following section. A full listing can be found in OMIM (http://www.ncbi.nlm.nih.gov/htbin-post/omim/dispmim?176640).

As has been found for other genes, in familial prion diseases, codons including a CpG are particularly vulnerable to mutation. This is because in the vertebrate genome the cytosine of a CpG is often methylated, and methyl-cytosine is vulnerable to deamination to thymine, with the result that CpG becomes TpG. Two examples of this mechanism in human prion disease are the P102L mutation linked to GSS and the E200K mutation associated with a familial form of CJD. These mutations have been independently isolated from a number of unrelated families from around the world and it is, therefore, likely that most have arisen independently and do not result from a common ancestor.

Some forms of familial CJD give rise to a slowly progressive dementing illness, which has, on occasions, been mistaken for familial AD. These diseases often have abundant

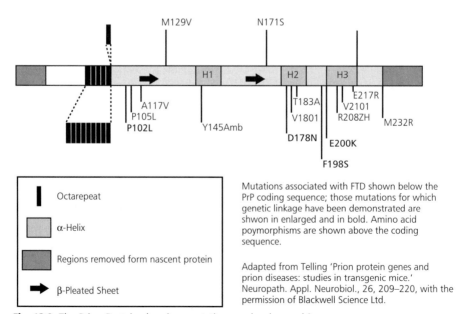

Mutations associated with FTD shown below the PrP coding sequence; those mutations for which genetic linkage have been demonstrated are shwon in enlarged and in bold. Amino acid poymorphisms are shown above the coding sequence.

Adapted from Telling 'Prion protein genes and prion diseases: studies in transgenic mice.' Neuropath. Appl. Neurobiol., 26, 209–220, with the permission of Blackwell Science Ltd.

Fig. 13.8 The Prion Protein showing mutations and polymorphisms.

deposits of amyloid, which are immunostained with antisera to PrP rather than with antisera to β-amyloid, as would be the case in AD. A number of these pedigrees have been shown to have a A117V mutation in the PrP gene. The F198S and Q217R mutations are associated respectively with familial CJD and GSS, both occurring with PrP^{SC} containing amyloid deposits and NFTs, which, like those found in AD, stain with antisera to Tau protein.

The D178N mutation is particularly interesting because when it occurs in cis with the Valine allele of the codon 129 polymorphism, familial CJD results; but when it occurs in cis with the Methionine allele of the polymorphism, fatal familial insomnia results. The codon 129 polymorphism is known to reside in one of the short stretches of beta-sheet found in PrP^C and, as such, is thought to occupy a region which may be critical for the transformation of the α-helix rich PrP^C to the β-sheet rich PrP^{SC}. It is possible that the effect of the D178N mutation on the tertiary structure of PrP^C is subtly different between the Met and Val alleles and that this would give rise to a difference in the conformation of the PrP^{SC}, which would, in turn, impact on the clinical phenotype. Certainly the proteinase-K resistant fragments associated with the two conditions are of different sizes: that characteristic of FFI being 19 kDa, whereas that found in this form of familial CJD is 21 kDa.

The M129V polymorphism and susceptibility to CJD

The M129V polymorphism appears to have a strong influence on the susceptibility to acquired and iatrogenic CJD. The data are summarized in Table 13.5. The main conclusion is that homozygosity, particularly for the Met allele, at codon 129 confers increased susceptibility to CJD. Work carried out on archival material from sufferers of

Table 13.5 Homozygosity at Codon 129 of the Prion Gene appears to influence susceptibility to Prion disease

Disease	Nos. Tested	Met/Met (%)	Met/Val (%)	Val/Val (%)	Homozygous (%)
Healthy Controls	261	96 (37)	136 (52)	29 (11)	125 (48)
Sporadic CJD	73	57 (78)	9 (12)	7 (10)	64 (88)
Iatrogenic CJD	63	38 (60)	7 (11)	18 (29)	56 (89)
vCJD	102	102 (100)	0	0	100
Fore Kuru	15	8 (53)	4 (27)	3 (20)	11 (73)
Fore Survivors	17	0	11 (65)	6 (35)	6 (35)
Non-Fore controls	27	4 (15)	15 (56)	8 (29)	12 (44)

Data from Johnson and Gibbs, 1998, supplemented with data from Lee *et al.* 2001. The Chi-squared test suggests significance at the 0.0001 level for codon 129 homozygotes versus heterozygotes for sporadic CJD and iatrogenic CJD compared individually against healthy controls. The Chi-squared statistic is significant at the 5% level for the Fore Kuru group compared against the Fore Survivors. The Chi-squared statistic is not significant at the 5% level for the Fore Kuru and Fore survivor groups compared separately with the Non-Fore controls

Kuru has been particularly instructive. First, it appears that homozygosity at codon 129 (particularly for Met) was associated with an earlier age of onset and a shorter duration of illness than was heterozygosity, but that the clinical picture was similar across all genotypes (Cervenakova *et al.* 1998). Second, by employing a smaller subset of the archival material which originated from males, who would only have been exposed to infection between the ages of 1 and 10, it was possible to show that a higher prevalence of M/M homozygosity amongst those males that developed Kuru before the age of 20. Furthermore, the survivors of this exposed cohort had a very low frequency of M/M homozygotes, indicating that individuals with this genotype had experienced negative selection (Lee *et al.* 2001). Because, so far, all victims of vCJD have been M/M homozygotes, it appears that the findings for Kuru may also be applicable to the vCJD 'epidemic'. If this is so, it suggests that we will only be able to estimate the likely duration of the 'epidemic' when victims with V/V and M/V genotypes start to appear.

Conclusions and future directions

The last 10 years have seen great advances in our understanding of the genetics and biology of rare forms of dementia caused by mutations in single genes. Increasingly attention is being turned to the more common and genetically complex forms. The genetic epidemiological data on VaD are not conclusive and more work is needed. However, it seems likely that there is an important environmental component which should be amenable to reduction by the Public Health measures already aimed at vascular disease. The evidence for a genetic influence on the commonest form of dementia, late onset AD, is somewhat stronger. Moreover, the various FAD mutations strongly implicate abnormal metabolism of APP and the over production of Aβ as likely to be central to the pathogenesis of the disorder. However, in view of recent findings in the tauopathies, it would seem premature to exclude an important role for Tau in the pathogenesis of AD, and understanding how mis-metabolism of APP leads to abnormalities of Tau should now be a priority for biological studies.

Molecular genetic studies have clearly implicated APOE4 as a risk factor for late onset AD in almost all populations studied. Genetic epidemiological studies suggest that several other loci of similar effect size also exist (Daw *et al.* 1999) and recently allele-sharing studies in affected sibling pairs have implicated a number of chromosomal regions (Kehoe *et al.* 1999; Myers *et al.* 2000; Pericak-Vance *et al.* 2000). Even with access to the working draft of the human genome sequence, identification of the susceptibility loci may take time, given the large size of the regions concerned. However, parallel advances in the biology of AD should increasingly allow plausible positional candidates to be nominated and this, together with advances in high throughput genotyping, research initiatives to ascertain powerful samples for association studies and improved understanding of patterns of LD in the human genome offer grounds for cautious optimism (Chapter 17).

References

Abrahamson M. and Grubb A. (1994) Increased body temperature accelerates aggregation of the Leu-68 → Gln mutant cystatin C, the amyloid-forming protein in hereditary cystatin C amyloid angiopathy. *Proc Natl Acad Sci USA* **91**:1416–20.

Abrahamson M., Grubb A., Olafsson I. *et al.* (1987) Molecular cloning and sequence analysis of cDNA coding for the precursor of the human cysteine proteinase inhibitor cystatin C. *FEBS Lett* **216**:229–33.

Ancolio K., Dumanchin C., Barelli H. *et al.* (1999) Unusual phenotypic alteration of beta amyloid precursor protein (betaAPP) maturation by a new Val-715 → Met betaAPP-770 mutation responsible for probable early-onset Alzheimer's disease. *Proc Natl Acad Sci USA* **96**:4119–24.

Anderton, B. H., Dayanandan, R., Killick, R. *et al.* (2000) Does dysregulation of the Notch and wingless/Wnt pathways underlie the pathogenesis of Alzheimer's disease? *Mol Med Today* **6**:54–9.

Arrasate M., Perez M., Armas-Portela R. *et al.* (1999) Polymerization of tau peptides into fibrillar structures. The effect of FTDP-17 mutations. *FEBS Lett* **446**:199–202.

Arriagada P. V., Growdon J. H., Hedley-Whyte E. T. *et al.* (1992) Neurofibrillary tangles but not senile plaques parallel duration and severity of Alzheimer's disease. *Neurology* **42**:631–9.

Artiga M. J., Bullido M. J., Frank A. *et al.* (1998) Risk for Alzheimer's disease correlates with transcriptional activity of the APOE gene. *Hum Mol Genet* **7**:1887–92.

Asada T., Yamagata Z., Kinoshita T. *et al.* (1996) Prevalence of dementia and distribution of ApoE alleles in Japanese centenarians: an almost-complete survey in Yamanashi Prefecture, Japan. *J Am Geriatr Soc* **44**:151–5.

Baker M., Litvan I., Houlden H. *et al.* (1999) Association of an extended haplotype in the tau gene with progressive supranuclear palsy. *Hum Mol Genet* **8**:711–15.

Bales K. R., Verina T., Dodel R. C. *et al.* (1997) Lack of apolipoprotein E dramatically reduces amyloid beta-peptide deposition. *Nat Genet* **17**:263–4.

Bamford K. A., Caine E. D., Kido D. K. *et al.* (1995) A prospective evaluation of cognitive decline in early Huntington's disease: functional and radiographic correlates. *Neurology* **45**:1867–73.

Barrett-Connor E. (1998). Rethinking estrogen and the brain. *J Am Geriatr Soc* **46**:918–20.

Beck E., Daniel P. M., Alpers M. *et al.* (1966) Experimental 'kuru' in chimpanzees. A pathological report. *Lancet* **2**:1056–9.

Bennett P., Bonifati V., Bonuccelli U. *et al.* (1998) Direct genetic evidence for involvement of tau in progressive supranuclear palsy. European Study Group on Atypical Parkinsonism Consortium. *Neurology* **51**:982–5.

Bergem A. L., Engedal K., Kringlen E. (1997) The role of heredity in late-onset Alzheimer disease and vascular dementia. A twin study. *Arch Gen Psychiatry* **54**:264–70.

Betard C., Robitaille Y., Gee M. *et al.* (1994) Apo E allele frequencies in Alzheimer's disease, Lewy body dementia, Alzheimer's disease with cerebrovascular disease and vascular dementia. *Neuroreport* **5**:1893–6.

Bird T. D., Nemens E. J., Kukull W. A. (1993) Conjugal Alzheimer's disease: is there an increased risk in offspring? *Ann Neurol* **34**:396–9.

Blansjaar B. A., Thomassen R., and Van Schaick H. W. (2000) Prevalence of dementia in centenarians. *Int J Ger Psychiatry* **15**:219–25.

Bonifati V., Joosse M., Nicholl D. J. *et al.* (1999) The tau gene in progressive supranuclear palsy: exclusion of mutations in coding exons and exon 10 splice sites, and identification of a new intronic variant of the disease-associated H1 haplotype in Italian cases. *Neurosci Lett* **274**:61–5.

Borchelt D. R., Ratovitski T., van Lare J. *et al.* (1997) Accelerated amyloid deposition in the brains of transgenic mice coexpressing mutant presenilin 1 and amyloid precursor proteins. *Neuron* **19**:939–45.

Braak H. and Braak E. (1991) Neuropathological stageing of Alzheimer-related changes. *Acta Neuropathol (Berl)* **82**:239–59.

Brass L. M., Isaacsohn J. L., Merikangas K. R. *et al.* (1992) A study of twins and stroke. *Stroke* **23**:221–3.

Breitner J. C. (1991) Clinical genetics and genetic counseling in Alzheimer disease. *Ann Intern Med* **115**:601–6.

Breitner J. C., Silverman J. M., Mohs R. C. *et al.* (1988) Familial aggregation in Alzheimer's disease: comparison of risk among relatives of early- and late-onset cases, and among male and female relatives in successive generations. *Neurology* **38**:207–12.

Breitner J. C., Welsh K. A., Gau B. A. *et al.* (1995) Alzheimer's disease in the National Academy of Sciences-National Research Council Registry of Aging Twin Veterans. III. Detection of cases, longitudinal results, and observations on twin concordance. *Arch Neurol* **52**:763–71.

Breitner J. C., Wyse B. W., Anthony J. C. *et al.* (1999) APOE-epsilon4 count predicts age when prevalence of AD increases, then declines: the Cache County Study. *Neurology* **53**:321–31.

Brown J., Ashworth A., Gydesen S. *et al.* (1995) Familial non-specific dementia maps to chromosome 3. *Hum Mol Genet* **4**:1625–8.

Brun A. and Gustafson L. (1993) The Lund Longitudinal Dementia Study: a 25-year perspective on neuropathology, differential diagnosis and treatment, in B. Corain, K. Iqbal, M. Nicolini *et al.* (eds) *Alzheimer's Disease: Advances in Clinical and Basic Research*. New York: John Wiley & Sons, Inc.

Buxbaum J. D., Liu K. N., Luo Y. *et al.* (1998) Evidence that tumor necrosis factor alpha converting enzyme is involved in regulated alpha-secretase cleavage of the Alzheimer amyloid protein precursor. *J Biol Chem* **273**:27765–7.

Cai H., Wang Y., McCarthy D. *et al.* (2001) BACE1 is the major beta-secretase for generation of Abeta peptides by Neurons. *Nat Neurosci* **4**:233–4.

Campion D., Dumanchin C., Hannequin D. *et al.* (1999) Early-onset autosomal dominant Alzheimer disease: prevalence, genetic heterogeneity, and mutation spectrum. *Am J Hum Genet* **65**:664–70.

Carmelli D., Swan G. E., Robinette D. *et al.* (1992) Genetic influence on smoking—a study of male twins. *N Engl J Med* **327**:829–33.

Cattaneo E., Rigamonti D., Goffredo D. *et al.* (2001) Loss of normal huntingtin function: new developments in Huntington's disease research. *Trends Neurosci* **24**:182–8.

Cervenakova L., Goldfarb L. G., Garruto R. *et al.* (1998) Phenotype-genotype studies in kuru: implications for new variant Creutzfeldt-Jakob disease. *Proc Natl Acad Sci USA* **95**:13239–41.

Cha J. H. (2000) Transcriptional dysregulation in Huntington's disease. *Trends Neurosci* **23**:387–92.

Chan W., Fornwald J., Brawner M. *et al.* (1996) Native complex formation between apolipoprotein E isoforms and the Alzheimer's disease peptide A beta. *Biochemistry* **35**:7123–30.

Chong S. S., Almqvist E., Telenius H. *et al.* (1997) Contribution of DNA sequence and CAG size to mutation frequencies of intermediate alleles for Huntington disease: evidence from single sperm analyses. *Hum Mol Genet* **6**:301–9.

Clark L. N., Poorkaj P., Wszolek Z. *et al.* (1998) Pathogenic implications of mutations in the tau gene in pallido-ponto-nigral degeneration and related neurodegenerative disorders linked to chromosome 17. *Proc Natl Acad Sci USA* **95** 13103–07.

Collee J. G. and Bradley R. (1997a) BSE: a decade on—Part 2. Lancet **349**:715–21.

Collee J. G. and Bradley R. (1997b) BSE: a decade on—Part I. Lancet **349**:636–41.

Conrad C., Andreadis A., Trojanowski J. Q. *et al.* (1997) Genetic evidence for the involvement of tau in progressive supranuclear palsy. *Ann Neurol* **41**:277–81.

Corder E. H., Saunders A. M., Strittmatter W. J. *et al.* (1995) Apolipoprotein E, survival in Alzheimer's disease patients, and the competing risks of death and Alzheimer's disease. *Neurology* **45**:1323–8.

Coughlan A. (2001) BSE: the chaos continues. *New Scientist* **172**:14–16.

Craft S., Teri L., Edland S. D. *et al.* (1998) Accelerated decline in apolipoprotein E-epsilon4 homozygotes with Alzheimer's disease. *Neurology* **51**:149–53.

Creutzfeldt H. G. (1920) Uber eine eigenartige herdformige Erkrankung des Zentralnervensystems. *Zeitschrift fur die gesamte Neurologie und Psychiatrie* **57**:1–18.

Cruts M., Hendriks L., Van Broeckhoven C. (1996) The presenilin genes: a new gene family involved in Alzheimer disease pathology. *Hum Mol Genet 5 Spec No* 1449–55.

Cuille J. and Chelle P. L. (1936) Pathologie animale. La maladie dite de la 'tremblante' du mouton est-elle innoculable? *Comptes Rendus de l'Academie des Sciences, Paris* **26**:1552–4.

Cummings C. J., Mancini M. A., Antalffy B. *et al.* (1998) Chaperone suppression of aggregation and altered subcellular proteasome localization imply protein misfolding in SCA1. *Nat Genet* **19**:148–54.

Cummings C. J. and Zoghbi H. Y. (2000) Fourteen and counting: unraveling trinucleotide repeat diseases. *Hum Mol Genet* **9**:909–16.

D'Arcangelo G., Homayouni R., Keshvara L. *et al.* (1999) Reelin is a ligand for lipoprotein receptors. *Neuron* **24**:471–9.

Daw E. W., Heath S. C., Wijsman E. M. (1999) Multipoint oligogenic analysis of age-at-onset data with applications to Alzheimer disease pedigrees. *Am J Hum Genet* **64**:839–51.

Delacourte A. (1999) Biochemical and molecular characterization of neurofibrillary degeneration in frontotemporal dementias. *Dement Geriatr Cogn Disord* **10**(Suppl 1):75–9.

Dickson D. W. (1997) The pathogenesis of senile plaques. *J Neuropathol Exp Neurol* **56**:321–39.

Drachman D. A. (1994). If we live long enough, will we all be demented? *Neurology* **44**:1563–5.

Duff K., Eckman C., Zehr C. *et al.* (1996) Increased amyloid-beta42(43) in brains of mice expressing mutant presenilin 1. *Nature* **383**:710–13.

Dumanchin C., Camuzat A., Campion D. *et al.* (1998) Segregation of a missense mutation in the microtubule-associated protein tau gene with familial frontotemporal dementia and parkinsonism. *Hum Mol Genet* **7**:1825–9.

Eckman C. B., Mehta N. D., Crook R. *et al.* (1997) A new pathogenic mutation in the APP gene (I716V) increases the relative proportion of A beta 42(43) *Hum Mol Genet* **6**:2087–9.

Ezquerra M., Pastor P., Valldeoriola F. *et al.* (1999) Identification of a novel polymorphism in the promoter region of the tau gene highly associated to progressive supranuclear palsy in humans. *Neurosci Lett* **275**:183–6.

Farlow M. R., Cyrus P. A., Nadel A. *et al.* (1999) Metrifonate treatment of AD: influence of APOE genotype. *Neurology* **53**:2010–16.

Farlow M. R., Lahiri D. K., Poirier J. *et al.* (1998) Treatment outcome of tacrine therapy depends on apolipoprotein genotype and gender of the subjects with Alzheimer's disease. *Neurology* **50**:669–77.

Farrer L. A., Cupples L. A., Haines J. L. *et al.* (1997) Effects of age, sex, and ethnicity on the association between apolipoprotein E genotype and Alzheimer disease. A meta-analysis. APOE and Alzheimer Disease Meta Analysis Consortium. *JAMA* **278**:1349–56.

Farrer L. A., Cupples L. A., van Duijn C. M. *et al.* (1995) Apolipoprotein E genotype in patients with Alzheimer's disease: implications for the risk of dementia among relatives. *Ann Neurol* **38**:797–808.

Farrer L. A., Myers R. H., Connor L. *et al.* (1991) Segregation analysis reveals evidence of a major gene for Alzheimer disease. *Am J Hum Genet* **48**:1026–33.

Farrer L. A., Myers R. H., Cupples L. A. *et al.* (1990) Transmission and age-at-onset patterns in familial Alzheimer's disease: evidence for heterogeneity. *Neurology* **40**:395–403.

Farrer L. A., O'Sullivan D. M., Cupples L. A. *et al.* (1989) Assessment of genetic risk for Alzheimer's disease among first-degree relatives. *Ann Neurol* **25**:485–93.

Feany M. B. and Dickson D. W. (1996) Neurodegenerative disorders with extensive tau pathology: a comparative study and review. *Ann Neurol* **40**:139–48.

Feany M. B., Mattiace L. A., Dickson D. W. (1996) Neuropathologic overlap of progressive supranuclear palsy, Pick's disease and corticobasal degeneration. *J Neuropathol Exp Neurol* **55**:53–67.

Ferrucci L., Guralnik J. M., Salive M. E. *et al.* (1996) Cognitive impairment and risk of stroke in the older population. *J Am Geriatr Soc* **44**:237–41.

Finkenstaedt M., Szudra A., Zerr I. *et al.* (1996) MR imaging of Creutzfeldt-Jakob disease. *Radiology* **199**:793–8.

Fitch N., Becker R., Heller A. (1988) The inheritance of Alzheimer's disease: a new interpretation. *Ann Neurol* **23**:14–19.

Gajdusek D. C., Gibbs C. J., Alpers M. (1966) Experimental transmission of a Kuru-like syndrome to chimpanzees. *Nature* **209**:794–6.

Gale C. R., Martyn C. N., Cooper C. (1996) Cognitive impairment and mortality in a cohort of elderly people. *BMJ* **312**:608–11.

Games D., Adams D., Alessandrini R. *et al.* (1995) Alzheimer-type neuropathology in transgenic mice overexpressing V717F beta-amyloid precursor protein. *Nature* **373**:523–7.

Gatz M., Pedersen N. L., Berg S. *et al.* (1997) Heritability for Alzheimer's disease: the study of dementia in Swedish twins. *J Gerontol A Biol Sci Med Sci* 52:M117–25.

Gearing M., Mirra S. S., Hedreen J. C. *et al.* (1995) The Consortium to Establish a Registry for Alzheimer's Disease (CERAD) Part X. Neuropathology confirmation of the clinical diagnosis of Alzheimer's disease. *Neurology* **45**:461–6.

Giordani B., Berent S., Boivin M. J. *et al.* (1995) Longitudinal neuropsychological and genetic linkage analysis of persons at risk for Huntington's disease. *Arch Neurol* **52**:59–64.

Goate A., Chartier-Harlin M. C., Mullan M. *et al.* (1991) Segregation of a missense mutation in the amyloid precursor protein gene with familial Alzheimer's disease. *Nature* **349**:704–6.

Goedert M., Spillantini M. G., Crowther R. A. *et al.* (1999) Tau gene mutation in familial progressive subcortical gliosis. *Nat Med* **5**:454–7.

Goldberg Y. P., Andrew S. E., Theilmann J. *et al.* (1993) Familial predisposition to recurrent mutations causing Huntington's disease: genetic risk to sibs of sporadic cases. *J Med Genet* **30**:987–90.

Goode B. L. and Feinstein S. C. (1994) Identification of a novel microtubule binding and assembly domain in the developmentally regulated inter-repeat region of tau. *J Cell Biol* **124**:769–82.

Graffagnino C., Gasecki A. P., Doig G. S. *et al.* (1994) The importance of family history in cerebrovascular disease. *Stroke* **25**:1599–604.

Grilli M., Diodato E., Lozza G. *et al.* (2000) Presenilin-1 regulates the neuronal threshold to excitotoxicity both physiologically and pathologically. *Proc Natl Acad Sci USA* **97**:12822–7.

Growdon J. H., Locascio J. J., Corkin S. *et al.* (1996) Apolipoprotein E genotype does not influence rates of cognitive decline in Alzheimer's disease. *Neurology* **47**:444–8.

Gusella J. F. and MacDonald M. E. (2000) Molecular genetics: unmasking polyglutamine triggers in neurodegenerative disease. *Nat Rev Neurosci* **1**:109–15.

Gusella J. F., Wexler N. S., Conneally P. M. *et al.* (1983) A polymorphic DNA marker genetically linked to Huntington's disease. *Nature* **306**:234–8.

Gutekunst C. A., Norflus F., Hersch S. M. (2000) Recent advances in Huntington's disease. *Curr Opin Neurol* **13**:445–50.

Haas C., Cazorla P., Miguel C. D. *et al.* (1997) Apolipoprotein E forms stable complexes with recombinant Alzheimer's disease beta-amyloid precursor protein. *Biochem J* 325(Pt 1):169–75.

Haass C. and Steiner H. (2001) Protofibrils, the unifying toxic molecule of neurodegenerative disorders? *Nat Neurosci* **4**:859–60.

Hademenos G. J., Alberts M. J., Awad I. *et al.* (2001) Advances in the genetics of cerebrovascular disease and stroke. *Neurology* **56**:997–1008.

Hadlow W. J. (1959). Scrapie and Kuru. *Lancet* **2**:289–90.

Hardy J. and Allsop D. (1991). Amyloid deposition as the central event in the aetiology of Alzheimer's disease. *Trends Pharmacol Sci* **12**:383–8.

Hardy J. A., Owen M. J., Goate A. M. *et al.* (1989) Molecular genetics of Alzheimer's disease. *Biochem Soc Trans* **17**:75–6.

Harper J. D., Wong S. S., Lieber C. M. *et al.* (1997) Observation of metastable A beta amyloid protofibrils by atomic force microscopy. *Chem Biol* **4**:119–25.

Hartley D. M., Walsh D. M., Ye C. P. *et al.* (1999) Protofibrillar intermediates of amyloid beta-protein induce acute electrophysiological changes and progressive neurotoxicity in cortical neurons. *J Neurosci* **19**:8876–84.

Haywood A. M. (1997) Transmissible spongiform encephalopathies. *N Engl J Med* **337**:1821–8.

Hegde R. S., Mastrianni J. A., Scott M. R. *et al.* (1998) A transmembrane form of the prion protein in neurodegenerative disease. *Science* **279**:827–34.

Heller D. A., de Faire U., Pedersen N. L. *et al.* (1993) Genetic and environmental influences on serum lipid levels in twins. *N Engl J Med* **328**:1150–6.

Henderson A. S., Easteal S., Jorm A. F. *et al.* (1995) Apolipoprotein E allele epsilon 4, dementia, and cognitive decline in a population sample. *Lancet* **346**:1387–90.

Hendrie H. C., Ogunniyi A., Hall K. S. *et al.* (2001) Incidence of dementia and Alzheimer disease in 2 communities: Yoruba residing in Ibadan, Nigeria, and African Americans residing in Indianapolis, Indiana. *JAMA* **285**:739–47.

Hendrie H. C., Osuntokun B. O., Hall K. S. *et al.* (1995) Prevalence of Alzheimer's disease and dementia in two communities: Nigerian Africans and African Americans. *Am J Psychiatry* **152**:1485–92.

Heston L. L. and Mastri A. R. (1977) The genetics of Alzheimer's disease: associations with hematologic malignancy and Down's syndrome. *Arch Gen Psychiatry* **34**:976–81.

Heston L. L., Mastri A. R., Anderson V. E. *et al.* (1981) Dementia of the Alzheimer type. Clinical genetics, natural history, and associated conditions. *Arch Gen Psychiatry* **38**:1085–90.

Heutink P. (2000). Untangling tau-related dementia. *Hum Mol Genet* **9**:979–86.

Hiesberger T., Trommsdorff M., Howell B. W. *et al.* (1999) Direct binding of Reelin to VLDL receptor and ApoE receptor 2 induces tyrosine phosphorylation of disabled-1 and modulates tau phosphorylation. *Neuron* **24**:481–9.

Higgins J. J., Litvan I., Pho T. T. *et al.* (1998) Progressive supranuclear gaze palsy is in linkage disequilibrium with the tau and not the alpha-synuclein gene. *Neurology* **50**:270–3.

Hill A. F., Zeidler M., Ironside J. *et al.* (1997) Diagnosis of new variant Creutzfeldt-Jakob disease by tonsil biopsy. *Lancet* **349**:99–100.

Hirono N., Yasuda M., Tanimukai S. *et al.* (2000) Effect of the apolipoprotein E epsilon4 allele on white matter hyperintensities in dementia. *Stroke* **31**:1263–8.

Hoenicka J., Perez M., Perez-Tur J. *et al.* (1999) The tau gene A0 allele and progressive supranuclear palsy. *Neurology* **53**:1219–25.

Hofman A., Schulte W., Tanja T. A. *et al.* (1989) History of dementia and Parkinson's disease in 1st-degree relatives of patients with Alzheimer's disease. *Neurology* **39**:1589–92.

Holcomb L., Gordon M. N., McGowan E. *et al.* (1998) Accelerated Alzheimer-type phenotype in transgenic mice carrying both mutant amyloid precursor protein and presenilin 1 transgenes. *Nat Med* **4**:97–100.

Holmes C., Cairns N., Lantos P. *et al.* (1999) Validity of current clinical criteria for Alzheimer's disease, vascular dementia and dementia with Lewy bodies. *Br J Psychiatry* **174**:45–50.

Holtzman D. M., Bales K. R., Tenkova T. *et al.* (2000) Apolipoprotein E isoform-dependent amyloid deposition and neuritic degeneration in a mouse model of Alzheimer's disease. *Proc Natl Acad Sci USA* **97**:2892–7.

Horwich A. L. and Weissman J. S. (1997) Deadly conformations—protein misfolding in prion disease. *Cell* **89**:499–510.

Houlden H., Baker M., Adamson J. *et al.* (1999) Frequency of tau mutations in three series of non-Alzheimer's degenerative dementia. *Ann Neurol* **46**:243–8.

Hsiao K., Chapman P., Nilsen S. *et al.* (1996) Correlative memory deficits, Abeta elevation, and amyloid plaques in transgenic mice. *Science* **274**:99–102.

Hsiao K., Meiner Z., Kahana E. *et al.* (1991) Mutation of the prion protein in Libyan Jews with Creutzfeldt-Jakob disease. *N Engl J Med* **324**:1091–7.

Hsich G., Kenney K., Gibbs C. J. *et al.* (1996) The 14–3–3 brain protein in cerebrospinal fluid as a marker for transmissible spongiform encephalopathies. *N Engl J Med* **335**:924–30.

Huntington's Disease Collaborative Research Group (1993) A novel gene containing a trinucleotide repeat that is expanded and unstable on Huntington's disease chromosomes. *Cell* **72**:971–83.

Huang D. Y., Goedert M., Jakes R. *et al.* (1994) Isoform-specific interactions of apolipoprotein E with the microtubule-associated protein MAP2c: implications for Alzheimer's disease. *Neurosci Lett* **182**:55–8.

Huff F. J., Auerbach J., Chakravarti A. *et al.* (1988) Risk of dementia in relatives of patients with Alzheimer's disease. *Neurology* **38**:786–90.

Hutton M., Lendon C. L., Rizzu P. *et al.* (1998) Association of missense and 5'-splice-site mutations in tau with the inherited dementia FTDP-17. *Nature* **393**:702–5.

Hyman B. T., Van Hoesen G. W., Damasio A. R. (1990) Memory-related neural systems in Alzheimer's disease: an anatomic study. *Neurology* **40**:1721–30.

Iglesias S., Chapon F., Baron J. C. (2000) Familial occipital calcifications, hemorrhagic strokes, leukoencephalopathy, dementia, and external carotid dysplasia. *Neurology* **55**:1661–7.

Ironside J. W., Hilton D. A., Ghani A. *et al.* (2000) Retrospective study of prion-protein accumulation in tonsil and appendix tissues. *Lancet* **355**:1693–4.

Iwatsubo T., Odaka A., Suzuki N. *et al.* (1994) Visualization of A beta 42(43) and A beta 40 in senile plaques with end-specific A beta monoclonals: evidence that an initially deposited species is A beta 42(43) *Neuron* **13**:45–53.

Jakob A. (1921) Uber eigenartige Erkrankungen des Zentralnerven-systems mit bemerkenswerten anotomischen Befunde. *Zeitschrift fur die gesamte Neurologie und Psychiatrie* **64**:147–228.

Jarvik G. P., Wijsman E. M., Kukull W. A. *et al.* (1995) Interactions of apolipoprotein E genotype, total cholesterol level, age, and sex in prediction of Alzheimer's disease: a case-control study. *Neurology* **45**:1092–6.

Jellinger K., Danielczyk W., Fischer P. *et al.* (1990) Clinicopathological analysis of dementia disorders in the elderly. *J Neurol Sci* **95**:239–58.

Jen J., Cohen A. H., Yue Q. *et al.* (1997) Hereditary endotheliopathy with retinopathy, nephropathy, and stroke (HERNS) *Neurology* **49**:1322–30.

Joachim C. L., Mori H., Selkoe D. J. (1989) Amyloid beta-protein deposition in tissues other than brain in Alzheimer's disease. *Nature* **341**:226–30.

Johnson R. T. and Gibbs C. J. Jr. (1998) Creutzfeldt-Jakob disease and related transmissible spongiform encephalopathies. *N Engl J Med* **339**:1994–2004.

Jordan B. D., Relkin N. R., Ravdin L. D. *et al.* (1997) Apolipoprotein E epsilon4 associated with chronic traumatic brain injury in boxing. *JAMA* **278**:136–40.

Joutel A., Corpechot C., Ducros A. *et al.* (1996) Notch3 mutations in CADASIL, a hereditary adult-onset condition causing stroke and dementia. *Nature* **383**:707–10.

Joutel A., Vahedi K., Corpechot C. *et al.* (1997) Strong clustering and stereotyped *Nature* of Notch3 mutations in CADASIL patients. *Lancet* **350**:1511–15.

Kalimo H., Viitanen M., Amberla K. *et al.* (1999) CADASIL: hereditary disease of arteries causing brain infarcts and dementia. *Neuropathol Appl Neurobiol* **25**:257–65.

Kase C. S. (1991). Epidemiology of multi-infarct dementia. *Alzheimer Dis Assoc Disord* **5**:71–6.

Kase C. S., Wolf P. A., Chodosh E. H. *et al.* (1989) Prevalence of silent stroke in patients presenting with initial stroke: the Framingham Study. *Stroke* **20**:850–2.

Kehoe P., Wavrant-De Vrieze F., Crook R. *et al.* (1999) A full genome scan for late onset Alzheimer's disease. *Hum Mol Genet* **8**:237–45.

Khaw K. T. and Barrett-Connor E. (1986) Family history of heart attack: a modifiable risk factor? *Circulation* **74**:239–44.

Kirkwood S. C., Siemers E., Bond C. *et al.* (2000) Confirmation of subtle motor changes among presymptomatic carriers of the Huntington disease gene. *Arch Neurol* **57**:1040–4.

Kirkwood S. C., Siemers E., Stout J. C. *et al.* (1999) Longitudinal cognitive and motor changes among presymptomatic Huntington disease gene carriers. *Arch Neurol* **56**:563–8.

Kivipelto M., Helkala E. L., Hanninen T. *et al.* (2001) Midlife vascular risk factors and late-life mild cognitive impairment: a population-based study. *Neurology* **56**:1683–9.

Klatzo I., Gajdusek D. C., Zigas V. (1959) The pathology of Kuru. *Laboratory Investigation* **8**:799–847.

Kosunen O., Soininen H., Paljarvi L. *et al.* (1996) Diagnostic accuracy of Alzheimer's disease: a neuropathological study. *Acta Neuropathol (Berl)* **91**:185–93.

Kubota M., Yamaura A., Ono J. *et al.* (1997) Is family history an independent risk factor for stroke? *J Neurol Neurosurg Psychiatry* **62**:66–70.

Kumar-Singh S., De Jonghe C., Cruts M. *et al.* (2000) Nonfibrillar diffuse amyloid deposition due to a gamma(42)-secretase site mutation points to an essential role for N-truncated A beta(42) in Alzheimer's disease. *Hum Mol Genet* **9**:2589–98.

Kurz A., Egensperger R., Haupt M. *et al.* (1996) Apolipoprotein E epsilon 4 allele, cognitive decline, and deterioration of everyday performance in Alzheimer's disease. *Neurology* **47**:440–3.

Laccone F. and Christian W. (2000) A recurrent expansion of a maternal allele with 36 CAG repeats causes Huntington disease in two sisters. *Am J Hum Genet* **66**:1145–8.

LaDu M. J., Pederson T. M., Frail D. E. *et al.* (1995) Purification of apolipoprotein E attenuates isoform-specific binding to beta-amyloid. *J Biol Chem* **270**:9039–42.

Lammich S., Kojro E., Postina R. *et al.* (1999) Constitutive and regulated alpha-secretase cleavage of Alzheimer's amyloid precursor protein by a disintegrin metalloprotease. *Proc Natl Acad Sci USA* **96**:3922–7.

Launer L. J., Andersen K., Dewey M. E. *et al.* (1999) Rates and risk factors for dementia and Alzheimer's disease: results from EURODEM pooled analyses. EURODEM Incidence Research Group and Work Groups. European Studies of Dementia. *Neurology* **52**:78–84.

Lautenschlager N. T., Cupples L. A., Rao V. S. *et al.* (1996) Risk of dementia among relatives of Alzheimer's disease patients in the MIRAGE study: what is in store for the oldest old? *Neurology* **46**:641–50.

Lee H. S., Brown P., Cervenakova L. *et al.* (2001) Increased susceptibility to Kuru of carriers of the PRNP 129 methionine/methionine genotype. *J Infect Dis* **183**:192–6.

Leeflang E. P., Zhang L., Tavare S. *et al.* (1995) Single sperm analysis of the trinucleotide repeats in the Huntington's disease gene: quantification of the mutation frequency spectrum. *Hum Mol Genet* **4**:1519–26.

Leissring M. A., Akbari Y., Fanger C. M. *et al.* (2000) Capacitative calcium entry deficits and elevated luminal calcium content in mutant presenilin-1 knockin mice. *J Cell Biol* **149**:793–8.

Lemere C. A., Blusztajn J. K., Yamaguchi H. *et al.* (1996) Sequence of deposition of heterogeneous amyloid beta-peptides and APO E in Down syndrome: implications for initial events in amyloid plaque formation. *Neurobiol Dis* **3**:16–32.

Lendon C. L., Ashall F., Goate A. M. (1997) Exploring the etiology of Alzheimer disease using molecular genetics. *JAMA* **277**:825–31.

Levy E., Carman M. D., Fernandez-Madrid I. J. *et al.* (1990) Mutation of the Alzheimer's disease amyloid gene in hereditary cerebral hemorrhage, Dutch type. *Science* **248**:1124–6.

Lewis J., Dickson D. W., Lin W. L. *et al.* (2001) Enhanced neurofibrillary degeneration in transgenic mice expressing mutant tau and APP. *Science* **293**:1487–91.

Li Y. M., Lai M. T., Xu M. *et al.* (2000b) Presenilin 1 is linked with gamma-secretase activity in the detergent solubilized state. *Proc Natl Acad Sci USA* **97**:6138–43.

Li Y. M., Xu M., Lai M. T. *et al.* (2000a) Photoactivated gamma-secretase inhibitors directed to the active site covalently label presenilin 1. *Nature* **405**:689–94.

Liao Y. C., Lebo R. V., Clawson G. A. *et al.* (1986) Human prion protein cDNA: molecular cloning, chromosomal mapping, and biological implications. *Science* **233**:364–7.

Liberski P. P. (1995) Prions, beta-sheets and transmissible dementias: is there still something missing? *Acta Neuropathol (Berl)* **90**:113–25.

Liddell M. B., Bayer A. J., Owen M. J. (1995) No evidence that common allelic variation in the Amyloid Precursor Protein (APP) gene confers susceptibility to Alzheimer's disease. *Hum Mol Genet* **4**:853–8.

Lim A., Tsuang D., Kukull W. *et al.* (1999) Clinico-neuropathological correlation of Alzheimer's disease in a community-based case series. *J Am Geriatr Soc* **47**:564–9.

Luo Y., Bolon B., Kahn S. *et al.* (2001) Mice deficient in BACE1, the Alzheimer's beta-secretase, have normal phenotype and abolished beta-amyloid generation. *Nat Neurosci* **4**:231–2.

MacDonald M. E., Vonsattel J. P., Shrinidhi J. *et al.* (1999) Evidence for the GluR6 gene associated with younger onset age of Huntington's disease. *Neurology* **53**:1330–2.

Mahley R. W. (1988) Apolipoprotein E: cholesterol transport protein with expanding role in cell biology. *Science* **240**:622–30.

Mann D. (2000) Frontotemporal dementia. *CNS* **2**:19–23.

Mann D. M. (1989) Cerebral amyloidosis, ageing and Alzheimer's disease; a contribution from studies on Down's syndrome. *Neurobiol Aging* **10**:397–9.

Marsh R. F. and Bessen R. A. (1993) Epidemiologic and experimental studies on transmissible mink encephalopathy. *Dev Biol Stand* **80**:111–18.

Martin J. B. (1999) Molecular basis of the neurodegenerative disorders. *N Engl J Med* **340**:1970–80.

Martin R. L., Gerteis G., Gabrielli W. F. Jr (1988) A family-genetic study of dementia of Alzheimer type. *Arch Gen Psychiatry* **45**:894–900.

Mayeux R., Ottman R., Maestre G. *et al.* (1995) Synergistic effects of traumatic head injury and apolipoprotein-epsilon 4 in patients with Alzheimer's disease. *Neurology* **45**:555–7.

Mayeux R., Ottman R., Tang M. X. *et al.* (1993) Genetic susceptibility and head injury as risk factors for Alzheimer's disease among community-dwelling elderly persons and their first-degree relatives. *Ann Neurol* **33**:494–501.

Mayeux R., Sano M., Chen J. *et al.* (1991) Risk of dementia in first-degree relatives of patients with Alzheimer's disease and related disorders. *Arch Neurol* **48**:269–73.

Mayeux R., Saunders A. M., Shea S. *et al.* (1998) Utility of the apolipoprotein E genotype in the diagnosis of Alzheimer's disease. Alzheimer's Disease Centers Consortium on Apolipoprotein E and Alzheimer's Disease. *N Engl J Med* **338**:506–11.

McGuffin P., Owen M. J., O'Donovan M. C. *et al.* (1994) *Seminars in Psychiatric Genetics.* Gaskell: London.

McKhann G., Drachman D., Folstein M. *et al.* (1984) Clinical diagnosis of Alzheimer's disease: report of the NINCDS-ADRDA Work Group under the auspices of Department of Health and Human Services Task Force on Alzheimer's Disease. *Neurology* **34**:939–44.

Meyer M. R., Tschanz J. T., Norton M. C. *et al.* (1998) APOE genotype predicts when–not whether–one is predisposed to develop Alzheimer disease. *Nat Genet* **19**:321–2.

Mohs R. C., Breitner J. C., Silverman J. M. *et al.* (1987) Alzheimer's disease. Morbid risk among first-degree relatives approximates 50 per cent by 90 years of age. *Arch Gen Psychiatry* **44**:405–8.

Morgan D., Diamond D. M., Gottschall P. E. *et al.* (2000) A beta peptide vaccination prevents memory loss in an animal model of Alzheimer's disease. *Nature* **408**:982–5.

Morris H. R., Janssen J. C., Bandmann O. *et al.* (1999) The tau gene A0 polymorphism in progressive supranuclear palsy and related neurodegenerative diseases. *J Neurol Neurosurg Psychiatry* **66**:665–7.

Morris J. C., McKeel D. W. Jr., Fulling K. *et al.* (1988) Validation of clinical diagnostic criteria for Alzheimer's disease. *Ann Neurol* **24**:17–22.

Motte J. and Williams R. S. (1989) Age-related changes in the density and morphology of plaques and neurofibrillary tangles in Down syndrome brain. *Acta Neuropathol (Berl)* **77**:535–46.

Myers A., Holmans P., Marshall H. *et al.* (2000) Susceptibility locus for Alzheimer's disease on chromosome 10. *Science* **290**:2304–5.

Myers A., De-Vrieze F., Holmans P. *et al.* (2002, in press). A full genome screen for Alzheimer's disease: stage two analysis. *Am J Med Genet (Neuropsychiatric Genetics).*

Myers R. H., Leavitt J., Farrer L. A. *et al.* (1989) Homozygote for Huntington disease. *Am J Hum Genet* **45**:615–18.

Myers R. H., Schaefer E. J., Wilson P. W. *et al.* (1996) Apolipoprotein E epsilon 4 association with dementia in a population-based study: the Framingham study. *Neurology* **46**:673–7.

Nacharaju P., Lewis J., Easson C. *et al.* (1999) Accelerated filament formation from tau protein with specific FTDP-17 missense mutations. *FEBS Lett* **447**:195–9.

Nance M. A., Mathias-Hagen V., Breningstall G. *et al.* (1999) Analysis of a very large trinucleotide repeat in a patient with juvenile Huntington's disease. *Neurology* **52**:392–4.

Neary D., Snowden J. S., Gustafson L. *et al.* (1998) Frontotemporal lobar degeneration: a consensus on clinical diagnostic criteria. *Neurology* **51**:1546–54.

Nilsberth C., Westlind-Danielsson A., Eckman C. B. *et al.* (2001) The 'Arctic' APP mutation (E693G) causes Alzheimer's disease by enhanced Abeta protofibril formation. *Nat Neurosci* **4**:887–93.

Nishimura M., Yu G., George-Hyslop P. H. (1999) Biology of presenilins as causative molecules for Alzheimer disease. *Clin Genet* **55**:219–25.

Ogunniyi A., Baiyewu O., Gureje O. *et al.* (2000) Epidemiology of dementia in Nigeria: results from the Indianapolis-Ibadan study. *Eur J Neurol* **7**:485–90.

Oliva R., Tolosa E., Ezquerra M. *et al.* (1998) Significant changes in the tau A0 and A3 alleles in progressive supranuclear palsy and improved genotyping by silver detection. *Arch Neurol* **55**:1122–4.

Olson J. M., Goddard K. A., Dudek D. M. (2001) The amyloid precursor protein locus and very-late-onset Alzheimer disease. *Am J Hum Genet* **69**:895–9.

Onda H., Kasuya H., Yoneyama T. *et al.* (2001) Genomewide-linkage and haplotype-association studies map intracranial aneurysm to chromosome 7q11. *Am J Hum Genet* **69**:804–19.

Ott A., Breteler M. M., van Harskamp F. *et al.* (1995) Prevalence of Alzheimer's disease and vascular dementia: association with education. The Rotterdam study. *BMJ* **310**:970–3.

Ott A., Breteler M. M., van Harskamp F. *et al.* (1998) Incidence and risk of dementia. The Rotterdam Study. *Am J Epidemiol* **147**:574–80.

Payami H., Montee K., Kaye J. (1994a) Evidence for familial factors that protect against dementia and outweigh the effect of increasing age. *Am J Hum Genet* **54**:650–7.

Payami H., Montee K. R., Kaye J. A. *et al.* (1994b) Alzheimer's disease, apolipoprotein E4, and gender. *JAMA* **271**:1316–17.

Pearson R. C.A. and Powell T. P.S. (1989) The neuroanatomy of Alzheimer's disease. *Reviews in the Neurosciences* **2**:101–22.

Pericak-Vance M. A., Bass M. P., Yamaoka L. H. *et al.* (1997) Complete genomic screen in late-onset familial Alzheimer disease. Evidence for a new locus on chromosome 12. *JAMA* **278**:1237–41.

Pericak-Vance M. A., Grubber J., Bailey L. R. *et al.* (2000) Identification of novel genes in late-onset Alzheimer's disease. *Exp Gerontol* **35**:1343–52.

Plassman B. L. and Breitner J. C. (1996) Recent advances in the genetics of Alzheimer's disease and vascular dementia with an emphasis on gene-environment interactions. *J Am Geriatr Soc* **44**:1242–50.

Poirier J., Delisle M. C., Quirion R. *et al.* (1995) Apolipoprotein E4 allele as a predictor of cholinergic deficits and treatment outcome in Alzheimer disease. *Proc Natl Acad Sci USA* **92**:12260–4.

Polvikoski T., Sulkava R., Myllykangas L. *et al.* (2001) Prevalence of Alzheimer's disease in very elderly people: a prospective neuropathological study. *Neurology* **56**:1690–6.

Poorkaj P., Bird T. D., Wijsman E. *et al.* (1998) Tau is a candidate gene for chromosome 17 frontotemporal dementia. *Ann Neurol* **43**:815–25.

Prince M., Lovestone S., Cervilla J. *et al.* (2000) The association between APOE and dementia does not seem to be mediated by vascular factors. *Neurology* **54**:397–402.

Prusiner S. B. (1982) Novel proteinaceous infectious particles cause scrapie. *Science* **216**:136–44.

Prusiner S. B. (1993) Genetic and infectious prion diseases. *Arch Neurol* **50**:1129–53.

Prusiner S. B. (1997) Prion diseases and the BSE crisis. *Science* **278**:245–51.

Raaymakers T. W. (1999) Aneurysms in relatives of patients with subarachnoid hemorrhage: frequency and risk factors. MARS Study Group. Magnetic resonance angiography in relatives of patients with subarachnoid hemorrhage. *Neurology* **53**:982–8.

Raiha I., Kaprio J., Koskenvuo M. *et al.* (1996) Alzheimer's disease in Finnish twins. *Lancet* **347**:573–8.

Ranen N. G., Stine O. C., Abbott M. H. *et al.* (1995) Anticipation and instability of IT-15 (CAG)n repeats in parent-offspring pairs with Huntington disease. *Am J Hum Genet* **57**:593–602.

Reiman E. M., Caselli R. J., Yun L. S. *et al.* (1996) Preclinical evidence of Alzheimer's disease in persons homozygous for the epsilon 4 allele for apolipoprotein E. *N Engl J Med* **334**:752–8.

Rizzu P., Van Swieten J. C., Joosse M. *et al.* (1999) High prevalence of mutations in the microtubule-associated protein tau in a population study of frontotemporal dementia in the Netherlands. *Am J Hum Genet* **64**:414–21.

Roks G., Dermaut B., Heutink P. *et al.* (1999) Mutation screening of the tau gene in patients with early-onset Alzheimer's disease. *Neurosci Lett* **277**:137–9.

Rosenberg N. K., Sorensen S. A., Christensen A. L. (1995) Neuropsychological characteristics of Huntington's disease carriers: a double blind study. *J Med Genet* **32**:600–4.

Rubinsztein D. C., Leggo J., Chiano M. *et al.* (1997) Genotypes at the GluR6 kainate receptor locus are associated with variation in the age of onset of Huntington disease. *Proc Natl Acad Sci USA* **94**:3872–6.

Rubinsztein D. C., Leggo J., Coles R. *et al.* (1996) Phenotypic characterization of individuals with 30–40 CAG repeats in the Huntington disease (HD) gene reveals HD cases with 36 repeats and apparently normal elderly individuals with 36–39 repeats. *Am J Hum Genet* **59**:16–22.

Rumble B., Retallack R., Hilbich C. *et al.* (1989) Amyloid A4 protein and its precursor in Down's syndrome and Alzheimer's disease. *N Engl J Med* **320**:1446–52.

Saunders A. M. (2000) Apolipoprotein E and Alzheimer disease: an update on genetic and functional analyses. *J Neuropathol Exp Neurol* **59**:751–8.

Saunders A. M. and Roses A. D. (1993) Apolipoprotein E4 allele frequency, ischemic cerebrovascular disease, and Alzheimer's disease. *Stroke* **24**:1416–17.

Schellenberg G. D., D'Souza I., Poorkaj P. (2000) The genetics of Alzheimer's disease. *Curr Psychiatry Rep* **2**:158–64.

Schenk D., Barbour R., Dunn W. *et al.* (1999) Immunization with amyloid-beta attenuates Alzheimer-disease-like pathology in the PDAPP mouse. *Nature* **400**:173–7.

Schievink W. I., Parisi J. E., Piepgras D. G. (1997) Familial intracranial aneurysms: an autopsy study. *Neurosurgery* **41**:1247–51.

Schievink W. I., Schaid D. J., Michels V. V. *et al.* (1995) Familial aneurysmal subarachnoid hemorrhage: a community-based study. *J Neurosurg* **83**:426–9.

Selkoe D. J. (1991) The molecular pathology of Alzheimer's disease. *Neuron* **6**:487–98.

Selkoe D. J. (1999) Translating cell biology into therapeutic advances in Alzheimer's disease. *Nature* **399**:A23–31.

Selkoe D. J. (2000) Notch and presenilins in vertebrates and invertebrates: implications for neuronal development and degeneration. *Curr Opin Neurobiol* **10**:50–7.

Sergeant N., Wattez A., Delacourte A. (1999) Neurofibrillary degeneration in progressive supranuclear palsy and corticobasal degeneration: tau pathologies with exclusively 'exon 10' isoforms. *J Neurochem* **72**:1243–9.

Seshadri S., Wolf P. A., Beiser A. *et al.* (1997) Lifetime risk of dementia and Alzheimer's disease. The impact of mortality on risk estimates in the Framingham Study. *Neurology* **49**:1498–504.

Shiwach R. (1994) Psychopathology in Huntington's disease patients. *Acta Psychiatr Scand* **90**:241–6.

Shiwach R. S. and Norbury C. G. (1994) A controlled psychiatric study of individuals at risk for Huntington's disease. *Br J Psychiatry* **165**:500–5.

Siest G., Pillot T., Regis-Bailly A. *et al.* (1995) Apolipoprotein E: an important gene and protein to follow in laboratory medicine. *Clin Chem* **51**:1068–86.

Silverman J. M., Li G., Zaccario M. L. *et al.* (1994) Patterns of risk in first-degree relatives of patients with Alzheimer's disease. *Arch Gen Psychiatry* **51**:577–86.

Silverman J. M., Smith C. J., Marin D. B. *et al.* (1999) Identifying families with likely genetic protective factors against Alzheimer disease. *Am J Hum Genet* **64**:832–8.

Silverman J. M., Smith C. M., Marin D. B. (2000) Has familial aggregation in Alzheimer's disease been overestimated? *Int J Ger Psychiatry* **15**:631–7.

Sisodia S. S. (1998) Nuclear inclusions in glutamine repeat disorders: are they pernicious, coincidental, or beneficial? *Cell* **95**:1–4.

Skoog I. (1998) Status of risk factors for vascular dementia. *Neuroepidemiology* **17**:2–9.

Skoog I., Hesse C., Aevarsson O. *et al.* (1998) A population study of apoE genotype at the age of 85: relation to dementia, cerebrovascular disease, and mortality. *J Neurol Neurosurg Psychiatry* **64**:37–43.

Skoog I., Palmertz B., Andreasson L. A. (1994) The prevalence of white-matter lesions on computed tomography of the brain in demented and nondemented 85-year-olds. *J Geriatr Psychiatry Neurol* **7**:169–75.

Slooter A. J., Bronzova J., Witteman, J. C. *et al.* (1999) Estrogen use and early onset Alzheimer's disease: a population-based study. *J Neurol Neurosurg Psychiatry* **67**:779–81.

Slooter A. J., Tang M. X., van Duijn C. M. *et al.* (1997) Apolipoprotein E epsilon4 and the risk of dementia with stroke. A population-based investigation. *JAMA* **277**:818–21.

Snowdon D. A., Greiner L. H., Mortimer J. A. *et al.* (1997) Brain infarction and the clinical expression of Alzheimer disease. The Nun Study. *JAMA* **277**:813–17.

Sparkes R. S., Simon M., Cohn V. H. *et al.* (1986) Assignment of the human and mouse prion protein genes to homologous chromosomes. *Proc Natl Acad Sci USA* **83**:7358–62.

Spillantini M. G. and Goedert M. (1998) Tau protein pathology in neurodegenerative diseases. *Trends Neurosci* **21**:428–33.

Stern Y., Brandt J., Albert M. *et al.* (1997) The absence of an apolipoprotein epsilon4 allele is associated with a more aggressive form of Alzheimer's disease. *Ann Neurol* **41**:615–20.

Stewart R. (1998) Cardiovascular factors in Alzheimer's disease. *J Neurol Neurosurg Psychiatry* **65**:143–7.

Strittmatter W. J., Saunders A. M., Goedert M. *et al.* (1994) Isoform-specific interactions of apolipoprotein E with microtubule-associated protein tau: implications for Alzheimer disease. *Proc Natl Acad Sci USA* **91**:11183–6.

Strittmatter W. J., Saunders A. M., Schmechel D. *et al.* (1993) Apolipoprotein E: high-avidity binding to beta-amyloid and increased frequency of type 4 allele in late-onset familial Alzheimer disease. *Proc Natl Acad Sci USA* **90**:1977–81.

Tandon A., Rogaeva E., Mullan M. *et al.* (2000) Molecular genetics of Alzheimer's disease: the role of beta-amyloid and the presenilins. *Curr Opin Neurol* **13**:377–84.

Tang M. X., Jacobs D., Stern Y. *et al.* (1996) Effect of oestrogen during menopause on risk and age at onset of Alzheimer's disease. *Lancet* **348**:429–32.

Tanzi R. E. (1991) Gene mutations in inherited amyloidopathies of the nervous system. *Am J Hum Genet* **49**:507–10.

Telling G. C. (2000) Prion protein genes and prion diseases: studies in transgenic mice. *Neuropathol Appl Neurobiol* **26**:209–20.

ter Berg H. W., Dippel D. W., Limburg M. *et al.* (1992) Familial intracranial aneurysms. A review. *Stroke* **23**:1024–30.

Tolnay M. and Probst A. (1999) Review: tau protein pathology in Alzheimer's disease and related disorders. *Neuropathol Appl Neurobiol* **25**:171–87.

Tomlinson B. E. (1982) Plaques, tangles and Alzheimer's disease. *Psychol Med* **12**:449–59.

Tomlinson B. E., Blessed G., Roth M. (1970) Observations on the brains of demented old people. *J Neurol Sci* **11**:205–42.

Tournier-Lasserve E., Joutel A., Melki J. *et al.* (1993) Cerebral autosomal dominant arteriopathy with subcortical infarcts and leukoencephalopathy maps to chromosome 19q12. *Nat Genet* **3**:256–9.

Town T., Paris D., Fallin D. *et al.* (1998) The -491A/T apolipoprotein E promoter polymorphism association with Alzheimer's disease: independent risk and linkage disequilibrium with the known APOE polymorphism. *Neurosci Lett* **252**:95–8.

Trottier Y., Devys D., Imbert G. *et al.* (1995) Cellular localization of the Huntington's disease protein and discrimination of the normal and mutated form. *Nat Genet* **10**:104–10.

Van Broeckhoven C., Haan J., Bakker E. *et al.* (1990) Amyloid beta protein precursor gene and hereditary cerebral hemorrhage with amyloidosis (Dutch) *Science* **248**:1120–2.

van Duijn C. M., de Knijff P., Cruts M. *et al.* (1994) Apolipoprotein E4 allele in a population-based study of early-onset Alzheimer's disease. *Nat Genet* **7**:74–8.

van Duijn C. M., Farrer L. A., Cupples L. A. *et al.* (1993) Genetic transmission of Alzheimer's disease among families in a Dutch population based study. *J Med Genet* **30**:640–6.

van Duijn C. M., Havekes L. M., Van Broeckhoven C. *et al.* (1995) Apolipoprotein E genotype and association between smoking and early onset Alzheimer's disease. *BMJ* **310**:627–31.

Walsh D. M., Hartley D. M., Kusumoto Y. *et al.* (1999) Amyloid beta-protein fibrillogenesis. Structure and biological activity of protofibrillar intermediates. *J Biol Chem* **274**:25945–52.

Walsh D. M., Lomakin A., Benedek G. B. *et al.* (1997) Amyloid beta-protein fibrillogenesis. Detection of a protofibrillar intermediate. *J Biol Chem* **272**:22364–72.

West M. J., Coleman P. D., Flood D. G. *et al.* (1994) Differences in the pattern of hippocampal neuronal loss in normal ageing and Alzheimer's disease. *Lancet* **344**:769–72.

Westaway D., Telling G., Priola S. (1998) Prions. *Proc Natl Acad Sci USA* **95**:11030–1.

Wexler N. S., Young A. B., Tanzi R. E. *et al.* (1987) Homozygotes for Huntington's disease. *Nature* **326**:194–7.

Wilcock G. K. and Esiri M. M. (1982) Plaques, tangles and dementia. A quantitative study. *J Neurol Sci* **56**:343–56.

Wilcock G. K., Lilienfeld S., Gaens E. (2000) Efficacy and safety of galantamine in patients with mild to moderate Alzheimer's disease: multicentre randomised controlled trial. Galantamine International-1 Study Group. *BMJ* **321**:1445–9.

Wilkins R. H. and Brody I. A. (1969) Alzheimer's disease. *Arch Neurol* **21**:109–10.

Wolfe M. S., Xia W., Ostaszewski B. L. *et al.* (1999) Two transmembrane aspartates in presenilin-1 required for presenilin endoproteolysis and gamma-secretase activity. *Nature* **398**:513–17.

Wyss-Coray T. and Mucke L. (2000) Ibuprofen, inflammation and Alzheimer disease. *Nat Med* **6**:973–4.

Zareparsi S., Wirdefeldt K., Burgess C. E. *et al.* (1999) Exclusion of dominant mutations within the FTDP-17 locus on chromosome 17 for Parkinson's disease. *Neurosci Lett* **272**:140–2.

Zigas V. and Gajdusek D. C. (1957) Kuru: clinical study of a new syndrome resembling paralysis agitans in natives of the Eastern Highlands of Australian New Guinea. *Medical Journal of Australia* **2**:745–54.

Zubenko G. S., Huff F. J., Beyer J. *et al.* (1988) Familial risk of dementia associated with a biologic subtype of Alzheimer's disease. *Arch Gen Psychiatry* **45**:889–93.

Zurutuza L., Verpillat P., Raux G. *et al.* (2000) APOE promoter polymorphisms do not confer independent risk for Alzheimer's disease in a French population. *Eur J Hum Genet* **8**:713–16.

Part 4

Applications and implications

Chapter 14

Psychopharmacogenetics

Rob W. Kerwin and Maria J. Arranz

Introduction

Pharmacogenetic research is dedicated to the discovery of genetic alterations that affect drug action with the aim of producing safer and more effective clinical treatment. Although initially pharmacogenetics was consigned to the study of mutations in single genes affecting drug metabolism, it has been extended to cover single and multigenic variation in drug metabolism or action. The field of pharmacogenetics originated 50 years ago with the discovery of differences in the metabolism of therapeutic drugs between individuals. Biochemical research into drug metabolizing enzymes combined with molecular genetics lead to the identification of mutations related to poor metabolizing of drugs. A few successes have also been achieved in correlating genetic variation and drug response in psychiatric treatment. However, it is now with the advancement of high throughput of genetic data and the realisation of the potential of pharmacogenomics by the pharmaceutical companies that an explosion of data leading to individualization of treatment is expected.

Although pharmacogenetics is aimed at improving patient care rather than at acquiring knowledge into disease genes, additional knowledge on the aetiology of mental disorders may be obtained. Genetic components may be easier identified within subgroups than when confounded in the general patient population. Indeed, the subdivision of the patient population into responders and non-responders to a particular treatment may be a valid strategy to associate genetic variation with particular phenotypes.

The aim of this chapter is to inform of the findings of genetic variability associated with interindividual differences in psychiatric treatment and to review the estate of pharmacogenetic studies in the area. Historically, research started into drug metabolizing enzymes and their dose related effects. Garrod first suggested that genetically controlled enzymes responsible for the detoxification of foreign compounds may be lacking in some individuals (Garrod 1931). Werner Kalow succeeded in associating enzyme abnormality (serum cholinesterase) with drug sensitivity (succinylcholine) (Kalow 1962). During the 1960s and 1970s, Harris matched structural gene mutations with physiological and pathological data in hemobloglobinopathies and enzymopathies (1976). However, interindividual response variation could not be explained on the basis of metabolizing polymorphisms only and the research field was extended to include drugs' sites of action. It is likely that a combination of variation in both metabolizing enzymes and drug targets fully explains the heterogeneity in response to psychiatric treatment.

Strategies for pharmacogenetic research

The strategies used for the detection of genetic variability relevant to clinical response are based in classical genetic association studies. Natural mutations in candidate genes are investigated in unrelated individuals presenting variation in clinical response. Whereas this strategy has been successful in detecting genes with even minor influences in clinical response, its major drawback is the difficulty of replicating positive findings. Differences in sample size, ethnic origin, and assessment criteria can explain most of the discrepancies. However, these differences can be minimized by looking at extreme phenotypes. For example, when investigating candidate genes in relation to levels of improvement with a particular treatment, comparison of very good responders versus very poor responders will increase the chances of finding genetic factors influencing responsiveness. Such a strategy will also minimize the sample differences caused by the assessment criteria. In addition, standardized rules for pharmacogenetic studies have been proposed to increase the chances of replications (Cichon *et al.* 2000). More accurate strategies also include the use of trios (proband and parents) for TDT studies (see Chapter 3) as they are less prone to false positive findings.

Most of the pharmacogenetic findings reported in this chapter have been obtained by time-consuming association studies. However, with the development of high throughput technologies for mutation detection and genotyping of specific genes, research has been expanded to investigate multiple loci of interest in large samples. In a few years time, study of variants in single genes on small sized samples will become obsolete, as the term 'pharmacogenetics' will be finally substituted by 'pharmacogenomics'. Hopefully this will lead to a clear improvement on the clinical treatment of psychiatric disorders, as the possibility of designing the most beneficial treatment at the appropriate dose for each individual according to their genetic profile becomes a reality. However, before reaching this point, more research into genetic factors determining drug metabolizing rates and drug activity is required.

Variability in metabolizing enzymes

Since Kalow's (1962) and Harris' (1975) findings, a large number of mutations in metabolizing enzymes have been described which are known to contribute to interindividual differences in the pharmacokinetics of many drugs. External compounds have to follow a succession of oxidation reactions (Phase I) and conjugations (Phase II) by metabolizing enzymes to be assimilated and then secreted by an organism. Disruptive mutations in metabolizing enzymes can affect the incorporation or elimination of foreign compounds, resulting in their toxic accumulation or rapid elimination from the organism. Therefore a toxic response or lack of responsiveness will be observed after treatment with otherwise therapeutic doses. Although these polymorphisms may not directly influence the drug therapeutic value, the metabolizing rate will be affected and the therapeutic doses will have to be adjusted to the patient's phenotype to achieve maximum efficacy. Correlations between mutations in phase I enzymes (i.e. cytochrome P450) and individual responsiveness have already been described and will be summarized here. However, although genetic variability in phase II enzymes (N-acetyltransferases, UDP-glucuronosyltransferases, and thiopurine methyltransferases) is abundant, its relation with treatment response is still unclear.

Cytochrome P450 enzymes

Twelve families of cytochrome P450 (CYP) enzymes have been described of which four of them (CYP1–CYP4) are directly involved in drug metabolism (Bertilsson and Dahl 1996). They constitute the best studied family of xenobiotic-metabolizing enzymes. Mutations in CYP2D6, CYP2C9, and CYP2C19 have already been shown to be the cause of altered drug pharmacokinetics (Lin *et al.* 1996; Caraco 1998; Kidd *et al.* 1999).

The CYP2D6 debrisoquine/sparteine polymorphism was discovered in the early 1970s when individuals treated with the drug who suffered a drastic fall in blood pressure were found to possess a defective enzyme (Eichelbaum and Evert 1996). Four different phenotypes have been observed in the population according to their debrisoquine/sparteine metabolizing rate: poor metabolizers (PM), intermediate metabolizers (IM), extensive metabolizers (EM), and ultra rapid metabolizers (UM). PMs develop an adverse reaction when treated with CYP2D6-metabolized drugs whereas UMs need high doses for optimal therapy. There are also important interethnic differences: Caucasians have the highest rates of PMs (7 per cent) and East Asians the lowest (1 per cent) (Lin *et al.* 1996; Caraco 1998). Mutations in the *CYP2D6* gene have been found to be responsible for these phenotype variations. The gene coding for the CYP2D6 enzyme is highly polymorphic: 48 polymorphisms have been described in white Caucasians of European origin (Marez *et al.* 1997). Two of these polymorphisms, CYP2D6A and CYP2D6B, and a gene deletion are responsible for 95 per cent of the population PMs, which were shown to be homozygous for these mutations. The frequency of these alleles is 27 per cent in Europeans and 6 per cent in Asians, which explains the interethnic differences observed (Bertilsson and Dahl 1996). This is of great importance for psychiatric treatment, as many antipsychotic and antidepressant (most tricyclic antidepressants, see Table 14.1) drugs are metabolized by CYP2D6 (Linder *et al.* 1997). Individuals with *CYP2D6* mutated alleles are specially at risk of developing adverse reactions after treatment with antidepressants (Lennard 1993; Dahl *et al.* 1996). Recent studies suggest that defective CYP2D6 variants are associated with abnormal movements due to neuroleptic use (Ellingrod *et al.* 1999). Although clozapine was found to be partially metabolized by CYP2D6 *in vitro*, the drug response is not affected by variation in the enzyme (Dahl *et al.* 1994; Arranz *et al.* 1995a). It was later found that CYP1A2 and CYP3A4 are the major metabolizing pathways of clozapine.

Table 14.1 List of antidepressants and antipsychotic drugs metabolized by the major CYP metabolic enzymes

Type of drug	CYP1A2	CYP2C19	CYP2D6	CYP3A4
Antidepressants				
	Amitriptyline	Amitriptyline	Amitriptyline	Amitriptyline
	Clomipramine	Citalopram	Clomipramine	Clomipramine
	Fluvoxamine	Clomipramine	Desipramine	Imipramine
	Imipramine	Imipramine	Fluvoxamine	Venlafaxine
		Moclobemide	Imipramine	
			Mianserin	
			Nortriptyline	
			Paroxetine	
			Venlafaxine	
Antipsychotics				
	Clozapine	Clozapine	Clozapine	Clozapine
	Olanzapine		Haloperidol	Risperidone
			Olanzapine	Quetiapine
			Perphenazine	
			Remoxipride	
			Risperidone	
			Sertindole	

The CYP2C19 is the second best characterized drug metabolism polymorphism in humans. The CYP2C19 polymorphism was revealed by the deficient metabolism of the anticonvulsant drug mephenytoin. Two mutations in the CYP2C19 gene (636-G/A and 681-G/A) were revealed to be the cause of 87 per cent of Caucasian PMs and 99 per cent of Oriental PMs associated with this polymorphism (De Morais *et al*. 1994; Ferguson *et al*. 1998). Several other less frequent mutations also associated with a defective CYP2C19 enzyme have been reported (Ferguson *et al*. 1998; Ibeanu *et al*. 1998). As is the case for CYP2D6, PCR methods for the rapid detection of these mutations are available.

The CYP1A2 subfamily has been implicated in the metabolism of clozapine (Jerling *et al*. 1994) and olanzapine (Ring *et al*. 1996) among others. Although several mutations have been described in the *CYP1A2* gene (Sachse *et al*. 1999; Nakajima *et al*. 1999; Huang *et al*. 1999) no clear association with antipsychotic response has yet been reported. The possible effect of polymorphisms in CYP1A2 on clozapine response is still under investigation.

Recent investigations have revealed other less frequent polymorphisms in metabolizing enzymes. A deficient CYP2C9 enzyme is responsible for the phenytoin/tolbutamide poor metabolizer phenotype (Sullivan-Klose *et al*. 1996; Kidd *et al*. 1999). Two amino acid substitutions (Arg144Cys and Ile359Leu) in the CYP2C9 protein have been revealed to be responsible for the decrease in enzyme activity (Gill *et al*. 1999). However, the frequency of the genotype associated with the decreased enzyme activity is less than 2 per cent (Yasar *et al*. 1999; Gill *et al*. 1999). A —290-G/A polymorphism in the promoter region of the *CYP3A4* gene, the most abundant CYP form in human liver, has been described. However, this mutation was not found to be associated to the differential expression of CYP3A4 observed in humans (Westlind *et al.* 1999), although this genetic variant had been associated with prostate carcinogenesis (Rebbeck *et al.* 1998).

Phase II metabolizing enzymes

Phase II enzymes including N-acetyltransferase (NAT), UDP-glucuronosyltransferase (UGT), and thiopurine methyltransferase (TPMT) enzymes metabolize conjugations of xenobiotics. Altered activity of these enzymes may cause cytotoxic reactions due to accumulation of drug metabolites.

Interindividual variation in drug acetylation has been known for more than 40 years. Three different genes (NAT1, NAT2 and the pseudogene NATP) have been described in humans for N-acetyltransferase enzymes (Blum *et al.* 1990). Initially, the NAT1 gene was thought to be monomorphic whereas several variants of NAT2 were described to be associated with the slow acetylator phenotype (Meyer and Zanger 1997). The frequency of slow acetylators is about 40–60 per cent in white Caucasians and 10–30 per cent in Asian populations (Zielinska *et al.* 1997; Xie *et al.* 1997; Cascorbi *et al.* 1995; Lin *et al.* 1993). More than seven allelic variants have been described in the NAT2 gene of which at least five (Arg64Gln, Ile114Thr, Arg197Gln, Lys268Arg, and Gly286Glu) account for more than 90 per cent of the slow acetylators (Vatsis *et al.* 1991; Lin *et al.* 1993). Several allelic variants have been described suggesting that the NAT1 enzyme may also have polymorphic activity (Vatsis and Weber 1993; Doll *et al.* 1997). It was later demonstrated that the enzyme was defective in about 8 per cent of the individuals (Butcher *et al.* 1998) and two structural polymorphisms (Arg64Trp and Gln187Glu) were associated with slow acetylator phenotype.

In addition, a number of polymorphisms in the TPMT and UDG1 genes causing loss of enzyme activity have been reported (De la Moureyre *et al.* 1998; Clarke *et al.* 1997; Ciotti *et al.* 1997). However, the direct relation of polymorphisms in phase 2 metabolizing enzymes and mental disorders is still under investigation. Although a clear relation between polymorphisms in these enzymes and psychiatric drugs has not yet been established, they may play important roles in toxic effects causing cytotoxic accumulation of drug metabolites. Future genotype screenings for the safety of drugs must consider these variations.

Predictive value of polymorphisms in metabolizing enzymes

Because of the high frequency in the population of mutations in metabolizing enzymes, their importance in the success of therapeutic treatment cannot be underestimated. Although metabolizing polymorphisms may not be directly related to the aetiology of the disease, they determine the patient's predisposition to adverse reactions due to poor/fast drug metabolism. Therapeutic doses of drugs will have to be adjusted to the individuals' metabolizing rates. Previous knowledge of the individual's metabolizing rates will be important for medications metabolized by a single polymorphic enzyme, whereas the effect of rates on drugs metabolized by more than one enzyme will be minor. Many pharmaceutical companies screen their products to determine if they are metabolized by polymorphic enzymes. The existence of rapid PCR methods for the detection of most mutations related to defective enzymes facilitates the prediction of individuals' metabolizing rates. DNA chips for the detection of CYP2D6 and CYPC19 mutations are being developed for the identification of potential poor metabolizers (Hodgson and Marshall 1998).

Variability in drug targets

Initially the field of pharmacogenetics dealt only with variability in drug metabolism. However, even when metabolism is not affected by polymorphic enzymes, differences in the response to psychiatric treatment is observed. Therefore, other factors may be affecting treatment response. Pharmacogenetic research started looking into drugs' targets as another source of genetic variation that could affect chemical therapy. Mutations altering neurotransmitter receptor systems targeted by antipsychotics/antidepressants may play an important role in treatment outcome. Investigation into the drug site of action intends not only to find genetic variability related to responsiveness, but also to determine which drug targets are of therapeutic value. This is of special relevance in antipsychotic treatment as many newly developed drugs display high affinity for a variety of neuroreceptors with unclear therapeutic roles. In addition, polymorphisms in neurotransmitter systems may contribute to the aetiology of the treated disease.

Research has been mainly directed to neurotransmitter systems thought to be involved in the aetiology of mental disorders and also the target of psychiatric drugs. Therefore, the major candidate genes for investigation are those of the dopaminergic and serotonergic system which are central to the classical hypothesis of the origin of schizophrenia and affective disorders. Special attention has been given to those receptors targeted by the atypical antipsychotic clozapine in the hope of discovering the key of the superior efficacy of this drug in the treatment of refractory schizophrenia. In addition, other receptors targeted by drugs have been screened for mutations of potential pharmacogenetic value.

Dopamine receptors

Dopamine receptors have been the subject of extensive research as candidates for involvement in mental disorders. Neuroleptic treatment is based on the blockade of dopamine receptors which results in symptom improvement as well as production of extrapyramidal side-effects (EPS). According to the previous hypothesis, mutations affecting expression or functioning of dopamine receptors could influence the therapeutic action of antipsychotics targeting them such as haloperidol and chlorpromazine. Dopamine receptors are divided in two main groups, the dopamine 1 type (D1 and D5) and the dopamine 2 type (D2, D3 and D4). The later group of receptors, in particular, are classical targets for antipsychotic treatment (see Table 14.2).

Table 14.2 Receptor binding affinities (K_i nM) for the typical antipsychotic haloperidol and the atypical clozapine, risperidone, olanzapine and amperozide. Adapted from Sodhi and Murray (1997)

Receptor	Haloperidol	Clozapine	Risperidone	Olanzapine	Amperozide
D_1	120	141	75	31	260
D_2	1.3	83	3.1	11	140
D_3	3.2	200			
D_4	2.3	20	7	27	
5-HT_{1A}	> 1000	6.5	488	> 1000	> 1000
5-HT_{2A}	78	2.5	0.16	5	20
5-HT_{2C}	> 1000	8.6	25.8	11.3	440
5-HT_3	> 1000	95	> 1000	57	
H_1	> 1000	23	155	7	730
M_1	> 1000	1.9	> 1000	1.9	
α_1	46	3.9	2	19	130
α_2	360	11.6	3	228	590

D = dopamine, 5-HT = serotonin, H = histamine, M = muscarinic, α = alpha-adrenergic

Two polymorphisms, -141C *Ins/Del* and Ser311Cys, and several other rare variants have been described in the gene coding for the D2 receptor (Itokawa *et al.* 1993; Arinami *et al.* 1997). However, their value as predictors of response to dopamine-blockade treatment is still undetermined. Surprisingly few studies have been performed investigating the influence of these polymorphisms on neuroleptic treatment, although associations have been reported with disorganized symptomatology in schizophrenia (Serreti *et al.* 1998). Although clozapine affinity for striatal D2 receptors is relatively low, SPET scans have revealed that this drug displays limbic selectivity for D2 receptors (Pilowsky *et al.* 1997). However, association studies between the functional -141C *Ins/Del* polymorphism and clozapine response have provided inconsistent results (Arranz *et al.* 1998a; Malhotra *et al.* 1999).

One of the most exciting findings regarding dopamine receptors concerns the D3 subtype. A Ser9Gly polymorphism in the D3 gene has been implicated in both the aetiology of schizophrenia and response to clozapine. Association between this polymorphism and schizophrenia has been reported by several groups (Shaikh *et al.* 1996; Ebstein *et al.* 1997) and confirmed by a meta-analysis (Williams *et al.* 1998). Individuals homozygous for the Ser allele were poorer responders to clozapine treatment (Shaikh *et al.* 1996; Scharfetter *et al.* 1998) and good responders to neuroleptics (Krebs *et al.* 1998) although these findings were not confirmed by Malhotra and collaborators (1998a). In addition, evidence for association between this polymorphism and tardive dyskinesia (TD) has been provided by two recent studies (Basile *et al.* 1999; Segman *et al.* 1999). Although the associations reported were modest, these findings suggest that D3 receptors are implicated in psychotic phenotypes and targeting them could be of therapeutic value.

Particular attention has been paid to the highly variable D4 receptor since it was reported that clozapine displayed high binding affinities for this subtype (Van Tol *et al.* 1991). Additionally, D4 receptors variants have been related to behavioural traits (Ebstein *et al.* 1996; Benjamin *et al.* 2000) and ADHD (Swanson *et al.* 1998, 2000). However, inconclusive results have been obtained from studies of several D4 polymorphisms on clinical samples of clozapine treated patients (Shaikh *et al.* 1997; Sanyal and Van Tol 1997). In contrast, modest associations have been reported between a 48bp repeat polymorphism in exon 3 of the D4 gene and response to treatment with classical antipsychotics (Hwu *et al.* 1998; Cohen *et al.* 1999). A newly reported functional polymorphism in the promoter region of the D4 gene (Okuyama *et al.* 1999) may help to clarify the involvement of D4 receptors in antipsychotic response.

Few studies have been done on the D1 and D5 receptors, and none of them reported any important implication of these receptors in mental disorders or treatment response. Although systematic screening of the D1 and D5 genes revealed several variants (Cichon *et al.* 1996; Feng *et al.* 1998), these were not found to be relevant for psychiatric genetics. Only a marginal association of a *Dde* I polymorphism in the 5' untranslated region of D1 polymorphism with clozapine response in a small sample was determined by FDG PET scan (Potkin *et al.* 1998). However, failure to find response-associated polymorphisms does not rule out the potential therapeutic value of these receptors.

In general, clearer associations were observed between dopamine receptor variants and neuroleptic response rather than with atypical antipsychotics. Being the main targets of classical antipsychotics, inherited variants in dopamine receptors may have a greater influence on neuroleptics than on those atypical antipsychotics which target a wider variety of neurotransmitter receptors. However, further studies are required to discern the real value of polymorphisms in the dopaminergic system as predictors of antipsychotic response.

Serotonin receptors

Serotonin (5-HT) receptors have been extensively investigated in relation to psychiatric disorders and antipsychotic treatment. In contrast to classical neuroleptics, which target dopamine receptors, atypical antipsychotics preferentially target 5-HT receptors. High occupancy of brain 5-HT receptors is achieved by clozapine, risperidone, olanzapine and other newly developed antipsychotics (see Table 14.2). Therefore, it has been hypothesized that these receptors are responsible for the superior efficacy of atypical antipsychotics in the treatment of refractory schizophrenia and their lack of extra-pyramidal side effects. Pharmacogenetic investigations are needed to identify which 5-HT receptor subtypes are responsible for the therapeutic action of atypical antipsychotics.

According to their pharmacology and molecular composition, serotonin receptors have been divided into seven families (5-HT1–5-HT7). The 5-HT2 family has been directly implicated in the therapeutic efficacy of clozapine and has been extensively investigated for association with clinical response. A silent 102-T/C polymorphism in the 5-HT2A gene was found to be associated with clozapine (Arranz *et al.* 1995b, 1998b) and neuroleptic response (Joober *et al.* 1999). However, other investigators failed to replicate these findings (Masellis *et al.* 1995; Nöthen *et al.* 1995; Nimgaonkar *et al.* 1996; Malhotra *et al.* 1996a). Differences in sample size, ethnicity, response assessment and treatment duration were blamed for the contrasting results. Nevertheless, a meta-analysis of all published studies on clozapine treated samples found strong associations between allele 102-C and poor clinical response (Arranz *et al.* 1998c). The significance of this finding was unclear as the 102-T/C was a silent change with no influence in the amino acid sequence of the receptor. However, a polymorphism in the promoter region of the gene, -1438-G/A, which may influence 5-HT2A expression levels, was found to be in complete linkage disequilibrium with the 102-T/C polymorphism (Spurlock *et al.* 1998). This result was corroborated by similar findings of association between other 5-HT2A polymorphism, His452Tyr and poor response to clozapine (Arranz *et al.* 1997; Masellis *et al.* 1998). These results constitute the strongest evidence of variation in drug targets influencing therapeutic action. Moreover, genetic variation in the 5-HT2C subtype, a Cys23Ser polymorphism amino acid substitution, was associated with marked improvement after clozapine treatment (Sodhi *et al.* 1995), although this association has yet to be confirmed (Malhotra *et al.* 1996b; Rietschel *et al.* 1997; Masellis *et al.* 1998). Taken altogether, these results highlight the importance of 5-HT2 receptors as therapeutic targets and support the hypothesis that their genetic variations can influence treatment with drugs targeting them such as clozapine and other antipsychotics. Therefore previous knowledge of the 5-HT2 genotype status of a patient may provide valuable information as to the likely efficacy of treatment with atypical antipsychotics. However, variation in 5-HT2 receptors cannot fully explain the heterogeneity in response to clozapine. This can be a reflection of the wide variety of neuroreceptors targeted by clozapine. Although 5-HT2 receptors may be the most influential as suggested by their clozapine high affinity, other receptors targeted by the drug may contribute to determine response. Following these results, investigations of 5-HT3, 4, 5, 6 and 7, and receptors of the adrenergic, histaminergic, and muscarinic systems, all of them targeted by clozapine, were initiated.

Research into other 5-HT receptors has failed to provide definite evidence of association with antipsychotic response. Several polymorphisms have been described in the 5-HT3A receptor gene after completion of the screening of 90 per cent of the coding region. However, none of these polymorphisms constituted amino acid changes, nor were they associated with clozapine response (unpublished data). 5-HT4, 5 and 7 receptor variants have also been investigated on clozapine clinical samples (Joseph Birkett, unpublished data) but failed to provide any information on the contribution of these receptors to antipsychotic action. The only relevant results regard a -12-A/T poly-morphism in 5-HT5A which showed a trend towards association with clozapine (Birkett *et al.* 2000) and a silent 5-HT6 267-T/C variant which showed modest association with neuroleptic response (Yu *et al.* 1999). If confirmed by replication studies, these results sug-gest that 5-HT5A and 5-HT6 receptors have a minor contributing role in clinical response. The recently cloned 5-HT3B (Davies *et al.* 1999) has yet to be screened for mutations, and may prove to be a more informative gene.

Adrenergic and histaminergic receptors

Adrenergic receptors have been associated with cognitive functions and depression and are also targeted by atypical antipsychotics. The alpha-2 subfamily, in particular, is targeted by clozapine. We have performed association studies of 3 polymorphisms described in the alpha-1A Arg492Cys (Shibata *et al.* 1996) and alpha-2A (-1291-C/G and -261-G/A) (Lario *et al.* 1997) on clozapine treated patients but failed to find any relation. Kalkman and collaborators (1998) also investigated a variant of the alpha-2B adrenergic receptor and found no association with clozapine responsiveness. However, the studies on these genes have just started and should not be discarded as possible therapeutic targets.

Interesting results have been reported by Orange and colleagues regarding a poly-morphism in the histamine 2 receptor, 649-A/G, which was reported to be associated with schizophrenia (Orange *et al.* 1996). However, this polymorphism was not found in clinical samples after extensive screening of the gene by other investigators. Other infrequent mutations have been reported in the coding and promoter regions of Histamine type 1 and type 2 (H1 and H2) genes. However, none of them were found to be related to clinical response, except a H2 -1018-G/A polymorphism that was found to be marginally associated with schizophrenia (Mancama *et al.* 2000). It is also possi-ble that the recently cloned Histamine 3 (H3) subtype for which clozapine displays high affinity may have a major influence in determining response. Systematic screening of the H3 gene for mutations and future association studies will clarify its role.

Investigations into muscarinic receptors, also targets of atypical antipsychotics, are in the early stages and no relation to responsiveness has been reported yet. Other neuro-receptors implicated in mental disorders are the glutamate and NMDA receptors. However, occupancy of these receptors by antipsychotic drugs is low and unlikely to be related to therapeutic efficacy.

Neurotransmitter transporters

Serotonin and dopamine transporters (5-HTT and DAT, respectively) play an impor-tant role in maintaining neurotransmitter levels. Variations in these transporters can alter the availability of serotonin and dopamine in the brain and influence the efficacy of drugs targeting these systems.

A VNTR (variable number of tandem repeats) polymorphism first described by Vandenbergh and colleagues (1992) and several silent mutations have been detected in the DAT gene (Grünhage *et al.* 1998). The importance of these polymorphisms on drug response remains to be investigated in spite of reports of association with psychiatric phenotypes (Daly *et al.* 1998).

VNTR polymorphisms have been described in the promoter and intronic regions of the 5-HTT which were found to be associated with the efficacy of the antidepressant fluvoxamine (Smeraldi *et al.* 1998) and with psychosis in schizophrenic patients (Malhotra *et al.* 1998b). We have observed a trend towards association between those polymorphisms and clozapine suggesting a minor contribution of the 5-HTT in determining response (Arranz *et al.* 2000a). These results could be taken as further evidence of the influence of variation in drug targets on responsiveness. Being more directly related to the transporter, fluvoxamine efficacy is more strongly influenced by 5-HTT polymorphisms than clozapine's, which has a wider variety of targets.

Predictive value of polymorphisms in neurotransmitter systems

Polymorphisms in neurotransmitter systems are directly related to lack of drug activity. Independently of their metabolizing rate, drugs may show lack of efficacy because of alterations at the target level. Therefore, even if the right doses of a drug are being used, polymorphisms rendering the targeted receptors inactive or affecting their expression will contribute to treatment failure. Finding receptor polymorphisms determining response to a particular treatment will facilitate the selection of the most appropriated therapy for each individual according to their genotype (Roses 2000).

Polymorphisms predicting clinical response can also inform the therapeutic value of the receptor. Presence of response-determining mutations in a receptor protein indicates mediation in treatment therapeutic activity. However, lack of response-associated variants in a particular neuroreceptor does not necessarily mean lack of therapeutic value. Therapeutic action may still be mediated through receptors lacking influential genetic mutations. Although of no use for genetic response prediction, such receptors cannot be discarded as therapeutic targets.

Individualization of treatment

From the evidence reviewed in this chapter, it is likely that a combination of metabolizing and drug target polymorphisms will produce the best prediction for the selection of the optimal dose and optimal drug as a function of the individual's genetic profile. Pharmaceutical companies have started work on rapid methods for the prediction of metabolic rates. DNA chips for the detection of the individual's CYP2D6 and CYP2C19 genotypes are already available, and kits for the accurate prediction of metabolizing rates will surely follow. However, research into selection of the most beneficial drug is still in its early stages.

Selection of the most beneficial drug will also require the combination of genotypes at several loci. Although strong associations have been reported between serotonin receptors and clozapine, it is obvious that genetic factors in other relevant loci contribute to determining response. We have developed a protocol for the prediction of response to clozapine which includes genotyping of 6 polymorphisms located in 4 neurotransmitter receptors targeted by the drug (Arranz *et al.* 2000b). This method can predict clozapine response with 80 per cent certainty. These results suggest the way forward for the individualisation of treatment. Hopefully, the prediction values can be improved with the detection of additional genetic variation contributing to variation in response. Similar research into genetic factors determining antipsychotic treatment should be encouraged.

Future of pharmacogenetics: pharmacogenomics

The final goal of pharmacogenetics it to help clinicians to choose the best treatment for each individual patient. Progress in pharmacogenetics has been achieved by establishing links between specific single genes and particular phenotypes by association studies. However, pharmaceutical companies are developing new techniques that will speed up the research into genes implicated in disease or treatment. Methods for the detection of single nucleotide polymorphisms (SNPs) at large scale have already been developed by several companies (Marshall and Hodgson 1998) and SNPs maps of the human genome are already available (Boehnke 2000; Roses 2000). Expressed sequence tag (EST) sequencing will speed up the identification of new genes (Kennedy 1997). Kits for the diagnosis of several human disorders including cystic fibrosis, beta-thalasemia, and Tay-sachs disease among others, using genetic information are already available (Persidis 1998). All these advances will lead to the rapid development of new diagnostic methods and therapeutic products using genomic information and, hopefully, to the improvement of patient care.

References

Arinami T., Gao M., Hamaguchi H. *et al.* (1997) A functional polymorphism in the promoter region of the dopamine D2 receptor gene is associated with schizophrenia. *Human Molecular Genetics* **6**:577–82.

Arranz M. J., Dawson E., Shaikh S. *et al.* (1995a) Cytochrome P4502D6 genotype does not determine response to clozapine. *British Journal of Clinical Pharmacology* **39**:417–20.

Arranz M. J., Collier D., Sodhi M. *et al.* (1995b) Association between clozapine response and allelic variation in the 5-HT2A receptor gene. *Lancet* **346**:281–2.

Arranz M. J., Collier D. A., Munro J. *et al.* (1997) Analysis of a structural polymorphism in the 5-HT2A receptor and clinical response to clozapine. *Neuroscience Letters* **217**:177–8.

Arranz M. J., Li T., Munro J. *et al.* (1998a) Lack of association between a polymorphism in the promoter region of the dopamine-2 receptor gene and clozapine response. *Pharmacogenetics* **8**:481–4.

Arranz M. J., Munro J., Owen M. J. *et al.* (1998b) Evidence for association between polymorphisms in the promoter and coding regions of the 5-HT2A receptor gene and response to clozapine. *Molecular Psychiatry* **3**:61–6.

Arranz M. J., Munro J., Sham P. *et al.* (1998c) Meta-analysis of studies on genetic variation in 5-HT2A receptors and clozapine response. *Schizophrenia Research* **32**:93–9.

Arranz M. J., Bolonna A. A., Munro J. *et al.* (2000a) The serotonin transporter and clozapine response. *Molecular Psychiatry* **5**:124–30.

Arranz M. J., Munro J., Birkett J. *et al.* (2000b) Pharmacogenetic prediction of clozapine response. *Lancet* **355**:1615–16.

Basile V. S., Masellis M., Badri F. *et al.* (1999) Association of the Msc I polymorphism of the dopamine D3 receptor gene with tardive dyskinesia in schizophrenia. *Neuropsychopharmacology* **21**:17–27.

Benjamin J., Osher Y., Kotler M. *et al.* (2000) Association between tridimensional personality questionnaire (TPQ) traits and three functional polymorphisms: dopamine receptor D4 (DRD4), serotonin transporter promoter region (5-HTTLPR) and catechol O-methyltransferase (COMT) *Molecular Psychiatry* **5**:96–100.

Bertilsson L. and Dahl M. L. (1996) Polymorphic drug oxidation. *CNS Drugs* 5(3):200–23.

Birkett J. T., Arranz M. J., Munro J. *et al.* Association analysis of the 5-HT5A gene in depression, psychosis and antipsychotic response. *Neuroreport 2000* **11**(9):2017–20.

Blum M., Grant D. M., McBride W. *et al.* (1990) Human arylamine N-acetyltransferease genes: isolation, chromosomal localisation and functional expression. *DNA Cell Biology* **9**:193–203.

Boehnke M. (2000) A look at linkage disequilibrium. *Nature Genetics* **25**:246–7.

Butcher N. J., Ilett K. F., Minchin R. F. (1998) Functional polymorphism of the human arylamine N-acetyltransferase type 1 gene caused by (CT)-T-190 and G(560)A mutations. *Pharmacogenetics* **8**:67–72.

Caraco Y. (1998) Genetic determinants of drug responsiveness and drug interactions. *Therapeutic Drug Monitoring* **20**:517–24.

Cascorbi I., Drakoulis N., Brockmoller J. *et al.* (1995) Arylamine N-acetyltransferase (NAT2) and their allelic linkage in unrelated Caucasian individuals-correlation with phenotypic activity. *American Journal of Medical Genetics* **57**:581–92.

Cichon S., Nöthen M. M., Stöber G. *et al.* (1996) Systematic screening for mutations in the 5′-regulatory region of the human dopamine D-1 receptor (DRD1) gene in patients with schizophrenia and bipolar affective disorder. *American Journal of Medical Genetics* **67**:424–8.

Cichon S., Nöthen M. M., Rietschel M. *et al.* (2000) Pharmacogenetics in schizophrenia. *American Journal of Medical Genetics* **97**:98–106.

Ciotti M., Marrone A., Potter C. *et al.* (1997) Genetic polymorphism in the human UGT1A6 (planar phenol) UDP-glucuronosyltransferase: pharmacological implications. *Pharmacogenetics* **7**:485–95.

Clarke D. J., Moghrabi N., Monaghan G. *et al.* (1997) Genetic defects of the UDP-glucuronosyltransferase-1 (UGT-1) gene that cause familial non-haemolytic unconjugated hyperbilirubinaemias. *Clinica Chimica Acta* **266**:63–74.

Cohen B. M., Ennulat D. J., Centorrino F. *et al.* (1999) Polymorphisms of the dopamine D-4 receptor and response to antipsychotic drugs. *Psychopharmacology* **141**:6–10.

Dahl M. L., Llerena A., Bondesson U. *et al.* (1994) Disposition of clozapine in man: lack of association with debrisoquine and S-mephenytoin hydroxylation polymorphisms. *British Journal of Clinical Pharmacology* **37**:71–4.

Dahl M. L., Bertilsson L., Nordin C. (1996) Steady state plasma levels of nortriptyline and its 1.0-hydroxy metabolite: Relationship to the CYP2D6 genotype. *Psychopharmacology* **123**:315–19.

Daly G., Hawi Z., Fitzgerald M. *et al.* (1998) Attention deficit hyperactivity disorder: association with the dopamine transporter (DAT1) but no with the dopamine receptor (DRD4) *American Journal of Medical Genetics* **81**:501.

Davies P. A., Pistis M., Hanna M. C. *et al.* (1999) The 5-HT3B subunit is a major determinant of serotonin-receptor function. *Nature* **397**:359–63.

De la Moureyre C. S.V., Debuysere H., Mastain B. *et al.* (1998).Genotypic and phenotypic analysis of the polymorphic thiopurine S-methyltransferase gene (TPMT) in a European population. *British Journal of Pharmacology* **125**:879–87.

De Morais S. M. F., Wilkinson G. R., Blaisdell J. *et al.* (1994) The major genetic-defect responsible for the polymorphism of S-mephenytoin metabolism in humans. *Journal of Biological Chemistry* **269**:15419–22.

Doll M. A., Jiang W., Deitz A. C. *et al.* (1997) Identification of a novel allele at the human NAT1 acetyltransferase locus. *Biochemical and Biophysical Research Communications* **233**:584–91.

Ebstein R. P., Novick O., Umansky R. *et al.* (1996) Dopamine D4 receptor (D4DR) exon III polymorphism associated with the human personality trait of novelty seeking. *Nature Genetics* **12**:78–80.

Ebstein R. P., Macciardi F., Heresco-Levi U. *et al.* (1997) Evidence for association between the Dopamine D3 receptor gene DRD3 and schizophrenia. *Human Heredity* **47**:6–16.

Eichelbaum M. and Evert B. (1996) Influence of pharmacogenetics on drug disposition and response. *Clinical and Experimental Pharmacology and Psychology* **23**:983–5.

Ellingrod V. L., Schultz S. K., Arndt S. *et al.* (1999) Association between cytochrome P4502D6 (CYP2D6) genotype, neuroleptic exposure, and abnormal involuntary movement scale (AIMS) score. *Schizophrenia Research* **36**(1–3): 90.

Feng J. N., Sobell J. L., Heston L. L. *et al.* (1998) Scanning of the dopamine D1 and D5 receptor genes by REF in neuropsychiatric patients reveals a novel missense change at a highly conserved amino acid. *American Journal of Medical Genetics* **81**:172–8.

Ferguson R. J., De Morais S. M. F., Benhamou S. *et al.* (1998) A new genetic defect in human CYP2C19: Mutation of the initiation codon is responsible for poor metabolism of S-mephenytoin. *Journal of Pharmacology and Experimental Therapeutics* **284**:356–61.

Garrod A. E. (1931) *Inborn Factors in Disease: An Essay*. Oxford University Press: New York.

Gill H.J., Tjia J. F., Kitteringham N. R. *et al.* (1999) The effect of genetic polymorphisms in CYP2C9 on sulphamethoxazole N-hydroxylation. *Pharmacogenetics* **9**:43–53.

Grünhage F., Rietschel M., Albus M. *et al.* (1998) Systematic search for variation in the human dopamine transporter gene. *American Journal of Medical Genetics* **81**:500.

Harris H. (1976) Enzyme variants in human populations. *Johns Hopkins Medical Journal* **138**:245–52.

Hodgson J. and Marshall A. (1998) Pharmacogenomics: will the regulators approve? *Nature Biotechnology* **16**(supp):13–15.

Huang J. D., Guo W. C., Lai M. D. *et al.* (1999) Detection of a novel cytochrome P-450 1A2 polymorphism (F21L) in Chinese. *Drug Metabolism and Disposition* **27**:98–101.

Hwu H. G., Hong C. J., Lee Y. L. *et al.* (1998) Dopamine D4 receptor gene polymorphisms and neuroleptic response in schizophrenia. *Biological Psychiatry* **44**:483–7.

Ibeanu G. C., Goldstein J. A., Meyer U. *et al.* (1998) Identification of new human CYP2C19 alleles (CYP2C19*6 and CYP2C19*2B) *Journal of Pharmacology an Experimental Therapeutics* **286**:1490–5.

Itokawa M., Arinami T., Futamura N. *et al.* (1993) A structural polymorphism of human dopamine D2 receptor (SER(311)-CYS). *Biochemical and Biophysical Research Communications* **3**:1369–75.

Jerling M., Lindstrom L., Bondesson U. *et al.* (1994) Fluvoxamine inhibition and carbamazepine induction of the metabolism of clozapine: evidence from a therapeutic drug monitoring service. *Therapeutic Drug Monitoring* **16**:368–74.

Joober R., Benkelfat C., Brisebois K. *et al.* (1999) T102C polymorphism in the 5HT2A gene and schizophrenia: relation to phenotype and drug response variability. *Journal of Psychiatry and Neuroscience* **24**:141–6.

Kalkman H. O., Charara N., Kuntzelmann G. *et al.* (1998) Structural polymorphism in the alpha (2B) adrenoceptor: No association with schizophrenia or clozapine responsiveness. *American Journal of Medical Genetics* **81**:510.

Kalow W. (1962) *Pharmacogenetics—Heredity and the Response to Drugs.* Philadelphia: W. B. Saunders Company.

Kennedy G. C. (1997) Impact of genomics on therapeutic drug development. *Drug Development Research* **41**:112–19.

Kidd R. S., Straughn A. B., Meyer M. C. *et al.* (1999) Pharmokinetics of chlorpheniramine, phenytoin, glipizide and nifedipine in an individual homozygous for the CYP2C9*3 allele. *Pharmacogenetics* **9**:71–80.

Krebs M. O., Sautel F., Bourdel M. C. *et al.* (1998) dopamine D3 receptor gene variants and substance abuse in schizophrenia. *Molecular Psychiatry* **3**:337–41.

Lario S., Calls J., Cases A. *et al.* (1997) Short report on DNA marker at candidate locus. *Clinical Genetics* **51**:129–30.

Lennard, M. S. (1993) Genetically-determined adverse drug-reactions involving metabolism. *Drug Safety* **9**:60–77.

Lin H. J., Han C. Y., Lin B. K. *et al.* (1993) Slow acetylator mutations in the human polymorphic N-acetyltransferase gene in 786 Asians, Blacks, Hispanics, and Whites—application to metabolic epidemiology. *American Journal of Human Genetics* **52**:827–34.

Lin K.-M., Poland R. E., Wan Y.-J. *et al.* (1996) The evolving science of pharmacogenetics: clinical and ethnic perspectives. *Psychopharmacology Bulletin* **32**:205–17.

Linder M. W., Prough R. A., Valdes R. (1997) Pharmacogenetics: a laboratory tool for optimising therapeutic efficiency. *Clinical Chemistry* **43**:254–66.

Malhotra A.K, Goldman D., Ozaki N. *et al.* (1996a) Lack of association between polymorphisms in the 5-HT$_{2A}$ receptor gene and antipsychotic response to clozapine. *American Journal of Psychiatry* **153**:1092–4.

Malhotra A. K., Goldman D., Ozaki N. *et al.* (1996b) Clozapine response and the 5HT(2C)Cys(23)Ser polymorphism. *Neuroreport* **7**:2100–2.

Malhotra A. K., Goldman D., Buchanan R. W. *et al.* (1998a) The dopamine D3 receptor (DRD3) Ser9Gly polymorphism and schizophrenia: a haplotype relative risk study and association with clozapine response. *Molecular Psychiatry* **3**:72–5.

Malhotra A. K., Goldman D., Mazzanti C. *et al.* (1998b) A functional serotonin transporter (5-HTT) polymorphism is associated with psychosis in neuroleptic-free schizophrenics. *Molecular Psychiatry* **3**:328–32.

Malhotra A. K., Buchanan R. W., Kim S. *et al.* (1999) Allelic variation in the promoter region of the dopamine D2 receptor gene and clozapine response. *Schizophrenia Research* **36**(1–3):91.

Mancama D., Arranz M. J., Munro J. *et al.* (2000 in press) The histamine 1 and histamine 2 receptor genes—candidates for schizophrenia and clozapine drug response. *Gene Screen.*

Marez D., Legrand M., Sabbagh N. *et al.* (1997) Polymorphism of the cytochrome P450 CYP2D6 gene in a European population: characterisation of 48 mutations and 53 alleles, their frequencies and evolution. *Pharmacogenetics* **7**:193–202.

Marshall A. and Hodgson J. (1998) DNA chips: an array of possibilities. *Nature Biotechnology* **16**:27–31.

Masellis M., Paterson A. D., Badri F. *et al.* (1995) Genetic variation of the 5-HT$_{2A}$ receptor and response to clozapine. *Lancet* **345**:908.

Masellis M., Basile V., Meltzer H. Y. *et al.* (1998) Serotonin subtype 2 receptor genes and clinical response to clozapine in schizophrenia patients. *Neuropsychopharmacology* **19**:123–32.

Meyer U. A. and Zanger U. M. (1997) Molecular mechanisms of genetic polymorphisms of drug metabolism. *Annual Review of Pharmacological Toxicology* **37**:269–96.

Nakajima M., Yokoi T., Mizutani M. *et al.* (1999) Genetic polymorphism in the 5′- flanking region of human CYP1A2 gene: effect on the CYP1A2 inducibility in humans. *Journal of Biochemistry* **125**:803–8.

Nimgaonkar V. L., Zhang X. R., Brar J. S. *et al.* (1996) 5-HT$_2$ receptor gene locus: association with schizophrenia or treatment response not detected. *Psychiatric Genetics* **6**:23–7.

Nöthen M. M., Rietschel M., Erdmann J. *et al.* (1995) Genetic variation of the 5-HT$_{2A}$ receptor and response to clozapine. *Lancet* **345**:908.

Okuyama Y., Ishiguro H., Toru M. *et al.* (1999) A genetic polymorphism in the promoter region of DRD4 associated with expression and schizophrenia. *Biochemical and Biophysical Research Communications* **258**:292–5.

Orange P. R., Heath P. R., Wright S. R. *et al.* (1996) Individuals with schizophrenia have an increased incidence of the H$_2$R$_{649G}$ allele for the histamine H2 receptor gene. *Molecular Psychiatry* **1**:466–9.

Persidis A. (1998) Pharmacogenomics and diagnostics. *Nature Biotechnology* **16**:791–2.

Pilowsky L. S., Mulligan R. S., Acton P. D. *et al.* (1997) Limbic selectivity of clozapine. *Lancet* **350**:490–1.

Potkin S. G., Kennedy J., Badri F. *et al.* (1998) A genetic PET scan study: D1 alleles predict clinical response to clozapine and corresponding brain metabolism. *American Journal of Medical Genetics* **81**:496–7.

Rebbeck T. R., Jaffe J.M., Walker A. H. *et al.* (1998) Modification of clinical presentation of prostate tumours by a novel genetic variant in CYP3A4. *Journal of the National Cancer Institute* **90**:1225–9.

Rietschel M., Naber D., Fimmers R. *et al.* (1997) Efficacy and side-effects of clozapine not associated with variation in the 5-HT2C receptor. *Neuroreport* **8**:1999–2003.

Ring B. J., Catlow J., Lindsay T. J. *et al.* (1996) Identification of the human cytochromes P450 responsible for the in vitro formation of the major oxidative metabolites of the antipsychotic agent olanzapine. *Journal of Pharmacology and Experimental Therapeutics* **276**:658–66.

Roses A. D. (2000) Pharmacogenetics and the practice of medicine. *Nature* **405**:857–65.

Sachse C., Brochmoller J., Bauer S. *et al.* (1999) Functional significance of a C- > A polymorphism in intron I of the cytochrome P450 CYP1A2 gene tested with caffeine. *British Journal of Clinical Pharmacology* **47**:445–9.

Sanyal S. and Van Tol H. H. M. (1997) Review of the role of dopamine D4 receptors in schizophrenia and antipsychotic action. *Journal of Psychiatric Research* **31**:219–32.

Scharfetter J., Chaudhri H. R., Hornik K. *et al.* (1998) Association of dopamine D3 receptor gene polymorphism and response to clozapine. *American Journal of Medical Genetics* **81**:499.

Segman R., Neeman T., Heresco-Levy U. *et al.* (1999) Genotypic association between the dopamine D3 receptor and tardive dyskinesia in chronic schizophrenia. *Molecular Psychiatry* **4**:247–53.

Serretti A., Macciardi F., Smeraldi E. (1998) Dopamine receptor D2 Ser Cys311 variant associated with disorganized symptomatology of schizophrenia. *Schizophrenia Research* **34**:207–10.

Shaikh S., Collier D. A., Sham P. C. *et al.* (1996) Allelic association between a Ser-9-Gly polymorphism in the dopamine D3 receptor gene and schizophrenia. *Human Genetics* **97**:714–19.

Shaikh S., Makoff A., Collier D. *et al.* (1997) Dopamine D4 receptors: potential therapeutic implications in the treatment of schizophrenia. *CNS Drugs* **8**:1–11.

Shibata K., Hirasawa A., Moriyama N. *et al.* (1996) a1a-Adrenoceptor polymorphism: pharmacological characterization and association with benign prostatic hypertrophy. *British Journal of Pharmacology* **118**:1403–8.

Smeraldi E., Zanardi R., Benedetti F. *et al.* (1998) Polymorphism within the promoter of the serotonin transporter gene and antidepressant efficacy of fluvoxamine. *Molecular Psychiatry* **3**:508–11.

Sodhi M. S., Arranz M. J., Curtis D. *et al.* (1995) Association between clozapine response and allelic variation in the 5-HT2C receptor gene. *Neuroreport* **7**:169–72.

Sodhi M. S. and Murray R. M. (1997) Future therapies for schizophrenia. *Expert Opinion on Therapeutic Patents* **7**:151–65.

Spurlock G., Heils A., Holmans P. *et al.* (1998) A family based association study of T102c polymorphism in 5HT2A and schizophrenia plus identification of new polymorphisms in the promoter. *Molecular Psychiatry* **3**:42–9.

Sullivan-Klose T. H., Ghanayem B. I., Bell D. A. *et al.* (1996) The role of the CYP2C9 leu-359 allelic variant in the tolbutamide polymorphism. *Pharmacogenetics* **6**:341–9.

Swanson J. M., Sunohara G. A., Kennedy J. L. *et al.* (1998) Association of the dopamine receptor D4 (DRD4) gene with a refined phenotype of attention deficit hyperactivity disorder (ADHD): a family based approach. *Molecular Psychiatry* **3**:38–41.

Swanson J., Oosterlaan J., Murias M. *et al.* (2000) Attention deficit/hyperactivity disorder children with a 7-repeat allele of the dopamine receptor D4 gene have extreme behaviour but normal performance on critical neuropsychological tests of attention. *Proceedings of the National Academy of Sciences of the United States of America* **97**:4754–9.

Van Tol H. H. M., Bunzow J. R., Guan H. C. *et al.* (1991) Cloning of the gene for a human dopamine D4-receptor with high-affinity for the antipsychotic clozapine. *Nature* **350**:610–14.

Vanderbergh D. J., Persico A. M., Uhl G. R. (1992) A human dopamine transporter cDNA predicts reduced glycosylation, displays a novel repetitive element and provides racially-dimorphic TaqI RFLPs. *Molecular Brain Research* **15**:161–6.

Vatsis K. P., Martell K. J., Weber W. W. (1991) Diverse point mutations in the human gene for polymorphic N-acetyltransferase. *Proceedings of the National Academy of Sciences of the USA* **88**:6333–7.

Vatsis K. P. and Weber W. W. (1993) Structural heterogeneity of Caucasian N-acetyltransferase at the NAT1 gene locus. *Archives of Biochemistry and Biophysics* **301**:71–6.

Westlind A., Lofberg L., Tindberg N. *et al.* (1999) Interindividual differences in hepatic expression of CYP3A4: Relationship to genetic polymorphism in the 5′- upstream regulatory region. *Biochemical and Biophysical Research Communications* **259**:201–5.

Williams J., Spurlock G., Holmans P. *et al.* (1998) A meta-analysis and transmission disequilibrium study of association between the dopamine D3 receptor gene and schizophrenia. *Molecular Psychiatry* **3**:141–9.

Xie H. G., Xu Z. H., Ouyang D. S. *et al.* (1997) Meta-analysis of phenotype and genotype of NAT2 deficiency in Chinese populations. *Pharmacogenetics* **7**:503–14.

Yasar U., Eliasson E., Dahl M. L. *et al.* (1999) Validation of methods for CYP2C9 genotyping: frequencies of mutant alleles in a Swedish population. *Biochemical and Biophysical Research Communications* **254**:628–31.

Yu Y. W.-Y., Tsai S.-J., Lin C.-H. *et al.* (1999) Serotonin-6 receptor variant (C267T) and clinical response to clozapine. *Neuroreport* **10**:1231–3.

Zielinska E., Niewiarowski W., Bodalski J. *et al.* (1997) Arylamine N-acetyltransferase (NAT2) gene mutations in children with allergic diseases. *Clinical Pharmacology and Therapeutics* **62**:635–42.

Chapter 15

Genetic counselling

Jane Scourfield and Michael J. Owen

Introduction: the process of counselling

Genetic counselling is a process which aims to facilitate informed and autonomous decision making, appreciation of the inheritance of a genetic condition, and improvement in the emotional well-being of those affected or their family members (Biesecker *et al.* 1999). Harper (1993) has defined it as:

> The process by which patients or relatives at risk of a disorder that may be hereditary are advised of the consequences of this disorder, the probability of developing or transmitting it, and of the ways in which this may be prevented or ameliorated

Currently, most genetic counselling is focused on rare Mendelian diseases; however, advances are likely to mean that in years to come genetic counselling will include improved empirical risk predictions for common diseases, such as the majority of psychiatric disorders, which are only partly genetic in origin. Despite this hoped-for improvement in risk prediction, considerable uncertainty will remain. For most common diseases it is likely that the proportion of cases attributable to any particular susceptibility genotype will be small (as in the case of breast cancer and BRCA1 and 2). Other, non-genetic, factors will have a substantial role, so that a negative result from any genetic test would give little reassurance that an individual will remain disease free. For those with a positive genetic test result Holzman and Marteau (2000), in their examination of the likely impact of genetic susceptibility tests, show that only with a relative risk of 20 and a population risk of 5 per cent will the positive predictive value of a positive test result exceed 50 per cent.

Another important element of genetic counselling is often the correction of mistaken beliefs, for example that all the offspring of a parent with serious mental illness necessarily have the same 'hereditary taint' or that all disorders with a genetic component are necessarily untreatable. The case of phenylketonuria provides us with an example of a genetic condition, which is treatable by manipulation of the environment (the removal of phenylalanine from the diet). For most psychiatric disorders it is assumed that a vulnerability is inherited and that this interacts with environmental stresses which are also required for the disorder to become manifest. Note, however, that the 'environment' in complex diseases might include any random biological processes during development, such as changes in gene structure or expression, epigenetic processes, as well as psycho-social or physical adversity (McGuffin *et al.* 1994; Petronis and Gottesman 2000).

Individuals seeking information from genetic counselling are usually relatives of psychiatric patients who want to know the risk of that disorder developing in themselves or their children. Less commonly, third parties such as legal representatives or professionals involved in adoption may seek advice. An important principle of the counselling process is that it aims to be non-directive, helping individuals and families make their own decisions rather than prescribing a particular course. Any psychiatrist seeing patients who wish to discuss genetic issues surrounding psychiatric disease should have in mind the following goals (Clarke 1994: 1–29):

1 Listening; find out from the clients what their questions are, if they relate to a specific disease or to an individual within the family who is affected with a particular disorder.

2 Be sure that an accurate diagnosis has been made and determine a complete and accurate pedigree. This is fundamental to any counselling process if meaningful information is to be exchanged.

3 Communicate information in a non-directive way appropriate to the clients' level of knowledge.

The estimated risk of recurrence needs to be weighed against the burden of the disorder, a difficult measurement to make. Burden may be considered the cost to a family of recurrence of a disorder, and once the risk/burden ratio is understood it is for the client to decide how acceptable this is. It is often useful to summarize in writing the main points which have been raised in a genetic counselling session and to provide the client with a copy of the summary.

Estimating risk

The assessment of risk in genetic counselling is based on three types of information (Murphy and Chase 1975)

1 *Empirical information*, consisting of estimates based on available research data about the recurrence risk of a disorder in various categories of relatives. This is the type of information used when the mode of inheritance is unknown.

2 *Modular information*, which depends on a clear understanding of the mode of inheritance of the disorder.

3 *Particular information*, which is a compilation of all the data that can be used in assessing the risks to a particular family. This would include, if available, the results of DNA tests.

For most psychiatric disorders the specific mode of inheritance is unknown, so in general the information given to clients in genetic counselling will be of the empirical type. The empirical data on which risk estimates are based come from family, twin and adoption studies, and these are summarized in the relevant chapters for each disorder.

Genetics and clinical practice

With the exception of certain rare familial dementias (Chapter 13, and below), there are currently no molecular genetic tests to predict psychiatric disorders, but future developments should lead to greater predictive accuracy. Predispositional tests are being carried out already for some gene mutations which are risk factors for disease but do not present a certainty of developing the disease (for example genetic testing for hereditary breast and ovarian cancer). Such tests will become the main type of genetic test offered in the future as genes predisposing to common disorders (including psychiatric disorders) are identified (Bell 1998).

Huntington's disease (HD) illustrates some of the issues surrounding genetic prediction of late-onset diseases, although common psychiatric disorders do not show its simple Mendelian pattern of inheritance and are more treatable. Predictive testing for HD has been available for several years, and since the mutation in the HD gene has been identified (Huntington's Disease Collaborative Research Group 1993) those who will or will not develop the disease can be identified with near certainty. Testees undergo careful pre- and post-test counselling which has been well evaluated and has resulted in very few adverse reactions (Quaid 1993; Harper *et al.* 1996). In fact, there is now evidence that test results are not the only factors influencing individual responses. Not surprisingly, the amounts of social support people have, along with their coping resources, have an important influence as does the degree of distress before the test (Marteau and Croyle 1998).

However, despite careful assessment and counselling procedures, various unforeseen problems have arisen. These include unintentional risk alteration e.g. testing of offspring resulting in parents knowing their own genetic status without wishing to, including the not-uncommon event of discovering non-paternity. Inappropriate referral is another problem, with some requests for genetic testing having been received without the consent of the individual at risk. Predictive genetic testing is not an innocuous investigation on a par with other laboratory tests, because of its wider implications and its effect on relatives. Personal, family, insurance and career prospects are profoundly affected by a positive test result and the decision whether or not to proceed with testing is one for the individual to make for him/herself. People should not be referred for testing without their knowledge.

The disclosure of genetic information is another difficult area. Whilst the results of genetic tests are subject to the confidentiality requirements of any other medical investigation, when there are implications for relatives the situation is not straightforward. There is potential for conflict within the health care professional-patient relationship should the patient refuse to warn at-risk relatives about relevant genetic information. The American Society for Human Genetics (1998) and Nuffield Council for Bioethics (1993) recommend that in exceptional circumstances genetic test results may be disclosed to relatives against the patient's wishes. Such circumstances would include risk of imminent or forseeable serious harm to relatives, and situations where harm from failing to disclose would outweigh harm from disclosure. Genetic testing of children is another issue raised by HD testing programmes. If a child will benefit from the result of a genetic test—for example if medical treatment will be modified—then the choice is a relatively straightforward one. However, when predictive tests for an adult-onset disorder are requested in children the results will be of no immediate advantage to the child and indeed, the child will have lost his/her right to choose not to be tested, as is the case with many adults who prefer not to know their genetic risk status. A child found to be at high risk of a late onset disease would be subject to very different expectations in terms of health, education, career and personal development that might jeopardize personal happiness. There is international consensus that childhood predictive genetic testing for late onset disease is not appropriate and international regulatory bodies have advised strongly against it. (World Federation of Neurology Research Committee. Research Group on Huntington's Disease 1989; American Society of Human Genetics: 1995). Such issues may be generalized, to some extent, to the future development of predictive genetic tests for other late onset disorders, such as psychiatric disorders. However, the issues are complicated by the more complex and only partially understood patterns of inheritance. Also, the progressive and untreatable course of HD means that parallels with the majority of psychiatric disorders are not perfect.

Genetic counselling and clinical practice: Alzheimer's disease as an example

In contrast to HD where it is possible to provide *modular* risk estimates for patients' relatives and a *particular* risk once DNA testing has been completed, risk estimates for common psychiatric disorders derive mainly from *empirical* data based on family studies which examine rates of disorder among various classes of relatives. In genetic counselling for psychiatric disorders, therefore, it is at present possible only to talk in general terms of risks to relatives and not to estimate precise risks for specific families or individuals.

An important exception is the case of dementia, where the genes for a number of extremely rare, autosomal dominant forms have been identified. Also we now have clear evidence that APOE genotype is an important risk factor for the more typical, non-autosomal dominant forms of Alzheimer's disease (AD). This provides a good illustration of the issues relating to predispositional genetic testing for other psychiatric disorders.

Autosomal dominant Alzheimer's disease

Families with autosomal dominant forms of AD are fortunately extremely rare, but the occurrence of numerous individuals with early onset disease (<55 years) in several generations is highly suggestive, as is the presence of many affected individuals in one generation. It has been recommended that families containing three members with a history of early onset AD occurring before the age of 60 should, if they request advice, be referred to a clinical geneticist (Liddell *et al.* 2001). Predictive or prenatal testing is potentially feasible in FAD families, but will depend upon the availability of DNA from at least one affected individual in order to establish which mutation is causing disease in that family. However, in 30 per cent of cases no clearly causative mutations are detected (Campion *et al.* 1999) and there has been one report of apparent non-penetrance of a PSEN1 mutation in a healthy 68-year-old member of a FAD pedigree (Rossor *et al.* 1996). Notwithstanding these caveats, in the absence of any preventative treatment, the demand for predictive testing is likely to be low, as has been the case in HD (Binedell *et al.* 1998). However, some families will wish for leukocyte DNA to be collected from their affected relative to facilitate future testing, a procedure that can best be undertaken by, and after consultation with, the local department of clinical genetics.

Non-Mendelian Alzheimer's disease

The findings from family studies (Chapter 13) can be used to advise relatives in only the broadest terms. One can do little more than say that, on balance, the risk to the first-degree relatives of AD cases, who developed the disorder at any time up to the age of 85, is increased three- to fourfold relative to the risk in controls. This would seem to translate to a risk of developing AD of between one in five and one in six, which, although an improvement on a risk of one-in-two, can be alarming. In order to put this risk into better perspective, it may be preferable to present the risk estimates in a graphical form as is done in the paper of Breitner and colleagues (1988). If this is done, clients can see that even at 78 years, the age of greatest risk, the actual predicted risk is only some 3 per cent and that the risk at age 70 is under 1 per cent. Many find this reassuring – partly because their worries were about getting AD in their fifties or sixties, thus preventing them from enjoying retirement (Liddell *et al.* 2001).

In the case of AD patients who became demented late in old age, say by the late eighties, relatives probably run the same 30 to 50 per cent risk of developing dementia as anyone else who lives to the age of 90 and beyond. Like other disorders that reflect the combined action of several genes, risk to relatives drops rapidly as the degree of genetic relatedness falls. Data are limited but the risk to second-degree relatives, such as grandchildren, is probably less than twice the population levels (Heston *et al.* 1981).

There is rather worrying evidence from one study which suggests that where *both* parents have a history of AD, the risk of their children developing the disorder is substantially increased. Bird and colleagues (Bird *et al.* 1993) studied 31 couples where both had a history of AD and their 87 children. Fourteen of the thirty (47 per cent) who survived to age 65 and beyond developed dementia. Further stratification by age showed that the proportion of demented children increased markedly with age. Thus, 13/19 (68 per cent) who survived to age 70 or longer were affected; by age 75 the proportion was 12/16 (75 per cent); and, by age 80, the proportion was 7/8 (88 per cent). These findings should be regarded as preliminary since the high risks observed may have resulted at least in part from ascertainment biases and the sample was relatively small. However, there is a case for referring worried offspring from such families for genetic counselling.

APOE genotyping in AD

The risk estimates in Table 15.1 suggest that, at most, some 50 per cent of APOE e4 homozygotes will develop AD within their lifetime. These predictions agree well with the general population based study of Henderson *et al.*, which suggested that the risk of being demented by age 90 in APOE e4 homozygotes was about 50 per cent (Henderson *et al.* 1995).

This illustrates the difficulty in using molecular information to give meaningful risk estimates for complex genetic diseases. From the previous section it is clear that knowledge of a person's APOE genotype is of little more use in predicting their chances of succumbing to AD than is knowledge of their family history of dementia. Furthermore, it seems that even individuals with the APOE e4/APOE e4 genotype have, on average, a greater than 50 per cent chance of escaping the disease. Therefore, APOE genotyping currently has no role in predicting the risk of developing AD.

Some claims have been made that APOE genotyping may be an aid to diagnosing AD. The study by Mayeux *et al.* (1998) suggested that for patients referred to specialised AD assessment centres, APOE genotyping used in combination with clinical criteria might improve the specificity of the diagnosis. The data show that whereas the demonstration of one or more APOE e4 allele in a person suspected of suffering with dementia slightly increases the accuracy of a clinical diagnosis of AD, the absence of an APOE e4 allele has little value in either endorsing or refuting a clinical diagnosis of AD. Thus, even in a selected group of patients, presumably all with a high a priori chance of AD, APOE genotyping seems to confer negligible diagnostic benefit.

Recommendations

The following guidelines from Liddell *et al.* 2001, are based on the reports of several groups, in the UK and USA. (Brodaty *et al.* 1996; The American College of Genetics Statement on use of apolipoprotein E testing for Alzheimer disease 1995; Alzheimer's Association 1998; Tunstall and Lovestone 1999).

Table 15.1 Lifetime risk of AD according to APOE e4 allele number

Sex	Male (%)	Female (%)
APOE e4 status unknown	6.3	12
No APOE e4	4.6	9.3
APOE e4 Heterozygote	12	23
APOE e4 Homozygote	35	53

Adapted from Liddell *et al.* 2001, based on data from Seshadri *et al.* 1995, 1997

Families where there is evidence of FAD (or indeed any familial early onset dementia) should be referred to a specialist centre with experience in the genetics of complex disorders.

Counselling for such families should follow the process established for HD. We would add that for AD even more than for HD it is essential to establish the causative mutation in that particular family, if one can be identified, before proceeding further. When a mutation is established in the family this opens the possibility of predictive testing for that family, although we expect the demand for such testing to be low in the absence of effective treatments for the condition.

For late onset AD all the groups are agreed that there is no role for APOE genotyping in prediction or risk-assessment. There is a commercial interest in APOE genotyping as part of the diagnostic process and it is possible that a role may develop in the future. However the UK Alzheimer's Disease Genetics Consortium agreed that, on the basis of the data available in 1999, there was no such role at the present.

Early onset familal dementia is therefore straightforward – refer to a medical geneticist, experienced in dementia. For late onset AD there is wide agreement that there is no practically useful genetic test available. But genetic testing is not the same as genetic counselling. The absence of a test should not mean that relatives are denied information and the opportunity to discuss their concerns. So, when asked 'Am I likely to get AD?' by the relative of a patient with late-onset AD, how should one answer? In most cases of AD it is only possible to advise relatives of their risk in the broadest terms. Extrapolating from family history studies it is possible to say that the risk to the children is in the region of one-in-five to one-in-six.

Occasionally psychiatrists encounter families with several siblings affected by late-onset AD and a history of dementia in previous generations. Such families almost certainly exhibit high genetic loading, but there are few reliable data on whether and to what extent risk increases with the number of affected relatives, although studies of other common disorders suggest that this is likely to be the case. There is also preliminary evidence for substantially increased risk in the offspring of parents who both have a diagnosis of AD. However, it should be borne in mind that the number of affected relatives may well depend on personal and public health factors affecting longevity as well as on the degree of genetic loading for AD, and it may not be advisable or possible to base discussions on risk estimates that are increased when several relatives have been affected. Nevertheless, in such pedigrees, it may be worth taking a detailed family history and consulting with a clinical geneticist.

Conclusions

Genetic testing in psychiatric disorders is a complex area that raises a number of ethical issues which are discussed in Chapter 16. With genetic research regularly the target of media attention, the resulting increase in public awareness of genetic risk for disease makes it likely that those with a family history of psychiatric disorder will request genetic counselling. Within psychiatry the modes of inheritance of major disorders are complex, and although molecular genetic research is advancing rapidly, it is unlikely that any form of molecular genetic testing will be available in the near future. In many instances the predictive value of tests for such genes will be low. This applies, as we have seen, to APOE testing for late-onset Alzheimer's disease. Indeed, even when all susceptibility genes for a given disorder have been identified, it will still not be possible to predict the development of disease with certainty until the relevant environmental risk factors have also been identified and the nature of the various interactions understood. Such interactions may be as complex as chaotic systems like the weather, which is notoriously unpredictable over even the relatively short term (Owen *et al.* 2000).

HD testing has shown us that despite up to 84 per cent of at risk individuals having expressed interest in testing prior to its development, the numbers actually coming forward have been much lower than expected (Harper *et al.* 1996; Binedell *et al.* 1998). It seems that in fact the majority of at risk adults prefer to live with uncertainty. Any genetic testing in complex disorders such as common psychiatric disorders could not give the precise predictions possible in HD testing programmes. Although there is a perception that at-risk individuals are thirsty for knowledge, experience suggests that those coming forward for testing would be a minority.

However, on a more optimistic note, genetic testing could have other roles that are likely to be of more value to patients and clinicians, for example, in helping to optimize treatment choices by testing for genes that are found to influence treatment responses in psychiatric disorders, leading to a greater individualization of treatment.

References

Alzheimers Assoc, Natl Inst Aging (1998) Consensus report of the Working Group on Molecular and Biochemical Markers of Alzheimer's Disease. *Neurobiol Aging* **19**:109–16.

The American College of Genetics (1995) Statement on use of apolipoprotein E testing for Alzheimer disease. American College of Medical Genetics/American Society of Human Genetics Working Group on APOE and Alzheimer disease. *JAMA* **274**:1627–9.

American Society of Human Genetics (1995) Ethical, legal and psychosocial implications of genetic testing in children and adolescents. *Am J Hum Genet* **57**:1233–41.

American Society of Human Genetics Subcommittee on Familial Disclosure (1998) Professional disclosure of familial genetic information. *Am J Hum Genet* **62**:474–83.

Bell J. (1998) The new genetics in clinical practice. *British Medical Journal* **316**:618–20.

Biesecker B. B. and Marteau T. M. (1999) The future of genetic counseling: an international perspective. *Nature Genetics* **22**:133–7.

Binedell J., Soldan J. R., Harper P. S. (1998) Predictive testing for Huntington's disease: 1. Predictors of uptake in South Wales. *Clinical Genetics* **54**:477–88.

Bird T. D., Nemens E. J., Kukull W. A. (1993) Conjugal Alzheimer's disease: is there an increased risk in offspring? *Annals of Neurology* **34**:396–9.

Breitner J. C. S., Murphy E. A., Silverman J. M. *et al.* (1988) Age-dependent expression of familial risk in Alzheimers Disease. *American Journal of Epidemiology* **128**:536–48.

Brodaty H., Conneally M., Gauthier S. *et al.* (1996) Medical and scientific committee ADI Consensus statement on predictive testing. *Alzheimer Dis Assoc Disord* **9**:182–7.

Campion D., Dumanchin C., Hannequin D. *et al.* (1999) Early-onset autosomal dominant Alzheimer's disease: prevalence, genetic heterogeneity, and mutation spectrum. *Am J Hum Genet*, **65**:664–70.

Clarke A. (1994) *Genetic Counselling, Practice and Principles*. London: Routledge.

Harper P. (1993) *Practical Genetic Counselling*. London: Butterworth.

Harper P. S., Soldan J., Tyler A. (1996) Predictive tests in Huntington's disease, in P. Harper *Huntington's Disease*. London: W. B. Saunders.

Henderson A. S., Easteal S., Jorm A. F. *et al.* (1995) Apolipoprotein E allele μ4, dementia, and cognitive decline in a population sample. *Lancet* **346**:1387–90.

Heston L. L., Mastri A. R., Anderson E. *et al.* (1981) Dementia of the Alzheimer type: clinical genetics, natural history and associated conditions. *Archives of General Psychiatry* **38**:1058–90.

Holzman N. A. and Marteau T. M. (2000) Will genetics revolutionise medicine? *New England Journal of Medicine* **343**:141–4.

Huntington's Disease Collaborative Research Group (1993) A novel gene containing a trinucleotide repeat that is expanded and unstable on Huntington's disease chromosomes. *Cell* **72**:971–83.

Liddell M. B., Lovestone S., Owen M. J. (2001) Genetic risk of Alzheimer's disease: advising relatives. *British Journal of Psychiatry* **178**:7–11.

Marteau T. M. and Croyle R. T. (1988) Psychological responses to genetic testing. *British Medical Journal* **316**:693–6.

McGuffin P., Asherson P., Owen M. *et al.* (1994) The strength of the genetic effect. Is there room for an environmental influence in the aetiology of schizophrenia? *British Journal of Psychiatry* **164**:593–9.

Murphy E. A. and Chase G. A. (1975) *Principles of Genetic Counselling*. Chicago: Yearbook Publishers.

Mayeux R., Saunders A. M., Shea S. *et al.* (1998) Utility of the apolipoprotein E genotype in the diagnosis of Alzheimer's disease. *New England Journal of Medicine* **338**:506–11.

Nuffield Council on Bioethics (1993) *Genetic Screening: Ethical Issues*. Nuffield Council on Bioethics: London.

Owen M. J., Cardno A. G. and O'Donovan M. C. (2000) Psychiatric genetics: back to the future. *Molecular Psychiatry* **5**:22–31.

Petronis A., Gottesman I. I., Crow T. J. *et al.* (2000). Psychiatric epigenetics: a new focus for the new century. *Molecular Psychiatry* **5**:342–6.

Quaid K. (1993) Presymptomatic testing for Huntington's Disease in the United States. *Am J Hum Genet* **53**:785–7.

Rossor M. N., Fox N. C., Beck J. *et al.* (1996) Incomplete penetrance of familial Alzheimer's disease in a pedigree with a novel presenilin-1 gene mutation. *Lancet* **347**:1560.

Seshadri S., Drachman D. A., Lippa C. F. (1995) Apolipoprotein E (4 allele and the lifetime risk of Alzheimer's disease: what physicians know and what they should know. *Archives of Neurology* **52**:1074–9.

Seshadri S., Wolf P. A., Beiser A. *et al.* (1997) Lifetime risk of dementia and Alzheimer's disease: the impact of mortality on risk estimates in the Framingham study. *Neurology* **49**:1498–504.

Tunstall N. and Lovestone S. (1999) UK Alzheimer's disease genetics consortium. *Int J Geriatr Psychiatry* 14(9):789–91.

World Federation of Neurology Research Committee. Research Group on Huntington's Disease, W. (1989) Ethical issues policy statement on Huntington's disease molecular genetics predictive test. *Journal of Neurological Science* **94**:327–32.

Chapter 16

Ethical considerations in psychiatric genetics

Anne Farmer

Introduction

The inclusion of a chapter on the ethical aspects of genetic research into mental disorders and behavioural traits, not only in this, but also in other recent textbooks (Crusio and Gerlai 1999), indicates the importance of ethical concerns in this rapidly developing branch of neuroscience. However the debate into the ethical aspects of genetic research is not one that should only take place within the research community. Accounts of recent technological advances and genetic discoveries described in the lay press have suddenly and sometimes sensationally captured the public imagination. The cloning of Dolly the sheep (Wilmut *et al.* 1997) was greeted in the *Independent* newspaper of 26 February 1997, with the statement 'In the past few days we have been through a change in our condition as momentous as the Copernican revolution, or the splitting of the atom'. The landmark significance of the scientific breakthrough had not been lost on other newspaper journalists, and both establishment media and tabloids featured Dolly in their front-page headlines.

In January 2001, the print and video media featured pictures of a baby Rhesus monkey named 'Andi'—his name coming from 'i DNA' for inserted DNA. Andi was the result of the insertion of jellyfish genes into a Rhesus monkey embryo. The ethical issues associated with the births of Andi and Dolly need to be considered by the general public, who will have obtained most of their information regarding this research from exciting media accounts.

The importance of public opinion and its potential for shaping scientific endeavour has recently been demonstrated in the debate about genetically modified (GM) crops. A massive 'anti-GM' lobby led to all of the national supermarket chains in the UK vying with each other to be the first to claim that they had removed all products containing GM from their shelves. Although there has been much less public disquiet in the US or most other English speaking countries, public protests have led to the limitation of the scale of experimentation into GM crops in the UK. Public awareness about scientific endeavour in general, and genetic research in particular, is also heightened by the ready availability of information on the World Wide Web, often unfiltered for validity.

If the public is to continue supporting genetic research, both via the taxes that (mainly) pay for it and by direct participation as subjects, the ethical, legal and social aspects of the research need frank and honest debate. Those engaged in genetic research need to pay particular attention to advising both the public and research foundations about the strengths, weaknesses, opportunities and threats posed by the possible outcomes of their endeavours. This is particularly important in respect of genetic research into mental disorders and behavioural traits, where past abuses based on politicized, bad and distorted science, perpetrated in several countries in the name of eugenics, have cast a long shadow.

The rise of eugenics and its impact on psychiatric genetics

The term *genetics* was applied for the first time in 1905 by William Bateson to describe a new branch of science, following the 'rediscovery' of the work of Gregor Mendel, whose laws of inheritance based on research into pea plants had been largely ignored when first published in 1866. Around the same time as Mendel was undertaking his experiments, Francis Galton was developing his ideas, independently of solid genetic knowledge, about the inheritance of intellectual ability first set out in detail in the monograph, 'Hereditary Genius'. Elsewhere Galton introduced the word 'eugenics' (coming from the Greek meaning 'well-born') and applied this to a new branch of biology, devoted to the improvement of the human stock by analogy to agricultural breeding practices. Galton believed that individuals with 'undesirable' qualities should be discouraged from breeding, while those with 'good' characteristics should be encouraged to have lots of children. He believed that this would ensure that certain diseases would be eradicated by being bred out of the population (negative eugenics), while good qualities would increase among the population (positive eugenics). Such ideas would now be regarded by population geneticists as simplistic, but they became rapidly popular and were espoused not just, as is sometimes thought, by the politically right wing, but also by thinkers on the left such as the Fabian society in the UK, and by social reformers concerned with maternal and public health in other industrialized countries. Adherents included socialist intellectuals such as George Bernard Shaw and H. G. Wells. Thus, although the 'eugenics movement' is, in the minds of most, mainly associated with the Third Reich and its fascist policies, some socialist thinkers in the 1930s considered that it was quite reasonable that the 'feeble minded' should be prevented from having children, because they could 'cost other people a lot of money' (Mitchison 1997).

Galton was also among the first to suggest that examining how diseases run in families could give clues to their aetiologies, and that studies of twins could help to tease apart the influence of genetic and environmental risk factors. However it was not until later that the difference between monozygotic and dizygotic twins was understood, and not until 1928 that the twin method was applied to severe mental disorders by Luxenburger of the Munich School of psychiatric geneticists. This was after his Chairman, Ernst Rudin undertook the first systematic family study of schizophrenia in

1916. Consequently many of the fundamental discoveries related to the genetic basis of major mental disorders, derived from twin family and adoption methodologies, had largely been made by 1970. However, much of this early work became discredited by Rudin's later involvement with the Nazi party, and his work fell into disrepute. During the Third Reich, Nazi doctors undertook 'medical experiments' on thousands of concentration camp prisoners that caused unbelievable suffering, mutilation and death. In 1947, after the war had ended, many of these doctors were tried by the Allies for 'crimes against humanity' in a 'Doctors' Trial' at Nuremberg. Although Rudin was arrested and tried at a local Denazification Court at around the same time as the Nuremberg Trials, he was judged to have been a 'fellow traveller' and not a major contributor to Nazi crimes. Whether this is the truth or an overly lenient view remains to be seen. However even recent attempts to objectively review an English translation of Rudin's original work on schizophrenia in a leading journal led to an impassioned series of letters in the correspondence columns (Kendler and Zerbin-Rudin 1996; Faraone *et al.* 1997; Gejman 1997; Gottesman and Bertelsen 1996; Gershon 1997; Lerer and Segman 1997).

Rüdin's provision of a scientific and academic façade for Nazi eugenic policies blurred the distincton between the policies in Germany and the emerging field of psychiatric genetics. This was despite a number of leading psychiatric geneticists vehemently opposing such practices. Under the Nazi regime the eugenic policies that were implemented were undoubtedly repugnant, but also astonishingly naïve. The 'mentally ill' were forcibly sterilized and later policies included the 'euthanasia' of psychiatric patients who were exterminated alongside those also considered to leading 'lives unworthy of being lived' such as epileptics, the deaf and blind, Jews and Gypsies. More recent decades have seen other cruel regimes attempting to 'ethnically cleanse' all individuals from a particular racial group or cultural origin from various countries, including Rwanda and Serbia.

However, the introduction of eugenic principles has not been the sole province of dictators and despots. As described above, many groups and individuals, including those who would regard themselves as liberal and socialist, have also considered that it is reasonable to limit migration, or prevent those individuals considered inferior in some respect from reproducing. Examples include emigration policy in the United States, influenced by Charles Davenport the first Director of the Cold Spring Harbor laboratories, that favoured emigration from certain 'desirable countries' (e.g. Scandinavia, Germany and the British Isles) and discouraged migrants from other countries such as south and eastern Europe. Eugenic philosophies also influenced Scandinavian policies on the sterilization of the mentally handicapped. Similarly the Australian government introduced a policy of removing young aboriginal children from their families to be adopted into white families as a means of 'breeding them white', with the intent of eliminating Aboriginal peoples as a distinct ethnic group (Scott-Clark and Levy 1997). It is only very recently that the Australian government has admitted this policy, and compensation claims are beginning to reach the courts.

The 'guilt by association' between eugenics and the psychiatric genetics lingered long after the cessation of the Second World War and Rüdin's trial. Formerly flourishing university departments of genetics dwindled to a few enthusiasts, such as Eliot Slater and James Shields at the Institute of Psychiatry in London, Stromgren, Odegaard and Essen-Moller in Scandinavia and Kallmann in the USA, most of whom had studied in Munich before the War. The resurgence of psychiatric genetics as a scientific discipline attractive to young academics had to wait until the late 1960s.

Before examining the current ethical practices as they relate to psychiatric genetic research, it is first necessary to consider national and international codes of practice that govern any research involving human subjects introduced since the ending of the Second World War and the Nuremberg Doctors' Trials. The first of these, the Nuremberg Code, was devised from the judges' report after the trials.

The development of international and national ethical codes for scientific research on human subjects

The Nuremberg Code

The first international agreement regarding the use of human subjects in scientific research was the Nuremberg Code, produced in the aftermath of the Nuremberg Doctors' trials following the Second World War. Although not the first example of harmful research carried out on unwilling subjects, the 'experiments' undertaken by Nazi doctors during the Second World War on concentration camp prisoners were notoriously abhorrent in their scope, degree of harm and suffering inflicted on the participants.

The Nuremberg Code has served as the first set of ethical principles guiding medical researchers to receive international acceptance. It has guided governments and professionals since the 1950s. The Code's 10 points are shown in Box 1. The main points are that the voluntary consent of the subject is essential, that all unnecessary risks should be avoided and the degree of risk to the individual should not exceed the 'humanitarian importance of the problem'. Also, participants should always be free to withdraw from the experiment at any time.

The Declaration of Helsinki

The Declaration of Helsinki, Ethical Principles for Medical Research using Human Subjects, published in 1964, was the World Medical Association's effort to update the Nuremberg code (www.wma.net/e/policy/). There have been subsequent amendments to the Declaration, the most recent being in 2000 (Christie 2000). The Declaration took the ethical debate a stage further than the Nuremberg Code, as it set out in detail the general principles for all medical research, and the duty of the physician involved in the research study, in particular. The basic principles of the Declaration of Helsinki are as follows. First, all medical research involving human subjects must be scientifically

Box 1 The Nuremberg Code

1 The voluntary consent of the human subject is absolutely essential. This means that the person involved should have legal capacity to give consent; should be so situated as to be able to exercise free power of choice, without the intervention of any element of force, fraud, deceit, duress, over-reaching or other ulterior form of constraint or coercion; and should have sufficient knowledge and comprehension of the elements of the subject matter involved as to enable him to make an understanding and enlightened decision. The latter element requires that before the acceptance of an affirmative decision by the experimental subject there should be made known to him the nature, duration and purpose of the experiment; the methods and means by which it is to be conducted; all inconveniences and hazards reasonably to be expected; and the effects upon his health or person which may possibly come from his participation in the experiment.

   The duty and responsibility for ascertaining the quality of the consent rests upon each individual who initiates, directs or engages in the experiment. It is a personal duty and responsibility, which may not be delegated to another with impunity.

2 The experiment should be such as to yield fruitful results for the good of society, unprocurable by other methods or means of study, and not random and unnecessary in nature.

3 The experiment should be so designed and based on the results of animal experimentation, and a knowledge of the natural history of the disease or other problem under study, that the anticipated results will justify the performance of the experiment.

4 The experiment should be so conducted as to avoid all unnecessary physical and mental suffering and injury.

5 No study should be conducted where there is an a priori reason to believe that death or disabling injury will occur, except, perhaps, in those experiments where the experimental physicians also serve as subjects.

6 The degree of risk to be taken should never exceed that determined by the humanitarian importance of the problem to be solved by the experiment.

7 Proper preparations should be made and adequate facilities provided to protect the experimental subject against even remote possibilities of injury, death or disability.

8 The experiment should be conducted only by scientifically qualified persons. The highest degree of skill and care should be required through all stages of the experiment of those who conduct or engage in the experiment.

Box 1 *(continued)*

9 During the course of the experiment the human subject should be at liberty to bring the experiment to an end if he has reached the physical and mental state where continuation of the experiment seems to him to be impossible.

10 During the course of the experiment the scientist in charge must be prepared to terminate the experiment at any stage, if he has probable cause to believe, in the exercise of good faith superior skill and careful judgement that a continuation of the experiment is likely to result in injury, disability or death to the experimental subject.

sound, conducted by scientifically competent persons, under the supervision of a clinically competent medical person, who takes responsibility for the experimental subject. Second, the experimental procedure must be clearly formulated, and each potential subject adequately informed of the aims, methods, sources of funding and conflicts of interest of the researcher, as well as the risks and benefits of the research to him or her. Third, the Declaration stated that approval for the research should be obtained from an independent ethical review committee that conforms to the laws and regulations of the country where the research will take place. Also, such committees have a right to continue to monitor the research and the researchers must provide the necessary reports to facilitate monitoring, including serious adverse events, (this mainly relates to trials of new drugs). Fourth, the Declaration commented in respect of individuals who were unable to give informed consent because of mental or physical incapacity, or where the research was being undertaken on children, in those with mental retardation or dementia. Research in these groups should only be conducted when the health of the population represented by the participants will be promoted, and when it cannot be performed on legally competent subjects. Such research also requires consent by the subject's legally authorized representative, although the subject themselves may be able to 'assent' to the research procedures. Lastly, the Declaration of Helsinki made it clear that publishers and authors also have ethical obligations, to provide accurate and publicly available results of the research and not to publish any research that does not adhere to these principles of the Declaration.

The Declaration of Helsinki has provided clear guidance regarding the ethical checks and balances required for any research involving human subjects. The idea that independent ethics committees should approve the ethics of a research proposal remains the guiding principle in most countries. However in the US, several national scandals involving the abuse of research subjects have added to the debate, and have shaped the ethical and legal constraints placed on research activity. While the principles of the Declaration of Helsinki underpin US ethical guidelines, there are additional safeguards to protect the individual research subject imposed as government policy. Arguably, the US has the most stringent and detailed ethical approval system in the world. The history of the ethical principles guiding research involving human subjects, including genetic research, in the US will now be reviewed.

Ethical codes in the US: The Belmont Report, The Common Rule and the National Bioethics Advisory Commission

The Belmont report was produced in the United States in 1979, by a National Commission for the Protection of Human Subjects of Biomedical and Behavioral Research established by Congress in response to public concerns about several well-publicized biomedical abuses (details available at http://cme.nci.nih.gov). These included the Tuskegee Syphilis study where 400 black men with syphilis were compared with 200 individuals who did not have this disease, in a long-term outcome study. The men were recruited without informed consent and were also misinformed about some of the procedures done for the research. Although penicillin was found to be effective in the treatment of syphilis in the 1940s, none of the infected subjects were informed or offered treatment. In fact, the study continued until the 1970s, despite the study indicating significant differences in the death and complications rate in subjects compared to the controls as early as 1936. The public outrage erupting when the accounts of the study appeared in the national press led to the appointment of an advisory panel to give advice on how to ensure that such experiments would never be conducted in the future.

Two other studies conducted in the US that were undertaken without freely given and informed consent were the Jewish Chronic Disease Hospital Study and the Willowbrook School Study. The Jewish Hospital study involved injecting cancer cells into patients with chronic debilitating diseases, in an attempt to understand whether the body's inability to reject cancer cells was due to cancer or debilitation. Not only was the oral consent that was allegedly obtained not documented, but also the patients were not told that they would be receiving cancer cells for fear of frightening them unnecessarily. At the subsequent enquiry, the researchers involved were found guilty of fraud, deceit and unprofessional conduct.

In the Willowbrook School study, a series of mentally retarded children newly admitted to the school, were deliberately infected with hepatitis in order to examine the natural history of the infection. The researchers argued that the school's unsanitary and crowded conditions led to most of the children developing the infection anyway. Over time, the school stopped accepting admissions with the exception of the research unit. Consequently parents were forced to agree to their child's participation in the study as a condition of admission.

As a result of these well-publicized cases of unethical research, the Belmont Report—'Ethical Principles and Guidelines for the Protection of Human Subjects' was published in 1979 (http://cme.nci.nih.gov). This provides the fundamental ethical principles guiding all human research in the US up to the present time.

The guiding ethical principles of the Belmont Report are *respect for persons, benefi-cence* and *justice*. 'Respect for persons' requires that individuals should be treated as autonomous agents i.e. that he or she is 'capable of deliberation about personal goals and of acting under such deliberation. To respect autonomy is to give weight to the autonomous person's considered opinions and choices while refraining from obstruct-ing their actions' (Belmont Report). The second requirement under 'respect for per-sons' concerns 'persons with diminished autonomy' who 'may need additional protections'. The Belmont Report states that even subjects with diminished capacity to give informed consent should nonetheless still be given the opportunity of choosing whether or not to participate in the research, whenever possible. It is also important that the judgement that an individual does not have the capacity to give informed con-sent should be re-evaluated periodically.

The second principle of the Belmont Report is that of 'beneficence'. Beneficence requires the researcher to maximize possible benefits and minimize possible harm to the individual taking part in the research. In many instances the main benefit is to soci-ety in general rather than the individual. However, the principle of beneficence also requires that the benefits to society should not be obtained at the expense of injury to the individual.

The third principle of the Belmont Report, namely that of 'justice', states that the risks and benefits of the research should be spread equally. This means that certain groups of subjects are not being preferentially selected to participate in the research merely because they are available, in a compromised position or otherwise vulnerable. The principle of justice also requires that reasons for excluding certain groups from participation in the research are also carefully examined, e.g. where the research design includes only individuals from certain ethnic groups or from one gender.

In 1981 the Department of Health and Human services and the Food and Drug Administration published regulations based on the Belmont report, and by 1991 several federal departments and agencies agreed to adopt the basic human participant protec-tions that are referred to as the 'Common Rule', the abbreviated description of the 'Code of Federal Regulations, Title 45, Public Welfare, Part 46 Protection of Human Subjects' (Department of Health and Human Services, National Institutes of Health, Office for Protection from Research Risk 1991) (details available at: http://cme.nci.nih.gov).

The Common Rule federal regulations include four subparts. Subpart A provides the basic policy for the protection of human research subjects and includes who the policy applies to, the constitution of the Institutional Review Boards (IRBs) and their remit in overseeing all the research undertaken in a specific institution. Subparts B, C and D provide additional protection for research involving foetuses, pregnant women and human *in vitro* fertilization (B), that required when undertaking research on prisoners, (C), and the use of children as research subjects (D).

In 1995, President Clinton ordered the creation of a National Bioethics Advisory Commission (NBAC)(http://bioethics.gov/general.html) in order to further 'enhance human subject protections'. The commission consisted of 15 members appointed by the president. Their remit was to 'provide advice and make recommendations' to the National Science and Technology Council and other bodies on the appropriateness of policies and regulations relating to bio-ethical issues arising from research on human biology and behaviour, as well as the applications of that research. The NBAC charter expired 3 October 2001 with many issues still unresolved (see *http://bioethics.george-town.edu/nbac/*). Its valuable reports and recommendations on a number of issues, such as the need for a 'federal oversight system' to replace the rather piecemeal legislation operating at state level, remain in limbo. A number of further recommendations relate to the need for a research agenda into research ethics and for education, certification and accreditation, managing conflicts of interest, several issues relating to IRBs, research related compensation and the need for resources to undertake all this work.

These highly comprehensive and detailed ethical procedures may be being driven, in part, by the large number of professional ethicists and lawyers per head of the population in the US compared to other countries (see the professional journal online at *http://ajobonline.com*). Whether this is conducive or obstructive to research aimed at a better understanding of disease and/or relief of suffering has been debated. Levine (2001), a contributor to the Belmont Report, has expressed sharp concern over the crisis in confidence over the effectiveness of the review boards in actually protecting human research subjects.

In the UK the government's legislative response has been more measured, and has taken note of public consultation exercises rather more than has been the case in the US. The independent and governmental bodies examining the ethical issues relating both to research involving human subjects and to genetics in the UK will be described next.

The ethical advisory bodies in the UK: The Nuffield Council on Bioethics, the Human Genetics Commission

The UK has not experienced the same widespread public outrage about human subject research as the Tuskagee syphilis study, the Willowbrook school study and the Jewish Hospital study have caused in the US. Consequently, there has been no UK equivalent of the Belmont Report or the Common Rule, and until recently the Declaration of Helsinki principles were considered sufficient guidance for human subject research, when the impact of new genetic discoveries hit the newspaper headlines.

Ethical debate in the UK has involved the public rather than professional ethicists and lawyers, far more than it apparently has in the US. The Nuffield Council on Bioethics (*http://www.nuffieldbioethics.org/home/index.asp*) is an independent body that has produced a number of reports after wide consultation and a number of surveys of public opinion.

The Nuffield Council on Bioethics was established in 1991, to examine the ethical issues raised by new developments in medicine and biology. Unlike the official governmental bodies that have created and shaped the ethical framework for research in the

US, the UK's Nuffield Council provides an opportunity for public opinion to participate in its recommendations. The Council is funded jointly, by the Nuffield Foundation and the Wellcome Trust, both of which are independent charity trusts, and the UK government funded Medical Research Council. The terms of reference of the Nuffield Council are shown in Box 2.

The Council has issued a number of reports since its establishment in 1991. These are shown in Box 3.

Later in this chapter the Nuffield Council Report on Mental disorders and genetics will be discussed in more detail, as well as the forthcoming report on the ethical context of genetics and human behaviour (the working party on this last topic is consulting as this chapter is being written).

A comprehensive review of the regulatory and advisory framework for biotechnology, by the UK government in 1999, led to the establishment of the Human Genetics

---

Box 2 Terms of Reference of the Nuffield Council on Bioethics

To identify and define the ethical questions raised by recent advances in biological and medical research in order to respond to and anticipate public concern.

To make arrangements for examining and reporting on such questions with a view to promoting public understanding and discussion; this may lead, where needed, to the formulation of new guidelines by the appropriate regulatory or other body;

In the light of the outcome of its work, to publish reports and to make representations as the Council may judge appropriate.

Box 3 Reports produced by the Nuffield Council on Bioethics

- Genetic screening: ethical issues (1993)
- Human tissue: ethical and legal issues (1995)
- Animal to human transplantation: the ethics of xenotransplantation (1996)
- Mental disorders and genetics: the ethical context (1998)
- Genetically modified crops: the ethical and social issues (1999)
- The ethics of clinical research in developing countries—a discussion paper (1999)
- Stem Cell therapy: the ethical issues—a discussion paper (2001)

Commission (HGC). The HGC has a similar role in the UK as the NBAC in the US, namely to advise government. The HGC's remit according to its website (*www.hgc.gov.uk*) is 'to give Ministers strategic advice on the 'big picture' of human genetics, with particular focus on social and ethical issues'. The HGC also interacts with a number of other bodies that also provide advice directly to the government including the Medical Research Council, the National Institute for Clinical Excellence, the Genetics and Insurance Committee, and the National Health Service Executive bodies for the countries of the UK.

Clearly, in the US, UK, Germany and doubtless other countries too, the pace of new discoveries and technologies generated by genetic research has driven governments' needs to obtain expert advice in order to shape policy decisions. Issues such as reproductive and therapeutic cloning have complex legal ethical and social considerations (*Focus Magazine* 2001), yet legislation to regulate experimentation in these areas is required very rapidly to keep pace with the technological capabilities. As one newspaper commentator stated in relation to the birth of Dolly, 'Dolly's here now, the ethics will have to cope' (Jones 1997). Many consider that what is also required is more research into the ethical legal and social issues relating to genetic research, especially as it relates to human subjects. In the US, more funding has been assigned to research into the ethical, legal and social implications of genetic research (ELSI), than any other country. The ELSI bioethics research programme will be discussed next.

Ethical, legal and social implications (ELSI) of genetic research

The development of a research programme to specifically examine the ethical legal and social issues relating to genetic research, was due to the foresight of the first Director of the National Institutes of Health, Genetic Research Programme in the USA, Nobel Laureate James Watson. In 1988, Watson announced that 3 per cent of the budget for the Human Genome project would be devoted to supporting a programme of research and discussion on the ethical, legal and social implications of the new genetic knowledge. The ELSI bioethics programme addresses issues related to all genetic research, not just that relating to psychiatric disorder and behavioural traits. The annual ELSI budget has risen over the years to a current figure (at the time of writing) of around 7 million US dollars. The reason cited by Watson for apportioning such a large sum of money to bioethical research, was that he wished to oppose the notion that he was in any way 'a closet eugenicist, having as my real long term goal the unambiguous identification of genes, that lead to social and occupational stratification, as well as genes justifying racial discrimination' (Watson 1997). A valuable reference guide online to the intersection of ethics and human genetics can be found at *http://bioethics.georgetown.edu/nirehg.html*. Watson's own Cold Spring Harbor Laboratory hosts an archive online of the American Eugenics Movement, including photographs and reproductions of documents, at *http://vector.cshl.org/eugenics/list_topics.pl*.

France (Butler 1997), Germany (Abbott 1997), Sweden (Rider 2000) and the UK have also developed bioethical initiatives, although these have been funded on a more modest scale than ELSI. As mentioned above, in the UK, the Nuffield Council on Bioethics has already published recommendations of a Working Group into the ethical aspects of the genetics of psychiatric disorders, and recently, another Working Party has been meeting to examine the ethics of research into genetics and human behaviour.

Current ethical questions related to modern psychiatric genetics studies

Arguably, the first ethical question that arises in relation to examination of the genetics of mental disorders is whether the research should be undertaken at all. The second is that if the research should be done what ethical guidelines needs to be put in place? The conclusions of the Nuffield Council on Bioethics Working Group on mental illness and genetics were that psychiatric genetic research is extremely important in the quest for improved treatments for disorders and the reduction of suffering. The Working Group on mental disorders reviewed much evidence, but ultimately concluded that the outcomes of psychiatric genetic research are likely to be predominately beneficial, and that there were no substantial ethical objections to such research continuing.

If it is accepted that psychiatric genetic research into mental disorder should continue, the next question that is posed relates to what ethical constraints should be placed upon such research. The specific issues relating to psychiatric genetic research have been addressed elsewhere (Farmer and McGuffin 1999) and in relation to the ELSI programme, (Meslin 1997), and will be considered again now. They relate to the following:

1 Informed consent;

2 The role of IRBs/ECs;

3 The role of individuals and families as participants in genetic research;

4 Privacy, confidentiality and disclosure of genetic test results to third parties; and

5 The commercialization of products of genetic research.

Informed consent

As outlined above and enshrined in the Nuremberg Code, research subjects need to understand the benefit, foreseeable risks and alternatives to participation in research. Physical harm related to obtaining the sample of DNA required for genetic testing is minimal. Whilst whole blood provides a much higher yield of DNA, sufficient samples can also be obtained via a sample of buccal mucosa cells from the inside of the cheek. These can be obtained by a mouthwash or cotton swabs. The main ethical concerns about informed consent relate to 'non-physical harm'. For example, research subjects may fail to understand the difference between clinical diagnosis and treatment, and participation in a research study designed solely to generate scientific data. Individuals may think that in participating in a genetic study they will find out more about their own risk for a genetic disorder. For most psychiatric disorders, e.g. schizophrenia or bipolar disorder, the results of DNA testing for an individual subject are of no tangible benefit at the present time. This is not the case for rare disorders such as Huntington's disease and the rare autosomal dominant forms of Alzheimer's disease. In these disorders DNA testing can determine whether a subject who is currently well is at risk or not of the disorder, with near 100 per cent certainty. However, for all disorders, whether or not DNA testing can determine an individual's risk of developing the disorder or not, the view of most institutional review boards (IRBs) and ethics committees (ECs) is that it should be made clear to research subjects that the results of DNA testing will not be revealed, either to them or any third party. Also that any benefits that ensue as a result of the research are purely those of advancing medical and scientific knowledge. In the US research institutions can also obtain Confidentiality Certificates which protect them from forced disclosure to third parties. However if information about participation in a research study appears in the clinical record submitted to a health insurance company for payment of a claim, the subjects' confidentiality may also be breached.

Capacity to provide informed consent may be an important issue for those with mental disorders. Capacity is a clinical term that can be defined as a person having the cognitive and other psychological skills/understanding necessary to be capable of coming to an informed decision on whether to consent or not (Royal College of Psychiatrists 2001). It is now considered that unlike the legal definition of the term 'competence' that is either present or absent, capacity is not categorical; rather skills and understanding are dimensional and task and time specific. Subjects can also be given time and helped to understand the informed consent procedure. Consequently, even subjects who are deemed legally not competent to give informed consent, should be helped to understand what is involved and to participate in the decision making process (Royal College of Psychiatrists 2001). This is particularly true in the case of children, those with learning disability/mental retardation, or disorders such as dementia. Although consent will need to be obtained from the subject's legal representative or next of kin, the subjects themselves should engage in the decision and be helped to give their 'assent' to the procedures.

A more difficult problem is determining capacity to give informed consent in an acutely ill subject, with psychotic symptoms for example. Such a person might not be able to understand the issues of the research when ill but may be quite capable to giving informed consent when well. The general advice from IRBs/ECs is that recruiting such subjects into a research project should wait until they are well enough to give informed consent. In the case of genetic studies DNA can be obtained at any time, so this should not pose too much of a problem. Sometimes, the opinion of an independent clinician may be required, to advise the research team about whether the person has capacity or not. Despite these complications in the consenting procedure, it is nonetheless important that such individuals have an opportunity to participate in research studies. Under the 'justice' principle of the Belmont code, omitting subjects from participation in research because they are sometimes too ill to give informed consent would not only be 'unjust' but would also be discriminatory and unethical.

Ethics Committees (ECs) and Institutional Review Boards (IRBs)

There are two main problems in respect of ECs/IRBs examination of the ethics of psychiatric genetic research. The first of these is that the research is taking place in a rapidly expanding and technically complex field. Members of local committees assessing a particular application may not have the necessary expertise to assure themselves that the 'experiment will yield fruitful results for the good of society' (see Nuremberg Code—Box 1). However, research funded by major grant awarding bodies will have already been 'peer reviewed' and the scientific merit and feasibility of the proposal carefully assessed. Consequently, ECs/IRBs may feel reassured that the study has already received sufficient scientific scrutiny. However, other studies that are not funded in this way may need to be assessed for this aspect of its ethics by an external reviewer.

A second problem for both the IRBs/ECs and the research teams involved is that psychiatric genetic research requires the recruitment of large number of subjects and their families. As has been discussed elsewhere in this book, calculations show that to achieve sufficient statistical power to detect linkage or association, sample sizes in the order of thousands of subjects is sometimes required. This usually means that the research has to be undertaken in several sites, either within one country or in several countries, in order to obtain sufficient numbers of participants. This adds to the complexity and time taken to obtain ethical approval from all the relevant ECs/IRBs, since different committees have there own documentation and may well insist on different requirements. The latter can sometimes be contradictory, with one EC/IRB insisting on a methodological change that is not allowed by another.

In the UK efforts have been made to stream line the process by having special regional ethics committees that review all multicentre applications (i.e. more than three sites). These are called MRECs or multicentre research ethics committees. The introduction of MRECs was meant to get around the problem of researchers undertaking large multisite studies, having to apply to multiple local ECs (LRECs). It also meant that the committee could draw on a wider geographical area for its membership, and was therefore more able to obtain sufficient expertise among committee members to be able to examine the scientific merit of applications, without recourse to an external reviewer. However, what appears to have happened is that the new system has merely added another bureaucratic layer for researchers to get through before a study can commence. Instead of a single MREC approval being all that is required, this is now necessary *in addition* to local approval, so principal investigators have to apply to all relevant LRECs as well. The time this process takes can mean that project start dates have to be postponed by several months.

The role of ECs/IRBs in providing ethical approval for research has increased over recent years. Some have argued that their power to influence the type and methodology of research that is undertaken is excessive (Nichols 2000). Furthermore, Nichols (2000) has suggested that ethics committees may have become barriers to ethical research that could improve healthcare and that this, in itself, is immoral and unethical. There has also been concern expressed in *Annals of Internal Medicine* (Levine 2001), *Lancet* (Lancet 1997) and *Nature* (Wadman *et al.* 1997) that the ethical debate is being taken over by professional ethicists and lawyers. It has been suggested that this has led to the creation of a 'bioethics industry' which appears at times to be self-serving; that is, that the ethicists themselves are concerned with their own prestige and moral authority, rather than only the public good. As outlined above this is probably more likely the case in the US rather than the UK or elsewhere in Europe.

Individual research subjects and their families

Most genetic research involves not only individuals who have a particular disorder but also their unaffected relatives. Most ECs/IRBs insist that relatives may only be recruited via the affected person to whom they have been identified. Clearly it is unethical to approach relatives when the index subject withholds permission. Similarly, approaching relatives directly, rather than via the index case is also deemed unethical, with some exceptions.

One of the major problems relates to whether the family's perception of the illness suffered by one or more of its members is altered by the family becoming involved in a genetics study. This again relates to 'non-physical' harm. Unaffected family members may not be aware that a disorder such as depression has a familial component. In addition, there are still old fashion notions regarding 'hereditary taint' which may be raised when family members are approached regarding participation. Such issues have to be carefully considered when individual families are approached. Stigma for mental disorders is commonplace in all nations, a reality that impacts on employment, insurability, and marriage prospects for both probands and their unaffected relatives.

Confidentiality

Problems of privacy and confidentiality to the individuals participating in genetic research is a major problem for psychiatric disorders. Some family members may reveal symptoms to a researcher that they have concealed from their relatives, and clearly in such circumstances they should expect confidentiality to be strictly respected. Genotyping has been known to unearth non-paternity, with consequent protection of privacy and the recognition of a threat to the validity of genetic analysis.

A more general problematic issue is that third parties, such as employers or insurance companies, may seek the information revealed by genetic research. The Human Genetics Council in the UK has examined this issue and provided guidance for insurance companies. The companies themselves have also set up a UK Forum for Genetics and Insurance where such matters are discussed with interested parties, including the Royal College of Psychiatrists. As well as holding conferences to hear the latest genetic research findings, this group also considers public and political opinion, reinsurance mechanisms, self-regulation, codes of conduct and legislation. The majority of mental disorders are polygenic and multifactorial, and the presence of a specific disease-related gene is not going to change the actuarial risk of an individual developing a particular disorder. Consequently, genetic testing for disorders such as schizophrenia, bipolar disorder and depression, are unlikely to provide further information regarding risk in the individual than is already known. However for other conditions, such as Huntington's disease, genetic testing has much greater diagnostic significance. In this case, insurers may wish to know the results of any testing that has been undertaken, since the test result will give almost 100 per cent certainty that the individual will or will not develop Huntington's disease. Despite the fact that genetic testing is likely to be uninformative to insurance companies, the general view is that researchers should insist on a strict embargo on divulging any information resulting from genetic research to any third party. Indeed, as mentioned above, in the US research institutions can also obtain Confidentiality Certificates that protect them legally from forced disclosure to third parties.

Commercialization of the products of psychiatric genetic research

There has been considerable ethical debate about the patenting of genes responsible for Mendelian disorders as well as those contributing to the complex genetic diseases discussed throughout this volume. Already a number of genes have been patented in the United States, where it is accepted that the discoverer of a genetic mutation is its sole owner. Whether this is justifiable considering that a genetic mutation is not an invention, but a spontaneously occurring biological anomaly, is questionable. Also, the discovery requires someone's DNA. At present in North America the individuals whose DNA has led to the discovery, has no financial rights, unlike the case of a prospector who discovers oil or gold on some else's land. In this case, the landowner certainly has some claim to financial reward.

There is currently much interest, particularly from pharmaceutical companies, in obtaining large DNA banks to further the development of drugs. Clearly those who agree to participate in research should be informed that there is an industrial partner who will have access to their DNA, even if all other personal details are omitted from the database.

The government of Iceland has recently agreed to sell the DNA from the entire population to a commercial company, deCode Genetics (Berger 1999). In addition medical and geneological information will also be made available on the total population (approximately 270,000 individuals). There has been much criticism of this decision (Andersen and Arnason 1999), in particular that a commercial enterprise can take over an entire nation's DNA. From a genetic prospective the population of Iceland is highly informative, since there were a small number of founders who arrived about a thousand years ago. Consequently it is more likely that the genes associated with common disorders are going to be found in such a population than in countries where the population is older and more genetically intermixed. While it is possible to feel uncomfortable about a pharmaceutical company obtaining such a large amount of personal, and genetic information about one country's population, it may lead to the discovery of better and more targeted treatments for common psychiatric disorders, which could be a price worth paying.

Ethical issues related to the genetics of human behaviour

Despite some opposition it is now generally agreed that research into genetic research of psychiatric disorder is both desirable and ethical. Clearly mental disorders cause much suffering and impairment, as well as considerable cost both to the individual and society. The case has been made elsewhere (Farmer and Owen 1996) that research into the genetic aspects of psychiatric disorders will ultimately lead to improved understanding of such disorders, as well as improved and more individually targeted treatments (see also chapter 17).

However, while psychiatric geneticists may consider that discovering the genetic basis for mental disorders will 'enhance diagnostic accuracy, improve treatment and radically alter clinical practice' (Farmer and Owen 1996), many members of the general public may have derived the opinion that our scientific endeavours are more to do with creating Frankenstein's monster. Indeed, in relation to genetic research into variation within the normal population, the public are generally in agreement that genetics should not be used to develop 'enhancement technologies', (reproductive cloning) that would allow us to choose the most desirable embryos selected for certain traits or qualities, e.g. gender, beauty, intelligence. Since only the rich would be able to afford to choose their offspring in this way, the social inequalities of society associated with wealth would convert to genetic inequalities. Instead of a eugenics movement based on the fanaticism of an individual despot or government policy, the selection of foetuses for such 'enhanced qualities' is eugenics by another route, i.e. individual private parental selection. Some have even argued that eugenics in this sense is not immoral (Caplan *et al.* 1999). Indeed, the rich can already buy better schools or cosmetic surgery as means of enhancing themselves or their children, although these advantages are not heritable.

The remit of the Genetics and Human Behaviour working group of the Nuffield Council on Bioethics is to define and consider ethical legal and social issues arising from the study of the genetics of variation within the normal range of behavioural characteristics. The behavioural traits that will be considered will include antisocial personality, intelligence, sexual orientation, and substance abuse and addiction. While there is consistent evidence that these traits and characteristics are heritable, what there is less agreement about is whether it is appropriate to investigate this evidence. Some respected geneticists are quite vehemently opposed to such research (Harper 1997), on the basis that the outcome will lead to discrimination, and a resurgence of eugenic abuses.

Whatever the conclusions of the Nuffield working group regarding the ethics of such research, it is nonetheless already being undertaken (see for example Fisher *et al.* 1999).

Those involved in such research consider that examining individual differences in normal traits will enhance understanding of the abnormal, pathological or deviant. What ever the final report from the working group recommends, there are also other temporal changes operating, in relation to what is considered ethical and what is not. Over recent years, it has become apparent that ethical values appear to be relative rather than absolute; there being a gradual and almost imperceptible shift in opinion as scientific knowledge increases. For example, organ transplantation and *in vitro* fertilization were quite controversial when they first became possible, as were mass inoculations for polio, fluorides in drinking water, and the addition of iron and vitamins to milk and bread, and there was much consideration about their ethics. However, such techniques are now considered almost commonplace. Similarly debates that are exercising ethicists (and indeed the President of the US), at present, such as the use of human embryo-derived stem cells for therapeutic cloning to treat disorders such as Parkinson's disease (Wertz 2002), may cease to be so important if it becomes clear to the public that the techniques have led to widespread relief of suffering.

References

Abbott A. (1997) Germany's past still cast a long shadow. *Nature* **389**:647.

Andersen B. and Arnason E. W. (1999) Iceland's database is ethically questionable. *British Medical Journal* **318**:1565.

Berger A. (1999) Private company wins rights to Icelandic gene database. *British Medical Journal.* **318**:11.

Butler D. (1997) France reaps benefit and costs of going by the book. *Nature* **389**:661–2.

Caplan A. L., McGee G., Magnus D. (1999) What is immoral about eugenics? *British Medical Journal* **7220**:1284–5.

Christie B. (2000) Doctors revised declaration of Helsinki *British Medical Journal* **321**:913.

Crusio W. E. and Gerlai R. T. (1999) *Handbook of Molecular-genetic Techniques for Brain and Behavior Research*. Amsterdam: Elsevier Science B. V.

Independent, The (1997) Galileo, Copernicus and now Dolly! 24 February p. 17.

Faraone S. W., Gottesman I. I., Tsuang M. T. (1997) Fifty years of the Nuremberg code: a time for retrospection and introspection. *American Journal of Medical Genetics* **74**:345–7.

Farmer A. E. and Owen M. J. (1996) Genomics: the next psychiatric revolution? *British Journal of Psychiatry* **169**:135–8.

Farmer A. E. and McGuffin P. (1999) Ethical issues and psychiatric genetics, in W. E Crusio and R. T. Gerlai (eds) *Handbook of Medical Genetic Techniques for Brain and Behaviour Research. Techniques in the Behavioural and Neural Sciences*, Vol. 13. Amsterdam: Elsevier Science B. V.

Fisher P. J., Turic D., McGuffin P. *et al.* (1999) DNA pooling identifies QTLs for general cognitive ability in children on chromosome 4. *Hum Mol Genet* **8**:915–22.

Focus Magazine (2001) Science versus morality—special report, August, pp. 92–3.

Gottesman I. I. and Bertelsen A. (1996) Legacy of German psychiatric genetics: hindsight is always 20/20. *American Journal of Medical Genetics(Neuropsychiatric Genetics)* **67**:343–6.

Gejman P. V. (1997) Ernst Rudin and Nazi euthanasia: another stain on his career. *American Journal of Medical Genetics(Neuropsychiatric genetics)* **74**:455–6.

Gershon E. S. (1997) Ernst Rudin, a Nazi psychiatrist and geneticist. *American Journal of Medical Genetics(Neuropsychiatric Genetics)* **74**:457–8.

Harper P. (1997) Huntingdon's disease and the abuse of genetics, in P. Harper and A. Clarke(eds) *Genetics, Society and Clinical Practice*. London: Bios Scientific Publications.

Jones S. (1997) Dolly's here now, the ethics will have to cope. The *Daily Telegraph*, 5 March, p. 22.

Kendler K. and Zerbin-Rudin E. (1996) abstract and review of 'Zur Erbapathologie der schizophrenie' (Contribution to the genetics of schizophrenia) 1916. *Am J Med Genet(Neuropsychiatric Genetics)* **67**:338–42.

Lancet, The (1997) The ethics industry. *Lancet* 350:897.

Lerer B. and Segman R. H. (1997) Correspondence regarding German psychiatric genetics and Ernstrudin. *Am J Med Genet(Neuropsychiatric Genetics)* **74**:459–60.

Levine R. J. (2001) Institutional review boards: A crisis in confidence. *Annals of Internal Medicine* **134**:161–3.

Mayer S. (2001) Commons votes for human embryo stem cell research. *British Medical Journal* **322**:7.

Meslin E. M. (1997) Ethical, legal and social implications of research in psychiatric genetics: thoughts from the ELSI research programme at the US National Human Genome Research Institute. *Am J Med Genet* 74:6.

Mitchison N. (1997) An outline for boys and girls. Cited in: when Britain was good for breeding, The *Sunday Times* 31 August.

Nicholl J.(2000) The ethics of research ethics committees. *British Medical Journal* 1217.

Rider S. (2000) *Report on ELSA Activities in Sweden. Research Concerning Ethical, Legal and Social Aspects of Genome Research*. Uppsala: Universitetstryckeriet Ekonomikum.

Royal College of Psychiatrists (2001) *Guidelines for Researchers and for Research Ethics Committees on Psychiatric Research Involving Human Participants*. Council Report CR82. London: Gaskell.

Scott-Clark C. and Levy A. (1997) Little white lies. The *Sunday Times Magazine*, 31 August pp. 14–22.

Wadman M., Levitin C., and Abbott A. (1997) Business booms for guides to biology's moral maze. *Nature* **389**:658–9.

Watson J. D. (1997) Genes and politics. *Journal of Molecular Medicine* **75**:624–36.

Wertz D. C. (2002) Embryo and stem cell research in the United States: history and politics. *Gene Therapy* **9**(11):674–8.

Wilmut T., Schnieke A. K., McWhir J. *et al.* (1997) Viable offspring derived from foetal and adult mammalian cells. *Nature* **385**:810–13.

Chapter 17

The future and post-genomic psychiatry

Michael J. Owen, Peter McGuffin, and Irving I. Gottesman

Introduction

Research in psychiatric genetics is not for the fainthearted. Mental disorders pose great challenges to genetic analyses because, in the majority of cases, they result from the predisposing effects of alleles at an unknown number of different genes, as well as environmental influences. Further complexities include the likelihood of non-additive genetic effects, including gene–gene interactions (epistasis), and potential gene–environment (physical and experiential) interactions. Genetic complexity is compounded by possible epigenetic factors (Gottesman 2001; Petronis 2001), stochastic effects (McGuffin *et al.* 1994; Woolf 1997) and nosological complexity and uncertainty. It is seldom possible to validate psychiatric diagnoses on the basis of physical examination or laboratory tests, or even to confirm them *post mortem*, and for many disorders we have little idea of pathogenic mechanisms. It also seems likely that there is, possibly considerable, genetic heterogeneity within individual psychiatric syndromes which will serve to diminish the detectability of individual loci, and that there is some overlap between the risk alleles predisposing to different psychiatric phenotypes.

Current state of progress

Genetic epidemiology

In spite of these difficulties, genetic epidemiological studies using both simple and complex approaches (Rao and Province 2001) have shown convincingly that genes play an important role in the syndromes treated in the previous chapters, but, with a few rare exceptions, they are not sufficient to produce disease. Increasingly, heritability estimates are based on a process of biometrical model fitting (Chapter 2) which allows formal comparison of hypotheses concerning whether, and to what extent, genetic and environmental factors contribute to liability to a disorder. Genetic epidemiologists have begun to explore the extent to which there is overlap in genetic and environmental risk factors between disorders, whether genes influence clinical heterogeneity within disorders, and the relationship between genetic susceptibility to disorders and putative environmental risk factors.

Patterns of familial co-aggregation have been important in helping us to extend diagnostic boundaries. Examples include the schizophrenic spectrum and autistic 'spectrum' disorders that are mild, often subclinical phenotypes associated with genetic liability to schizophrenia and autism respectively (Chapters 7 and 10). There is also evidence of a considerable overlap between the genetic contributions to depression and anxiety (Chapters 9 and 12). Such findings have important implications for the traditional classifications of psychiatric disorders, and for establishing the most useful phenotypes and endophenotypes for molecular genetic studies.

Within a particular diagnostic category there is often considerable clinical hetero-geneity which may result in large differences in the optimal treatments and variation in the likely prognoses for patients. It is often assumed that genetic influences on clinical heterogeneity are primarily related to the degree or nature of genetic liability to the disorder. This issue has been investigated in a range of studies of age at onset in schizo-phrenia. There is consistent evidence for a substantial genetic influence on age at onset in schizophrenia (Chapter 10). However, it is likely that susceptibility genes for schizo-phrenia have only a small effect at best on age at onset, and that most of the effect is probably due to modifying genes and idiosyncratic epigenetic factors. A similar pattern is emerging for the major groups of schizophrenia symptoms (Chapter 10).

The status of putative environmental risk factors for psychiatric disorders has become more complex in the light of genetic epidemiological studies. For example, episodes of depressive illness are often preceded by a cluster of adverse life events, which are tradi-tionally regarded as environmental risk factors. However, there is increasing evidence that the experience or perception of life events themselves has a partly genetic basis (Chapters 2 and 9). Furthermore, people who suffer from depressive illnesses may not experience more life events than their unaffected relatives (McGuffin et al. 1988). Therefore, the association between life events and depression may be due at least partly to the co-inheritance of susceptibility to both (Owen and McGuffin 1997). Another example comes from the field of alcohol dependence, where age at first drink is tradi-tionally thought of as an environmental risk factor. However, there is evidence from data on twins that the relationship is due to common familial factors, and that early drinking does not have a causal effect on alcohol dependence (Prescott and Kendler 1999). The implication is that interventions aimed at delaying age of first drinking are unlikely to be successful in reducing alcohol dependence later in life.

It is often suggested that gene–environment interactions are likely to be important in many psychiatric disorders. However, it has been difficult to identify clear examples because, for most conditions, we have not yet confidently identified specific risk factors. There is some evidence, albeit unconfirmed, from adoption studies that individuals who are at relatively high genetic risk of disorders such as schizophrenia (Tienari et al. 1994) and adolescent conduct disorder (Cadoret et al. 1995) may be affected to a greater extent by dysfunction and stress within the family environment than individuals at lower genetic risk for these disorders. There is also preliminary evidence that possession of the apolipoprotein E (APOE) 4 allele is a risk factor for the presence and severity of chronic traumatic brain injury in boxers (Chapter 13). It seems highly likely that genetic epidemiological studies will play an increasingly important role in disentangling the complexities of disorders with multifactorial aetiologies as more specific risk factors are identified.

Molecular genetics

Linkage studies

The first wave of molecular genetic studies in psychiatry focused on large multiply-affected families and was based upon the hope that such pedigrees, or at least a proportion of them, result from the segregation of genes of major effect. This approach has been highly successful in identifying single gene forms of some common disorders, such as breast cancer and non-insulin dependent diabetes. In psychiatry there have been some notable successes also. Major genes have been identified for various forms of dementia (Chapter 13), mental retardation (Chapter 5) and most recently for a rare form of specific language impairment (Chapter 6).

Difficulties in replicating findings

Unfortunately it seems that, for the majority of psychiatric disorders, monogenic forms are at best extremely rare and quite possibly non-existent. As a consequence, the results of linkage studies have seemed to some to be disappointing, with positive studies often falling short of being compelling and failures to replicate being abundant. However, it is worth making three points. First, similar difficulties are being faced for other common diseases and so far the recent identification of NOD 2 as a susceptibility locus for Crohn's disease (Hugot et al. 2001; Ogura et al. 2001) and Calpain 10 as type II Diabetes susceptibility locus (Horikawa et al. 2000) are the only successful examples of positional cloning for a complex disorder. Second, replicated, positive linkages to several chromosomal regions are accumulating for a number of complex psychiatric disorders such as late-onset Alzheimer's disease, schizophrenia, bipolar disorder, autism and dyslexia. Third, we should expect true positive linkages to be difficult to replicate in disorders that are predisposed to by the combined action of several genes of moderate effect (Flint and Mott 2001; Suarez et al. 1994). However, the problem is that, in most instances, it is difficult to know whether the statistical evidence for linkage is sufficiently strong to warrant large-scale and expensive efforts aimed at cloning putative linked loci. One way of resolving this issue will be to study much larger samples, of say 600–800 nuclear families, which should be sufficient to detect susceptibility genes of moderate effect size, and studies of this sort should now become a priority (Chapter 3, Owen et al. 2000). Of course, the proof that a positive linkage is correct comes when the disease gene is identified, and a number of groups are already attempting to identify disease genes in candidate regions. These efforts should be encouraged by findings in Crohn's disease and diabetes as well as by converging linkage evidence for several disorders.

Candidate gene association studies

Allelic association studies offer a powerful means of identifying genes of modest to small effect (i.e. those conferring a relative risk of less than 2) in samples of realistic size. Until recently, these have been restricted to studies of functional or positional candidate genes. The most successful application of this approach in psychiatric genetics has been the identification of the e4 allele of the gene encoding apolipoprotein E (APOE) as a risk factor for Alzheimer's disease (Chapter 13). Promising findings have also been found for polymorphisms in DRD4 and DAT1 for ADHD (Chapter 7). However, in other psychiatric disorders our relative ignorance of pathophysiology and the lack of strong and circumscribed evidence for linkage have meant that compelling candidate genes have been hard to find. There have been some positive findings, but once again conflicting results have been reported. It has usually been assumed that these reflect the tendency of association studies to produce false positives (Chapter 3). However, we should not discount the possibility of false negatives since, where small effects (ORs approximately 1.2–1.5) are concerned, larger samples than those used in many studies may be required to have adequate power to achieve replication. Thoughtful reductionism of phenotypes to endophenotypes may also further the progress. On the other hand, putative associations should be assumed to be false until replicated, even if the p value is low. Other potential causes of conflicting findings are heterogeneity between patient samples due to differences in ethnicity, ascertainment or clinical methodology (Owen *et al.* 2000).

The requirement for large sample sizes has practical implications for sampling strategies and study design. While in some instances, particularly very common childhood disorders, family-based association samples might be almost as easy to collect as case control samples; in general, large sample sizes will be difficult to collect. The place for family-based studies is as part of a first or even second replication tier where the lower power, both intrinsic and due to smaller sample-size (due to difficulty of collection), will be offset by higher prior odds, the ability to make specific predictions concerning the specific associated allele and the direction of effect, and the fact that few markers will reach this stage of the sequential process.

The genome sequence and psychiatric genetics

Large-scale, systematic association studies

Recently there has been increasing interest in the possibility of systematic association studies covering the whole genome, or at least large regions implicated by linkage analysis (Risch and Merikangas 1996; Owen *et al.* 2000). Optimism has been fuelled by three factors. First, the most abundant form of genetic variation, the single nucleotide polymorphism (SNP), is usually bi-allelic and potentially amenable to binary, high throughput genotyping assays. Second, completion of a working draft of the human genome (Lander *et al.* 2001) has made it possible to identify or predict the presence of all or most genes in the genome, which can then be analysed for the presence of variation in disease populations by targeted resequencing. Third, coordinated large-scale resequencing efforts have allowed the construction of dense, genome-wide SNP maps (White *et al.* 2001).

Essentially there are two types of association study, direct and indirect. In the former, association is sought between and variants that can alter the structure, function or expression of a gene or genes. In contrast, indirect studies seek associations between markers and disease that are due to linkage disequilibrium between the markers and susceptibility variants. The hope is that if dense enough marker maps can be applied, then large regions of the genome can be systematically screened for evidence of LD without the requirement of actually identifying and screening every single functional SNP in the region. Concerns have been expressed on theoretical grounds that the distance over which useful levels of LD will be maintained in outbred populations (5 kB) is too small for this approach to be feasible (Kruglyak 1999). However, recent data from empirical studies are more promising and suggest that blocks or 'islands' of LD averaging between 10 and 100 kB occur across the genome in outbred, European-derived populations (Daly *et al.* 2001; Jeffreys *et al.* 2001) and that, even in outbred populations, these islands can be tagged by relatively few haplotypes (Johnson *et al.* 2001).

However, several problems remain. First, direct studies depend upon the identification of all functionally significant polymorphisms. While this is feasible in principle with regard to non-synonymous coding SNPs that are common in the target population, an exhaustive catalogue is still economically impossible except for circumscribed regions of the genome. Furthermore functional polymorphisms outside the coding sequence, such as those in regulatory regions that alter gene expression or those that effect splicing or RNA editing, might be very hard to find. Indeed, it is quite plausible that variants that influence gene expression in quite subtle ways are important in psychiatric disorders, and identification of these should be a priority (Owen *et al.* 2000).

Another important unknown is the extent to which the genes predisposing to psychiatric disorders possess few, relatively common susceptibility alleles conferring small relative risks such as APOE e4 (the common disease/common variant hypothesis) or from a larger number of rarer alleles with individually larger effect sizes (the allelic heterogeneity hypothesis). The safest assumption based upon considerations from population genetics and from empirical data is that both sorts of variant play a role (Wright and Hastie 2001), but it is not possible to predict which is likely to be operating in a given instance, and there are important differences in the optimum strategies to detect the two types. The success of indirect association studies is predicated upon the assumption that a common ancestral variant is enriched in the disease population. If allelic heterogeneity is great then there will be not be detectable LD between marker alleles and disease and the indirect approach simply will not work. The study of genetically isolated subpopulations where the small number of founders limits the number of different disease alleles may offer a partial solution.

The degree of allelic heterogeneity also has important implications for the choice of markers. Common variants are more likely to be detected by generic SNP harvesting approaches and to be available in current databases, though many will not. Rarer ones will require resequencing of disease cases and preferably those with a strong family history. However, both will require large samples to detect association. Rarer variants of larger effect will be more readily detected if the samples can be stratified on the basis of genetically valid variables such as family history, age at onset, severity and symptom profile, though the latter three might reflect modifying loci rather than being an index of loading at a specific susceptibility locus.

Finally, we are still some way from having a sufficiently rapid, cheap and accurate method of genotyping SNPs. Micro-array technology is advancing quickly, but several hurdles have to be overcome before fast, accurate genotyping of tens of thousands of SNPs is possible (Lander 1999). In the meantime methods based on DNA pooling may well be applicable. Pooling has been used successfully with microsatellites in model experiments (Barcellos *et al.* 1997; Daniels *et al.* 1998; Kirov *et al.* 2000) and to detect evidence of putative QTLs for cognitive ability (Fisher *et al.* 1999). Pooled genotyping is even more readily applicable to SNPs where problems relating to 'stutter' and differential amplification are irrelevant, and permit very accurate estimates of allele frequencies (Hoogendoorn *et al.* 2000; Norton *et al.* 2002).

Given the above considerations, it seems clear that the era of genome-wide association studies for psychiatric disorders, direct or indirect, is not yet at hand. Instead, studies in the next few years should probably focus mainly upon the direct approach utilizing non-synonymous SNPs in a wide range of functional and positional candidate genes. There are a number of chromosomal regions in several disorders where the evidence for linkage is sufficiently compelling for this to have a reasonable chance of success. Functional candidate studies should preferably dissect complete systems by a combination of the application of sensitive methods for mutation detection, followed by association studies in appropriately sized of samples. At present although the indirect approach is not widely applicable at a genome-wide level, smaller scale studies focusing upon specific regions indicated by the results of linkage studies may allow loci to be mapped.

Functional genomics

It is possible that new genes and pathways might be implicated by emerging transcriptomic and proteomic approaches, and preliminary findings using both approaches have already been reported (Edgar *et al.* 1999; Hakak *et al.* 2001; Mirnics *et al.* 2001). The problem here is that human studies will be hampered by the many confounding variables associated with *post mortem* studies of brains, while animal studies suffer from the difficulties inherent in extrapolating from animal behaviour to a complex human psychiatric disorder (Lewis and Levitt, 2002).

Future challenges

Refining phenotypes

It seems to be a widely held view, particularly among biological psychiatrists who are not geneticists, that an important reason for our failure so far to identify genes for many psychiatric disorders is that, by classifying patients according clinical diagnostic criteria only, we are studying the wrong phenotypes. The first thing to note is that, as we have seen, the commonly used diagnostic criteria define phenotypes which genetic epidemiology has shown to have moderate to high heritability. In principle, therefore, it should be possible to identify the genes predisposing to these syndromes if sufficiently large samples are studied. However, perhaps genetic validity could be improved by focusing upon aspects of clinical variation such as age of onset or symptom profiles, or by identifying biological markers both through brain imaging and drug responses that predict degree of genetic risk or define more homogeneous subgroups.

Unfortunately, despite much work, it has not been possible to identify genetically distinct subtypes within the major diagnostic categories for most mental disorders. Instead, clinical variation is likely to reflect in part at least a combination of quantitative variation in genetic risk for the disorder and the effect of modifying genes which influence illness expression rather than the risk of illness per se. Examples of such phenomena, relating to age at onset and symptom pattern in schizophrenia, were mentioned in Chapter 10 and are clearly seen whenever one encounters a pair of identical twins either completely or partially discordant for a psychiatric disorder.

The search for trait markers aims to move genetic studies beyond the clinical syndrome by identifying indices of genetic risk that can be measured in asymptomatic relatives of probands and/or by identifying markers of pathophysiological processes that are closer to the primary effects of susceptibility genes than clinical symptoms; so-called endophenotypes or intermediate phenotypes. Work in this area is developing fast; for example, candidate trait markers for schizophrenia include schizotypal personality traits, measures of cognitive processing, brain-evoked potentials and abnormalities in eye movements (De Lisi 1999). It is also hoped that advances in brain imaging will lead to the identification of genetically valid trait markers. However, there are still problems to overcome. First, we will need reliable measures that can be applied practically to a sufficient number of families or unrelated patients and second, we will need to ensure that the traits identified substantially reflect genetic rather than environmental factors.

Another approach is to view psychiatric disorders as extremes of traits that are continuously distributed in the population (and examples of this are given throughout Parts 1 and 2 of this book). These can be used in quantitative trait locus (QTL) linkage and association studies. This approach has been used to map a locus for reading disability on chromosome 6 (Chapter 6) and on chromosome 18 (Fisher *et al.* 2001), and similar approaches have been applied, for example, to linkage studies of alcohol dependence, and symptom severity in schizophrenia. This strategy has also been used to investigate a possible association between the dopamine transporter gene (DAT) and severity of attention deficit hyperactivity disorder (ADHD) (Waldman *et al.* 1998).

A further approach that may facilitate gene mapping in the future is the use of animal models. These are particularly difficult to construct for psychiatric disorders, which are predominantly defined in terms of subjective experiences described by patients. However, it may be possible to identify behavioural correlates of some more objective aspects of psychiatric disorders, or of physiological trait markers. Perhaps the most promising example is currently rodent emotionality (or 'anxiety'), which has been discussed in Chapter 12. Other examples of psychiatric problems and disorders for which plausible animal models currently exist include eating disorders (Chapter 12), substance abuse (Chapter 11), aggression and hyperactivity (Chapter 8).

Disorders that predominantly involved higher cognitive function such as schizophrenia are likely to prove more difficult to model in animals. However, there are features of the human phenotype such as subtle abnormalities of cell migration, enlarged cerebral ventricles and information processing abnormalities including defects in pre-pulse inhibition, that can be detected in animals. In fact, a possible approach to producing a mouse model for at least some of the schizophrenia phenotype is suggested by the finding that some people with velocardiofacial syndrome (VCFS) due to deletions of chromosome 22q11 also have schizophrenia (Chapter 10). Mice that are heterozygously deleted for a subset of the genes that are deleted in VCFS show sensorimotor gating and memory impairments, both of which have been implicated as endophenotypes in schizophrenia (Paylor *et al.* 2001). Further studies of models of this sort may well allow the genetic basis for the behavioural and psychiatric phenotypes in VCFS to be determined, which in turn might point to genetic pathways relevant to schizophrenia. Animal models for human psychiatric disorder might also emerge from large mutagenesis programmes such as that being carried out at the MRC Mammalian Genetics Unit, Harwell, UK (http://www.mgc.har.mrc.ac.uk). This is aimed at generating large numbers of new mouse phenotypes, many of which will carry disorders that model human genetic disease. The challenge here will be to develop rapid phenotypic screening protocols, allowing models of possible relevance to psychiatry to be selected for more detailed behavioural analysis.

Identifying modifying genes

Genes affecting the clinical features of a disease or that influence treatment response in many instances may simply reflect pleiotropic effects of aetiological risk factors or pharmacogenomic status. That is, there may be specific genes that influence clinical features and treatment response independent of those affecting liability. Thus another potentially fruitful line of inquiry might be to design studies aimed at seeking modifying rather than causative genes, since these may themselves allow novel drug targets to be identified. An example here is the association between the low activity allele of a polymorphism in COMT and rapid cycling in bipolar affective disorder (Kirov *et al.* 1998). An obvious potential example is the possibility that there are genes such as APOE4 for age of onset in Alzheimer's disease. A treatment that delayed the mean age of onset of AD by 10 years would virtually eliminate it from the general population. It is also important that we design and implement appropriate pharmacogenetic studies. However, as for trait markers, a phase of clinical genetic epidemiology is in most cases indicated prior to embarking upon large-scale pharmacogenetic studies. It should also be borne in mind that there is no a priori reason why response to behavioural and psychological treatments should be less influenced by genetic factors than pharmacological treatments.

Implications of identifying genes

Functional studies

What would be the implications of identifying genetic risk factors for the major psychiatric disorders? The most important, and the most obvious, is that it will inspire a new wave of neurobiological studies from which new and more effective therapies will hopefully emerge. However, while the unequivocal identification of associated genetic variants will represent a great advance, many years of work will still be required before this is likely to translate to the bedside. An early problem will be to determine exactly which genetic variation amongst several in LD in a given gene is actually responsible for the functional variation. This is clearly illustrated by work on the gene encoding the angiotensin converting enzyme (Rieder *et al.* 1999). Even where a specific variant within a gene with can be identified as the one of functional importance, functional analysis, in terms of effect at the level of the organism is likely to be particularly difficult for behavioural phenotypes in the absence of animal models. An extra level of complexity is that we will need to be able to produce model systems, both *in vivo* and *in vitro*, that allow gene–gene and gene–environment interaction to be studied.

Genetic nosology

While the development of new therapies will take time, it is likely that the identification of susceptibility genes will have an earlier impact on psychiatric nosology. By correlating genetic risk factors with clinical symptoms and syndromes it should be possible to study heterogeneity and comorbidity in order to improve the diagnosis and classification of mental disorders. The prospects will also be enhanced for identifying clinically useful biological markers of psychiatric disorders as an aid to diagnosis, thus moving beyond the current situation of making diagnoses based entirely on clinical signs and symptoms. Improvements in the validity of psychiatric diagnosis will clearly facilitate all avenues of research into these disorders. However, for our purposes, perhaps we should limit ourselves to speculating that improvements in diagnosis and classification will enhance our ability to detect susceptibility factors genetic and other-wise and thus a positive feedback between nosology, epidemiology and molecular genetics can be envisaged.

Molecular epidemiology

The identification of genetic risk factors should allow us to investigate the ways in which genes and environment interact. Studies of this kind will require large epidemi-ologically based samples together with the collection of relevant environmental data. This work could start now with DNA being banked for future use, although in many instances the identification of plausible environmental measures might require clues from the nature of the genetic risk factors yet to be identified. A major theme in rela-tion to this will be the bringing together of methodologies from genetics and epidemi-ology which have traditionally adopted somewhat differing analytical approaches (Skoultchi et al. 1997). Treating susceptibility alleles as risk factors in an epidemiological context will allow estimates of effect sizes within a population to be made. Accounting for specific genetic effects will also facilitate the search for independent environmental factors, and the investigation of potential gene–environment interactions. Scientific validity is likely to be enhanced by ensuring as far as possible that control samples are drawn from the same base population as controls. In addition, the use of incident cases, rather than cases who happen to have the disorder at the time of a study, should guard against the risk of identifying loci related only to chronicity of illness rather than susceptibility. Phenotypic assessment is likely to benefit from prospective studies, to counteract the tendency of patients to forget historical details, and the difficulty of making observed ratings retrospectively from case records. However, the price of improved scientific rigour is likely to be considerably more expensive studies, due to the longer period and larger number of investigators that will be required to ascertain the detailed data on the thousands of subjects which will probably be required.

Genetic testing

A further implication concerns genetic testing. This is a complex area that raises a num-ber of ethical issues which have been discussed elsewhere (Wang et al. 1998). However, the potential for predictive testing has probably been overstated, given that susceptibility to most psychiatric disorders almost certainly depends upon the combined effects of pre-disposing and protective alleles at a number of loci (cf. Gottesman and Erlenmeyer-Kimling 2001). Consequently, in many instances the predictive value of tests for such genes will be low. This applies, for example, to APOE testing for late-onset Alzheimer's disease, and has led to a recommendation not to perform such testing in asymptomatic individuals (Tienari et al. 1994). Indeed, even when all susceptibility genes for a given dis-order have been identified, it will still not be possible to predict the development of dis-ease with certainty until the relevant environmental risk factors have also been identified and the nature of the various interactions (epistatic and epigenetic) understood. Such interactions may be as complex as chaotic systems like the weather, which is notoriously unpredictable over even the relatively short term (Owen *et al.* 2000). However, on a more optimistic note, genetic testing could have other roles that are likely to be of more value to patients and clinicians, for example, in helping to optimize treatment choices by testing genes that are found to influence treatment responses in psychiatric disorders, leading to a greater individualization of treatment (Chapter 14).

Conclusion

Despite the sobering complexities of psychiatric genetic research, significant advances have been made in genetic epidemiology and in gene finding, most notably in the field of dementias. The prospects for determining the genetic basis of other common psychiatric disorders are good, but important difficulties still have to be overcome. Technical and methodological advances are certainly required. However, the ultimate identification of genetic risk factors for mental disorders and the translation of this into improved patient care will depend as much upon the traditional disciplines of clinical description and epidemiology and our ability to combine these with genetic, psychological and sociological approaches in the integrative science of psychiatry.

References

Barcellos L. F., Klitz W., Field L. *et al.* (1997) Association of mapping of disease loci by use of a pooled DNA genomic screen. *Am J Hum Genet* **61**:734–47.

Barden N., Morissette J., Shink E. *et al.* (1998) Confirmation of bipolar affective disorder susceptibility locus on chromosome 12 in the region of the Darier disease gene. *Am J Med Genet* **81**:475.

Berrettini W. H., Ferraro T. N., Goldin L. R., *et al.* (1994) Pericentromeric chromosome 18 DNA markers and manic-depressive illness: evidence for a susceptibility gene. *Proc Natl Acad Sci* **91**: 5918–21.

Biomed European Bipolar Collaborative Group (1997) No association between bipolar disorder and alleles at a functional polymorphism in the COMT gene. *British Journal of Psychiatry* **170**:526–8.

Blacker D., Wilcox M. A., Laird N. M. *et al.* (1998) Alpha-2 macroglobulin is genetically associated with Alzheimer disease. *Nat Genet* **19**:357–60.

Blackwood D. H. R., He L., Morris S. W. *et al.* (1996) A locus for bipolar affective disorder on chromosome 4p. *Nat Genet* **12**:427–30.

Blouin J. L., Dombroski B. A., Nath S. K. *et al.* (1998) Schizophrenia susceptibility loci on chromosomes 13q32 and 8p21. *Nat Genet* **20**:70–3.

Cadoret R. J., Yates W. R., Troughton E. *et al.* (1995) Gene-environment interaction in the genesis of aggressivity and conduct disorder. *Archives of General Psychiatry* **52**:916–24.

Cardno A. G., Marshall E. J., Coid B. *et al.* Heritability Estimates for Psychotic Disorders. *Archives of General Psychiatry* **56**:162–8.

Cardon L. R., Smith S. D., Fulker D. W. *et al.* (1994) Quantitative trait locus for reading disability on chromosome 6. *Science* **266**:276–9.

Cargill M., Altshuler D., Ireland J. *et al.* (1999) Characterization of single-nucleotide polymorphisms in coding regions of human genes. *Nature Genetics* **22**:231–8.

Collier D. A., Arranz M. J., Sham P. *et al.* (1996) The serotonin transporter is a potential susceptibility factor for bipolar affective disorder. *Neuroreport* **7**:1675–9.

Collins F. S., Guyer M. S., Chakravarti A. Variation on a theme: Cataloging human DNA sequence variation. *Science* **278**:1580–1.

Cook E. H., Stein M. A., Krasowski M. D. *et al.* (1995) Association of attention-deficit disorder and the dopamine transporter gene. *Am J Hum Genet* **56**:993–8.

Corder E. H., Saunders A. M., Strittmatter W. J. *et al.* (1993) Gene dose of apolipoprotein E type-4 allele and the risk of Alzheimer's disease in late-onset families. *Science* **261**:921–3.

Daly M. J., Rioux J. D., Schaffner S. F. *et al.* (2001) High-resolution haplotype structure in the human genome. *Nature Genetics* **29**:229–31.

Daniels J., Holmans P., Williams N. *et al.* (1998) A simple method for analyzing microsatellite allele image patterns generated from DNA pools and its application to allelic association studies. *Am J Hum Genet* **62**:1189–97.

DeLisi L. E. (1999) A critical overview of recent investigations into the genetics of schizophrenia. *Curr Opin Psychiatry* **12**:29–39.

Edgar P. F., Schonberger S. J., Dean B. *et al.* (1999) A comparative proteome analysis of hippocampal tissue from schizophrenic and Alzheimer's disease individuals. *Mol Psychiatry* **4**:173–8.

Ewald H., Mors O., Flint T. *et al.* A possible locus for manic depressive illness on chromosome 16p13. *Psychiatr Genet* **5**:71–81.

Fisher S. E., Francks C., Marlow A. J. *et al.* (2001) Independent genome-wide scans identify chromosome 18 quantitative-trait locus influencing dyslexia. *Nature Genetics* **30**:86–91.

Fisher P. J., Turic D., Williams N. M. *et al.* (1999) DNA pooling identifies QTLs for general cognitive ability in children on chromosome 4. *Human Molecular Genetics* **8**:915–22.

Flint J. and Mott R. (2001) Finding the molecular basis for quantitative traits: successes and pitfalls. *Nature Reviews: Genetics* **2**:437–45.

Flint J. (1997) Freeze! *Nat Genet* **17**:250–1.

Flint J., Corley R., DeFries J.C. *et al.* (1995) A simple genetic-basis for a complex psychological trait in laboratory mice. *Science* **269**:1432–5.

Ginns E. I., Ott J., Egeland J. A. (1996) A genome-wide search for chromosomal loci linked to bipolar affective disorder in the Old Order Amish. *Nat Genet* **12**:431–5.

Gottesman I. I. (2001) Psychopathology through a life span-genetic prism. *Am Psychol* **56**:867–78.

Hakak Y., Walker J. R., Li C. *et al.* (2001) Genome-wide expression analysis reveals dysregulation of myelination-related genes in chronic schizophrenia. *Proc Natl Acad Sci USA* **98**:4746–51.

Hardy J., Gwinn-Hardy K. (1998) Genetic classification of primary neurodegenerative disease. *Science* **282**:1075–9.

Hardy J. and Israël A. (1999) Alzheimer's disease: in search of g-secretase. *Nature* **398**:466–7.

Hoogendoorn B., Owen M. J., Oefner P. J. *et al.* (1999) Genotyping single nucleotide polymorphisms by primer extension and high performance liquid chromatography. *Human Genetics* **104**:89–93.

Horikawa Y., Oda N., Cox N. J. *et al.* (2000) Genetic variation in the gene encoding calpain-10 is associated with type 2 diabetes mellitus. *Nat Genet* **26**:163–75.

Houwen R. H. J., Baharloo S., Blankenship K. *et al.* (1994) Genome screening by searching for shared segments: mapping a gene for benign recurrent intrahepatic cholestasis. *Nat Genet* **8**:380–6.

Hugot J. P., Chamaillard M., Zouali H. *et al.* (2001) Association of NOD2 leucine-rich repeat variants with susceptibility to Crohn's disease. *Nature* **411**:599–603.

International Molecular Genetic Study of Autism Consortium (1998) A full genome screen for autism with evidence for linkage to a region on chromosome 7q. *Hum Mol Genetics* **7**:571–8.

Jeffreys A. J., Kauppi L., Neumann R. (2001) Intensely punctate meiotic recombination in the class II region of the major histocompatibility complex. *Nature Genetics* **29**:217–22.

Johnson G. C. L., Esposito L., Barratt B. J. *et al.* (2001) Haplotype tagging for the identification of common disease genes. *Nature Genetics* **29**:233–7.

Jorde L. B., Watkins W. S., Carlson M. *et al.* Linkage disequilibrium predicts physical distance in the adenomatous polyposis coli region. *Am J Hum Genet* 54:884–898.

Kehoe P. G., Carsten R., McIlroy S. *et al.* (1999) Variation in DCP1, encoding ACE, is associated with susceptibility to Alzheimer's disease. *Nat Genet* **21**:71–2.

Kehoe P. G., Wavrant De Vrieze F., Crook R. *et al.* (1999) A full genome scan for late onset Alzheimer's disease. *Hum Mol Genet* **8**:237–45.

Kelsoe J. R., Loetscher E., Spence M. A. *et al.* (1998) A genome survey of bipolar disorder indicates a susceptibility locus on chromosome 22. *Am J Med Genet* **81**:461–2.

Kendler K. S., Karkowski-Shuman L., O'Neill F. A. *et al.* (1997) Resemblance of psychotic symptoms and syndromes in affected sibling pairs from the Irish study of high-density schizophrenia families: evidence for possible etiologic heterogeneity. *Am J Psychiatry* **154**:191–8.

Kendler K. S., MacLean C. J., Ma Y. L. *et al.* (1999) Marker-to-Marker linkage disequilibrium on chromosomes 5q, 6p and 8p in the Irish high-density schizophrenia pedigrees. *Am J Med Genet* **88**:29–33.

Kendler K. S., Neale M. C., Kessler R. C. *et al.* (1992) Major depression and generalized anxiety disorder: same genes, (partly) different environments? *Archives of General Psychiatry* **49**:716–22.

Kendler K. S., Neale M., Kessler R. *et al.* (1993c): A twin study of recent life events and difficulties. *Archives of General Psychiatry* **50**:789–96

Kendler K. S., Pedersen N. L., Neale M. C. *et al.* (1995) A pilot Swedish twin study of affective illness including hospital- and population-ascertained subsamples: results of model fitting. *Behav Genet* **3**:217–32.

Kendler K. S., Tsuang M. T., Hays P. (1987) Age at onset in schizophrenia: a familial perspective. *Archives of General Psychiatry* **44**:881–90.

Kendler K. S., Walters E. E., Neale M. C. *et al.* (1995) The structure of the genetic and environmental risk factors for six major psychiatric disorders in women: phobia, generalized anxiety disorder, panic disorder, bulimia, major depression and alcoholism. *Archives of General Psychiatry* **52**: 374–83.

Kirov G., Murphy K. C., Arranz M. J. *et al.* (1998). Low activity allele of catecho-o-methyl transferase gene associated with rapid cycling bipolar affective disorder. *Molecular Psychiatry* **3**:342–5.

Kirov G., Williams N., Sham P. *et al.* (2000) Pooled genotyping of microsatellite markers in parent-offspring trios. *Genome Res* **10**:105–15.

Kruglyak L. (1999) Prospects for whole-genome linkage disequilibrium mapping of common disease genes. *Nat Genet* **22**:139–44.

Kruglyak L. and Lander E. S. (1995) Complete multipoint sib-pair analysis of qualitative and quantitative traits. *Am J Hum Genet* **57**:439–54.

LaHoste G. J., Swanson J. M., Wigal S. B. *et al.* (1996) Dopamine D4 receptor gene polymorphism is associated with attention-deficit hyperactivity disorder. *Mol Psychiatry* **1**:121–4.

Lander E. S. (1999) Array of hope. *Nature Genet* **21**(suppl):3–4.

Lander E. and Kruglyak L. (1995) Genetic dissection of complex traits: guidelines for interpreting and reporting linkage results. *Nat Genet* **11**:41–7.

Lander E. S., Linton L. M., Birren B. *et al.* (2001) Initial sequencing and analysis of the human genome. *Nature* **409**:860–921.

Lendon C., Ashall F., Goate A. M. (1997) Exploring the etiology of Alzheimer disease using molecular genetics. *JAMA* **277**:825–31.

Lesch K. P., Bengel D., Heils A. *et al.* (1996) Association of anxiety-related traits with a polymorphism in the serotonin transporter gene regulatory region. *Science* **274**:1527–31.

Lewis D. A. and Levitt P. (2002) Schizophrenia as a disorder of neurodevelopment. *Annu Rev Neurosci* **25**:409–32.

McGuffin P., Asherson P., Owen M. *et al.* (1994) The strength of the genetic effect—is there room for an environmental influence in the aetiology of schizophrenia? *British Journal of Psychiatry* **164**:593–9.

McGuffin P., Katz R., Bebbington P. (1988) The Camberwell Collaborative Depression Study III. Depression and adversity in the relatives of depressed probands. *British Journal of Psychiatry* **152**:775–82.

McGuffin P. and Owen M. J. (1996) Molecular genetic studies of schizophrenia. *Cold Spring Harbor Symposia on Quantitative Biology* **61**:815–22.

McGuffin P., Owen M. J. O'Donovan M. C. *et al.* (1994) *Seminars in Psychiatric Genetics.* London: Gaskell.

Mirnics K. (2001) Molecular characterisation of schizophrenia viewed by microarray analysis of the prefrontal cortex. *Neuron* **28**:53–67.

Murphy K. C., Jones L. A., Owen M. J. (1999 in press) Elevated rates of schizophrenia in adults with velo-cardio-facial syndrome. *Archives of General Psychiatry.*

National Institute on Aging/Alzheimer's Association Working Group (1996) Apolipoprotein E genotyping in Alzheimer's disease. *Lancet* **347**:1091–5.

Neale M. C. and Cardon L. R. (1992) *Methodology for Genetic Studies of Twins and Families.* NATO ASI Series: Kluwer Academic Publishers.

Neale M. C., Eaves L. J., Hewitt J. K. *et al.* (1989) Analysing the relationship between age at onset and risk to relatives. *Am J Hum Genet* **45**:226–39.

Norton N., Williams N. M., Williams H. J. *et al.* (2002) Universal, robust, highly quantitative SNP allele frequency measurement in DNA pools. *Hum Genet* **110**:471–8.

Nuffield Council on Bioethics (1998) *Mental Disorders and Genetics: The Ethical Context.* London: Nuffield Council on Bioethics.

Ogilvie A. D., Battersby S., Bubb V. J. *et al.* (1996) Polymorphism in serotonin transporter gene associated with susceptibility to major depression. *Lancet* **347**:731–3.

Ogura Y., Bonen D. K., Inohara N. *et al.* (2001) A frameshift mutation in NOD2 associated with susceptibility to Crohn's disease. *Nature* **411**:603–6.

Owen M. J., (1992) Will schizophrenia become a graveyard for molecular geneticists? *Psychol Med* **22**:289–93.

Owen M. J. Holmans P., McGuffin P. (1997) Association studies in psychiatric genetics. *Mol Psychiatry* **2**:270–3.

Owen M. J. and McGuffin P. (1997) Genetics and psychiatry. *British Journal of Psychiatry* **171**:201–2.

Own M. J., Cardno A. G., O'Donocan M. C. (2000) Psychiatric Genetics: back to the Future. *Molecular Psychiatry* **5**:22–31.

Page G. P. and Amos C. I. (1999) Comparison of linkage-disequilibrium methods for localization of genes influencing quantitative traits in humans. *Am J Hum Genet* **64**:1194–205.

Pekkarinen P., Bredbacka P. E., Terwilliger J. *et al.* (1994) Evidence for a susceptibility locus for manic depressive disorder in Xq26. *Am J Hum Genet* **55**:133.

Pericak-Vance M. A., Bass M. L., Yamaoka L. H. *et al.* (1998) Complete genomic screen in late-onset familial Alzheimer's disease. *Neurobiol Aging* **19**:S39–42.

Peterson A. C., Di Rienzo A., Lehesjoki A.-E. *et al.* (1995) The distribution of linkage disequilibrium over annonymous genome regions. *Hum Mol Genet* **4**:887–94.

Petronis, A. (2001) Human morbid genetics revisited: the relevance of epigenetics. *Trends in Genetics* **17**:142–6.

Petruhkin K., Fischer S. G., Pirastu M. *et al.* (1993) Mapping cloning and genetic characterization of the region containing in Wilson disease gene. *Nature Genet* **5**:338.

Philippe A., Martinez M., Guilloud Bataille M. *et al.* (1999) Genome-wide scan for autism susceptibility genes. *Hum Mol Genet* **8**:805–12.

Prescott C. A. and Kendler K. S. (1999) Age at first drink and risk for alcoholism: a noncausal association. *Alcoholism—Clinical and Experimental Research* **23**:101–7.

Reich T., Edenberg H. J., Goate A. *et al.* (1998) Genome-wide search for genes affecting the risk for alcohol dependence. *Am J Med Genet* **81**:207–15.

Rieder M. J., Taylor S. L., Clark A. G. *et al.* (1999) Sequence variation in the human angiotensin converting enzyme. *Nat Genet* **22**(1):59–62.

Risch N., Merikangas K. (1996) The future of genetic studies of complex human diseases. *Science* **273**:1516–7.

Risch N., and Teng J. (1998) The relative power of family-based and case-control designs for linkage disequilibrium studies of complex human diseases: I. DNA pooling. *Genome Res* **8**:1273–88.

Schizophrenia Collaborative Linkage Group. (1996) A combined analysis of D22S278 marker alleles in affected sib-pairs: support for a susceptibility locus for schizophrenia at chromosome 22q12. *Am J Med Genet* **67**:40–5.

Schizophrenia Linkage Collaborative Group for Chromosomes 3:6, and 8. (1996) Additional support for schizophrenia linkage on chromosomes 6 and 8: a multicentre study. *Am J Med Genet* **67**:580–94.

Sham P. C. (1998) Statistical methods in psychiatric genetics. *Statistical Methods in Medical Research* **7**:279–300.

Skoultchi A. I., Puech A., Saint-Jore B., *et al.* (1997) Comparative mapping of the human and mouse VCFS/DGS syntenic region discloses the presence of a large internal rearrangement. *Am J Hum Genet* **61**:A296.

Stine O. C., Xu J., Koskela R. *et al.* (1995) Evidence for linkage of bipolar disorder to chromosome 18 with a parent-of-origin effect. *Am J Hum Genet* **57**:1384.

Straub R. E., Lehner T., Luo Y. *et al.* (1994) A possible vulnerability locus for bipolar affective disorder on chromosome 21q22.3. *Nat Genet* **8**:291–6.

Straub R. E., MacLean C. J., O'Neill F. A. *et al.* (1997) Support for a possible schizophrenia vulnerability locus in a region of 5q22–31 in Irish families. *Mol Psychiatry* **2**:148–55.

Suarez B. K., Hampe C. L., Van Eerdewegh P. (1994) Problems of replicating linkage claims in psychiatry, in E. S. Gershon and C. R. Cloninger (eds) *Genetic Approaches to Mental Disorders*. Washington, DC: American Psychiatric Press.

Thapar A. and McGuffin P. (1996) Genetic influences on life events in childhood. *Psychol Med* **26**:813–20.

Tienari P., Lyman C. W., Moring J. *et al.* (1994) The Finnish Adoptive Family Study of Schizophrenia: implications for family research. *British Journal of Psychiatry* **164**(suppl. 23):20–6.

Waldman I. D., Rowe D. C., Abramowitz A., *et al.* (1998) Association and linkage of the dopamine transporter gene and attention-deficit hyperactivity disorder in children: heterogeneity owing to diagnostic subtype and severity. *Am J Hum Genet* **63**:1767–76.

Wang D. G., Fan J. B., Siao C. J. *et al.* (1998) Large-scale identification, mapping, and genotyping of single-nucleotide polymorphisms in the human genome. *Science* **280**:1077–82.

White P .S., Kwok P. Y., Oefner P. *et al.* (2001) Third International Meeting on Single Nucleotide Polymorphism and Complex Genome Analysis: SNPs: 'some notable progress'. *Eur J Hum Genet* **9**(4):316–8.

Wildenauer D. B., Schwab S. G., Hallmayer J. *et al.* (1998) Genome scan for autosomal genes conferring risk to schizophrenia in a German/Israeli sample. *Am J Med Genet* **81**:454.

Williams J., McGuffin P., Nothen M. *et al.* (1997) Meta-analysis of association between the 5HT2a receptor T102C polymorphism and schizophrenia. *Lancet* **349**:1221.

Williams J., Spurlock G., Holmans P. *et al.* (1998) A meta-analysis and transmission disequilibrium study of association between the dopamine D3 receptor gene and schizophrenia. *Mol Psychiatry* **3**:141–9.

Williams J., Spurlock G., McGuffin P. *et al.* (1996) Association between schizophrenia and the T102C polymorphism of 5-hydroxytryptamine type 2a receptor gene. *Lancet* **347**:1294–6.

Appendix: A brief guide to internet addresses for psychiatric genetics and genomics

Irving I. Gottesman

Advances in the areas of research impacting upon the subject matter of this book are so rapid and so broad that it behooves both the beginning and advanced scientist to be aware of the rich resources, including full text of articles, chapters, and books e.g. www.AnnualReviews.org and an updated version of the best selling *GENES* by B. Lewin http://www.ergito.com/docs/start.htm available for keeping current (and a step ahead of the uninformed) that are available on the World Wide Web. More than 100 classic and neoclassic essays by the 'discoverers' of some of the most important ideas in genetics are accessible online. We cannot guarantee that the URLs will still be active at the time you try to access them as the information providers themselves are in a state of flux, both in governmental and commercial sectors. Some sites may require simple and free registration, or only be accessible through a university or medical school library connection. Often, major journals will publish landmark papers on the web even to non-subscribers, as witnessed by the items connected with the sequencing of the human genome in *Nature* http://nature.com.genomics/human and in *Science* http://www.sciencemag.org/feature/plus/sfg/special/#sequence. Be aware of your risk to becoming addicted to clicking, as one click will lead to another in information retrieval. You will encounter the working draft of the human genome at http://genome.ucsc.edu. and the 'mother of all URLs' can be accessed at http://www.ncbi.nih.gov/genome/guide/human; following the links on this page will take you into a wonderland of constantly updated information about topics of wide ranging interest, including the genomes of our evolutionary cousins. Tools for data mining including those for sequence analysis, links to your favourite gene, and human-mouse homology maps are readily available at the marvellous National Center for Bioinformatics Research (NCBI) page. It may be more convenient for European scientists to visit the useful collection of genomic sites maintained by the Medical Research Council at http://www.hgmp.mrc.ac.uk/GenomeWeb/. While in the neighbourhood, drop by the University of Wales College of Medicine's Human Gene Mutation Database (HGMD) at http://archive.uwcm.ac.uk/uwcm/mg/ hgmd0.html which is also in a SNP consortium with Celera Genomics http://www.celera.com/. Explore the European Bioinformatics Institute based at Hinxton for the UK gene-mapping mavens http://www.ebi.ac.uk/ as well as its backer the Wellcome Trust http:www.wellcome.ac.uk/en/genome/.

You might also add the page from Cold Spring Harbor http://vector/cshl.org/ to your 'favorites' list.

Mouse genome informatics has been highly refined and can be explored to advantage at http://www.informatics.jax.org/mgihome/ maintained by the Jackson Laboratory in Bar Harbor, Maine; treat yourself to a dozen mice while you are surfing. A monthly indicator of new discoveries, updates and changes to mendelizing loci in the famous Online Mendelian Inheritance of Man (OMIM), the brainchild of Victor McKusick of Johns Hopkins University fame can be seen at http://www.ncbi.nlm.nih.gov/htbin-post/Omim/ dispupdates. Understanding the language and terms encountered in genetics, genomics, post-genomics, bioinformatics, and proteomics while surfing the URLs already mentioned should not be a problem, as numerous online glossaries are available to help the novice at http://sciencemag.org/feature/plus/sfg/education/glossaries.shtml as well as at http://genomicglossaries.com, generously maintained and updated by Mary Chitty at the Cambridge Healthtech Institute (USA).

Yet another world of important and fascination information is provided online by The National Human Genome Research Institute (Francis Collins, Director) of the National Institute of Health at http://www.nhgri.nih.gov/ also at no cost. Links to dozens of relevant sites can be tapped for genome centres in the USA, France, Canada, and the UK, for chromosome maps of all the model organisms, and direct access to journals. It is also an easy way to enter the sites maintained by the Department of Energy (DOE) and the Center for Disease Control and Prevention (CDC) in Atlanta, Georgia which has its very active Office of Genetics and Disease Prevention showcased http://www.cdc.gov/genetics with weekly updates on many targets of interest and summaries of the worldwide literature.

Fans of history and the humanities will want to visit the page maintained by Nobel Prize Museum in Stockholm http://www.nobel.se where it is easy to retrieve the sordid history of the Foundation of the Kaiser Wilhelm Institute for Medical Research during the Nazi era, as well as the autobiographies of such Nobel Laureates as John Nash, Jr and William Shockley. Standard, favourite search engines can be used to advantage for any initial inquiry such as www.google.com and www.altavista.com; try searching on the word 'twins' for an embarrassment of riches that requires further specification. A remarkable collection of classic genetic pieces written between 350 A.D. and 1932 are provided at http://www.esp.org/foundations/genetics/classical. Teachers, students, and their parents can be directed to the Genetic Science Learning Center http://www.genetics.utah.edu/section5/sc5afrm.html at the University of Utah in the hope that it will aid in the recruitment of the next generation or of more talent to the long struggle to understand the causes of mental disorders and their prevention.

The SNP consortium can be found at http://snp.cshl.org/ and searches for protein or DNA sequence for human draft data can be tried at http://www.ensembl.org/Data/blast.html.

Interests in more applied uses of research on mental disorders such as genetic counselling http://www.pitt.edu/~edugene/resource/ or for patients to read themselves http://www.geneclinics.org are beginning to be developed. Consumer-maintained URLs such as www.schizophrenia.com and www.NAMI.org can be recommended on an ad hoc basis after you have confirmed their usefulness for a particular person or situation. Pages concerned with advocacy and genetic discrimination issues can be sampled at http://www.tgac.org and http://www.nationalpartnership.org/healthcare/genetic/coalition.htm. Also related to professional education are the following choices: http://www.nchpeg.org/ and http://www.fgec.org.

From the NHGRI page one can also enter sites for the famous ELSI (see Chapter 16 by A. Farmer) or Ethical, Legal, and Social Implications of the Human Genome Project, find what the CIDR or Center for Inherited Disease Research will do for you, and delve into policy and public affairs relevant to research in the genetics and genomics of mental disorders. A comprehensive guide to bioethics links is provided by the Center for Bioethics at the University of Virginia School of Medicine http://www.med. Virginia.edu/bioethics/links.htm. Other important sites dealing with ethical issues have been covered in Chapter 16 and include http://bioethics.georgetown.edu/nbac/http://www.nuffieldbioethics.org/home/index.asp/the British Human Genetics Commission www.hgc.gov.uk and Jim Watson's own Cold Spring Harbor Laboratory archive of the American Eugenics Movement http://vector.cshl.org/eugenics/list_topics.pl.

Finally, if you have not already done so, you may want to visit the International Society for Psychiatric Genetics (ISPG) at http://www.ispg.net/.

Index